Lecture Notes in Computer Science   120

Edited by G. Goos, J. Hartmani

Advisory Board: W. Brauer   D.

Springer
Berlin
Heidelberg
New York
Barcelona
Budapest
Hong Kong
London
Milan
Paris
Santa Clara
Singapore
Tokyo

Michel Deza   Reinhardt Euler
Ioannis Manoussakis   (Eds.)

Combinatorics and Computer Science

8th Franco-Japanese and
4th Franco-Chinese Conference
Brest, France, July 3-5, 1995
Selected Papers

 Springer

Series Editors

Gerhard Goos, Karlsruhe University, Germany

Juris Hartmanis, Cornell University, NY, USA

Jan van Leeuwen, Utrecht University, The Netherlands

Volume Editors

Michel Deza
LIENS, Department de Mathématiques et d'Informatique
45 Rue d'Ulm, F-75230 Paris Cedex 05, France

Reinhardt Euler
Faculté des Sciences, Laboratoire d'Informatique de Brest
6 Avenue Le Gorgeu, F-29285 Brest Cedex, France

Ioannis Manoussakis
Université de Paris-Sud, Laboratoire de Recherche en Informatique
Centre d'Orsay, Bâtiment 490, F-91405 Orsay Cedex, France

Cataloging-in-Publication data applied for

Die Deutsche Bibliothek - CIP-Einheitsaufnahme

Combinatorics and computer science : 8th Franco-Japanese and
4th Franco-Chinese conference, Brest, France, July 3 - 5, 1995 ;
selected papers / M. Deza ... (ed.). - Berlin ; Heidelberg ; New
York ; Barcelona ; Budapest ; Hong Kong ; London ; Milan ;
Paris ; Santa Clara ; Singapore ; Tokyo : Springer, 1996
  (Lecture notes in computer science ; Vol. 1120)
  ISBN 3-540-61576-8
NE: Deza, Michel [Hrsg.]; GT

CR Subject Classification (1991): G.2, G.1.6, C.2.4, D.1.3,F.1.2

ISSN 0302-9743
ISBN 3-540-61576-8 Springer-Verlag Berlin Heidelberg New York

© Springer-Verlag Berlin Heidelberg 1996
Printed in Germany

Typesetting: Camera-ready by author
SPIN 10513453   06/3142 - 5 4 3 2 1 0   Printed on acid-free paper

Preface

The 4th Franco-Chinese and 8th Franco-Japanese conference "Combinatorics and Computer Science" (CCS'95) took place in Brest from July 3 to 5, 1995. Eighty participants from 10 countries came together to discuss their latest results in the form of 54 presentations.

This volume contains the final versions of 33 papers accepted for publication after a regular review procedure.

We would like to express our gratitude to the members of the program committee for their advice in setting up the program, to the sponsoring institutions for their immediate support, to the participants for creating a very fruitful and friendly atmosphere, to Le Quartz and Laurent Lemarchand for their strong support in organizing the conference, to the referees for their invaluable assistance, and to Springer–Verlag for publishing these proceedings.

May 1996   Michel Deza, Reinhardt Euler, Yannis Manoussakis

Table of Contents

Graph Theory

Combinatorial Optimization

Selected Topics

Parallel and Distributed Computing

Equitable and m-Bounded Coloring of Split Graphs

Bor-Liang Chen[1], Ming-Tat Ko[2] and Ko-Wei Lih[3]

[1] Institute of Applied Mathematics, Tunghai University, Taiwan 40704
[2] Institute of Information Science, Academia Sinica, Nankang, Taipei, Taiwan 11529
[3] Institute of Mathematics, Academia Sinica, Nankang, Taipei, Taiwan 11529

Abstract. An equitable coloring of a graph is a proper coloring such that the sizes of color classes are as even as possible. An m-bounded coloring of a graph is a proper coloring such that the sizes of color classes are all bounded by a preassigned number m. Formulas for the equitable and m-bounded chromatic numbers of a split graph are established in this paper. It is proved that split graphs satisfy the equitable Δ-coloring conjecture in Chen, Lih and Wu [4].

1 Introduction

All graphs considered in this paper are finite, loopless, and without multiple edges. A graph $G = (V(G), E(G))$ is said to be *equitably k-colorable* if the vertex set $V(G)$ can be partitioned into k independent subsets $V_1, V_2, \cdots, V_k$ such that $||V_i| - |V_j|| \leq 1$ for all i and j. This partition constitutes an *equitable coloring* Θ of G and each V_i is called a *color class*. The smallest integer k for which G is equitably k-colorable is called the *equitable chromatic number* of G, denoted by $\chi_=(G)$. Let $|G|$ and $\Delta(G)$ denote the number of vertices and the maximum degree of the graph G, respectively. Recently, Chen, Lih and Wu [4] proposed the following.

The Equitable Δ-Coloring Conjecture. A connected graph G is equitably $\Delta(G)$-colorable if it is different from all complete graphs K_m, odd cycles C_{2m+1}, and complete bipartite graphs of type $K_{2m+1,2m+1}$.

This is a stronger conjecture than an earlier one made in Meyer [7] which asserts that $\chi_=(G) \leq \Delta(G)$ for any connected graph G which is neither a complete graph nor an odd cycle. An earlier result of Hajnal and Szemerédi [5] implies that $\chi_=(G) \leq \Delta(G) + 1$ for any graph G. In Lih and Wu [6], the equitable Δ-coloring conjecture was established for bipartite graphs. Chen, Lih and Wu [4] proved that the conjecture holds for graphs G satisfying $\Delta(G) \geq |G|/2$ or $\Delta(G) \leq 3$.

It is usually very difficult to determine for what numbers k the graph G is equitably k-colorable. We should note that G is not necessarily equitably k-colorable for any k greater than $\chi_=(G)$. Extending results of Bollobás and Guy

[1], Chen and Lih [3] solved this problem completely for the class of trees. In this paper we are going to solve this problem for another class of graphs.

A connected graph G is called a *split* graph if its vertex set can be partitioned into two nonempty subsets U and V such that U induces a complete graph and V induces an independent set. We denote a split graph G as $G[U;V]$ and always assume that no vertex in V is adjacent to all vertices in U. It is obvious that $\chi_=(G) < |G|$ under our convention. We will determine when a split graph G is equitably colorable in this paper. In particular, the equitable Δ-coloring conjecture holds for split graphs.

The notion of m-bounded coloring will be discussed in the final section.

2 Equitable Coloring

Let $G = G[U;V]$ be a split graph such that $U = \{u_1, u_2, \cdots, u_n\}$ induces a complete graph and $V = \{v_1, v_2, \cdots, v_r\}$ induces an independent set. We use $\alpha'(G)$ to denote the largest size of a matching (i.e., non-incident edges) in G. Let $\lfloor x \rfloor$ ($\lceil x \rceil$) denote the largest (least) integer not greater (smaller) than x.

We assign a family of bipartite graphs $BG(k), k \geq 1$, to the given split graph $G = G[U;V]$ in the following way. The vertex set of $BG(k)$ is $\{u_{ij} : 1 \leq i \leq n$ and $1 \leq j \leq k\} \cup V$ and $\{u_{ij}, v_t\}$ is defined to be an edge of $BG(k)$ if and only if u_i and v_t are non-adjacent in G. We denote the two parts of $BG(k)$ by U^* and V for convenience. Note that $BG(k)$ is a subgraph of $BG(k+1)$.

The coloring of a split graph $G[U;V]$ is closely related to matchings of the graphs $BG(k)$'s. For instance, any given matching in $BG(k)$ induces a partial coloring of G in the following standard way. We use the ith color to color u_i and all those vertices in V that are matched to some $u_{ij}, 1 \leq j \leq k$.

Lemma 1. *Let $G[U;V]$ be a split graph and $k \geq 1$. If G can be colored so that the color classes are of sizes at least $k + 1$, then $\alpha'(BG(k)) = kn$.*

Proof. Each u_i belongs to the same color class with at least k vertices $v_{i_1}, v_{i_2}, \cdots,$ v_{i_k} of V. In the graph $BG(k)$, we match u_{ij} with v_{i_j} for $1 \leq j \leq k$. This matching saturates every vertex in the part U^*, hence it is a maximum matching. □

Theorem 2. *Let $G[U;V]$ be a split graph. If G is equitably k-colorable, then G is equitably $(k+1)$-colorable.*

Proof. Let $C_1, C_2, \cdots, C_k$ be the k equitable color classes of G so arranged that their sizes are in non-increasing order. Note that $k \geq n$. We may also assume that $u_i \in C_i$ for each u_i in $U = \{u_1, u_2, \cdots, u_n\}$. We follow the cyclic order $C_1, C_2, \cdots, C_k, C_1, \cdots$ to take out one vertex form each $C_i \cap V$ sequentially. Sooner or later the removed vertices form a new class which together with the trimmed classes will give rise to an equitable $(k+1)$-coloring of G. □

Theorem 3. *let $G[U;V]$ be a split graph. Let $m = \max\{k : \alpha'(BG(k)) = kn\}$ if the set in question is nonempty; otherwise let m be zero. Then $\chi_=(G) = n + \lceil (r - \alpha'(BG(m+1)))/(m+2) \rceil$.*

Proof. Let Θ be an equitable coloring of G with $\chi_=(G)$ colors. By Lemma 1 and the maximality of m, each color class is of size $\leq m + 2$. Let $C_1, C_2, \cdots, C_n$ be the n color classes of G such that $u_i \in C_i$ for each u_i in $U = \{u_1, u_2, \cdots, u_n\}$. The size of each C_i is at most $q + 1 \equiv \lceil |G|/\chi_=(G) \rceil$. The canonical way of constructing matchings used in the proof of Lemma 1 will give rise to the following consequences. First, $\alpha'(BG(q-1)) = (q-1)n$, hence $q - 1 \leq m$ by the maximality of m. Second, there is a matching in $BG(q)$ such that all vertices in $C_i \cap V, i = 1, 2, \cdots n$, are saturated. Thus there are at least $r - \alpha'(BG(q))$ vertices in V which are outside of all C_i's. Since $BG(q)$ is a subgraph of $BG(m + 1)$, we have $r - \alpha'(BG(q)) \geq r - \alpha'(BG(m + 1))$. Therefore Θ needs at least $\lceil (r - \alpha'(BG(m + 1)))/(m + 2) \rceil$ colors to color vertices outside of all C_i's. This shows that $\chi_=(G) \geq n + \lceil (r - \alpha'(BG(m + 1)))/(m + 2) \rceil$.

To prove the reverse direction of the above inequality, we will construct an equitable coloring of G using $n + \lceil (r - \alpha'(BG(m + 1)))/(m + 2) \rceil$ colors. Let M_m and M_{m+1} be maximum matchings in $BG(m)$ and $BG(m+1)$, respectively. Since $BG(m)$ is a subgraph of $BG(m + 1)$, $M_m \cup M_{m+1}$ is a disjoint union of cycles and paths. All cycles are of even length. We will construct a maximum matching M^* in $BG(m + 1)$ as follows. A path of odd length in $M_m \cup M_{m+1}$ cannot begin with an edge of M_m. For otherwise we can interchange edges of M_m with edges of M_{m+1} on that path to construct a matching larger than M_{m+1}. First from every path of odd length in $M_m \cup M_{m+1}$, we take out edges of M_{m+1} and put them into M^*. Next from every cycle and every path of even length, we take out edges of M_m and put them into M^*. We finally obtain a matching M^* in $BG(m + 1)$ such that $|M^*| = |M_{m+1}| = \alpha'(BG(m + 1)) \geq \alpha'(BG(m)) = mn$ and all vertices of M_m are vertices of M^*. Therefore, in $BG(m)$, M^* matchs the m duplicates of each u_i to m vertices in V. Using M^* to induce a partial coloring of G in the standard way, the color classes are of size $m + 1$ or $m + 2$. There are $r - \alpha'(BG(m+1))$ vertices in V left uncolored. We use $\lfloor (r - \alpha'(BG(m+1)))/(m+2) \rfloor$ colors to color the remaining vertices such that each color class is of size $m + 2$. If there are still vertices left, then we use the cyclic trimming method of Theorem 2 to obtain an equitable coloring of G using $n + \lceil (r - \alpha'(BG(m + 1)))/(m + 2) \rceil$ colors. Thus $\chi_=(G) \leq n + \lceil (r - \alpha'(BG(m + 1)))/(m + 2) \rceil$. $\square$

Corollary 4. *Let $G[U; V]$ be a split graph and m be the number defined in Theorem 3. Then G is equitably k-colorable for any $k \geq n + \lceil (r - \alpha'(BG(m+1)))/(m + 2) \rceil$.*

Theorem 5. *Let $G[U; V]$ be a split graph. Then G is equitably $\Delta(G)$-colorable.*

Proof. Let m be the number defined in Theorem 3. It suffices to show that $\Delta(G) \geq \chi_=(G) = n + \lceil (r - \alpha'(BG(m + 1)))/(m + 2) \rceil$. Since the maximum degree occurs in the part U, we have $\Delta(G) > n - 1$. If $\Delta(G) = n$, then each vertex in U is adjacent to at most one vertex in V. Therefore $\Delta(G) \geq |G|/2$. It follows from Chen, Lih and Wu [4] that G is equitably n-colorable. Hence $\Delta(G) = n \geq \chi_=(G)$. It remains to prove the case for $\Delta(G) \geq n + 1$.

Since $\alpha'(BG(m+1)) \leq (m+1)n$, there exits at least one u_{ij} in the part U^* of $BG(m+1)$ which does not belong to a given maximum matching in $BG(m+1)$.

Those vertices of V which do not belong to the same maximum matching can not be adjacent to u_{ij}. For otherwise we could extend that matching to a larger matching. It follows that $\Delta(G) \geq deg_G(u_i) \geq n - 1 + r - \alpha'(BG(m + 1))$.

Now $\chi_=(G) = n + \lceil (r - \alpha'(BG(m + 1)))/(m + 2) \rceil > n + 1$ only when $r - \alpha'(BG(m + 1)) > m + 2 \geq 2$. But then $\Delta(G) \geq n - 1 + r - \alpha'(BG(m + 1)) \geq n + \lceil (r - \alpha'(BG(m + 1)))/(m + 2) \rceil$. □

3 m-Bounded Coloring

An m-bounded coloring is a different colorability concept which has close connections with equitable colorability. For a given positive integer m, if the vertex set $V(G)$ can be partitioned into k independent subsets $V_1, V_2, \cdots, V_k$ such that $|V_i| \leq m$ for all i, then the graph G is said to possess an *m-bounded coloring*. The *m-bounded chromatic number* $\chi_m(G)$ of G is the smallest number of colors required for an m-bounded coloring of G. The m-bounded chromatic number of a tree was determined in Chen and Lih [2].

Theorem 6. *Let $G[U; V]$ be a split graph and $m \geq 1$ a given integer. Then $\chi_m(G) = n + \lceil (r - \alpha'(BG(m - 1)))/m \rceil$.*

Proof. The proof is a slight modification of that of Theorem 3. Let Θ be an m-bounded coloring of G with $\chi_m(G)$ colors. Since each color class has at most m elements, Θ gives rise to a matching in $BG(m - 1)$. There are at least $r - \alpha'(BG(m - 1))$ vertices in V left unmatched. To find a lower bound for $\chi_m(G)$, we partition these vertices into $\lceil (r - \alpha'(BG(m - 1)))/m \rceil$ classes such that all are of size m except possibly one of smaller size. This shows that $\chi_m(G) \geq n + \lceil (r - \alpha'(BG(m - 1)))/m \rceil$. To prove the reverse direction, we fix a maximum matching in $BG(m - 1)$. This matching induces a partial coloring of G such that each color class is of size at most m. Then we do the same as before to use $\lceil (r - \alpha'(BG(m - 1)))/m \rceil$ colors to color the remaining vertices of V. An m-bounded coloring of G is so constructed which uses $n + \lceil (r - \alpha'(BG(m - 1)))/m \rceil$ colors. Therefore $\chi_m(G) \leq n + \lceil (r - \alpha'(BG(m - 1)))/m \rceil$. □

Remark. Since the complement of a split graph is also a split graph, the equitable and m-bounded coloring problems for split graphs can be equivalently formulated in terms of partitions into minimum number of complete subgraphs whose sizes are either as even as possible or bounded by the number m.

References

1. Bollobás, B., Guy, R. K.: Equitable and proportional coloring of trees. J. Comb. Theory(B) **34** (1983) 177–186
2. Chen, B. L., Lih, K. W.: A note on the m-bounded chromatic number of a tree. Europ. J. Comb. **14** (1993) 311–322
3. Chen, B. L., Lih, K. W.: Equitable coloring of trees. J. Comb. Theory(B) **61** (1994) 83–87

4. Chen, B. L., Lih, K. W., Wu, P. L.: Equtiable coloring and the maximum degree. Europ. J. Comb. **15** (1994) 443–447
5. Hajnal, A., Szemerédi, E.: Proof of a conjecture of Erdös. In Combinatorial Theory and Its Applications. Vol II (P. Erdös, A Rényi and V. T. Sós, eds), Colloq. Math. Soc. János Bolyai **4**, North-Holland, Amsterdam (1970) 601–623
6. Lih, K. W., Wu, P. L.: On equitable coloring of bipartite graphs. Discrete Math. (to appear)
7. Meyer, W.: Equitable coloring. Amer. Math. Monthly **80** (1973) 920–922

Four Coloring
for a Subset of Maximal Planar Graphs
with Minimum Degree Five

Philippe Rolland
IRIN
Université de Nantes
2, rue de la Houssinière
44072 Nantes cedex France
Rolland@Irin.Univ-Nantes.fr

Abstract. In this paper, we present some results on maximal planar graphs with minimum degree five, denoted by $MPG5$ graphs [6]. We consider a subset of $MPG5$ graphs, called the $\mathcal{Z}$ graphs, for which all vertices of degree superior to five are not adjacent. We give a vertex *four* coloring for every $\mathcal{Z}$ graph.

1 Introduction

We restrict ourselves to undirected, connected and planar graphs G which have no loops or multiple edges. Given a simple graph $G = (X, E)$ with node set X and edge set E, and given a positive integer k, a k-coloring of G is an assignment of a color to each node where no two nodes joined by an edge receive the same color. The smallest k for which there exists a k-coloring of G is the chromatic number, and is denoted $\chi(G)$. G is called planar when it is isomorphic to a graph $G(\pi)$ whose vertex set X is a point set in a plane π while the edges are Jordan curves in π such that two different edges have, at most, end points in common. It is possible to represent the graph in π in many isomorphic ways; each of them is called a planar representation. If we consider a planar graph with no loops or faces bounded by two edges, it may be possible to add a new edge to the given representation of G, such that these properties are preserved. When no such adjunction can be made, we call G maximal planar. Clearly, every representation of a graph is contained in a maximal planar graph with the same vertex set. A planar graph is called triangulated when all faces have three corners. In [8], it is proved that a planar graph is maximal if and only if it is triangulated. In [6], G is called an $MPG5$ graph if G is maximal, planar, with a minimum degree equal to five; $MPG5_n$ is the $MPG5$ graph set with n vertices.

The four color problem of maximal planar graphs will be easier if there exists some vertex x with $dg(x) < 5$, and this for the following reasons: If $dg(x) < 4$ then x can be colored by one of the four colors. If $dg(x)=4$ then by using Kempe chains [5], x can also be colored by one of the four colors. Recently Robertson, Sanders, Seymour and Thomas, in [7], have given a direct proof of the Four Color Theorem (briefly, the 4CT). Thus 4CT is not yet *dead* since F. Guthrie in 1852. Without dealing with the whole problem of $4CT$, which is too difficult, we propose one four-coloring for a particular class of maximal planar graphs.

A v-vertex is a vertex of degree v, a vertex x is minor if $dg(x) \leq 6$. Let x be a v-vertex, with at least $v - 2$ (in the even v case or $v - 1$ in the odd v case) consecutive 5-neighbours. Then x is reducible. For $v = 5, 6$ or 7, this can be improved by Birkhoff, Franklin, Bernhart and Winn, [1, 8, 9]. The general case is due to Errera [4]. It may be possible to prove directly the *four*-coloring of all $\mathcal{Z}$ graphs by reducible configurations using the Errera's generalization. Here we will propose another kind of proof.

The paper is organized as follows. The next section is devoted to the basic definitions, as for example, $\mathcal{Z}$, A and Zoe graphs. In section 3 (resp. 4) we propose an algorithm to construct every A type (Zoe) graphs. Section 5 (resp. 6) we give a vertex *four* coloring for every A type (Zoe) graphs. Actually, it implies that we have defined one vertex *four* coloring for every $\mathcal{Z}$ graphs.

2 Basic definitions

Let $G = (X, E)$ be a simple graph and x be a vertex. $N(x)$ defines the set of neighbors of node x. $N^i(x)$ represents the i-neighborhood of x, i.e. the vertices with a distance i from x; for example,

$$N^2(x) = \left\{ \bigcup_{y \in N(x)} N(y) \right\} - N(x) - \{x\}.$$

The degree of node x, denoted $dg(x)$, is defined by the size $card(N(x))$. Δ (resp. δ) is the maximum (minimum) degree of G.

Definition 1 *Consider a simple graph $G = (X, E)$. We denote*

$$X_{sup6} = \{x \in X / dg(x) \geq 6\},$$
$$X_{inf6} = \{x \in X / dg(x) < 6\}.$$

Definition 2 *Consider an $MPG5$ graph $G=(X,E)$ with $X_{sup6} \neq \emptyset$. G is called,*

- *$\mathcal{Z}$, if $\forall x \in X_{sup6}, N(x) \subset X_{inf6}$.*

- *A type, if $\forall x \in X_{sup6}, \{N(x) \cup N^2(x)\} \subseteq X_{inf6}$: see Figure 1.*

- *Zoe, if G is $\mathcal{Z}$ and not A: see Figure 2.*

Thus with the previous definition, we have the following partition of $\mathcal{Z}$.

Proposition 3 *Let $G = (X, E)$ be an $MPG5$ graph. G is a $\mathcal{Z}$ graph if and only if, either G is A or Zoe.*

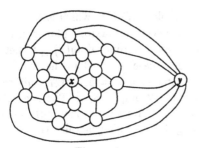

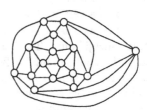

Figure 1:	Figure 2:
Example of an A type graph.	Example of a Zoe graph.

3   Generating an *A* type graph

We are going to give a *four* coloring constructive proof for every *A* type graph. For this we give a process to generate these graphs.

Proposition 4 *Let G be an MPG5 graph with n vertices. If $\exists x \in X_{sup6}$ with*

$$\{N(x) \cup N^2(x)\} \subseteq X_{inf6}$$

then G is an A type graph.

   Proof We propose a constructive proof. Suppose that we would like to build an *A* type graph *G*. Let *x* be a vertex in X_{sup6} with $dg(x) = k$. We propose three steps in order to generate an *A* type graph. We can start by drawing $N(x)$. Since *G* must be maximal, we have to draw a wheel with at least six vertices of degree three in the outer face. In the following example, see Figure 3, we have choosen $k = 7$.

Example: Build an A type graph with $\Delta(G) = 7$.

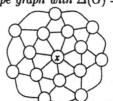

Figure 3:	Figure 4:
First step.	Second step.

   By definition, since $N(x) \subset X_{inf6}$, between $N(x)$ and $N^2(x)$ we must add exactly two edges for each vertex in $N(x)$. The fact that *G* is maximal and planar, implies that $card(N(x)) = card(N^2(x)) = k$ and all vertices in $N^2(x)$ *actually* have a degree four; see Figure 4.

   Since $N^2(x) \subset X_{inf6}$, between $N^2(x)$ and $N^3(x)$ we must add exactly one edge for each vertex in $N^2(x)$. *G* is maximal and planar, and so $N^3(x)$ is limited to a single vertex (called y) with degree k and $N^w(x) = \emptyset$ for $w > 3$; see Figure 1. □

By the previous proof, we can include the following observations:

- Let G be an A type graph with n vertices. n is even, and if we denote X_{sup6} by $\{x, y\}$ then $\Delta(G) = \frac{n-2}{2} = dg(x) = dg(y)$.

- The smallest A type graph G contains 14 vertices and $\Delta(G) = 6$. $MPG5_{14}$ contains only this graph.

- Let $G = (X, E)$ be an $MPG5$ graph and x be a vertex. There is no x such that

$$\left\{ \bigcup_{i=1}^{k} N^i(x) \right\} \subset X_{inf6} \text{ ,with } k > 3.$$

- All A type graphs with n vertices are isomorphic.

The algorithm (i) uses the previous method in order to generate an A type graph with $n = (2 \times k) + 2$ vertices.

```
(0)    Let x (resp. y) be a vertex;
(1)    Let {x₁, x₂, ..., xₖ} be an elementary cycle in clockwise order of k vertices
(2)         link each of them up a central vertex x
(3)    Do the same thing with a cycle {y₁, y₂, ..., yₖ} in counterclockwise order
(4)         and a central vertex y.
(5)    i = 1;
(6)    While i < k do
(7)         Add (xᵢ, yᵢ) and (xᵢ, yᵢ₊₁);
(8)         i = i + 1;
(9)    End while
(10)   Add (xₖ, yₖ) and (xₖ, y₁);
```

Algorithm (i): Generation of an A type graph.

4 A graph coloring

Theorem 5 *Every A type graph is four-colorable.*

Proof We propose the algorithm (ii) which four colors the vertices by the symbols 0, 1, 2 and 3. Let $G = (X, E)$ be an A type graph. We start with the following considerations: the vertices X are uncolored and $X_{sup6} = \{x, y\}$.

To prove the veracity of this algorithm we propose a Hamiltonian circuit on the dual of an A type graph. We use the notation in Figure 5:

$$N(x) = \{z, q_3, x_1^1, x_2^1 \ldots, x_r^1, t\} \tag{1}$$

$$N^2(x) = \{h, q_1, x_1^2, x_2^2, \ldots, x_r^2, q_2\} \tag{2}$$

$$N^3(x) = \{y\} \tag{3}$$

$$N(z) = \{h, q_1, q_3, x, t\} \tag{4}$$

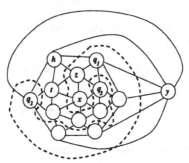

Figure 5: Hamiltonian circuit on the dual of an A type graph.

(1)	Color x by 0
(2)	Let $z \in N(x)$, starting from $z + 1$ in labelled clockwise order color alternatively each $t \in N(x)$ which is uncolored by the colors 2 and 3. *{t the last vertex colored}*
(3)	Let h be the common neighbour of z and t in $N^2(x)$
(4)	**If** t is colored by 2 **then**
(5)	Color h by 3 and y by 2
(6)	**else**
(7)	Color h by 2 and y by 3
(8)	**endif**
(9)	Color the uncolored vertices alternatively by 0 and 1

Algorithm (ii): A four-coloring for all A type graphs.

We consider the following paths,

$$P_1 = x, z, q_1, x_1^2, x_2^2 \ldots, x_i^2, \ldots, x_r^2, q_2 \tag{5}$$

$$P_2 = q_3, x_1^1, x_2^1 \ldots, x_i^1, \ldots, x_r^1, t, h, y \tag{6}$$

These two paths are independent and all vertices in X are visited. We could see this as a Hamiltonian circuit on the dual of G, where the internal face of this circuit will be defined by P_1. Thus, P_1 is 2-colourable, as P_2. In algorithm (ii), we have used the colors 0,1 for P_1 and the colors 2,3 for P_2. □

5 *Zoe* graph generation

As the *four* coloring A type graph proof, we are going to give a constructive prove of the *four* coloring for every *Zoe* graph. For this, we give in this section, a process to produce these graphs.

Definition 6 *Let $G = (X, E)$ be an $MPG5$ graph. Then we denote by X' the following vertex set:*

$$X' = \{x \in X/N(x) \subset X_{inf6}\}.$$

Definition 7 *A triangle T is an empty triangle, i.e. 3 corners labelled by vertices $a, b,$ and c. A pyramid W is a hexagon completely triangular, with six vertices on the external face labelled in clockwise order by a, c_1, b, a_1, c, b_1 and zero vertices on the internal face, for which, locally in W, $dg(a) = dg(b) = dg(c) = 2$ and $dg(a_1) = dg(b_1) = dg(c_1) = 4$. Vertices a, b, c are called the corners of the pyramid and of the triangle.*

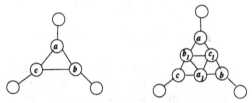

Figure 6: Polygon T, W.

We give a characteristic property for every *Zoe* graph.

Lemma 8 *Let $G = (X, E)$ be a Zoe graph. Then $\forall x \in X'$, $\exists y \in N^2(x) \cap X_{sup6}$.*

Proof Suppose $x \in X_{sup6}$. By using Proposition 4, since G is not A then $\exists y \in N^2(x) \cap X_{sup6}$. Suppose $x \in X_{inf6}$ with $\{N(x) \cup N^2(x)\} \subset X_{inf6}$. Then G is the smallest $MPG5$ graph, i.e. 12 vertices of degree five. This graph is not *Zoe* (since $X_{sup6} = \emptyset$), and there is a contradiction. □

The main result of this section is a vertex cover of $X_{inf6} - X'$ by a polygon set from T and W.

Corollary 9 *Let $G = (X, E)$ be a Zoe graph. There exists a vertex cover of $\{X_{inf6} - X'\}$ by a polygon set of T, W.*

In order to prove the previous corollary, we are going to describe the neighborhood by the polygons T and W for every vertex in X'. We distinguish two cases (every one gives a lemma): $x \in X_{sup6}$ and $x \in \{X' \cap X_{inf6}\}$. Before, we can observe the following results.

Lemma 10 *Consider $G = (X, E)$ as a Zoe graph, and $x, y \in X_{sup6}$ with $y \in N^2(x)$. There is a vertex v in the common second neighborhood of x and y with $v \in X'$ and a polygon p such that p separates these three previous vertices x, y, v.*

Proof Since G is Zoe, $\{N(x) \cup N(y)\} \subset X_{inf6}$, i.e. there is an edge (z,t) which bounds two triangles (x,z,t), (z,t,y). We consider a plane orientation defined by the clockwise order. Suppose $N(x)$ and $N(y)$ are described by:

$$N(x) = \{x_1, \ldots, x_k, t, z, x_{k+3}, \ldots, x_{dg(x)}\} \tag{7}$$

$$N(y) = \{y_1, \ldots, y_{k+2}, z, t, y_{k+5}, \ldots, y_{dg(y)}\} \tag{8}$$

Since G is a Zoe graph, $x_{k+3}, y_{k+2} \in X_{inf6}$. So there exists a vertex s such that $N(s) = \{\ldots, x_{k+4}, x_{k+3}, y_{k+2}, y_{k+1}\}$ in order to enclose x_{k+3}, y_{k+2}. We have two choices: either s is in X_{sup6} or in X_{inf6}.

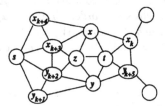

Figure 7: Neighborhood of T.

Suppose $s \in X_{sup6}$, see Figure 7. z is a corner of a polygon T. The external edge of T, i.e. (x_{k+3}, z), (z, y_{k+2}) and (y_{k+2}, x_{k+3}) bounds a triangle (x_{k+3}, z, x), (z, y_{k+2}, y) and (y_{k+2}, x_{k+3}, s). Since by hypothesis G is Zoe, $N(s) \subset X_{inf6}$, i.e. $s \in X'$. Of course, s belongs to the common second neighborhood of x and y; thus we can use s as v.

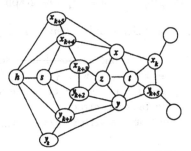

Figure 8a: Neighborhood of W.

Suppose $s \in X_{inf6}$, see Figure 8a. To enclose s, there exists a vertex h such that $N(h) = \{\ldots, x_{k+4}, s, y_{k+1}, \ldots\}$. Since $x_{k+4} \in N(x)$, we have $(h, x_{k+5}) \in E$, and then $N(x_{k+4}) = \{x_{k+5}, x, x_{k+3}, s, h\}$. Now, we use the same process around y_{k+1}, we have $N(y_{k+1}) = \{h, s, y_{k+2}, y, y_k\}$, and so to enclose the vertex y_{k+1} the edge $(h, y_k) \in E$.

Thus z is a corner of a polygon W. The external edge of W, i.e. (x_{k+3}, z), (z, y_{k+2}), (y_{k+2}, y_{k+1}), (y_{k+1}, s), (s, x_{k+4}) and (x_{k+4}, x_{k+3}) bounds a triangle: (x_{k+3}, z, x), (z, y_{k+2}, y), (y_{k+2}, y_{k+1}, y), (y_{k+1}, s, h), (s, x_{k+4}, h) and (x_{k+4}, x_{k+3}, x).

We have two cases. Suppose $h \in X_{sup6}$, then since G is Zoe, $h \in X'$. Suppose $h \in X_{inf6}$, see Figure 8b. In order to enclose h, there exists w, for which (w, x_{k+5}, y_k) is a T polygon.

By hypothesis $N(x) \subset X_{inf6}$ and $w, x_{k+5} \in N(x)$; $N(y) \subset X_{inf6}$ and $w, y_{k+1} \in N(y)$. Therefore $N(h) \subset X_{inf6}$. So, in these two cases $h \in X'$ and h belongs to the common second neighborhood of x and y; we can use h as v. $\square$

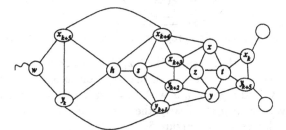

Figure 8b: Neighborhood of W: $h \in X_{inf6}$.

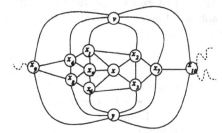

Figure 8c: Neighborhood of a vertex x in $X' \cap X_{inf6}$.

Lemma 11 *Let G be a Zoe graph and x be a vertex in $X' \cap X_{inf6}$. Then $N^2(x)$ contains exactly two vertices y, z in X_{sup6} and there is exactly one polygon W and one polygon T around x.*

Proof Let x be a vertex in $X' \cap X_{inf6}$. Suppose $\{N^2(x) \cap X_{sup6}\} = \emptyset$. Then $\{N(x) \cup N^2(x)\} \subset X_{inf6}$ and so by corollary 4, G is an A type graph. There is a contradiction since a Zoe graph can not be A type. Suppose $\{N^2(x) \cap X_{sup6}\} = \{y\}$. We consider the sub-graph H induced by $N(x) \cup \{x\}$; all vertices in the outer face have a degree four. Since by hypothesis, only y has a degree at least six, there are four consecutive vertices in outer face of H which must receive only one edge in G; there is a contradiction since $N(y)$ must be in X_{inf6} and the outer face is not a triangle. Suppose $card(\{N^2(x) \cap X_{sup6}\} \geq 3)$. Then since $dg(x) = 5$, there are two adjacent vertices in $N(x) \cap X_{sup6}$. There is a contradiction since G is a Zoe graph. Finally, suppose $\{N^2(x) \cap X_{sup6}\} = \{y, v\}$. Since G is Zoe, y and v are not adjacents. We use the notation proposed in Figure 8c: in clockwise order $N(x) = \{x_1, x_2, x_3, x_4, x_5\}$, $N^2(x) = \{x_6, v, x_7, y, x_8\}$, $N^3(x) = \{x_9, \ldots, x_{10}, \ldots\}$, and finally $N(v) = \{x_1, x_6, x_9 \ldots, x_{10}, x_2\}$. Since x_9, x_{10} belong to the first neighborhood of y and v, they also belong to X_{inf6}. Therefore we have one polygon W and one polygon T around $N(x)$; i.e. $x_9, x_6, x_1, x_5, x_4, x_8$ defining W and x_2, x_7, x_3 defining T. $\square$

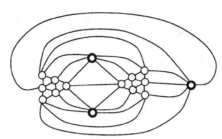

Figure 8d: The Zoe graph with 3 vertices of degree six.

Lemma 10 describes the *left* side of $[z, t; x, y]$. We can observe the same results on the *right*; i.e. there is a vertex v' in the common second neighborhood of x and y with $v' \in X'$ and a polygon p' with t as corner such that p' separate these three previous vertices x, y, v'. Clearly, there is $v = v'$ only with $|X_{sup6}| = 3$; i.e three vertices of degree six, see Figure 8d.

Lemma 8, 10, and 11, imply that there exists a vertex cover of $X_{inf6} - X'$ by a polygon set of T, W; and so we have proved Corollary 9.   □

Let G be a *Zoe* graph, see Figure 9. We call P the polygon set of T and W in G. We consider the algorithm *(iii)*.

(1)	Let G be a *Zoe* graph with n vertices
(2)	**While** $X' \neq \emptyset$ **do**
(3)	Let x be a vertex in X'
(4)	Remove x
(5)	**End while**
(6)	Let Q be this current graph
(7)	**While** $P \neq \emptyset$ **do**
(8)	Let p be a polygon in P
(9)	Replace p by a single vertex v
(10)	**End while**
(11)	Let K be this current graph

Algorithm (iii): A 3-regular planar graph.

The graph Q is the subgraph of G defined by the induced vertex set $\{X_{inf6} - X'\}$. By using Corollary 9, Q is a connected planar graph with $\delta = 3$ and $\Delta = 4$ if there is at least one W, otherwise $\Delta = 3$; see Figure 10. Since a polygon p is defined by 3 corners of degree 3, the vertex v in line (9) is such that $dg(v) = 3$; and so K is a 3-regular connected planar graph; see Figure 11.

We have some observations about K. We consider G and x a vertex in X'. The minimum of polygons T, W around x is two. This minimum is defined by two cases:

1. $dg(x) = 6$ and we have two W or three T around x.

2. $dg(x) = 5$, we have one T, one W.

If G contains some vertices in the previous two cases, K is not simple (excepted when $dg(x) = 6$ with three T); more precisely, for each of these vertices we have a double edge in K. Thus, to generate a Zoe graph we can use the three following steps:

1. Generate a 3-regular planar graph K with or without double edges.

2. Build Q by replacing every vertex in K by either T or W; but by respecting the previous constraints 1,2.

3. Build G by putting in Q, a vertex v in every non triangular face f. Draw a wheel by addition of edges, between v and the vertices which define f.

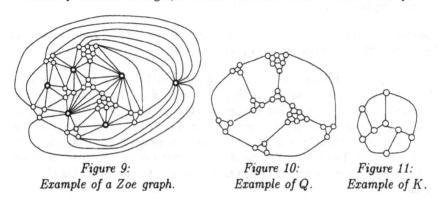

| *Figure 9:* | *Figure 10:* | *Figure 11:* |
| *Example of a Zoe graph.* | *Example of Q.* | *Example of K.* |

In order to see the previous three steps, we can associate the first step with Figure 11, the second step with Figure 10, and the third with Figure 9.

6 Zoe graph coloring

We recall Brooks theorem:

Theorem 12 *[2] Let $G = (X, E)$ be a connected simple graph with $\Delta(G) = d$. Then G is d-colorable if and only if $G \neq K_{d+1}$.*

Theorem 13 *Every Zoe graph is four colorable.*

Proof We present some observations about the coloring of G, Q, and K in the algorithm (iii).

– The vertex set X' is one-colorable, i.e. there exists a k-coloring of G such that all the vertices of X' use the same color, for example we could color the vertex set X' by 3.

– Suppose $\{X_{in f6} - X'\}$ contains only triangles T; then K is a simple 3-regular planar graph, and so $K \neq K_4$. By using the theorem of Brooks, K is 3 colorable.

Let G be a *Zoe* graph without polygon W. By the two previous observations, there exists a four coloring of G such that X' is one-colorable (we will suppose by the color 3).

Consider a polygon T. We use the notation of Figure 6. Suppose that we use the following coloring: $(a, 0)$, $(b, 1)$, $(c, 2)$. Now, in graph G, we replace the polygon T by a polygon W. Without conflicts of coloring, we can color the polygon W as follows $(a, 0)$, $(b, 1)$, $(c, 2)$ and $(a_1, 0)$, $(b_1, 1)$, $(c_1, 2)$.

Continue this process, until the graph G will be without polygon T (step by step, each polygon T has been replaced with a polygon W). So, we have proved the following theorem:

Theorem 14 *Let $G = (X, E)$ be a Zoe graph. There exists a four-coloring of G for which the vertex set defined by X' is 1-colorable.*

We propose a different way of proving theorem 14. We consider the *one*-coloring of X', and we propose another 3-coloring of K. We consider a vertex set S defined as a maximal independent vertex set in K. We can prove that the subgraph induced by the vertex set $\{vertices(K) - S\}$ is a spanning tree. Since a tree is 2-colorable, and an independent vertex set is *one*-colorable, then K is 3-colorable. This decomposition of K by a maximal independent vertex set can be interesting in order to find generalizations of the four coloring of the $\mathcal{Z}$ graphs.

By theorems 5 and 13 we have given a vertex *four*-coloring of all $\mathcal{Z}$ graphs, and so we have proved the following theorem,

Theorem 15 *All $\mathcal{Z}$ graphs are four-colorable.*

7   Conclusion

We have given a *four*-coloring for every $\mathcal{Z}$ graph. This result can be more interesting if we consider another result about the generation of $MPG5$ graphs as given in [3]. Start on the smallest A type graph and use a local transformation called T. Obviously, an open problem is to use T and the precedent coloring, and to find a generalization of $\mathcal{Z}$ graphs. We can observe that during a private communication, Paul Seymour has given another proof for the *four* coloring of the $\mathcal{Z}$ graphs, by using the Kempe argument, Brooks theorem and D-reducibility. This proof states that we can always one-color the vertices in X_{sup6} and three color those in X_{inf6}. But in the perspective of finding a generalization using our results from [3] it was interesting to give a method to realize a vertex coloring for all $\mathcal{Z}$ graphs.

References

1. A. Bernhart. *Another reducible edge configuration*, volume 70, pages 144–146. Amer. J; Math., 1948.

2. R. L. Brooks. On coloring the nodes of the network. In *Proc. Cambridge Philos. Soc.*, volume 37, pages 194–197, 1941.

3. J. Hardouin Duparc and P. Rolland. Transformations for maximal planar graphs with minimum degree five. In *Computing and Combinatorics*, volume 959, pages 366–371. LNCS Springer Verlag, 1995.

4. A. Errera. Une contribution au problème des quatre couleurs. *Bull. de la soc. Math. de France*, 53:52–74, 1925.

5. A. B. Kempe. *On the geographical problem of the four colours*, volume 2, pages 193–200. Amer. J. Math., 1879.

6. C.A. Morgenstern and H.D. Shapiro. *Heuristics for Rapidly Four-Coloring Large Planar Graphs*, pages 869–891. Algorithmica. Springer-Verlag New York, 1991.

7. P. Seymour N. Robertson, D. Sanders and R. Thomas. The four colour problem, 1994.

8. Oystein Ore. *The four-color problem*. Academic Press, 1967.

9. Thomas L. Saaty and Paul. C. Kainen. *The four-color problem, assaults and conquest*. Mc Graw-Hill International Book Company, 1977.

Enumeration Algorithm for the Edge Coloring Problem on Bipartite Graphs

Yasuko MATSUI[1] and Tomomi MATSUI[2]

[1] Tokyo Metropolitan University , Tokyo 192-03, Japan
[2] University of Tokyo, Tokyo 113, Japan

Abstract. In this paper, we propose an algorithm for finding all the edge colorings in bipartite graphs. Our algorithm requires $O(T(n, m, \Delta) + K \min\{n^2 + m, T(n, m, \Delta)\})$ time and $O(m\Delta)$ space, where n denotes the number of vertices, m denotes the number of edges, Δ denotes the number of maximum degree, $T(n, m, \Delta)$ denotes the time complexity of an edge coloring algorithm, and K denotes the number of edge colorings.

1 Introduction

The finding of all objects that satisfy a specified property is a fundamental problem in combinatorics, computational geometry, and operations research. This paper deals with the edge coloring problem in bipartite graphs. There are many applications for the edge coloring problem. For example, scheduling problems and timetable problems are discussed in [1, 3, 8, 14]. In this paper, we propose an algorithm for finding all the minimum edge colorings in bipartite graphs.

Let us consider a bipartite graph $B = (U, V, E)$ with vertex sets U, V and edge set E. We denote $|U \cup V|$ by n and $|E|$ by m. Δ denotes the maximum degree of any vertex in $U \cup V$. In this paper, we allow a graph with parallel edges and assume that $n \leq m$.

An *edge coloring* of a graph associates a color with each edge in the graph in such a way that no two edges of the same color have a common endpoint. A *minimum edge coloring* is an edge coloring which uses the fewest number of colors as possible. The edge coloring problem finds a minimum edge coloring. In 1916, König proved the following theorem [10].

Theorem 1. *Let B be a bipartite graph and let Δ be the maximum degree of any vertex. Then a minimum edge coloring of B uses exactly Δ colors.*

In this paper, $T(n, m, \Delta)$ denotes the time complexity of an algorithm for finding a minimum edge coloring in a bipartite graph. The currently best time bound is achieved by Cole and Hopcroft's algorithm [2]. They proposed an edge coloring algorithm whose time complexity is $O(\max\{m \log n, n \log n (\log \Delta)^3\})$.

Recently, we proposed an algorithm for finding all the minimum edge colorings in bipartite graphs [13], which requires $O(Knm)$ time and $O(nm^2)$ space, where K denotes the number of minimum edge colorings. However, our new algorithm is more efficient in time bound and space complexity. Our algorithm requires $O(T(n, m, \Delta) + K \min\{n^2 + m, T(n, m, \Delta)\})$ time and $O(m\Delta)$ space. So, our algorithm generates each additional edge coloring in $O(\min\{n^2 + m, T(n, m, \Delta)\})$ time.

2 Algorithm

In this section, we describe the main framework of our algorithm.

First, we give some definitions and notations. An edge subset $M \subseteq E$ is called a *matching* if each pair of edges in M are not incident with a common vertex. A matching is said to *cover* a vertex subset W, if every vertex in W is an endpoint of an edge in the matching. A degree of a vertex x, denoted by $degree(x)$, is the number of edges which are incident with x.

From Theorem 1, it is clear that a minimum edge coloring of a bipartite graph B corresponds to a set of Δ matchings which is a partition of the edge set E. In this paper, we identify a minimum edge coloring of B with a set of Δ matchings. For any edge e and any minimum edge coloring C, we denote $M(C, e)$ as a matching in C which contains the edge e. If B has exactly one minimum edge coloring, we say B is *uniquely edge colorable*. When the graph B has two distinct minimum edge colorings C and C', there exists an edge $e \in E$ satisfying that $M(C, e) \neq M(C', e)$. A matching M of B is called a *feasible matching*, when there exists a minimum edge coloring including M. Theorem 1 directly implies the following lemma.

Lemma 2. *A matching M of B is feasible if and only if M covers every vertex of B whose degree is equal to Δ.*

For any edge e, $\mathcal{F}(B, e)$ denotes the set of all feasible matchings of B including the edge e. For any edge e and for any minimum edge coloring C, it is clear that $M(C, e) \in \mathcal{F}(B, e)$. Given a feasible matching M of B, $\mathcal{C}(B, M)$ denotes the set of all minimum edge colorings of B containing the matching M. Clearly, the family $\{\mathcal{C}(B, M) | M \in \mathcal{F}(B, e)\}$ is a partition of all minimum edge colorings of B. When we have a feasible matching $M \in \mathcal{F}(B, e)$, the problem to enumerate all minimum edge colorings in $\mathcal{C}(B, M)$ is reduced to the problem to enumerate all the minimum edge colorings in the bipartite graph $B \setminus M \equiv (U, V, E \setminus M)$. So, we can divide the problem of enumerating all the minimum edge colorings of B into subproblems. This idea implies the following algorithm.

Algorithm $main(B)$

A1: **begin**
A2:   set $C := \emptyset$
A3:   find a minimum edge coloring C_t of B
A4:   call $find\_all\_colorings(B, C_t, C)$
A5: **end**

Subprocedure *find_all_colorings*(B, C'_t, C')

B1: **begin**

B2:   **if** C'_t is the unique minimum edge coloring of B **then** output $C' \cup C'_t$

B3:   **else**

B4:   **begin**

B5:   choose an edge e of B

B6:   generate all the matchings in $\mathcal{F}(B, e)$

B7:   **for each** $M \in \mathcal{F}(B, e)$ **do**

B8:   **begin**

B9:   find a minimum edge coloring C'' of $B \setminus M$

B10:   *find_all_colorings*$(B \setminus M, C'', C' \cup \{M\})$

B11:   **end**

B12:   **end**

B13:   **return**

B14: **end**

To complete this algorithm, we must solve the following three problems.

(1) The problem to determine whether a given graph is uniquely edge colorable or not at Line **B2**.
(2) The problem to choose an appropriate edge at Line **B5**.
(3) The problem for finding all the matchings in $\mathcal{F}(B, e)$ at Line **B6**.

In the rest of this paper, we discuss the above problems.

3   Uniquely edge colorable bipartite graph

In this section, we discuss the problems (1) and (2) described in the previous section.

When a given graph B has two distinct minimum edge colorings C and C', there exists an edge e satisfying that $M(C, e) \neq M(C', e)$. In our algorithm, we choose such an edge e at Line B5. Since $M(C, e) \neq M(C', e)$, set $\mathcal{F}(B, e)$ includes at least two feasible matchings. Therefore, if we choose the edge e at Line B5, the original enumeration problem is divided into at least two subproblems. From the above discussion, we need to solve the following problem in our algorithm.

Subproblem `uniquely-colorable`(B, C)

Input: a bipartite graph B and a minimum edge coloring C in B
Output: an edge e satisfying that $M(C, e) \neq M(C', e)$, if B has a minimum edge coloring C' different from C; else, say "uniquely edge colorable"

There are some previous studies on uniquely edge colorable graphs [9, 12]. In the following, we give a simple condition of uniquely edge colorable bipartite graph.

Theorem 3.   *A connected bipartite graph $B = (U, V, E)$ is not uniquely edge colorable if and only if there exist two vertices x and y ($x \neq y$) such that $degree(x) \geq 3, degree(y) \geq 2$.*

Proof. If the degree of each vertex is less than 3, then the graph B is a path or a cycle, and so B is uniquely edge colorable. Now suppose that there exists a vertex x such that $degree(x) \geq 3$ and the degrees of any other vertices are less than 2. Then the graph B becomes a star and so B is uniquely edge colorable.

Next we show the inverse implication. Assume that there exist two vertices $x, y (x \neq y)$ such that $degree(x) \geq 3$ and $degree(y) \geq 2$. Let v be a vertex with $degree(v) = \Delta \geq 3$. If every vertex adjacent to v is a leaf vertex, then B becomes a star and it is a contradiction. So there exists a vertex u adjacent to v such that $degree(u) \geq 2$. We denote the edge connecting u and v by e. Since $degree(u) \geq 2$, there exists an edge $e'_1 \neq e$ incident with u. Let C be a minimum edge coloring of B. Here we denote the feasible matching $M(C, e'_1)$ by M'. Since the degree of v is equal to Δ, by Theorem 1, the matching M' contains an edge e'_2 which is incident with v. The property $degree(v) \geq 3$ implies that there exists an edge e'' incident to v satisfying that $e'_2 \neq e'' \neq e$. The matching $M(C, e'')$ is denoted by M''. Let $\widetilde{B}$ be the graph induced by the edge set $M' \cup M''$. Clearly, each component of $\widetilde{B}$ is a path or a cycle and the component including the vertex v is not an edge. If $\widetilde{B}$ is disconnected, then $\widetilde{B}$ is not uniquely edge colorable and it directly implies that the original graph B is not uniquely edge colorable. Now consider the case that $\widetilde{B}$ is connected. Then $\widetilde{B}$ contains a path P connecting v and u and including the edges e'_1 and e'_2. The path P and the edge e forms a cycle, which is denoted by C. Obviously, the cycle C is an alternating cycle with respect to the feasible matching M'. Let M^* be the symmetric difference of C and M'. Then M^* is also a feasible matching of B and M^* is not contained in C. Lemma 2 directly implies that there exists a minimum edge coloring C^* which contains the feasible matching M^*. Since $M^* \not\subseteq C$, the minimum edge coloring C^* is different from C and so B is not uniquely edge colorable.

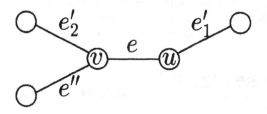

Fig. 1. Two vertices u and v such that $degree(u) \geq 2, degree(v) = \Delta \geq 3$.

The above theorem directly implies the following.

Corollary 4.   *A bipartite graph B is uniquely edge colorable if and only if B is a matching, a cycle, a path, or a star.*

From Theorem 3, it is clear that we can determine whether a given bipartite graph is uniquely edge colorable or not by checking the degree of every vertex. The proof of Theorem 3 gives an idea for solving the subproblem uniquely-colorable(B, C). When the subgraph $\widetilde{B}$ defined in the proof is disconnected, we can output arbitrary edge in $\widetilde{B}$ as a solution of the subproblem uniquely-colorable(B, C). If $\widetilde{B}$ is connected, we output the edge e connecting the vertices u and v. So, the above procedure solves the subproblem in $O(\min\{n + \Delta, n + m\})$time.

4 Finding all feasible matchings

In this section, we consider the problem to enumerate all the feasible matchings which contain a specified edge e.

Our algorithm partitions the set of feasible matchings including e into two subsets iteratively. More precisely, given two distinct feasible matchings M and M' in $\mathcal{F}(B, e)$, we choose an edge $f \in (M \cup M') \setminus (M \cap M')$ and partition the set $\mathcal{F}(B, e)$ into two subsets $\mathcal{M} = \{M'' \in \mathcal{F}(B, e) | f \in M''\}, \mathcal{M}' = \{M'' \in \mathcal{F}(B, e) | f \notin M''\}$. This partition implies that exactly one of $M, M' \in \mathcal{M}$(say M), while the other (say M') belongs to $\mathcal{M}'$. Clearly, the recursive application of the above procedure constructs a binary tree of subproblems. By using depth first rule, we can find all the feasible matchings without repetition. Due to solve the above subproblem, it is very natural to consider the following subproblem.

Subproblem matching-unique (B, M, I, D)
Input: a bipartite graph B, a feasible matching M of B and sets of edges I, D
Output: a feasible matching M' satisfying that $M' \supseteq I, M' \cap D = \emptyset$ and $M \neq M'$, if one exists; else, say "none exists".

We can drop the conditions $e \in M$ and $e \in M'$ by setting I with $e \in I$.

In the following, we describe an algorithm for solving the above subproblems. From Lemma 2, we can replace the condition that M and M' is a feasible matching of B with the condition that M and M' cover a given vertex subset W. If we delete all edges in D and all vertices covered by I (and edges incident with the vertices), the subproblem matching-unique is transformed to the following problem.

Subproblem find-matching (B, W, M)
Input: a bipartite graph B, a vertex subset W of B and a matching M which covers W.
Output: a matching M' which differs from M and covers W, if one exists; else, say "none exists"

The following lemma provides an idea for solving the subproblem find-matching.

Lemma 5. *Let $B = (U, V, E)$ be a bipartite graph, W a vertex subset and M a matching which covers W. There exists a matching M' which differs from M and covers W if and only if either (1) or (2) holds.*

(1) There exists an alternating cycle with respect to M.

(2) There exists an alternating path with respect to M which connects two vertices not in W.

Proof. One direction is trivial. For the other direction, suppose that there exists a matching M' which covers W and differs from M. Consider the graph B' induced by the edges $(M \cup M') \setminus (M \cap M')$. In this graph, each connected component is an alternating path or an alternating cycle with respect to M. When the graph B' contains an alternating cycle, we have done. Consider the case that B' has an alternating path. Then, the terminal vertices of the path are covered by one matching and not by another matching. If a terminal vertex is contained in W, it contradicts our assumption that each vertex in W is covered by both M and M'. So, the alternating path connects two vertices not in W.

Figure 2 (a1) corresponds to the condition (1) of Lemma 5 and (a2) to (2).

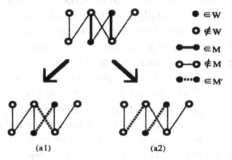

(a1) (a2)

Fig.2. Conditions (1),(2) of Lemma 5.

Now, we explain how to solve **find-matching**(B, W, M). Let $G'(B, M)$ be a directed graph obtained from B by directing the edges in $E - M$ from U to V, and the edges in M from V to U. Each directed elementary cycle in $G'(B, M)$ corresponds to an alternating cycle with respect to B and M. The inverse implication also holds. The depth first search method finds a directed elementary cycle in a given directed graph(see [11]). Figure 3 shows the directed graph for the example bipartite graph in Figure 2.

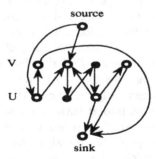

Fig.3. The network corresponding to a bipartite graph.

Next, we describe how to find an alternating path with respect to M which connects two vertices not in W. Let U''(respectively V'') be a set of vertices in U(respectively V) which are covered by M. A directed graph $G''(B, M)$ is a graph obtained from B by directing edges in $E - M$ from U to V, the edges in M from V to U, by adding one source vertex s and the edges from s to $(V'' \setminus W) \cup (U \setminus U'')$, and finally by adding one sink vertex t and the edges from $(V \setminus V'') \cup (U'' \setminus W)$ to t. Each directed s-t path in $G''(B, M)$ corresponds to an alternating path with respect to M which connects two vertices not in W. The inverse implication also holds. The depth first search method finds a directed elementary s-t path [11].

From the above, we can solve the subproblem matching–unique in $O(n+m)$ time and $O(n+m)$ space by employing the depth first search method [11] (where n denotes the number of vertices and m denotes the number of edges in B).

Our algorithm generates the feasible matchings by solving a sequence of subproblems iteratively. When we apply the problem dividing procedure described before, we obtain a binary tree structure of subproblems. At any inner node (non-leaf node) of the binary tree structure, the corresponding subproblem finds a new matching and a given enumeration problem is divided into two subproblems. At every leaf node of the binary tree structure, the corresponding subproblem says "none exists" and we output the current feasible matching. So, the number of nodes of the binary tree structure is equal to $2|\mathcal{F}(B, e)| - 1$. Thus, the total time complexity for finding all the feasible matchings in $\mathcal{F}(B, e)$ is $O((n + m)|\mathcal{F}(B, e)|)$, when we have one matching in $\mathcal{F}(B, e)$. Since we solve the subproblems recursively, we traverse the binary tree structure by using the depth first rule. So, the space requirement becomes $O(n + m)$.

5 Complexity

Finally, we discuss the computational complexity and memory requirements of our algorithm for finding all the minimum edge colorings.

At Line A3 of the algorithm *main*, we employ Cole and Hopcroft's method [2] and find a minimum edge coloring in $T(n, m, \Delta) = O(\max\{m \log n, n \log n(\log \Delta)^3\})$ time. In Section 3, we proposed a method to choose an edge e at Line B5 such that $\mathcal{F}(B, e)$ has at least two feasible matchings. Thus, the original enumeration problem is divided into at least two subproblems at Line B7. The algorithm *main* calls the subprocedure *find_all_colorings* recursively and generates a tree structure of the subprocedures. At every leaf node of the tree structure, we output a minimum edge coloring. At any inner node, we divide a given enumeration problem into at least two subproblems. Thus, the number of inner nodes of the tree structure is less than the number of minimum edge colorings in the original graph. So, the algorithm *main* calls the subprocedure *find_all_colorings* at most $2K - 1$ times, where K denotes the number of minimum edge colorings in the original graph. The computational time required at Lines B1-B5,B7,B8,B10-B14 is bounded by $O(m)$. At Line B6, we can generate all the feasible matchings in $\mathcal{F}(B, e)$ in $O(m|\mathcal{F}(B, e)|)$ time. Clearly, the subprocedure $find\_all\_colorings(B, C'_t, C)$ calls the subprocedure $find\_all\_colorings$ $|\mathcal{F}(B, e)|$ times at Line B10. Thus, the total time required at Line B6 is bounded by $O(mK)$. From the above, the total computational effort required at Lines B1-B8,B10-B14 is bounded by $O(mK)$. At Line B9, we need to find a new coloring of the graph $B \setminus M$. If we employ Cole and Hopcroft's method, it requires $O(\max\{m \log n, n \log n(\log \Delta)^3\})$ time. However, we have an edge coloring C'_t of B. Then the set $\{M' \setminus M : M' \in C'_t\}$ is an edge coloring of $B \setminus M$ using Δ' colors where Δ' is the maximum degree of any vertex in B. Since M is a feasible matching of B, the maximum degree of any vertex in $B \setminus M$ is $\Delta' - 1$. If we employ König's color flipping method [10], we can find an edge coloring of $B \setminus M$ from C'_t in $O(n^2 + m)$ time. Thus, the time complexity of Line B9 is bounded by $O(\min\{n^2 + m, T(n, m, \Delta)\})$. From the above, the total time complexity of the algorithm $main$ is bounded by $O(T(n, m, \Delta) + K \min\{n^2 + m, T(n, m, \Delta)\})$.

At last, we discuss the space requirement. If we generate all the feasible matchings in $\mathcal{F}(B, e)$ at Line B6, then we need a huge amount of memory space, since $\mathcal{F}(B, e)$ contains exponential number of matchings in the worst case. However, we can reduce the space requirement easily. At Line B6, we execute the matching enumeration algorithm proposed in Section 2 and at each time an additional matching is obtained, we give a pause to the matching enumeration algorithm and execute the loop B8-B11. Since the matching enumeration algorithm requires $O(m)$ space, the space complexity of the subprocedure $find\_all\_colorings$ is bounded by $O(m)$. Clearly, the height of the tree structure of the subprocedures generated by the algorithm $main$ is less than or equal to $\min\{\Delta, m\}$. Since the algorithm $main$ traverses the tree structure by using the depth first rule, the total memory requirement is bounded by $O(\min\{m\Delta, m^2\})$.

From the above discussions, our algorithm requires $O(T(n, m, \Delta) + K \min\{n^2 + m, T(n, m, \Delta)\})$ time and $O(\min\{m\Delta, m^2\})$ space. The bottleneck of the time complexity is Line B9. If we have an $O(T')$ time algorithm for finding an edge coloring using Δ colors (i.e., a minimum coloring) of a bipartite graph B from an edge coloring using $\Delta + 1$ colors, the time complexity of our algorithm is reduced to $O(T(n, m, \Delta) + K(m + T'))$ time.

References

1. Bondy,J.A., Murty,U.S.R.: Graph theory with applications. North-Holland (1976)
2. Cole,R., Hopcroft,J.: On Edge Coloring Bipartite Graphs. SIAM J.Comput. **11** (1982) 540-546
3. Dempster,M.A.H.: Two algorithms for the time-table problem, Combinatorial Mathematics and its Applications(ed. D.J.A.Welsh). Academic Press New York (1971) 63-85
4. Fukuda,K., Matsui,T.: Finding All the Minimum Cost perfect Matchings in bipartite Graphs. Networks **22** (1992) 461-468
5. Fukuda,K., Matsui,T.: Finding All the Perfect Matchings in Bipartite Graphs. Appl. Math. Lett. **7** 1 (1994) 15-18
6. Gabow,H.N.: Using Euler Partitions to Edge Color Bipartite Multigraphs. Information J. Comput. and Information Sciences **5** (1976) 345-355

7. Gabow,H.N., Kariv,O.: Algorithms for Edge Coloring Bipartite Graphs and Multi-graphs. SIAM J. Comput. **11** (1982) 117-129
8. Gonzalez,T., Sahni,S.: Open Shop Scheduling to Minimize Finish Time. J. ACM. **23** (1976) 665-679
9. Greenwell,D.L., Kronk,H.V.: Uniquely Line Colorable Graphs. Canad. Math. Bull. **16** (1973) 525-529
10. König,D.: Über Graphen und ihre Anwendung auf Determinantentheorie und Mengenlhere. Math. Ann. **77** (1916) 453-465
11. Tarjan,R.E.: Depth-first search and linear graph algorithms. SIAM J. Comput. **1** (1972) 146-160
12. Thomason,A.G.: Hamiltonian Cycles and Uniquely Edge Colourable Graphs. Ann. Discrete Math. **3** (1978) 259-268
13. Yoshida,Matsui.Y., Matsui,T.: Finding All the Edge Colorings in Bipartite Graphs. T.IEE. Japan 114-C **4** (1994) 444-449 (in Japanese)
14. de Werra,D.: On some combinatorial problems arising in scheduling. INFOR. **8** (1970) 165-175

On-Line Recognition of Interval Graphs
in O($m + nlog\ n$) Time

Wen-Lian Hsu

Institute of Information Science, Academia Sinica, Taipei, Taiwan, R.O.C.

Abstract. Since the invention of *PQ*-trees by Booth and Lueker in 1976 the recognition of interval graphs has been simplified dramatically. In [7], we presented a very simple linear-time recognition algorithm based on scanning vertices arranged in a special perfect elimination ordering. Our approach is to decompose a given interval graph into uniquely representable components whose models can be obtained by considering "strictly overlapping" pairs of intervals. This method, however, does not yield an efficient on-line algorithm since it uses the perfect elimination scheme, which is hard to maintain efficiently in an on-line fashion.

Utilizing the decomposition approach and an "abstract" interval representation we are able to design an O($m + nlog\ n$) time on-line recognition algorithm in this paper. The O($nlog\ n$) factor comes from the fact that we need to maintain a concatenable queue to search for certain minimal interval "cuts" in the abstract representation.

1. Introduction

Interval graphs have a wide range of applications (cf. [4]). Several linear time algorithms have been designed to recognize interval graphs [1,2,6,7,8]. Booth & Lueker [1] first used *PQ*-trees to recognize interval graphs in linear time as follows: obtain a perfect elimination ordering of the vertices of the given chordal graph. From such an ordering determine all maximal cliques. If the graph is an interval graph, then a linear order of the maximal cliques satisfying certain consecutive property can be obtained using *PQ*-trees and an interval model can be constructed. However, the data manipulation of *PQ*-trees is rather involved and the complexity analysis is also quite tricky. Korte and Möhring [8] simplified the operations on a *PQ*-tree by reducing the number of templates. Hsu and Ma [6] gave a simpler decomposition algorithm without using *PQ*-trees. All of these algorithms rely on the following fact (cf. [3]): a graph is an interval graph iff there exists a linear order of its maximal cliques such that for each vertex v, all maximal cliques containing v are consecutive. It should be noted that, maximal cliques are hard to be determined efficiently in an on-line fashion.

In 1991, Hsu [7] gave the first linear time recognition algorithm which directly places the intervals without considering maximal cliques. The consecutive property of maximal cliques will be satisfied in a natural fashion. The key is to find a good ordering of intervals to be placed. In this paper, we adopt some idea of [7] in designing an on-line recognition algorithm but drop the vertex ordering requirement. Our algorithm maintains a data structure that allows the intervals to be added one by one at a time and reports whether the current graph is an interval graph. The time complexity of our algorithm is $O(n\log n + m)$. However, the algorithm does not support the deletion of intervals.

For each interval model D we can group its endpoints into maximal consecutive set of left or right endpoints, called the endpoint *blocks* of D. A *cut* C is defined to be a pair $\{B_L, B_R\}$ of neighboring left and right blocks, where the right block B_R is to the right of the left block B_L. Thus, there is a one-to-one correspondence between cuts and blocks of the model. The *cut width* of C is defined to be the number of intervals of D intersecting with a vertical line separating endpoints in B_L and B_R. By the above definition, we can obtain a unique sequence of cut widths of D from left to right. These numbers are stored in a queue, which will be an important component for the on-line algorithm.

Below, we briefly review the basic idea of the algorithm in Hsu [7]. First of all, an important definition of strictly adjacent vertices was introduced in [7]. This enables one to find the modular decomposition tree efficiently based on finding connected components on strictly adjacent edges. If the given graph is prime, then there is a simple algorithm to construct a unique (normalized) interval model. The algorithm starts by finding the modular decomposition tree and then computing a special vertex ordering for each prime component based on a spanning tree of those connected components. The idea of finding the modular decomposition tree will be useful for the on-line algorithm. However, the special vertex ordering will no longer be necessary.

We now sketch the basic idea of our on-line algorithm. There are two data structures used in our algorithm: the substitution decomposition tree, and the concatenable queue of cut widths. The first part of the algorithm is the construction of the substitution decomposition tree of the current graph, which follows the approach in [7]; the second part is, for each prime component C, maintaining a concatenable queue that keeps track of the sequence of cut widths of any interval model of C. The substitution decomposition tree enables us to concentrate on the interval test for each prime component. However, in deciding whether an intermediate graph is an interval graph, the algorithm does not actually construct an interval model at every iteration (which would be very time-consuming). Instead, it constructs the block sequence for each prime component (which was affected by the newly-added interval) by checking some numbers associated with the concatenable queue of cuts.

When a new vertex v is added to the current graph, the tree is modified and the queues of those related prime components are concatenated. At each iteration, the tree modification can be carried out in $O(\deg(v))$ time using a simple vertex partitioning strategy; the updating of the queue can be done in $O(\log n + \deg(v))$ time.

2. Background

We shall review several notations used in [7]. The interval representation of an interval graph is usually far from unique. To eliminate uninteresting variations of the endpoint orderings , we shall consider the following restricted interval models:

Definition. *Define the open neighborhood $N(u)$ to be the set of vertices adjacent to u in G. Define the closed neighborhood $N[u]$ to be $N(u) \cup \{u\}$. An interval i is said to contain an interval j if every point of j is contained in i. An interval model for a graph G is said to be normalized (denoted by an N-model) if it satisfies that, for any two vertices u, v in G, the interval for u contains the interval for v if and only if $N[u]$ contains $N[v]$.*

Two vertices u, v are said to be *similar* if $N(u)-\{v\} = N(v)-\{u\}$. Two adjacent (resp. nonadjacent) similar vertices are said to form a *true (resp. false) twin*. We use the following lemma from [5].

Lemma 2.1 [5]. *If an interval graph G does not contain any twin, then there is an N-model for G.*

Two normalized interval models are said to be *equivalent* if the endpoint ordering of one is either identical to or is the reversal of that of the other. An interval graph G is said to have a *unique N-model* if all N-models for G are equivalent. Another way of saying this is that the left-to-right block sequence is unique. To obtain uniquely representable interval graphs we shall apply the substitution decomposition described below. For details the reader is advised to consult the paper by Spinrad [10].

A *module* in an undirected graph G is a set of vertices S such that for any vertex $u \in S$ and $v \notin S$, $(u,v) \in E$ iff $(u',v) \in E$ for every $u' \in S$. A module S is *nontrivial* if $1 < |S| < |V|$. A graph is *prime* if there exists no nontrivial module. A *substitution decomposition* (or *modular decomposition*) of a graph is to substitute a nontrivial module with a marker vertex and perform this recursively for the module as well as for the reduced graph containing that marker. The result of a substitution decomposition is often represented by a tree, where each subtree represents a nontrivial module marked by its root.

A vertex is *simplicial* if its neighbors form a clique. A module which forms a clique is called a *type I module* (or *S-module*). A module is called a *type II module* (or *N-module*) if it is connected but not type I. A module is called a *type III module* (or *P-module*) if it not connected. As shown in [6], there are only three different types of modules for interval graphs. For the graph G in Figure 1 we illustrate its modular decomposition tree in Figure 2, where N indicates a type II module, P indicates a type III module and S indicates a type I module). Note that all the vertices are represented as leaves in the decomposition tree.

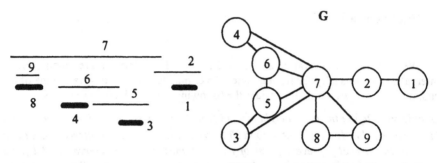

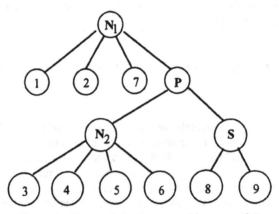

The corresponding interval graph G

Figure 1. The graph G

Figure 2. The modular decomposition tree of G

Note that type III modules are those independent sets (two sets are independent if no vertex in one is adjacent to any vertex in the other) with the same external neighborhoods, which can be easily found in linear time by partitioning the vertex set based on their neighborhoods. In the next section, we illustrate how to compute this tree on-line based on vertex partitioning.

3. On-Line Construction of the Substitution Decomposition Tree

Below, we assume that the given graph G is indeed an interval graph. If this is not the case, some violation would be discovered along the way and one could then terminate the recognition algorithm. We shall apply the vertex partitioning strategy described below to construct the current decomposition tree efficiently. At each iteration we need to keep the block sequence for the representative graph of each prime component. This sequence is stored as a linked list. For each vertex in G we record the blocks containing its two endpoints.

Theorem 3.1. *At each iteration (when a new vertex v is added), the substitution decomposition tree can be updated in $O(\deg(v))$ time.*

This theorem can be proved by following through the construction procedure described below.

Initially, the decomposition tree is empty. Let T be the current tree associated with the current graph G. Let v be the new vertex added to G. Mark all vertices of G that are adjacent to v. The nodes of T can be classified as follows.

1. A node of T is said to be *full* if every vertex in its corresponding submodule is adjacent to v.

2. A node is said to be *partial* if some but not all vertices in its corresponding submodule are adjacent to v.

3. A node is said to be *empty* if no vertex in the submodule it represents is adjacent to v.

A leaf node of T is either full or empty. Note that it suffices to mark the neighbors of v in G to identify all the partial and full nodes in a bottom up fashion. Hence the number of steps involved in the labeling will be proportional to $\deg(v)$. The ancestors of a partial node must all be partial.

Two intervals are said to *cross* each other if they overlap but none is contained in the other. A submodule M is said to *cross* the interval v and vice versa if (i) v overlaps with some but not all intervals in M; and (ii) v overlaps with some interval u (outside M) which does not overlap with any interval in M. A submodule M is said to be *contained in* v if every interval in M overlaps with v. A module M is said to *contain* v if (i) v overlaps with some interval in M; and (ii) for any u in $G\backslash M$, v overlaps u iff every interval in M overlaps with u.

The construction of the decomposition tree for $G \cup \{v\}$ is done recursively in a top-down fashion starting with the root module. At each partial submodule, we illustrate how to insert the endpoints of v into the block sequence of the representative graph of the submodule. The root module $M(G)$ must contain v. We shall first describe the change of the block structure in case both endpoints of v are inserted into the same component M. Consider the representative graph for M and its corresponding block sequence. Consider the following cases:

1. M is a type I module (a clique): split the node for M into two nodes representing two submodules M_1 and M_2 with $M_1 \cup M_2 = M$, where M_1 consists of all vertices of M that are adjacent to v and $M_2 = M - M_1$. Connect these two nodes to the parent of the node for M. Connect all vertices in M_1 to the node for M_1 and those in M_2 to the node for M_2. Construct a node for v and connect it to M_1. Construct a new node for v and connect this node to that for M_1. This is illustrated in Figure 3.

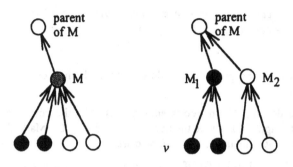

Figure 3. The split of a type I module

2. *M* is a type *II* module: Let B_L be the block containing the right-most left endpoint of neighbors of v and B_R the block containing the left-most right endpoint of neighbors of v. Consider two subcases:

(1) v is a simplicial vertex: B_L should be to the left of B_R. Thus, the endpoints of this new interval v should be inserted into a new block located in between B_L and B_R. The exact way to locate the position of this block will be described in Section 4. Consider the following three subcases:

    (a) v has a true twin, say v', in the representative graph: then a type *I* node consisting of $\{v,v'\}$ is created. We then replace the node for v' in M by this node and connect the nodes for v and v' to this new node.

    (b) v has a false twin, say v', in the representative graph: then a type II node consisting of $\{v,v'\}$ is created. We then replace the node for v' in M by this node and connect the nodes for v and v' to this new node.

    (c) v has neither a true twin nor a false twin: then v is not contained in any nontrivial submodule of $M \cup \{v\}$. Thus, we construct a new node for v and connect this node to the node for M.

(2) v is not a simplicial vertex: B_R should be to the left of B_L. The left endpoint of v should split B_R and the right endpoint of v should split B_L. In this case v cannot have a false twin. Consider the following subcases:

    (a) v has a true twin, say v', in the representative graph (with endpoints in the same left and right block as v does): then a type I node consisting of $\{v,v'\}$ is created. We then replace the node for v' in M by this node and connect the nodes for v and v' to this new node.

    (b) v does not have a true twin: then v is not contained in any nontrivial submodule of $M \cup \{v\}$. Thus, we construct a new node for v and connect this node to the node for M. Recursively insert the endpoints of v into any submodule of M crossing v.

3. *M* is a type *III* module: there can be at most two partial submodules in M. Group the full and partial submodules of M to form a type *II* module with v. The full submodules become simplicial vertices in the representative graph for the new type *II*

module. We illustrate this in Figure 4, where M_2 and M_4 are partial submodules and M_3 is a full submodule. For those partial modules that cross v, we shall discuss the subtree construction below.

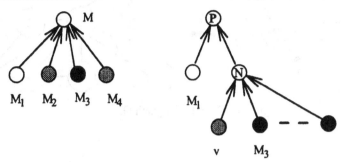

Figure 4. The merge of submodules M_2, M_3 and M_4 into a type II module

Next, we consider the case that v crosses a module M and only one endpoint of v is inserted into the block sequence for the representative graph of M. Without loss of generality, assume we are inserting the left endpoint of v. Note that in this case the node for v has already been connected to higher level nodes, so vertex v only split the lower level submodules. Consider the following cases:

1. M is a type I module (a clique): this case is analogous to the above.

2. M is a type II module: let B_R be the block containing the left-most right endpoint of neighbors of v. Then the left endpoint of v should split B_R. The new block sequence will then be merged with the one containing the right endpoint of v.

3. M is a type III module: Consider the following two cases:

(1) if v does not cross any submodule of M, then divide M into two submodules M_1 and M_2, where M_1 consists of all vertices of M that are adjacent to v. Replace the node for M by the node for M_1. Connect all submodules in M_1 to the node for M_1. Connect all submodules in M_2 to the parent node for v.

(2) if v crosses a submodule, say M', of M, then divide M into three submodules M_1, M' and M_2, where M_1 consists of all submodule of M that are contained in v, M_2 consists of all submodules of M that are independent of v. Submodules in M_1 and M_2 are processed as described in (1). Recursively insert the endpoint of v into M'.

Note that in the above operations, if it is found that some parent has only one child then this parent can be deleted and replaced by this child. We illustrate an example of our algorithm below. Suppose we add another vertex 10 to the graph G with neighbors 2, 7 and 9 (call the resulting graph G_1). We first insert both endpoints of v into the root module $N1$ using case 2(2)(b). Since v crosses the submodule P, we insert the left endpoint of v into the submodule P using case 3(2). Since v crosses the submodule S, we insert the left endpoint into the submodule S using case 1. The new type II decomposition tree is shown in Figure 6.

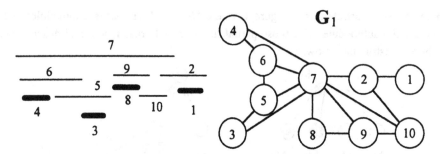

Figure 5. The interval graph G_1

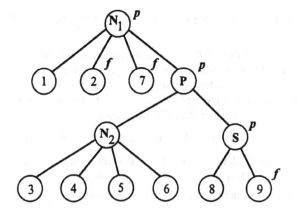

(i) The original decomposition tree of G

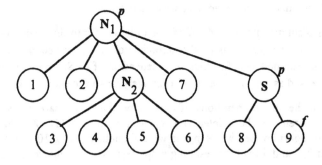

(ii) The tree obtained by eliminating the partial P-node

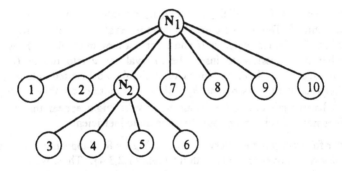

(iii) The final tree obtained by eliminating the partial S-node

Figure 6. The modular decomposition tree of $G \cup \{10\}$

4. Updating the Queue of Cut Widths

In addition to computing the block sequence for each prime component, we also construct a priority queue of cut widths. We illustrate the queues for the graph G in Figure 7.

The block sequences for G are

(a) the sequence for N_1 is $\{7,P\},\{P'\},\{2,\},\{7'\},\{1\},\{1',2'\}$

(b) the sequence for N_2 is $\{4,6\},\{4'\},\{5\},\{6'\},\{3\},\{3',5'\}$

The queues of cut widths are the same for both components

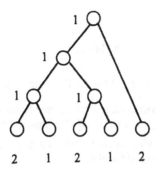

Figure 7. The queues of cut widths for the components of G

The main reason for keeping track of this queue is that they can be used more effectively for interval graph test. This can be explained as follows. Supposing the newly added vertex v is simplicial (namely, its neighbors form a clique) in some prime component H. Denote the current unique interval model of the representative graph for H by D. To test whether H and v form a new interval graph, we need to check whether it is possible to insert the interval for v into model D. By the discussion in Section 3, both endpoints of v should be inserted into new blocks created in between the blocks B_L and B_R. Since it is quite likely that intervals in H-$N[v]$ form a barrier preventing the inclusion of v, it might appear that we need to check vertices not in $N[v]$, which could become quite inefficient.

Consider for example, the interval model in Figure 8. Supposing we are adding a new interval v which overlaps only with intervals $\{1,2,3,4\}$. Thus, we need to locate the place for the new blocks containing the endpoints of v. However, without additional information, it seems that a search involving vertices not adjacent to v is unavoidable.

Inserting a new interval which only overlaps with intervals 1, 2, 3 and 4

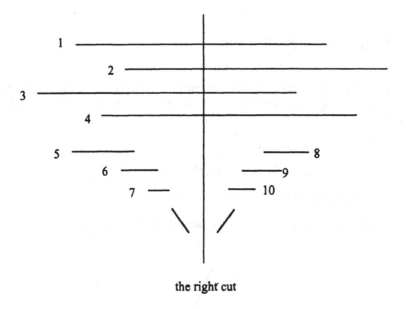

the right cut

Figure 8. Finding the right cut for the simplicial vertex

We resolve this problem by maintaining a concatenable queue of cut widths in D and checking whether the minimum cut width of D within the two blocks B_L and B_R equals $|N(v) \cap H|$. Once this cut width is found one can quickly identify the location of the cut by tracing down the queue and update it.

In general, we shall associate a queue of cut widths for each prime component of the graph in addition to the block sequence. Whenever a new interval is inserted we first determine the new decomposition tree as described in the last section and the change of the block sequences. Then we need to specify the change of the associated queues.

We first locate the two blocks B_L and B_R. If the new vertex v is a simplicial vertex, we use the queues to determine the location of the new blocks. Otherwise, the endpoints of v should split B_L and B_R, we can perform the normal tree spliting to update the queue.

Next, we discuss an efficient method for updating the queue when several prime components are combined into a new component $M \cup \{v\} = M_v$, where M is the smallest submodule of G containing v. The queue update is done iteratively in a top-down fashion, starting from the module M. We first determine the unique way to insert the endpoints of v into the block sequence of M in a manner described above. For any submodule, say M', split by v (there can be at most two of these) we repeat the above process with M' and so on. The final queue and block sequence for M_v is obtained by recursively substituting the new queues and block sequences of the submodules back to its immediate supermodules.

Now consider the cut widths associated with M_v. Along the path from a submodule M' to M, every intermediate submodule represents a simplicial vertex in the representative graph of its parent module. To update the cut width for the queue of M_v, we can precompute the cut width increment for every block in these components and determine the final cut width all at once. Initially, we simply concatenate the individual queues of neighboring node in the path using the current cut widths associated with the nodes. Each concatenation takes at most $O(\log n)$ time, the total number of concatenations equals the number of partial nodes in T. Since the total number of nodes ever created for the decomposition tree is at most $O(n)$ the total concatenation time is at most $O(n\log n)$.

If the above algorithm terminates without encountering any contradiction, then we conclude that the current graph is an interval graph; otherwise, it is not. Below, we illustrate the corresponding change of the priority queue for the cut widths of N_2 in Figure 9.

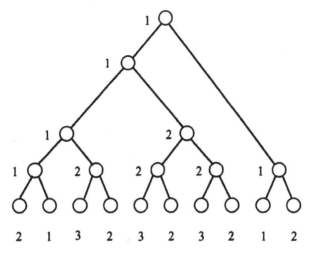

Figure 9. The cut width queue of N_1

References

1. K. S. Booth and G. S. Lueker , *Linear algorithms to recognize interval graphs and test for the consecutive ones property,* Proc. 7th ACM Symp. Theory of Computing, (1975), 255-265.
2. K. S. Booth and G. S. Lueker, *Testing for the consecutive ones property, interval graphs and graph planarity using PQ-tree algorithms,* J. Comput. Syst. Sci. 13, (1976), 335-379.
3. D. R. Fulkerson and O. A. Gross, *Incidence Matrices and Interval Graphs,* Pacific J. Math. 15, (1965), 835-855.
4. M. C. Golumbic, *Algorithmic Graph Theory and Perfect Graphs,* Academic Press, New York, 1980.
5. W. L. Hsu , *O(mn) Recognition and Isomorphism Algorithms for Circular-Arc Graphs,* SIAM J. Comput. 24, (1995), 411-439.
6. W. L. Hsu and C. H. Ma, *Fast and Simple Algorithms for Recognizing Chordal Comparability Graphs and Interval Graphs,* Lecture Notes in Computer Science 557, 52-60, (1991), to appear in SIAM J. Comput.
7. W. L. Hsu, *A simple test for interval graphs,* Lecture Notes in Computer Science 657, (1992), 11-16.
8. N. Korte and R. H. Möhring, *An incremental linear time algorithm for recognizing interval graphs,* SIAM J. Comput. 18, (1989), 68-81.
9. G. S. Lueker and K. S. Booth, *Interval graph isomorphism,* JACM 26, (1979), 195.
10. J. Spinrad, *On Comparability and Permutation Graphs,* SIAM J. Comput. 14 (1985), 658-670.

Connected Proper Interval Graphs and the Guard Problem in Spiral Polygons (Extended Abstract)

Chiuyuan Chen *† and Chin-Chen Chang

*Institute of Applied Mathematics, National Chiao Tung University,
Hsinchu 30050, Taiwan, R.O.C.*

1. Introduction

The main purpose of this paper is to study the hamiltonicity of proper interval graphs and applications of these graphs to the guard problem in spiral polygons. The *intersection graph* of a family F of nonempty sets is the graph derived by representing each set in F by a vertex and connecting two vertices by an edge if and only if their corresponding sets intersect. An *interval graph* is the intersection graph G of a family I of intervals on the real line. I is usually called the *interval model* for G. A *proper interval graph* is an interval graph with an interval model I such that no interval in I properly contains another.

A *walk* in a graph $G = (V, E)$ is a sequence $(v_1, v_2, ..., v_k)$ of vertices such that $v_i v_{i+1} \in E$ for $1 \leq i \leq k - 1$. A walk is *closed* if $v_1 = v_k$ and is *open* if $v_1 \neq v_k$. A *path* is a walk in which no vertices are repeated. A *Hamiltonian path* in G is a path that contains all the vertices of G. A *circuit* is a closed walk in which no vertices are repeated, except $v_1 = v_k$. A *Hamiltonian circuit* in G is a circuit that contains all vertices of G. G is said to be *Hamiltonian-connected* if for any two distinct vertices, there is a Hamiltonian path joining them.

The *Hamiltonian path (circuit) problem* is, given an undirected graph $G = (V, E)$, to determine whether G contains a Hamiltonian path (circuit). These two problems are well-known NP-complete problems [6,13]. They remain NP-complete for planar cubic 3-connected graphs [7] and bipartite graphs [15]. The Hamiltonian circuit problem also remains NP-complete for split graphs [10],

*This research was partially supported by the National Science Council of the Republic of China under grant No. NSC83-0208-M009-054.

† e-mail: cychen@cc.nctu.edu.tw.

edge graphs [1], planar bipartite graphs [12], and grid graphs [12]. Polynomial time algorithms for the Hamiltonian circuit problem are known only for 4-connected planar graphs [11], proper interval graphs [2], interval graphs [14,16], and circular-arc graphs [20].

In [2], Bertossi proved that a proper interval graph has a Hamiltonian path if and only if it is connected. He also gave a slightly complicated condition for a proper interval graph to have a Hamiltonian circuit. Using this, he proposed an algorithm for finding a Hamiltonian circuit in a proper interval graph.

In Section 2, we shall derive simple conditions for a proper interval graph to have a Hamiltonian path, have a Hamiltonian circuit, and be Hamiltonian-connected, respectively. From these conditions, algorithmic results are also given.

A polygon P is *simple* if no pair of non-consecutive edges sharing a point. All polygons discussed in this paper are assumed simple. A vertex v of P is *convex* (or *concave*) if its interior angle is less than (or greater than) 180 degrees. A *convex* (or *concave*) *chain* of P is a sequence of consecutive convex (or concave) vertices. P is a *spiral polygon* if it has exactly one concave subchain.

A point p in a polygon P is said to *see* or *cover* another point q if the line segment $\overline{pq}$ lies entirely within P; note that $\overline{pq}$ may touch the boundary of P. For example, in Figure 1(a), k, j, and p can see each other, but c can not see k. The set of points that cover P are called (*point*) *guards* of P. A guard who is also a vertex of P is called a *vertex guard*. Every vertex guard is a point guard, but the converse is not true. Point guards are usually more powerful than vertex guards [17]. For example, in Figure 1(a), $\{p, q\}$ is a set of point guards that cover P, but three vertex guards are required to cover P.

The *vertex visibility graph* of a polygon P is a graph whose vertices correspond to the vertices of P and two vertices of the graph are adjacent if and only if their corresponging vertices in P can see each other. Everett and Corneil [4] proved that the vertex visibility graph of a spiral polygon is an interval graph. However, the vertex visibility graph of a spiral polygon may not be a proper interval graph; Figure 1(b) shows an example of a vertex visibility graph that is not a proper interval graph since $\{c, a, i, e\}$ induces a $K_{1,3}$. It is shown in [19] that a proper interval graph contains no induced copy of $K_{1,3}$.

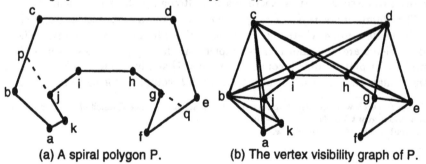

(a) A spiral polygon P.   (b) The vertex visibility graph of P.

Figure 1.

In Section 3, we consider the problem of *finding a minimum sized point guards that cover a spiral polygon* P. We shall prove that the class of stick intersection graphs associated with spiral polygons equals the class of nontrival connected proper interval graphs. We shall also give a linear time algorithm for solving the above quard problem by means of the properties of proper interval graphs.

2. Hamiltonian path, Hamiltonian circuit, and Hamiltonian-connectivity

This section studies the hamiltonicity of proper interval graphs. We start with some terminologies. The *connectivity* $\kappa(G)$ of a graph $G = (V, E)$ is the minimum number of vertices whose removal from G results in a trivial or disconnected graph. G is *k-connected* if $\kappa(G) \geq k$. For $S \subseteq V$, the graph $G - S$ is derived by removing from G all vertices in S and all edges incident with vertices in S. Let $c(G)$ denote the number of components of G. The *scattering number* $s(G)$ of a graph G is

$$\max \{c(G - S) - |S| \mid S \subseteq V \text{ and } c(G - S) \neq 1\}.$$

The *closed neighborhood* $N[v]$ of a vertex v is the set of vertices adjacent to v plus v it self. An ordering $[v_1, v_2, ..., v_n]$ of the vertices of G is a *consecutive ordering* if for every i, $N[v_i]$ is consecutive, i.e., $N[v_i] = \{v_j : i_1 \leq j \leq i_2\}$ for some $i_1 \leq i_2$. For example, in Figure 2, $N[v_1] = \{v_1, v_2, v_3\}$, $N[v_2] = \{v_1, v_2, v_3\}$, $N[v_3] = \{v_1, v_2, v_3, v_4, v_5\}$, $N[v_4] = \{v_3, v_4, v_5\}$, and $N[v_5] = \{v_3, v_4, v_5\}$. Therefore $[v_1, v_2, v_3, v_4, v_5]$ is a consecutive ordering of the graph in Figure 2.

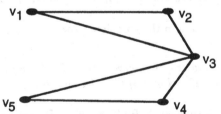

Figure 2. A proper interval graph.

Roberts [18] proved that G is a proper interval graph if and only if its augmented adjacency matrix, which is the adjacency matrix plus the identity matrix, satisfies the consecutive 1's property for columns; also see [3]. This fact can be restated as Theorem 1.

Theorem 1. *A graph $G = (V, E)$ is a proper interval graph if and only if G has a consecutive ordering.*
Proof. The proof appears in the complete paper. ∎

Note that $[v_1, v_2, ..., v_n]$ is a consecutive ordering of $G = (V, E)$ if and only if

$i < j < k$ and $v_i v_k \in E$ imply $v_i v_j \in E$ and $v_j v_k \in E$. .... (1)

A graph G is *triangulated* if every cycle of length greater than 3 possesses a chord, i.e., an edge joining two nonsecutive vertices of the cycle. A *clique* of G is a complete subgraph. A vertex v of G is called *simplicial* if the set of vertices adjacent to v induces a clique. A *perfect vertex elimination scheme* of G is an ordering $[v_1, v_2, ..., v_n]$ of the vertices such that v_i is a simplicial vertex of the subgraph induced by $\{v_i, v_{i+1}, ..., v_n\}$ for $1 \le i \le n$.

Note that $[v_1, v_2, ..., v_n]$ is a perfect vertex elimination scheme of $G = (V, E)$ if and only if

$i < j < k$ and $v_i v_j \in E$ and $v_i v_k \in E$ imply $v_j v_k \in E$. .... (2)

It is well-known that every interval graph is triangulated [9] and a graph is triangulated if and only if it has a perfect vertex elimination scheme [5]. Note that condition (1) implies condition (2).

Lemma 2. *If $[v_1, v_2, ..., v_n]$ is a consecutive ordering of G, then both $[v_1, v_2, ..., v_n]$ and $[v_n, v_{n-1}, ..., v_1]$ are perfect vertex elimination schemes of G.*

Now we are ready to study the hamiltonicity of proper interval graphs.

Theorem 3. *For any proper interval graph $G = (V, E)$ of $n \ge k + 1$ vertices, the following statements are equivalent for any positive integer k.*
(i) $s(G) \le 2 - k$.
(ii) G is *k-connected.*
(iii) *For any consecutive ordering $[v_1, v_2, ..., v_n]$ of G, $v_i v_j \in E$ for $1 \le |i - j| \le k$.*
Proof. The proof appears in the complete paper. ∎

Theorem 4. [2] *For any proper interval graph $G = (V, E)$ of $n \ge 2$ vertices, G has a Hamiltonian path if and only if G is 1-connected.*
Proof. ($\Rightarrow$) Obvious.

($\Leftarrow$) Suppose G is 1-connected. By Theorem 3, for any consecutive ordering $[v_1, v_2, ..., v_n]$ of G, $v_i v_j \in E$ for $|i - j| = 1$. Thus $\langle v_1, v_2, v_3, ..., v_n \rangle$ is a Hamiltonian path of G. ∎

Theorem 5. *For any proper interval graph $G = (V, E)$ of $n \ge 3$ vertices, G has a Hamiltonian circuit if and only if G is 2-connected.*
Proof. ($\Rightarrow$) Obvious.

($\Leftarrow$) Suppose G is 2-connected. By Theorem 3, for any consecutive ordering $[v_1, v_2, ..., v_n]$ of G, $v_i v_j \in E$ for $1 \le |i - j| \le 2$. Thus $\langle v_1, v_3, v_5, v_7, ..., v_{n-2}, v_n, v_{n-1}, v_{n-3}, v_{n-5}, ..., v_4, v_2, v_1 \rangle$ is a Hamiltonian circuit of G if n is odd and $\langle v_1, v_3, v_5, v_7, ..., v_{n-1}, v_n, v_{n-2}, v_{n-4}, ..., v_4, v_2, v_1 \rangle$ is a Hamiltonian circuit of G if n is even. ∎

Theorem 6. *For any proper interval graph $G = (V, E)$ of $n \geq 4$ vertices, G is Hamiltonian-connected if and only if G is 3-connected.*

Proof. ($\Rightarrow$) This part of proof appears in the complete paper.

($\Leftarrow$) Suppose G is 3-connected. By Theorem 3, For any consecutive ordering $[v_1, v_2, ..., v_n]$ of G, $v_i v_j \in E$ for $1 \leq |i - j| \leq 3$. Suppose v_l and v_m, $l < m$, are two arbitrary distinct vertices of G. A Hamiltonian path from v_l to v_m can be constructed as follows.

Case where l is odd and $m + n$ is even : $\langle v_l, v_{l-2}, v_{l-4}, v_{l-6}, ..., v_1, v_2, v_4, v_6, ..., v_{l-1}, v_{l+1}, v_{l+2}, v_{l+3}, ..., v_{m-1}, v_{m+1}, v_{m+3}, v_{m+5}, ..., v_{n-1}, v_n, v_{n-2}, v_{n-4}, v_{n-6}, ..., v_m \rangle$.

Case where l and $m + n$ are both odd : $\langle v_l, v_{l-2}, v_{l-4}, v_{l-6}, ..., v_1, v_2, v_4, v_6, ..., v_{l-1}, v_{l+1}, v_{l+2}, v_{l+3}, ..., v_{m-1}, v_{m+1}, v_{m+3}, v_{m+5}, ..., v_{n-2}, v_n, v_{n-1}, v_{n-3}, v_{n-5}, ..., v_m \rangle$.

Case where l and $m + n$ are both even : $\langle v_l, v_{l-2}, v_{l-4}, v_{l-6}, ..., v_2, v_1, v_3, v_5, v_7, ..., v_{l-1}, v_{l+1}, v_{l+2}, v_{l+3}, ..., v_{m-1}, v_{m+1}, v_{m+3}, v_{m+5}, ..., v_{n-1}, v_n, v_{n-2}, v_{n-4}, v_{n-6}, ..., v_m \rangle$.

Case where l is even and $m + n$ is odd : $\langle v_l, v_{l-2}, v_{l-4}, v_{l-6}, ..., v_2, v_1, v_3, v_5, v_7, ..., v_{l-1}, v_{l+1}, v_{l+2}, v_{l+3}, ..., v_{m-1}, v_{m+1}, v_{m+3}, v_{m+5}, ..., v_{n-2}, v_n, v_{n-1}, v_{n-3}, v_{n-5}, ..., v_m \rangle$.

Therefore G is Hamiltonian-connected. ∎

3. Guard problem in spiral polygons

In this section, we give a linear time algorithm to solve the guard problem in spiral polygons by using the properties of connected proper interval graphs.

We assume that a spiral polygon P is given in *standard form*: the vertices of P are listed as a concave chain $[u_1, u_2, ..., u_m]$ in clockwise order and a convex chain $[w_1, w_2, ..., w_n]$ in clockwise order such that u_1 and w_1 are adjacent and u_m and w_n are adjacent; see Figure 3(a).

An edge e of P is called *concave* if it contains at least one concave vertex. A *stick* s of P is a longest line segment containing a concave edge and lying inside P. Denote $s_1 = \overline{a_1 b_1}$, $s_2 = \overline{a_2 b_2}$, ..., $s_{m+1} = \overline{a_{m+1} b_{m+1}}$ the sticks containing concave edges $\overline{w_1 u_1}$, $\overline{u_1 u_2}$, $\overline{u_2 u_3}$, ..., $\overline{u_{m-1} u_m}$, $\overline{u_m w_n}$, respectively; see Figure 3(a) and (b). For each i, R_i denotes the region bounded by stick s_i and the convex chain of P and $\overset{\frown}{a_i b_i}$ denotes the boundary of R_i extending from a_i to b_i in clockwise order; see Figure 3(c). The following lemma is clear.

Lemma 7. *The intersection graph $G_P(S)$ of $S = \{s_1, s_2, ..., s_{m+1}\}$ equals the intersection graph $G_P(R)$ of $R = \{R_1, R_2, ..., R_{m+1}\}$ equals the intersection graph $G_P(B)$ of $B = \{\overset{\frown}{a_1 b_1}, \overset{\frown}{a_2 b_2}, ..., \overset{\frown}{a_{m+1} b_{m+1}}\}$.*

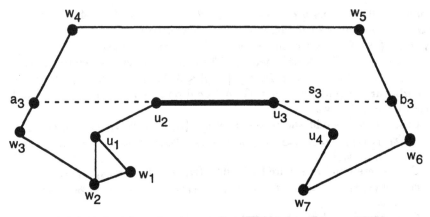

(a) Standard form, concave edge $\overline{u_2u_3}$, and stick $s_3 = \overline{a_3b_3}$.

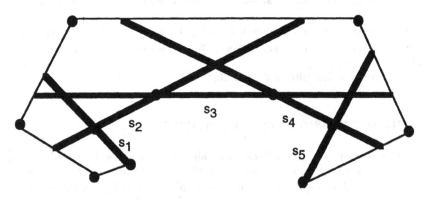

(b) Sticks.

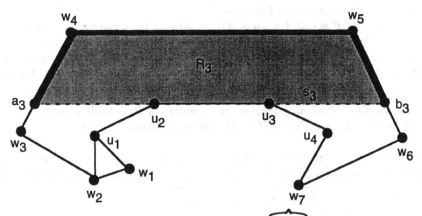

(c) Region R_3 (shaded) and boundary $\overset{\frown}{a_3b_3}$ (bolded).

Figure 3.

The intersection graphs $G_P(S), G_P(R), G_P(B)$, which are equal by Lemma 7, are called the *stick, region,* and *boundary intersection graph* associated with P, respectively. Note that the intersection graph of the sticks in Figure 3(b) equals the proper interval graph in Figure 2, in which the vertex v_i corresponds to the stick s_i for $1 \leq i \leq 5$. This is not an accident, since we have the following theorem.

Theorem 8. *The class of stick intersection graphs associated with spiral polygons equals the class of nontrivial connected proper interval graphs.*
Proof. The proof appears in the complete paper. ∎

An *independent set* of a graph $G = (V, E)$ is a subset $S \subseteq V$ such that no two vertices in S are adjacent. A *maximum independent set* of G is an independent set of maximum cardinality. A *clique cover* of G is a partition of the vertex set $V = A_1 + A_2 + ... + A_t$ such that each A_i induces a clique of G. A *minimum clique cover* of G is a clique cover of minimum cardinality.

In [8], Gavril proposed an algorithm for finding a maximum independent set and a minimum clique cover of a triangulated graph. Let $N_{R_i} = \{R_j \mid R_j$ intersects R_i and $j \geq i\}$. By Lemma 7 and the proof of Theorem 8, $G_P(R)$ is a proper interval graph in which $[R_1, R_2, ..., R_{m+1}]$ is a consecutive ordering. By Lemma 2, this ordering is a perfect vertex elimination scheme. Therefore we can slightly modify Gavril's algorithm to find a maximum independent set of $G_P(R)$ according to $[R_1, R_2, ..., R_{m+1}]$ as follows: Inductively define a sequence of regions $R_{i_1}, R_{i_2}, ..., R_{i_t}$ such that $R_{i_1} = R_1$; R_{i_j} is the first region in the sequence $[R_1, R_2, ..., R_{m+1}]$ which follows $R_{i_{j-1}}$ and which is not in $N_{R_{i_{j-1}}}$ (since P is spiral, if R_{i_j} is not in $N_{R_{i_{j-1}}}$, then R_{i_j} is not in $N_{R_{i_1}} \cup N_{R_{i_2}} \cup ... \cup N_{R_{i_{j-1}}}$); all regions following R_{i_t} are in $N_{R_{i_t}}$ (since P is spiral, if all regions following R_{i_t} are in $N_{R_{i_1}} \cup N_{R_{i_2}} \cup ... \cup N_{R_{i_t}}$, then they are in $N_{R_{i_t}}$). Hence $N_{R_{i_1}} \cup N_{R_{i_2}} \cup ... \cup N_{R_{i_t}} = \{R_1, R_2, ..., R_{m+1}\}$. We have the following lemma.

Lemma 9. *The set $\{R_{i_1}, R_{i_2}, ..., R_{i_t}\}$ is a maximum independent set of $G_P(R)$ and $\{N_{R_{i_1}}, N_{R_{i_2}}, ..., N_{R_{i_t}}\}$ is a minimum clique cover of $G_P(R)$.*
Proof. The proof appears in the complete paper. ∎

Lemma 10. *P requires at least t points guards.*
Proof. The proof appears in the complete paper. ∎

Lemma 11. *$\{b_{i_1}, b_{i_2}, ..., b_{i_t}\}$ is a minimum sized point guards for P.*
Proof. The proof appears in the complete paper. ∎

We now have an algorithm for finding a minimum sized point guards that cover a spiral polygon P.

Algorithm Minimum_Point_Guard.

Input: A spiral polygon P in standard form. (Assume that P has m concave vertices.)

Output: A set D, which is a minimum sized point guards that cover P.

Step 1. Find the $m+1$ sticks $s_1 = \overline{a_1 b_1}$, $s_2 = \overline{a_2 b_2}$, ..., $s_{m+1} = \overline{a_{m+1} b_{m+1}}$ of P.

Step 2. Find a sequence of points b_{i_1}, b_{i_2}, ..., b_{i_t} in the following way: $b_{i_1} = b_1$; b_{i_j} is the first point in $[b_1, b_2, ..., b_{m+1}]$ which follows $b_{i_{j-1}}$ such that stick s_{i_j} does not intersect stick $s_{i_{j-1}}$; b_{i_t} has the property that all sticks following stick s_{i_t} intersect with s_{i_t}. $D = \{b_{i_1}, b_{i_2}, ..., b_{i_t}\}$.

Theorem 13. *Algorithm Minimum_Point_Guard finds a minimum sized point guards that covers P and it runs in linear time.*
Proof. Correctness of the algorithm follows from Lemmas 10, 11 and 12. Since each step of this algorithm runs in linear time, this algorithm runs in linear time. ∎

Corollary 14. *The minimum number of point guards required to cover a spiral polygon P equals the cardinality of a maximum independent set of the stick intersection graph associated with P.*

Acknowledgement
The authors wish to thank Professor Gerard J. Chang for his reading the drafts and making many useful suggestions that greatly improve the presentation of this paper.

References
[1] A. A. Bertossi, The edge Hamiltonian path problem is NP-complete, *Inform. Process. Lett.* 13 (1982) 157-159.

[2] A. A. Bertossi, Finding Hamiltonian circuits in proper interval graphs, *Inform. Process. Lett.* 17 (1983) 97-101.

[3] G. Ding, Convering the edges with consecutive sets, *J. Graph Theory* 15 (1991) 559-562.

[4] H. Evertt and D. G. Corneil, Recognizing visibility graphs of spiral polygons, *J. Algorithms* 11 (1990) 1-26.

[5] D. R. Fulkerson and O. A. Gross, Incidence matrices and interval graphs, *Pacific J. Math.* 15 (1965) 835-855.

[6] M. R. Garey and D. S. Johnson, *Computers and Intractability: A Guide to the Theory of NP-completeness*, Freeman, San Francisco (1979).

[7] M. R. Garey, D. S. Johnson, and R. E. Tarjan, The planar Hamiltonian circuit problem is NP-complete, *SIAM J. Comput.* 5 (1976) 704-714.

[8] F. Gavril, Algorithms for minimum coloring, maximum clique, minimum covering by cliques, and maximum independent set of a chordal graph, *SIAM J. Comput.* 1 (1972) 180-187.

[9] P. C. Gilmore and A. J. Hoffman, A characterization of comparability graphs and of interval graphs, *Canad. J. Math.* 16 (1964) 539-548.

[10] M. C. Golumbic, *Algorithmic Graph Theory and Perfect Graphs*, Academic Press, New York (1980).

[11] D. Gouyou-Beauchamps, The Hamiltonian circuit problem is polynomial for 4-connected planar graphs, *SIAM J. Comput.* 11 (1982) 529-539.

[12] A. Itai, C. H. Papadimitriou, and J. L. Szwarcfiter, Hamiltonian paths in grid graphs, *SIAM J. Comput.* 11 (1982) 676-686.

[13] R. M. Karp, Reducibility among combinatorial problems, in: *Complexity of Computer Computation* (eds. R. E. Miller and J. W. Thatcher) Plenum Press, New York (1972) 85-103.

[14] J. M. Keil, Finding Hamiltonian circuits in interval graphs, *Inform. Process. Lett.* 20 (1985) 201-206.

[15] M. S. Krishnamoorthy, An NP-hard problem in bipartite graphs, *SIGACT News* 7 (1975) 26.

[16] G. K. Manacher, T. A. Mankus, and C. J. Smith, An optimum $\Theta(n\log n)$ algorithm for finding a canonical Hamiltonian path and a canonical Hamiltonian circuit in a set of intervals, *Inform. Process. Lett.* 35 (1990) 205-211.

[17] J. O'Rourke, *Art Gallery Theorems and Algorithms*, Oxford University Press, New York (1987).

[18] F. S. Roberts, *Representations of Indifference Relations*, Ph.D. thesis, Stanford University (1968).

[19] F. S. Roberts, Indifference graphs, in: *Proof Techniques in Graph Theory* (ed. F. Harary) Academic Press, New York (1969) 139-146.

[20] W. K. Shih, T. C. Chern, and W. L. Hsu, An $O(n^2\log n)$ algorithm for the Hamiltonian cycle problem on circular-arc graphs, *SIAM J. Comput.* 21 (1992) 1026-1046.

Weighted Connected Domination and Steiner Trees in Distance-Hereditary Graphs*

(Extended Abstract)

Hong-Gwa Yeh and Gerard J. Chang

Department of Applied Mathematics
National Chiao Tung University
Hsinchu 30050, Taiwan
Email: gjchang@math.nctu.edu.tw

The concept of domination can be used to model many location problems in operations research. In a graph $G = (V, E)$, a *dominating set* is a subset D of vertices such that every vertex in $V - S$ is adjacent to some vertex in D. A dominating set of G is *connected* if the subgraph $G[D]$ induced by D is connected. The *connected domination problem* is to find a minimum-sized connected dominating set of a graph. Suppose, moreover, that each vertex v in G is associated with a weight $w(v)$ that is a real number. The *weighted connected domination problem* is to find a connected dominating set D such that $w(D) \equiv \sum_{v \in D} w(v)$ is as small as possible.

The concept of Steiner trees originally concerned points in Euclidean spaces, but it is also closely related to connected domination in graphs. Suppose T is a subset of vertices in a graph $G = (V, E)$. The *Steiner tree problem* is to find a subset S of $V - T$ such that $G[S \bigcup T]$ is connected. S and T are called the *Steiner set* and *target set*, respectively. We can also consider the weighted version of the Steiner tree problem.

The connected domination and Steiner tree problems have the same complexity for many classes of graphs. For instance, they are both polynomially solvable for strongly chordal graphs [22], permutation graphs [5], cographs [16, 8], series-parallel graphs [10, 20, 21, 22], and distance-hereditary graphs [12, 2]; and they are $\mathcal{NP}$-complete for bipartite graphs [13, 19], split graphs [22, 17], chordal graphs [22, 17], and chordal bipartite graphs [18]. It is also known that the connected domination problem is polynomially solvable for k-trees (fixed k) [1] and 1-CUBs [7] and $\mathcal{NP}$-complete for k-CUBs ($k \geq 2$) [7]. The Steiner tree problem is polynomially solvable in homogeneous graphs [11].

The purpose of this paper is to present linear time algorithms for the

---

*Supported in part by the National Science Council under grant NSC84-2121-M009-023.

weighted connected domination problem with general weights and the Steiner tree problem with non-negative weights in distance-hereditary graphs. D'Atri and Moscarini [12] gave $O(|V||E|)$ algorithms for connected domination and Steiner tree problems in distance-hereditary graphs. Brandstädt and Dragan [2] presented a linear time algorithm for the connected r-domination and Steiner tree problems in distance-hereditary graphs.

We first give a brief introduction to distance-hereditary graphs. A graph is *distance-hereditary* if every two vertices have the same distance in every connected induced subgraph. Distance-hereditary graphs were introduced by Howorka [15]. The characterization and recognition of distance-hereditary graphs have been studied in [3, 12, 14]. Note that the class of distance-hereditary graphs is a subclass of all parity graphs [4] and a superclass of all cographs [6, 9].

Suppose A and B are two sets of vertices in a graph $G = (V, E)$. The *neighborhood* $N_A(B)$ of B in A is the set of vertices in A that are adjacent to some vertex in B. The *closed neighborhood* $N_A[B]$ of B in A is $N_A(B) \bigcup B$. For simplicity, $N_A(v)$, $N_A[v]$, $N(B)$, and $N[B]$ stand for $N_A(\{v\})$, $N_A[\{v\}]$, $N_V(B)$, and $N_V[B]$, respectively. The *distance* $d_G(x, y)$ or $d(x, y)$ between two vertices x and y in G is the minimum length of an x-y path in G. The *hanging* h_u of a connected graph $G = (V, E)$ at a vertex $u \in V$ is the collection of sets $L_0(u), L_1(u), \ldots, L_t(u)$ (or $L_0, L_1, \ldots, L_t$ if there is no ambiguity), where $t = \max_{v \in V} d_G(u, v)$ and $L_i(u) = \{v \in V : d_G(u, v) = i\}$ for $0 \le i \le t$. For any $1 \le i \le t$ and any vertex $v \in L_i$, let $N'(v) = N(v) \bigcap L_{i-1}$. For any $U \subseteq V$, a vertex $v \in U \bigcap L_i$ with $1 \le i \le t$ has a *minimal neighborhood* in L_{i-1} *with respect to* U if $N'(w)$ is not a proper subset of $N'(v)$ for any $w \in U \bigcap L_i$. When $U = V$ we omit the term U in the above definition.

Theorem 1 ([12]) *For a connected graph $G = (V, E)$ the following statements are equivalent.*

(1) G is a distance-hereditary graph.

(2) Every cycle of length at least five in G has two crossing chords.

(3) For every hanging $h_u = (L_0, L_1, \ldots, L_t)$ of G and every pair of vertices x, $y \in L_i$ $(1 \le i \le t)$ that are in the same component of $G[V - L_{i-1}]$, we have $N'(x) = N'(y)$.

Theorem 2 ([3]) *Suppose $h_u = (L_0, L_1, \ldots, L_t)$ is a hanging of a connected distance-hereditary graph at u. For any two vertices $x, y \in L_i$ with $i \ge 1$, $N'(x)$ and $N'(y)$ are either disjoint, or one of the two sets is contained in the other.*

Theorem 3 (Fact 3.4 in [14]) *Suppose $h_u = (L_0, L_1, \ldots, L_t)$ is a hanging of a connected distance-hereditary graph at u. For each $1 \le i \le t$, there exists a vertex $v \in L_i$ such that v has a minimal neighborhood in L_{i-1}. In addition, if v satisfies the above condition then for every pair of vertices x and y in $N'(v)$, we have $N_{V-N'(v)}(x) = N_{V-N'(v)}(y)$.*

The first result of this paper is to present a linear time algorithm for finding a minimum weighted connected dominating set of a connected distance-hereditary graph $G = (V, E)$ in which each vertex v has a weight $w(v)$ that is a real number. The following lemmas and theorem are the basis of our algorithm.

Lemma 4 *Suppose $G = (V, E)$ is a connected graph with a weight function w on V. Let V' be the set of vertices v with $w(v) < 0$ and w' be defined by $w'(v) = \max\{w(v), 0\}$ for all $v \in V$. If D is a minimum w'-weighted connected dominating set of G, then $D \bigcup V'$ is a minimum w-weighted connected dominating set of G.*

Lemma 5 *Suppose $h_u = \{L_0, L_1, \ldots, L_t\}$ is a hanging of a connected distance-hereditary graph at u. For any connected dominating set D and $v \in L_i$ with $2 \le i \le t$, $D \bigcap N'(v) \ne \emptyset$.*

Theorem 6 *Suppose $G = (V, E)$ is a connected distance-hereditary graph with a non-negative weight function w on vertices. Let $h_u = \{L_0, L_1, \ldots, L_t\}$ be a hanging at a vertex u of minimum weight. Consider the set $\mathcal{A} = \{N'(v) : v \in L_i$ with $2 \le i \le t$ and v has a minimal neighborhood in $L_{i-1}\}$. For each $N'(v)$ in $\mathcal{A}$, choose one vertex v^* in $N'(v)$ of minimum weight, and let D be the set of all such v^*. Then D or $D \bigcup \{u\}$ or some $\{v\}$ with $v \in V$ is a minimum weighted connected dominating set of G.*

By Lemma 4 and Theorem 6, we can design an efficient algorithm for the weighted connected domination problem in distance-hereditary graphs. To implement the algorithm efficiently, we do not actually find the set $\mathcal{A}$. Instead, we perform the following step for each $2 \le i \le t$. Sort the vertices in L_i such that

$$|N'(x_1)| \le |N'(x_2)| \le \cdots \le |N'(x_j)|.$$

We then process $N'(x_k)$ for k from 1 to j. At iteration k, if $N'(x_k) \bigcap D = \emptyset$, then $N'(x_k)$ is in $\mathcal{A}$ and we choose a vertex of minimum weight to put it into D; otherwise $N'(x_k) \notin \mathcal{A}$ and we do nothing.

Algorithm WCD-dh. Find a minimum weighted connected dominating set of a connected distance-hereditary graph.
Input: A connected distance-hereditary graph $G = (V, E)$ and a weight $w(v)$ of real number for each $v \in V$.
Output: A minimum weighted connected dominating set D of graph G.

begin

    $D \longleftarrow \emptyset$;

    let $V' = \{v \in V : w(v) < 0\}$;

    $w(v) \longleftarrow 0$ for each $v \in V'$;

    let u be a vertex of minimum weight in V;

    determine the hanging $h_u = (L_0, L_1, \ldots, L_t)$ of G at u;

    for $i = 2$ to t **do**

    begin

        let $L_i = \{x_1, \ldots, x_j\}$;

        sort L_i such that $|N'(x_{i_1})| \le |N'(x_{i_2})| \le \ldots \le |N'(x_{i_j})|$;

        for $k = 1$ to j **do**

            if $N(x_{i_k}) \bigcap D = \emptyset$ **then** $D \longleftarrow D \bigcup \{y\}$ where y is a vertex

            of minimum weight in $N(x_{i_k})$

    end

    if not ($L_1 \subseteq N[D]$ and $G[D]$ is connected) **then** $D \longleftarrow D \bigcup \{u\}$;

    for $v \in V$ that dominates V: **if** $w(v) < w(D)$ **then** $D \longleftarrow \{v\}$;

    $D \longleftarrow D \bigcup V'$

end

Theorem 7 *Algorithm WCD-dh gives a minimum weighted connected dominating set of a connected distance-hereditary graph in linear time.*

The second result of this paper is to present a linear time algorithm for finding a minimum weighted Steiner tree with respect to a target set $T \subseteq V$ in a connected distance-hereditary graph $G = (V, E)$ with a non-negative weight $w(v)$ for each $v \in V$. In this section $h_u = \{L_0, \ldots, L_t\}$ denotes a hanging of G at a vertex u in the target set T. The key to our algorithm for the weighted Steiner tree problem is the following theorem. The theorem is similar to Theorem 6, but even simpler.

Theorem 8 *Suppose $U = \{x \in V: x$ lies on a shortest u-v path for some v in $T\}$ and $\mathcal{B} = \{N'(x): x \in U \bigcap L_i$ has a minimal neighborhood in L_{i-1} relative to U and $N'(x) \bigcap T = \emptyset\}$. Then the set S formed by choosing a vertex x^* of minimum weight in each $N'(x) \in \mathcal{B}$ is a minimum weighted Steiner set with respect to T in G.*

Theorem 8 provides the basic idea for designing a good algorithm for the weighted Steiner tree problem. Similar to the implementation of WCD-dh, we do not actually find U and $\mathcal{B}$. Instead, at any L_i we sort $|N'(x)|$ for all $x \in S \bigcup T$ to find $\mathcal{B}$.

Algorithm WST-dh. Find a minimum weighted Steiner set of a distance-hereditary graph with nonnegative weights on vertices.

Input: A connected distance-hereditary graph $G = (V, E)$ with nonnegative weight $w(v)$ for each $v \in V$ and a subset $T \subseteq V$.

Output: A subset $S \subseteq V - T$ of minimum weight such that $G[S \bigcup T]$ is connected.

begin

        $S \longleftarrow \emptyset$;

        let u be a vertex of T;

        determine the hanging $h_u = (L_0, L_1, \ldots, L_t)$ of G at u;

        for $i = t$ **to** 2 **step** -1 **do**

        begin

            let $(S \bigcup T) \bigcap L_i = \{x_1, \ldots, x_p\}$;

            if $p \neq 0$ **then**

            begin

                sort $x_1, x_2, \ldots, x_p$ such that $|N'(x_{j_1})| \leq \ldots \leq |N'(x_{j_p})|$;

                for $k = 1$ **to** p **do**

                    if $N'(x_{j_k})$ has no vertices in $S \bigcup T$

                    then $S \longleftarrow S \bigcup \{y\}$ where y is a vertex of minimum

                        weight in $N'(x_{j_k})$

        end

        end

end

Theorem 9 *Algorithm WST-dh solves the weighted Steiner tree problem for a connected distance-hereditary graph with a non-negative weight function in linear time.*

      The weighted Steiner tree problem with arbitrary real weights on vertices remains open.

References

[1] S. Arnborg and A. Proskurowski, *Linear time algorithms for NP-hard problems restricted to partial k-trees*, Discrete Appl. Math. 23, pp. 11-24, 1989.

[2] A. Brandstädt and F. F. Dragan, *A linear-time algorithm for connected r-domination and Steiner tree on distance-hereditary graphs*, preprint, 1994.

[3] H. J. Bandelt and H. M. Mulder, *Distance-hereditary graphs*, J. Comb. Theory Series B, pp. 182-208, 1986.

[4] M. Burlet and J. P. Uhry, *Parity graphs*, Annals of Discrete Math. 21, pp. 253-277, 1984.

[5] C. J. Colbourn and L. K. Stewart, *Permutation graphs: connected domination and Steiner trees*, Discrete Math. 86, pp. 145-164, 1990.

[6] D. G. Corneil, H. Lerchs, and L. Stewart, *Complement reducible graphs*, Discrete Appl. Math. 3, pp. 163-174, 1981.

On Central Spanning Trees of a Graph

S. BEZRUKOV

Universität-GH Paderborn

FB Mathematik/Informatik

Fürstenallee 11

D–33102 Paderborn

F. KADERALI, W. POGUNTKE

FernUniversität Hagen

LG Kommunikationssysteme

Bergischer Ring 100

D–58084 Hagen

Abstract

We consider the collection of all spanning trees of a graph with distance between them based on the size of the symmetric difference of their edge sets. A central spanning tree of a graph is one for which the maximal distance to all other spanning trees is minimal. We prove that the problem of constructing a central spanning tree is algorithmically difficult and leads to an NP-complete problem.

1 Introduction

All the basic notions concerning graphs, which are not explained here, may be found in any introductory book on graph theory, e.g. [4]. In the whole paper we consider undirected connected graphs without loops, but maybe with multiple edges. For a graph G we denote by $V(G)$ and $E(G)$ its vertex and edge sets, respectively.

Let T_1 and T_2 be a pair of spanning trees of a graph G. We define the **distance** between T_1 and T_2 as

$$D(T_1,T_2) = \frac{1}{2} \cdot |(E(T_1) \cup E(T_2)) \setminus (E(T_1) \cap E(T_2))|,$$

i.e. the distance equals half of the symmetric difference between $E(T_1)$ and $E(T_2)$ (notice that the symmetric difference itself is always of even size).

For a fixed tree T there exists a polynomial algorithm to find a tree T' such that $D(T,T')$ is maximal. Indeed, assign weights to the graph edges. Each edge of T gets weight 1, and all the other edges get weight 0. Now apply an algorithm to find a minimal weight spanning tree in G. This results in a maximally distant tree T' with respect to T.

A pair of spanning trees T_1, T_2 of a graph G is called **maximally distant** if $D(T_1,T_2) \geq D(T_1',T_2')$ for any spanning trees T_1',T_2' of G. Maximally distant trees were studied in a number of papers (cf. [9, 10, 11]). An algorithm for finding a pair of maximally distant spanning trees is presented in [9] and requires polynomially many steps with respect to the number of vertices in the graph. A spanning tree T being part of a pair of maximally distant trees is called **extremal**. Extremal trees are useful in the mixed analysis of electrical circuits.

In our paper we are interested in the dual problem of finding a spanning tree T, such that $\max_{T'} D(T,T')$ is minimal. We call such a tree a **central** tree of G. The notion of a central tree was introduced in [2], and some applications of such trees to circuit analysis one can find in [7]. A central tree can also be useful for broadcasting messages in a communication network. Central trees were intensively studied in the literature (see [8, 12, 13, 14]), but presently we know only the paper [1] being devoted to the construction of central trees. Unfortunately the algorithm described in [1] contains a gap, and in [5] one can find a counterexample to this algorithm. Also no result is known to us concerning the complexity of the problem of constructing a central tree. This complexity aspect is the main point of our analysis here.

Given a graph G, let us define its **tree graph** $T(G)$. The vertices of $T(G)$ correspond to the spanning trees of G, and two vertices of $T(G)$ are adjacent iff the distance between the corresponding spanning trees is 1. Thus the notion of a central tree of G corresponds to a **central vertex** in the graph $T(G)$. The problem to find all central vertices in a graph is known to be polynomial with respect to the number of its vertices, but in our case the number of spanning trees of G may be exponentially large with respect to $|V(G)|$, and so the result concerning the central vertices cannot be directly applied to construct central trees.

Let H be a subgraph of a graph G. Denote by $\overline{H}$ the complement of H in G, i.e. the subgraph obtained by deletion of all the edges of H in G. Let $r(\overline{H})$ denote the rank of $\overline{H}$, i.e. the number of vertices of G minus the number of components of $\overline{H}$.

Proposition 1 (cf. [6]) *If T is a central tree, then $r(\overline{T}) \leq r(\overline{T'})$ for any other spanning tree T'.*

Therefore deletion of a central tree from G results in a maximal number of components in the remaining graph. Note that, dually, deletion of an extremal tree results in a minimal number of components in the remaining graph (cf. [6]). This is the property making extremal trees useful in circuit analysis.

Consider the following problem, which we call **Central Tree**:

INSTANCE: A graph G and an integer number k.
QUESTION: Is there a spanning tree T of G, such that the graph $\overline{T}$ consists of k components?

In the next section we prove that this problem is NP-complete.

2 The main result

Theorem 1 *The Central Tree problem is NP-complete*

We use transformation from the problem X3C (Exact Cover by 3-sets), which is known to be NP-complete (see the problem [SP2] in [3]) and is the following:

INSTANCE: A set E of $|E| = 3k$ elements and a collection F of 3-element subsets of E.
QUESTION: Does F contain an exact cover for E, i.e. a subcollection $F' \subseteq F$ such that every element of E occurs in exactly one member of F'?

In the proof we take an instance for the X3C problem and construct some graph, considered later as an input to the Central Tree problem. Hence, we consider the Central Tree problem only for the graphs obtained in such a way, thus showing that it is NP-complete even for this restricted class of graphs.

Let F be a collection of 3-subsets as an instance of the X3C problem. We represent F by a bipartite graph $G(F)$ with the bipartition sets E and S consisting of $3k$ and $|F|$

vertices respectively. The vertices of E correspond to the base set, and the vertices of S correspond to the 3-subsets in F. The edges of $G(F)$ are determined by the incidence structure of the set system.

We construct an instance for the Central Tree problem as follows. Take the graph $G(F)$ and add $\binom{3k}{2}$ edges, connecting the vertices of E, such that the subgraph of $G(F)$ induced by the vertex set E is a complete graph. Add an extra vertex v to the obtained graph, and connect v with each vertex of S. The resulting graph is denoted by $G(F)$.

In Fig. 1a an example of such a graph $G(F)$ for the case $|E| = 6$, $|F| = 4$ is shown. The sets $\{v_1\}$ and $\{v_2\}$ form an exact cover of the base set.

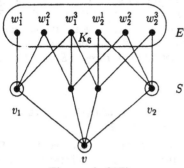
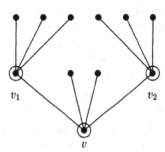

a. The graph $G(F)$. b. A spanning tree of $G(F)$.

Fig. 1

Now we show that an exact cover of the set E (consisting of k subsets) exists iff there exists a spanning tree in $G(F)$, which splits $G(F)$ into $k+2$ components. We assume that

$$|E| \geq 6 \quad \text{and} \tag{1}$$
$$|S| > k+1. \tag{2}$$

Observe that if one bounds these sets the X3C problem is polynomially solvable.

Indeed, let an exact cover exist. Denote by $v_1, ..., v_k$ the vertices of S, corresponding to the covering subsets (cf. Fig. 1a). Furthermore, for each v_i denote by w_i^1, w_i^2, w_i^3 its neighbors in E. Then the subsets $\{w_i^1, w_i^2, w_i^3\}$ are disjoint. Consider the spanning tree T in G induced by the edges of the form (v, u) with $u \in S$ and (v_i, w_i^1), (v_i, w_i^2), (v_i, w_i^3) for $i = 1, ..., k$. Then the components of $G \setminus T$ consist of $k+1$ single vertices $v, v_1, ..., v_k$ and the subgraph induced by the vertex set $E \cup (S \setminus \{v_1, ..., v_k\})$. Thus we have $k+2$ components.

We refer to Fig. 1b for $k = 2$. There a spanning tree of the graph $G(F)$ is shown and after deletion of it the vertices v_1, v_2, v form 3 single components, and the rest of the vertices form the 4th component.

We showed that the existence of an exact cover of size k in the set system F implies the existence of a spanning tree in the graph $G(F)$, after deletion of which we get a graph with $k+2$ components. Now we show the reverse direction. In fact we show that the structure of components must be exactly as described above.

So, let T be a spanning tree, which splits G into exactly $k+2$ components. We claim that all the vertices of E belong to the same component.

Suppose this is not true and there exists a component C such that

$$0 < |C \cap E| < |E|.$$

Denote $E' = C \cap E$. If $|E'| > 1$ and $|E \setminus E'| > 1$, then the edge cut separating E' and $E \setminus E'$ which is part of T necessarily contains a circle, contradicting that T is a tree.

Let $E' = \{w\}$ and assume the component containing w has one more vertex u. Without loss of generality we may assume that $u \in S$ and $(w, u) \in E(G(T))$. Now the edge cut separating u and w from $E \setminus E'$ has to contain a circle, which is again a contradiction.

Thus, we have a component consisting of the single vertex w. It is not difficult to see that all the vertices $E \setminus w$ belong to only one component, which we denote by D. Now each vertex $u \in S$ belongs to this component D, since otherwise the edge cut separating u and w from the set $E \setminus \{w\}$ contains a circle. Therefore, in this case the graph $G(F) \setminus T$ consists of at most 3 components, formed by the single vertices w and v and the subgraph induced by the vertex set $(E \setminus \{w\}) \cup S$. But (1) and $|E| = 3k$ imply that the number of components must be at least 4, a contradiction. Hence all the vertices of E must belong to the same component.

Consider the following two cases:

Case 1. Assume that the vertex v forms a single component in $G(F) \setminus T$. Then the spanning tree T includes all the edges of the form (v, u) with $u \in S$.

Case 1a. Assume that there exists a component, which contains at least 2 vertices of S. Then this component must contain at least one vertex from E (since otherwise it is not connected), and so by the above it must contain all the set E. Therefore, there exists at most one component containing at least 2 vertices of S, and we have determined already 2 components of the graph $G(F) \setminus T$. Thus all the other k components are formed by k single vertices (we denote them by $v_1, ..., v_k$) of S, and so the tree T contains the edges of the form (v_i, w_i^1), (v_i, w_i^2), (v_i, w_i^3) (here $w_i^1, w_i^2, w_i^3 \in E$ incident with v_i, and the sets of the form $\{w_i^1, w_i^2, w_i^3\}$ must be disjoint). Thus, the vertices $v_1, ..., v_k$ form an exact cover of the set E.

Case 1b. Assume that there is no component containing at least 2 vertices of S. This leads us to a contradiction, since in this case each vertex of S (maybe except one of them, which is connected to E) forms a single component, and so either $k = |S|$ or $k = |S| - 1$, which contradicts (2).

Case 2. Now assume that the component containing the vertex v (denote this component by K) also contains some vertices $u_1, ..., u_t \in S$. We show that this is, however, impossible.

Case 2a. Assume similarly to the above that there exists another component C ($C \neq K$), which contains at least 2 vertices of $S \setminus \{u_1, ..., u_t\}$. This leads us to the conclusion that C contains the whole set E, $E \cap K = \emptyset$ and all the other k components are formed by k single vertices of the set $S' = S \setminus (C \cup K)$ (with $|S'| = k$). Similarly to the above the vertices of S' have to form an exact cover of the set E. Consider the vertex u_1. Its neighborhood $W = \{w^1, w^2, w^3\}$ in E is covered by the vertices of S'. If there exists a vertex $v_1 \in S'$ such that its neighborhood in E contains at least 2 vertices of W, then the edge cut separating the vertices u_1 and v_1 contains a cycle, contradicting that T is a tree. Thus, there exist two vertices $v_1, v_2 \in S'$ adjacent with the vertices of W and the edge cut separating the vertices u_1, v_1, v_2, v contains a cycle, which is again a contradiction.

Case 2b. Assume that there is no other component containing at least 2 vertices of S. If $K \cap E = \emptyset$, then similarly to Case 2a we obtain a contradiction that T is not a tree.

Thus K has to contain the whole set E and each vertex from the set $S \setminus \{u_1, ..., u_t\}$ has to form a single component. There are $k + 1$ of these components, and they have to be separated by a tree. This means that the neighborhoods $\{w_i^1, w_i^2, w_i^3\}$ of them in E must be disjoint, which is impossible because $|E| = 3k$.

References

[1] Amoia A., Cottafava G.: *Invariance Properties of central trees*, IEEE Trans. Circuit Theory, vol. CT-18 (1971), 465–467.

[2] Deo D.: *A central tree*, IEEE Trans. Circuit Th., vol. CT-13 (1966), 439–440.

[3] Garey M.R., Johnson D.S.: *Computers and Intractability: A Guide to the Theory of NP-completeness*, Freeman 1979.

[4] Harary F.: *Graph theory*, Addison-Wesley Publ. Company, 1969.

[5] Kaderali F.: *A counterexample to the algoritm of Amoia and Cottafava for finding central trees*, preprint, FB 19, TH Darmstadt. June 1973.

[6] Kaderali F.: *Über zentrale und maximal entfernte Bäume*, unpublished manuscript.

[7] Kajitani Y., Kawamoto T., Shinoda S.: *A new method of circuit analysis and central trees of a graph*, Electron. Commun. Japan, vol. 66 (1983), No. 1, 36–45.

[8] Kawamoto T., Kajitani Y., Shinoda S.: *New theorems on central trees described in connection with the principal partition of a graph*, Papers of the Technical Group on Circuit and System theory of Inst. Elec. Comm. Eng. Japan, No. CST77-109 (1977), 63–69.

[9] Kishi G., Kajitani Y.: *Maximally distant trees and principal partition of a linear graph*, IEEE Trans. Circuit Theory, vol. CT-16 (1969), 323–330.

[10] Kishi G., Kajitani Y.: *Maximally distant trees in a linear graph*, Electronics and Communications in Japan (The Transactions of the Institute of Electronics and Communication Engineers of Japan), vol. 51 (1968), 35–42.

[11] Kishi G., Kajitani Y.: *On maximally distant trees*, Proceedings of the Fifth Annual Allerton Conference on Circuit and System Theory, University of Illinois, Oct. 1967, 635–643.

[12] Shinoda S., Kawamoto T.: *On central trees of a graph*, Lecture notes in Computer Sci., vol. 108 (1981), 137–151.

[13] Shinoda S., Kawamoto T.: *Central trees and critical sets*, in Proc. 14th Asilomar Conf. on Circuit, Systems and Comp., Pacific Grove, Calif., 1980, D.E. Kirk ed., 183–187.

[14] Shinoda S., Saishu K.: *Conditions for an incidence set to be a central tree*, Papers of the Technical Group on Circuit and System theory of Inst. Elec. Comm. Eng.Japan, No. CAS80-6 (1980), 41–46.

Complete Bipartite Decompositions of Crowns, with Applications to Complete Directed Graphs

Chiang Lin, Jenq-Jong Lin and Tay-Woei Shyu

ABSTRACT. For an integer $n \geq 3$, the crown S_n^0 is defined to be the graph with vertex set $\{a_1, a_2, \cdots, a_n, b_1, b_2, \cdots, b_n\}$ and edge set $\{a_i b_j : 1 \leq i, j \leq n, i \neq j\}$. We consider the decomposition of the edges of S_n^0 into the complete bipartite graphs, and obtain the following results.

The minimum number of complete bipartite subgraphs needed to decompose the edges of S_n^0 is n.

The crown S_n^0 has a $K_{l,m}$-decomposition (i.e., the edges of S_n^0 can be decomposed into subgraphs isomorphic to $K_{l,m}$) if $n = \lambda lm + 1$ for some positive integers λ, l, m. Furthermore, the l-part and m-part of each member in this decomposition can be required to be contained in $\{a_1, a_2, \cdots, a_n\}$ and $\{b_1, b_2, \cdots, b_n\}$, respectively.

Every minimum complete bipartite decomposition of S_n^0 is trivial if and only if $n = p + 1$ where p is prime (a complete bipartite decomposition of S_n^0 that uses the minimum number of complete bipartite subgraphs is called a minimum complete bipartite decomposition of S_n^0 and a complete bipartite decomposition of S_n^0 is said to be trivial if it consists of either n maximal stars with respective centers $a_1, a_2, \cdots, a_n$, or n maximal stars with respective centers $b_1, b_2, \cdots, b_n$).

The above results have applications to the directed complete bipartite decomposition of the complete directed graphs.

1. Introduction

The decomposition of the edges of a graph into complete bipartite subgraphs is called a *complete bipartite decomposition*. Let $\tau(G)$ be the minimum number of complete bipartite subgraphs in a complete bipartite decomposition of a graph G. We call $\tau(G)$ the *complete bipartite number* of G. The first graphs for which the complete bipartite number was studied were the complete graphs. Graham and Pollark [3] proved that $\tau(K_n) = n - 1$. Tverberg [12] and Peck [8] proved this result using linear and quadratic equations, and matrix ranks, respectively. The complete bipartite number of the complete multigraphs was investigated in [6,9]. That of the weak product of complete graphs was determined in [10]. The problem of decomposing the complete graph into complete bipartite subgraphs was also considered in [1,4,5].

For an integer $n \geq 3$, the crown S_n^0 was originally defined in [11] to be the ordered set $P = (X, <)$ where $X = \{a_1, a_2, \cdots, a_n, b_1, b_2, \cdots, b_n\}$ and $x < y$ in $P \iff x = a_i, y = b_j$ for some $i \neq j$. We use the same terminology and notation to denote the graph which is the Hasse diagram of the crown. That is to say, in this paper the crown S_n^0 is the graph with vertex set $\{a_1, a_2, \cdots, a_n, b_1, b_2, \cdots, b_n\}$ and edge set $\{a_i b_j : 1 \leq i, j \leq n, i \neq j\}$ (see Figure 1 for S_5^0). Equivalently, the crown S_n^0 is the graph obtained by deleting a perfect matching from the complete bipartite graph $K_{n,n}$. Hereafter $\{a_1, a_2, \cdots, a_n, b_1, b_2, \cdots, b_n\}$ always means the vertex set of S_n^0 defined above. In this paper we will investigate the complete bipartite decomposition problem of the crown S_n^0 and the related problem of the complete directed graphs.

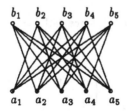

Figure 1 S_5^0

2. Complete Bipartite Decomposition of Crowns

Before entering the discussion of decomposition of the crowns, we consider that of the bipartite graphs.

Suppose that G is a bipartite graph with bipartition (X, Y), where $X = \{x_1, x_2, \cdots, x_m\}$ and $Y = \{y_1, y_2, \cdots, y_n\}$. We use $M(G)$ to denote the $m \times n$ matrix (e_{ij}) where

$$e_{ij} = \begin{cases} 1 & \text{if } x_i \text{ is adjacent to } y_j \\ 0 & \text{otherwise.} \end{cases}$$

Furthermore if H is a subgraph of G, we use $M(H)$ to denote the $m \times n$ matrix (e_{ij}) where

$$e_{ij} = \begin{cases} 1 & \text{if } x_i \text{ is adjacent to } y_j \text{ in } H \\ 0 & \text{otherwise.} \end{cases}$$

With these notations, it is easy to see that if the edges of G can be decomposed into subgraphs $H_1, H_2, \cdots, H_t$, then

$$M(G) = M(H_1) + M(H_2) + \cdots + M(H_t). \tag{1}$$

For a matrix M, we use $r(M)$ to denote the rank of M.

Lemma 1. *Suppose G is a bipartite graph. Then $\tau(G) \geq r(M(G))$.*

Proof. Suppose $\tau(G) = t$ and G is decomposed into complete bipartite subgraphs $H_1, H_2, \cdots, H_t$. Since each H_i is complete bipartite, $r(M(H_i)) = 1$. It follows from (1) that

$$r(M(G)) \leq r(M(H_1)) + r(M(H_2)) + \cdots + r(M(H_t)) = t.$$

This completes the proof. □

Now we determine $\tau(S_n^0)$. Note that [13]

$$\det \begin{bmatrix} a & b & \cdots & b \\ b & \ddots & \ddots & \vdots \\ \vdots & \ddots & \ddots & b \\ b & \cdots & b & a \end{bmatrix}_{n \times n} = \big(a + (n-1)b\big)\big(a-b\big)^{n-1}. \tag{2}$$

Note also that (2) holds for $n = 1$, since $\det[a] = a$.

Theorem 2. $\tau(S_n^0) = n$.

Proof. We see that

$$M(S_n^0) = \begin{bmatrix} 0 & 1 & \cdots & 1 \\ 1 & \ddots & \ddots & \vdots \\ \vdots & \ddots & \ddots & 1 \\ 1 & \cdots & 1 & 0 \end{bmatrix}_{n \times n}.$$

Using (2), we have
$$\det\big(M(S_n^0)\big) = (n-1)(-1)^{(n-1)} \neq 0,$$
which implies $r\big(M(S_n^0)\big) = n$. By Lemma 1, $\tau(S_n^0) \geq n$.

On the other hand, it is easy to see that S_n^0 can be decomposed into n complete bipartite graphs, namely, the maximal stars with centers at $a_1, a_2, \cdots, a_n$, respectively. Thus $\tau(S_n^0) \leq n$. This completes the proof. $\square$

Here is an alternative proof of $\tau(S_n^0) \geq n$. Suppose G is a graph with vertex set $\{x_1, x_2, \ldots, x_l\}$. Let $A(G)$ be the adjacency matrix of G, i.e., $A(G) = (m_{ij})$ is the $l \times l$ matrix where

$$m_{ij} = \begin{cases} 1 & \text{if } x_i \text{ is adjacent to } x_j \\ 0 & \text{otherwise.} \end{cases}$$

Let p and q be the number of positive and negative eigenvalues of $A(G)$, respectively (counted with multiplicity). It was shown [7,10] that $\tau(G) \geq \max\{p, q\}$. Since $V(S_n^0) = \{a_1, a_2, \ldots, a_n, b_1, b_2, \ldots, b_n\}$, we have

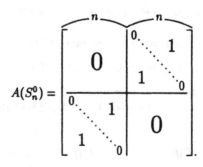

Hence

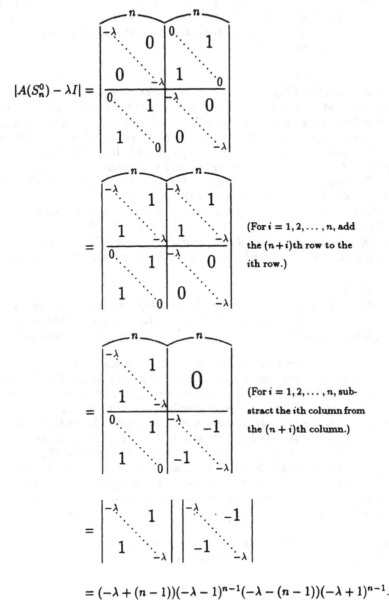

$$= (-\lambda + (n-1))(-\lambda - 1)^{n-1}(-\lambda - (n-1))(-\lambda + 1)^{n-1}.$$

So $A(S_n^0)$ has n positive eigenvalues and n negative eigenvalues, which implies that $\tau(S_n^0) \geq n$.

For disjoint nonempty sets A, B, we use $K(A, B)$ to denote the graph with vertex set $A \cup B$ and edge set $\{ab : a \in A, b \in B\}$. Further if $|A| = l$, $|B| = m$, we call A and B the l-part and the m-part of $K(A, B)$, respectively. If G and H are graphs such that the edges of G can be decomposed into subgraphs isomorphic to H, then we say that G has an H-decomposition.

Theorem 3. *Suppose $n = \lambda lm + 1 \geq 3$ for some positive integers λ, l, m. Then S_n^0 has a $K_{l,m}$-decomposition. Furthermore, the l-part and m-part of each member in this decomposition can be required to be contained in $\{a_1, a_2, \cdots, a_n\}$ and $\{b_1, b_2, \cdots, b_n\}$, respectively.*

Proof. Case 1. $\lambda = 1$.

We will decompose S_{lm+1}^0 into $lm + 1$ subgraphs each of which is isomorphic to $K_{l,m}$ with l-part in $\{a_1, a_2, \ldots, a_{lm+1}\}$ and m-part in $\{b_1, b_2, \ldots, b_{lm+1}\}$. Now we describe these subgraphs. For $i = 1, 2, \ldots, lm + 1$, first let $A_i = \{a_i, a_{i+1}, \ldots, a_{i+(l-2)}, a_{i+(l-1)}\}$ and $B_i = \{b_{i+l}, b_{i+2l}, \ldots, b_{i+(m-1)l}, b_{i+ml}\}$, where the subscripts are taken modulo $lm + 1$. Then let $H_i = K(A_i, B_i)$. It is obvious that each H_i is isomorphic to $K_{l,m}$ with l-part in $\{a_1, \ldots, a_n\}$ and m-part in $\{b_1, \ldots, b_n\}$. To show that S_n^0 is decomposed into $H_1, H_2, \ldots, H_n$, we label each edge of S_n^0 as follows. Each edge can be assumed to be $a_j b_k$ with $1 \leq j \leq lm + 1$ and $j < k < j + lm + 1$. We refer to this edge as an s-edge where $s = k - j$. We see that each H_i contains exactly one s-edge for $s = 1, 2, \ldots, lm$. For $1 \leq i < i' \leq lm + 1$, H_i and $H_{i'}$ are edge-disjoint, since the s-edge of H_i is distinct from that of $H_{i'}$ for $s = 1, 2, \ldots, lm$. Counting the edges in S_n^0 and the edges in H_i, we see that the decomposition is verified.

Case 2. $\lambda \geq 2$.

By Case 1, $S_{\lambda lm+1}^0$ can be decomposed into subgraphs isomorphic to $K_{\lambda l, m}$ with λl-part in $\{a_1, \ldots, a_n\}$ and m-part in $\{b_1, \ldots, b_n\}$. A natural way to decompose $K_{\lambda l, m}$ into subgraphs isomorphic to $K_{l,m}$ leads to the required decomposition. $\square$

If $\mathscr{D}$ is a complete bipartite decomposition of a graph G with $|\mathscr{D}| = \tau(G)$, then $\mathscr{D}$ is called a *minimum (complete bipartite) decomposition* of G. Consider the decomposition of crowns. The decomposition of S_n^0 consisting of either n maximal stars with centers at $a_1, a_2, \cdots, a_n$ respectively, or n maximal stars with centers at $b_1, b_2, \cdots, b_n$, respectively, is called a *trivial (complete bipartite) decomposition*. Since $\tau(S_n^0) = n$, the trivial decomposition of S_n^0 is a minimum decomposition. More generally, by letting $\lambda = 1$ in Theorem 3, we can decompose S_{lm+1}^0 into $lm+1$ complete bipartite subgraphs which are isomorphic to $K_{l,m}$; this decomposition is a minimum one. So for $n = lm + 1$ where $l, m \geq 2$, S_n^0 has a nontrivial minimum decomposition. In fact the converse is also true. To show this, we need the following technical lemma.

Lemma 4. *Suppose l, m, r are nonnegative integers such that $l \geq 1$ and $m + r \geq 2$. Let $A = (a_{ij})$ be a square matrix of order $n = l + m + r$ where*

$$
a_{ij} = \begin{cases} 0 & \text{if } i = j \\ 0 & \text{if } 1 \leq i \leq l \text{ and } l + m + 1 \leq j \\ 1 & \text{otherwise.} \end{cases}
$$

Then $\det A = (-1)^{n-1}(n - lr - 1)$.

Proof.

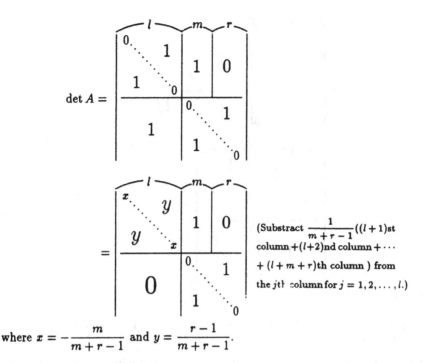

where $x = -\dfrac{m}{m+r-1}$ and $y = \dfrac{r-1}{m+r-1}$.

Applying (2), we obtain

$$\det A = (x + (l-1)y)(x - y)^{l-1}(m+r-1)(-1)^{m+r-1}$$
$$= \frac{-m + (l-1)(r-1)}{m+r-1}(-1)^{l-1}(m+r-1)(-1)^{m+r-1}$$
$$= (-m - l - r + lr + 1)(-1)^{l+m+r-2}$$
$$= (-1)^{n-1}(n - lr - 1). \quad \square$$

We give a necessary condition for the minimum decomposition of S_n^0.

Theorem 5. *Let H be a member in a minimum decomposition of S_n^0. Suppose H is isomorphic to $K_{l,r}$. Then $lr = n - 1$.*

Proof. Since $K_{l,r}$ is a subgraph of S_n^0, we have $l + r \le n$. Thus either $l \le n - 2$ or $r \le n - 2$ for $n \ge 3$. Without loss of generality assume $l \le n - 2$. We may also assume $H = K(A, B)$ where $A \subset \{a_1, a_2, \ldots, a_n\}$, $B \subset \{b_1, b_2, \ldots, b_n\}$, $|A| = l$ and $|B| = r$. Furthermore, without loss of generality, let $A = \{a_1, a_2, \ldots, a_l\}$, $B = \{b_{n-r+1}, b_{n-r+2}, \ldots, b_n\}$.

Let $\{H, H_2, H_3, \ldots, H_n\}$ be a minimum decomposition of S_n^0. Then, by (1),

$$M(S_n^0) = M(H) + M(H_2) + M(H_3) + \cdots + M(H_n).$$

Let $A = M(S_n^0) - M(H)$. Then $A = M(H_2) + M(H_3) + \cdots + M(H_n)$. Thus $r(A) \le r(M(H_2)) + r(M(H_3)) + \cdots + r(M(H_n)) = n - 1$, which implies $\det A = 0$. On the other hand,

$$A = M(S_n^0) - M(H)$$

$$= \left[\begin{array}{ccc} 0. & & 1 \\ & \ddots & \\ 1 & & 0. \end{array} \right] - \left[\begin{array}{c|c} 0 & 1 \\ \hline 0 & 0 \end{array} \right]$$

$$= \left[\begin{array}{ccc|cc|c} 0. & & 1 & & & \\ & \ddots & & 1 & 0 & \\ 1 & & 0 & & & \\ \hline & & & 0. & & 1 \\ & 1 & & & \ddots & \\ & & & 1 & & 0. \end{array} \right],$$

where $m = n - (l + r)$.

Note that $m + r = n - l \geq 2$. By Lemma 4, $\det A = (-1)^{n-1}(n - lr - 1)$. From $\det A = 0$, we have $lr = n - 1$. $\square$

Theorem 6. *Every minimum complete bipartite decomposition of S_n^0 is trivial if and only if $n = p + 1$ for some prime p.*

Proof. (Necessity) Suppose, on the contrary, that $n = lm + 1$ for some $l, m \geq 2$. By Theorem 3 ($\lambda = 1$), S_n^0 can be decomposed into copies of $K_{l,m}$. This decomposition has size n. Thus S_n^0 has a nontrivial minimum decomposition, a contradiction to the assumption. This confirms the necessity.

(Sufficiency) Now $n = p + 1$ where p is prime. Let $\mathscr{D}$ be a minimum decomposition of S_n^0. Suppose H is a member in $\mathscr{D}$ and H is isomorphic to $K_{l,r}$. By Theorem 5, $lr = n - 1 = p$. Since p is prime, we may assume $l = 1$, $r = p$. Thus each member in $\mathscr{D}$ is isomorphic to the star $K_{1,p} = K_{1,n-1}$. We can see that the centers of these stars must be either all in $\{a_1, a_2, \cdots, a_n\}$ or all in $\{b_1, b_2, \cdots, b_n\}$. Hence $\mathscr{D}$ is a trivial decomposition. $\square$

3. Applications: Directed Complete Bipartite Decompositions of Complete Directed Graphs

The results about the decomposition of crowns into complete bipartite subgraphs in Section 2 have applications to the decomposition of complete directed graphs into directed complete bipartite subgraphs.

Let DK_n denote the complete directed graph with vertex set $\{1, 2, \ldots n\}$ and arc set $\{\overrightarrow{xy} : 1 \leq x, y \leq n, x \neq y\}$. For disjoint nonempty sets S and T, let $DK(S, T)$ denote the directed complete bipartite graph with vertex set $S \cup T$ and arc set $\{\overrightarrow{st} : s \in S, t \in T\}$. If $|S| = l$ and $|T| = m$, $DK(S, T)$ is also denoted by $DK_{l,m}$.

For $n \geq 3$, we establish a one-to-one correspondance between the complete bipartite decompositions of S_n^0 and the directed complete bipartite decompositions of DK_n in the following way. For $S \subset \{a_1, a_2, \cdots, a_n\}, T \subset \{b_1, b_2, \cdots, b_n\}$, let $\overline{S} = \{i : a_i \in S\}, \overline{T} = \{i : b_i \in T\}$. We associate the complete bipartite subgraph $K(S, T)$ of S_n^0 with the directed complete bipartite subgraph $DK(\overline{S}, \overline{T})$ of DK_n. Thus corresponding to a complete bipartite decomposition $\mathscr{D}$ of S_n^0 is the directed complete bipartite decomposition $\overline{\mathscr{D}} = \{DK(\overline{S}, \overline{T}) : K(S, T) \in \mathscr{D}\}$ of DK_n. Note that $|\mathscr{D}| = |\overline{\mathscr{D}}|$. For a directed graph G, we also use $\tau(G)$ to denote the minimum number of directed complete bipartite subgraphs needed to partition the arcs of G. It was proved in [2] that $\tau(DK_n) = n$ using matrix factorization technique. The same result was proved in [9] using linear equation technique. Now we give a simple proof.

Theorem 7. $\tau(DK_n) = n$.

Proof. This follows from Theorem 2 and the above one-to-one correspondance. □

If G and H are directed graphs such that the arcs of G can be decomposed into subgraphs isomorphic to H, then we say that G has an H-decomposition. Here is an application of Theorem 3.

Theorem 8. $DK_n(n \geq 2)$ has a $DK_{l,m}$-decomposition if and only if $lm|n - 1$.

Proof. (Sufficiency) This is trivial if $n = 2$. Now let $n \geq 3$, and $n = \lambda lm + 1$. In the correspondance mentioned in this section, the subgraph of S_n^0 which is isomorphic to $K_{l,m}$ with l-part in $\{a_1, \cdots, a_n\}$ and m-part in $\{b_1, \cdots, b_n\}$ is associated with a subgraph of DK_n which is isomorphic to $DK_{l,m}$. Thus a $K_{l,m}$-decomposition of S_n^0 with l-part in $\{a_1, \ldots, a_n\}$ and m-part in $\{b_1, \ldots, b_n\}$ corresponds to a $DK_{l,m}$-decomposition of DK_n. The result follows from Theorem 3.

(Necessity) Let v be a vertex of DK_n. Each $DK_{l,m}$ with v in the l-part contributes m to the number of arcs incident from v. And each $DK_{l,m}$ with v in the m-part contributes l to the number of arcs incident to v. Thus outdeg(v) is a multiple of m, and indeg(v) is a multiple of l. Hence $m|n - 1$ and $l|n - 1$. Also, since the numbers of arcs of DK_n and $DK_{l,m}$ are $n(n - 1)$ and lm respectively, we have $lm|n(n - 1)$. Therefore $lm|n - 1$, since $\gcd(n, n - 1) = 1$. □

If $\mathscr{D}$ is a directed complete bipartite decomposition of a directed graph G with $|\mathscr{D}| = \tau(G)$, then $\mathscr{D}$ is called a *minimum* (*directed complete bipartite*) *decomposition* of G. The decomposition of DK_n consisting of either the directed stars $DK(\{i\}, \{1, 2, \ldots, n\} - \{i\})(i = 1, 2, \cdots, n)$ or the directed stars $DK(\{1, 2, \ldots, n\} - \{i\}, \{i\})(i = 1, 2, \cdots, n)$ is called a *trivial* (*directed complete bipartite*) *decomposition*.

As an application of Theorem 6, we have the following:

Theorem 9. *Every minimum directed complete bipartite decomposition of DK_n is trivial if and only if $n = p + 1$ for some prime p.*

Proof. Consider the correspondance mentioned in this section once again. For $i = 1, 2, \cdots, n$, the maximal star with center at a_i (respectively, at b_i) of S_n^0 is associated with the directed star $DK(\{i\}, \{1, 2, \ldots, n\} - \{i\})$ (respectively, the directed star $DK(\{1, 2, \ldots, n\} - \{i\}, \{i\}))$. So the trivial decompositions of S_n^0

correspond to the trivial decompositions of DK_n. Also, it is obvious that the minumum decompositions of S_n^0 correspond to the minumum decompositions of DK_n. Thus the condition that every minimum decomposition of DK_n is trivial is equivalent to that every minimum decomposition of S_n^0 is trivial. The theorem follows from Theorem 6. $\square$

References

1. D. De Caen and D. G. Hoffman, *Impossibility of decomposing the complete graph on n points into n−1 isomorphic complete bipartite graphs*, SIAM J. Disc. Math. **2** (1989), 48–50.
2. F. R. K. Chung, R. L. Graham, and P. M. Winkler, *On the addressing problem for directed graphs*, Graphs and Combinatorics **1** (1985), 41–50.
3. R. L. Graham and H. O. Pollak, *On embedding graphs in squashed cubes*, Springer Lecture Notes in Math. **303** (1973), 99–110.
4. A. Granville, A. Moisiadis, and R. Rees, *Bipartite planes*, Congr. Numer. **61** (1988), 241–248.
5. C. Huang and A. Rosa, *On the existence of balanced bipartite designs*, Utilitas Math. **4** (1973), 55–75.
6. Q.-X. Huang, *On complete bipartite decomposition of complete multigraphs*, Ars Combinatoria **38** (1994), 292–298.
7. T. Kratzke, B. Reznick, and D. West, *Eigensharp graphs: decomposition into complete bipartite subgraphs*, Trans. Amer. Math. Soc. **308** (1988), 637–653.
8. G. W. Peck, *A new proof of a theorem of Graham and Pollak*, Discrete Math. **49** (1984), 327–328.
9. D. Pritikin, *Applying a proof of Tverberg to complete bipartite decompositions of digraphs and multigraphs*, J. Graph Theory **10** (1986), 197–201.
10. B. Reznick, P. Tiwari, and D. B. West, *Decomposition of product graphs into complete bipartite subgraphs*, Discrete Math. **57** (1985), 189–193.
11. W. T. Trotter, *Dimension of the crowns S_n^k*, Discrete Math. **8** (1974), 85–103.
12. H. Tverberg, *On the decomposition of K_n into complete bipartite graphs*, J. Graph Theory **6** (1982), 493–494.
13. W. D. Wallis, *Combinatorial designs*, Marcel Dekker, Inc, New York, and Basel, 1988, p. 23.

Department of Mathematics, National Central University, Chung-Li, Taiwan, R.O.C.
E-mail address: lchiang@math.ncu.edu.tw

Finding an Antidirected Hamiltonian Path Starting with a Forward Arc from a Given Vertex of a Tournament.

E. Bampis[1] P. Hell[2] Y. Manoussakis[3] M. Rosenfeld[4]

[1] La.M.I., Université d'Evry, Boulevard des Coquibus, 91025 Evry Cedex, France
[2] School of Computing Science, Simon Fraser University, Burnaby, B.C., Canada V5A 1S6
[3] L.R.I, URA 410, Bât 490, Université de Paris Sud, 91405 Orsay Cedex, France
[4] Department of Mathematics, Pacific Lutheran University, Tacoma, Washington 98447, USA

Abstract. We prove (in two different ways) that in a sufficiently large tournament a vertex of outdegree at least two is always the beginning of an antidirected hamiltonian path starting with a forward arc. The proofs yield algorithms to find, if possible, an antidirected Hamiltonian path starting in a given vertex with an arc of a given direction. The first proof yields the theorem for all tournaments of order at least 19. The second proof only applies to somewhat larger tournaments, but leads to more efficient (sequential and parallel) algorithms.

1 Introduction

A *tournament* is an orientation of a complete graph, i.e. a digraph such that between every two vertices there is exactly one arc. The *order* of a tournament is its number of vertices, usually denoted by n. An *antidirected path (ADP)* in a tournament is a simple path in which every two adjacent arcs have opposite orientations. (In other words, no two adjacent arcs of the path form a directed path.) An *antidirected Hamiltonian path (ADHP)* in a tournament is an antidirected path containing all the vertices of the tournament. In a similar way, we define an *antidirected Hamiltonian cycle (ADHC)*.

Let x be a vertex of a tournament T. An *x-antidirected Hamiltonian path (x-ADHP)* in T is an ADHP in T with one endpoint x. An x-ADHP has a unique arc incident with x. If that arc is oriented away from x we call it a *forward x-antidirected Hamiltonian path (x-FADHP)*, and if it is oriented towards x we call it a *backward x-antidirected Hamiltonian path (x-BADHP)*.

According to [3, 8, 6, 9, 10] all tournaments of order at least 8 have an ADHP, all tournaments of order at least 11 have an x-ADHP from each vertex x, and all tournaments of even order at least 16 have an ADHC. The proofs of these results imply efficient sequential algorithms for finding ADHP's, x-ADHP's and ADHC's in tournaments, of complexities $O(n), O(n^2)$ and $O(n^2)$, respectively. Algorithms for paths in tournaments, oriented in any prescribed pattern of forward and backward arcs, are discussed in [4].

Recently, parallel algorithms for finding ADHP's, x-ADHP's, and ADHC's have been proposed [1], based on new proofs of their existence. The complexity of these algorithms is $O(\log n)$ time, $O(n/\log n)$ processors for ADHP's, $O(\log^2 n)$ time, $O(n^2/\log n)$ processors for x-ADHP's, and $O(\log^2 n)$ time, $O(n^2/\log n)$ processors for finding ADHC's.

In this paper, we study the following problem: Given a tournament T and a vertex x, find an x-FADHP (or an x-BADHP) in T, if one exists. If x has outdegree 0, then clearly such a path does not exist. If x has outdegree at least 2 and T is large enough, then according to our theorem an x-FADHP exists (and we can find it by the techniques inherent in the proof of the theorem). Finally, if x has outdegree 1, and y is its unique outneighbour, then a x-FADHP of T exists if and only if there exists a y-BADHP of $T \setminus x$ (and can be obtained from such a y-BADHP by adjoining the arc xy). Of course these results have analogues for x-BADHP. Note that to decide the existence of an x-FADHP in a large T we only need to test the outdegree of x: If it is at least 2 then there will be an x-FADHP; if it is 1 and its unique outneighbour is y then we recursively test the existence of a y-BADHP in $T \setminus x$; and if it is 0 then there will be no x-FADHP. (The existence of an x-FADHP and x-BADHP in small tournaments T can be tested by exhaustive search without affecting the overall complexity of the algorithm.) To actually find such paths we will use the technique of our second proof, below.

We will give two proofs of our result. The first proof has the advantage that it applies to all tournaments of order at least 19. The second proof only applies to tournaments of order at least 46, but yields a more efficient algorithm (of complexity $O(n)$) to actually find an x-FADHP, if one exists. In the last section we remark that this algorithm can be easily parallelized, yielding a parallel algorithm of complexity $O(\log n)$ time, $O(n/\log n)$ processors. The reason the second proof yields a more efficient algorithm is that it is based on finding antidirected Hamitonian paths; the first proof relies on finding antidirected Hamiltonian cycles. In both cases the complexity of the algorithm is dominated by finding such paths or cycles, and as noted above, the best currently known algorithm is $O(n)$ for the paths but $O(n^2)$ for the cycles.

We end this section with two observations which will be useful in the sequel. If T is a tournament of odd order which has an ADHP, then it has an ADHP $\leftarrow \rightarrow \leftarrow \ldots \rightarrow$ as well as an ADHP $\rightarrow \leftarrow \rightarrow \ldots \leftarrow$. Indeed, if, say, T contained an ADHP $x_1 \leftarrow x_2 \rightarrow \ldots \rightarrow x_n$, and if $x_1 \leftarrow x_n$, then $x_n \rightarrow x_1 \leftarrow x_2 \ldots \leftarrow x_{n-1}$ is also an ADHP in T; if on the other hand $x_1 \rightarrow x_n$, then we would take $x_1 \rightarrow x_n \leftarrow x_{n-1} \ldots \leftarrow x_2$.

We denote by TT_5 the ('transitive') tournament on five vertices $v_1, v_2, \ldots, v_5$ in which $v_i \rightarrow v_j$ if and only if $i < j$. The tournament TT_5 is a subtournament of any tournament of order at least 14 [7]. Here is an easy algorithm to find a TT_5 in any tournament T of order 16: Pick v_1 to be any vertex of T of outdegree at least

8. In the subtournament induced by 8 outneighbours of v_1, pick v_2 to be a vertex of outdegree at least 4. In the subtournament induced by 4 outneighbours of v_2, pick v_3 to have outdegree at least 2. Finally, let $v_4 \rightarrow v_5$ be two outneighbours of v_3 in this subtournament. In the sequel we shall only use this simple algorithm, and thus only use the result that any tournament of order at least 16 contains a TT_5, even though using the above result with 14 in place of 16 ([7]) would lower slightly the bound in the second proof.

2 The first proof

We begin with a lemma which explains how in a tournament T (of order divisible by four), which has an ADHC, we can find vertices x such that T has both an x-FADHP and an x-BADHP.

Lemma 2.1 *Let T be a tournament of order $4k$ and let $\{a_1 \rightarrow a_2 \leftarrow a_3 \rightarrow a_4 \leftarrow \cdots \rightarrow a_{4k} \leftarrow a_1\}$ be an ADHC in T. Then there exists an a_{2i}-FADHP and an a_{2j+1}-BADHP in T, for some subscripts i and j.*

Note that an a_{2i}-BADHP and an a_{2j+1}-FADHP are contained in the given ADHC.

Proof: Without loss of generality we may assume that $a_1 \rightarrow a_3$. (The subscripts below are reduced modulo $4k$.)
Case 1. $a_{2i-1} \leftarrow a_{2i+1}$ or $a_{2i+2} \leftarrow a_{2i}$, for some i.
If $a_4 \leftarrow a_2$, then $a_2 \rightarrow a_4 \leftarrow a_5 \rightarrow \cdots \rightarrow a_{4k} \leftarrow a_1 \rightarrow a_3$ is an ADHP with a_2 and a_3 the vertices indicated in the lemma. Hence we may also assume that $a_4 \rightarrow a_2$. If $a_5 \rightarrow a_3$ then the ADHP $a_4 \rightarrow a_2 \leftarrow a_1 \rightarrow a_{4k} \leftarrow a_{4k-1} \rightarrow \cdots a_5 \rightarrow a_3$ with the end vertices a_4 and a_3 will prove the claim. By repeating this argument, the result follows.
Case 2. $a_{2i+4} \leftarrow a_{2i+1}$, for some i.
If, say, $a_1 \rightarrow a_4$ then $a_3 \leftarrow a_1 \rightarrow a_4 \leftarrow a_5 \rightarrow a_6 \leftarrow \cdots \rightarrow a_{4k} \leftarrow a_2$ is an ADHP with a_2 and a_3 the indicated vertices. An identical argument holds for any edge of the form $a_{2i+1} \rightarrow a_{2i+4}$.
If neither of the two above cases occurs, then T has another ADHC, which consists of k "blocks" each of order 4: $a_1 \leftarrow a_4 \rightarrow a_2 \leftarrow a_3 \rightarrow a_5 \leftarrow a_8 \rightarrow a_6 \leftarrow a_7 \cdots \rightarrow a_{4t+1} \leftarrow a_{4t+4} \rightarrow a_{4t+2} \leftarrow a_{4t+3} \cdots \rightarrow a_{4k-3} \leftarrow a_{4k} \rightarrow a_{4k-2i} \leftarrow a_{4k-1} \rightarrow a_1$. This cycle contains an a_4-FADHP and an a_1-BADHP. $\square$

The next lemma is crucial for our proof:

Lemma 2.2 *Let T be a tournament of order $2k$ and let $t_1 \rightarrow t_2$ be an arc of T. Assume that $\{a_1 \rightarrow a_2 \leftarrow a_3 \rightarrow a_4 \leftarrow \cdots \rightarrow a_{2k-2} \leftarrow a_1\}$ is an ADHC in $T \setminus \{t_1, t_2\}$. Then one of the following must hold:*

1. T contains a t_i-BADHP with $i = 1$ or $i = 2$, or
2. $a_{2i} \to t_1$ and $t_1 \to a_{2i-1}$, for all i.

Proof: Assume that 2 does not hold. If, for some index i, $t_1 \to a_{2i}$, then the ADHP $t_2 \leftarrow t_1 \to a_{2i} \leftarrow a_{2i-1} \cdots$ shows that 1 holds. Hence we may assume that $a_{2i} \to t_1$. Since 2 does not hold, some index j has $a_{2j+1} \to t_1$. If one of the vertices a_{2j}, a_{2j+2} dominates t_2 (say $a_{2j} \to t_2$), then the path $t_2 \leftarrow a_{2j} \to t_1 \leftarrow a_{2j+1} \to a_{2j+2} \leftarrow \cdots$ is the desired ADHP. If t_2 dominates both vertices $\{a_{2j}, a_{2j+2}\}$ then the ADHP $t_1 \leftarrow a_{2j+1} \to a_{2j} \leftarrow t_2 \to a_{2j+2} \leftarrow a_{2j+3} \to \cdots$ is the desired path. (Subscripts are reduced modulo $2k - 2$.) $\square$

Remark. By applying the above lemma to T^{-1}, the reversal of T, and the arc $t_2 \to t_1$, it is easily seen that one of the following also must hold:

- T contains a t_i-FADHP, with $i = 1$ or $i = 2$, or
- $a_{2i} \leftarrow t_2$, and $t_2 \leftarrow a_{2i-1}$, for all i.

We are ready for our main result:

Theorem 2.3 *Let T be a tournament of order $n \geq 19$ and let v be a vertex of outdegree at least 2. Then T contains a v-FADHP.*

Proof: Without loss of generality, let $t_1 \to t_2$ be the two outneighbours of v. Assume first that n is odd. Let $T' = T \setminus v$. By our assumptions, $T' \setminus \{t_1, t_2\}$ is a tournament of even order, at least 16, and thus has an ADHC $a_1 \to a_2 \leftarrow a_3 \to \cdots \to a_{n-3} \leftarrow a_1$, [6]. We apply Lemma 2.2 to T' with the edge $t_1 \to t_2$: If alternative 1 holds, we obtain the desired ADHP $v \to t_i \leftarrow \cdots$. Thus we may suppose that alternative 2 holds. If $n - 3 = 4k$ for some integer k, then according to Lemma 2.1 there is an a_{2i-1}-BADHP in $T' \setminus \{t_1, t_2\}$, for some i, and thus a v-FADHP in T of the form $v \to t_2 \leftarrow t_1 \to a_{2i-1} \leftarrow \cdots$. If $n-3 = 4k+2$ for some integer k we proceed as follows: Since $k \geq 4$, and 2 holds, v has outdegree at least 9. Let $s_1 \to s_2$ be two vertices, other than t_2, dominated by t_1. Again, the tournament $T^* = T \setminus \{v, t_1, t_2, s_1, s_2\}$ has $4k \geq 16$ vertices and hence it has an ADHC. By an argument similar to the previous case, the desired v-FADHP in T is obtained as follows: Applying Lemma 2.1 to T' and the arc $s_1 \to s_2$, if alternative 1 holds then we take the path $v \to t_2 \leftarrow t_1 \to s_i \leftarrow \cdots$ and if alternative 2 holds then we take the path $v \to t_2 \leftarrow t_1 \to s_2 \leftarrow s_1 \to a_{2i-1} \leftarrow x \cdots$.

Note that by an identical argument, if T is a tournament of odd order, and v is a vertex with indegree at least 2 then T admits a v-BADH.

It remains to deal with the case of even n. If t_1 has indegree at least two in the tournament $T \setminus v$, which has odd order $n - 1 \geq 19$, then by the above proof $T \setminus v$ has a t_1-BADHP and thus T has a v-FADHP. Hence we may assume that t_1 has indegree at most one in $T \setminus v$. Note that $T \setminus \{v, t_1, t_2\}$ is also a tournament of odd order $n - 3 \geq 17$ and thus contains an ADHP $x_1 \leftarrow x_2 \rightarrow \cdots \rightarrow x_{n-3}$. We may assume without loss of generality that $x_1 \leftarrow x_{n-3}$. Since t_1 has indegree at most one, we have $t \rightarrow x_1$ or $t \rightarrow x_{n-3}$, and hence $v \rightarrow t_2 \leftarrow t_1 \rightarrow x_1 \leftarrow \rightarrow \cdots$ or $v \rightarrow t_2 \leftarrow t_1 \rightarrow x_{n-3} \leftarrow \rightarrow \cdots$ is the desired ADHP in T. □

Remark. There exists a tournament with $n = 10$ vertices and a vertex v of outdegree two, but without a v-FADHP: Simply take a tournament on 7 vertices which has no ADHP (the 'quadratic residue' tournament QR_7, cf. [8]) and add three new vertices, v, t_1, t_2, and arcs $v \rightarrow t_1, v \rightarrow t_i$ and $v \leftarrow x, t_i \rightarrow x$ for all $i = 1, 2$ and all vertices x of the seven vertex tournament QR_7.

The proof of our theorem does imply a sequential algorithm for finding a x-FADHP in a tournament, if one exists. The complexity of that algorithm is dominated by the complexity of finding an ADHC in a tournament, and thus is $O(n^2)$. In the next section, we give another proof of our theorem, which only applies to tournaments of order at least 46, but yields an $O(n)$ algorithm.

3 The second proof

Lemma 3.1 *Suppose T is a tournament of even order $n \geq 16$, and let $v_1, v_2, \ldots v_5$ be the vertices of a particular copy of TT_5 in T. Then T contains a v_i-ADHP for some $i \in \{1, 2, \ldots, 5\}$.*

Proof: (We have already observed that any tournament of order at least 16 contains at least one TT_5.) Let $c_1 \leftarrow c_2 \rightarrow \cdots \leftarrow c_{n-6} \rightarrow c_{n-5}$ be an ADHP in the tournament $T \setminus \{v_1, v_2, \ldots v_5\}$ (which is of order $n - 5 \geq 11$).

If $v_1 \rightarrow c_1$ then $v_3 \rightarrow v_4 \leftarrow v_2 \rightarrow v_5 \leftarrow v_1 \rightarrow c_1 \leftarrow \cdots$ is a v-ADHP in T with $v \in \{v_1, v_2, \ldots v_5\}$. A symmetric argument applies if $v_1 \rightarrow c_{n-5}$.

If $v_2 \rightarrow c_1$ then $v_1 \rightarrow v_4 \leftarrow v_3 \rightarrow v_5 \leftarrow v_2 \rightarrow c_1 \leftarrow \cdots$ is a desired path. A symmetric argument applies if $v_2 \rightarrow c_{n-5}$.

If $v_3 \rightarrow c_1$ then $c_{n-5} \leftarrow c_2 \rightarrow \cdots \rightarrow c_1 \leftarrow v_3 \rightarrow v_5 \leftarrow v_1 \rightarrow v_4 \leftarrow v_2$ is a desired path. A symmetric argument applies when $v_3 \rightarrow c_{n-5}$.

Therefore we assume that c_1 and c_{n-5} dominate the vertices v_1, v_2 and v_3. Without loss of generality, we may assume that $c_1 \rightarrow c_{n-5}$. Now, if $c_2 \rightarrow v_1$ then $v_1 \leftarrow c_2 \rightarrow \cdots \leftarrow c_{n-6} \rightarrow c_{n-5} \leftarrow c_1 \rightarrow v_3 \leftarrow v_2 \rightarrow v_5 \leftarrow v_4$ is a desired path, and if $c_2 \rightarrow v_3$ then $v_3 \leftarrow c_2 \rightarrow \cdots \leftarrow c_{n-6} \rightarrow c_{n-5} \leftarrow c_1 \rightarrow v_2 \leftarrow v_1 \rightarrow v_5 \leftarrow v_4$ is a desired path.

Otherwise, $c_2 \leftarrow v_1, c_2 \leftarrow v_3$, and we may take the path $c_3 \leftarrow c_4 \rightarrow \cdots \leftarrow c_{n-6} \rightarrow c_{n-5} \leftarrow c_1 \rightarrow v_2 \leftarrow v_1 \rightarrow c_2 \leftarrow v_3 \rightarrow v_5 \leftarrow v_4$. □
Observe that the ADHP claimed by the lemma can be found in time $O(n)$.

Theorem 3.2 *Let T be a tournament of order $n \geq 46$, and let x be a vertex of T of outdegree at least two. Then T contains an x-FADHP which can be found in time $O(n)$.*

Proof: Case 1: n is even. In this case assume only that $n \geq 42$. Find a vertex y in $T \setminus x$ which has indegree (in $T \setminus x$) at least 21. Let z any vertex with $z \to y$, and let t be any vertex with $x \to t$. If $z \to t$, then consider the tournament $T \setminus \{x, y, t, z\}$. Among the (at least 18) inneighbours of y in this tournament, find a TT_5 and, using Lemma 3.1, find a FADHP in the even tournament $T \setminus \{x, y, t, z\}$, starting in a vertex v of the TT_5. Then $x \to t \leftarrow z \to y \leftarrow v \to \cdots$ is an x-FADHP in T.

Let t_1, t_2 be two vertices with $x \to t_i$, and let z_1, z_2 be two other vertices with $y \leftarrow z_i$. Without loss of generality we may assume that $t_1 \leftarrow t_2, z_1 \leftarrow z_2$. We may also assume that $t_2 \to z_1$, by the preceding paragraph. Now consider the tournament $T \setminus \{x, y, t_1, t_2, z_1, z_2\}$. Among the (at least 16) inneighbours of y, find a TT_5. Again using Lemma 3.1, find a FADHP in this even tournament, starting in a vertex v of the TT_5. The $x \to t_1, \leftarrow t_2 \to z_1 \leftarrow z_2 \to y \leftarrow v \to \cdots$ is an x-FADHP in T.

Remark. By applying the same procedure in T^{-1}, the reversal of T, the following must hold: If T is a tournament of even order $n \geq 42$, then for each vertex x with indegree at least two there is an x-BADHP in T.

Case 2: n is odd. Now we will assume that $n \geq 46$. Find a vertex y in $T \setminus x$ of indegree at least 23. If there are vertices z, t with $z \to y, x \to t, z \to t$, then consider the tournament $T \setminus \{x, t_1, y'\}$. It is of even order $n - 3 \geq 43$ and hence contains a y-BADHP; this yields the following x-FADHP in T: $x \to t \leftarrow z \to y \leftarrow \cdots$. Otherwise we choose t_1, t_2, z_1, z_2 with $x \to t_i, y \leftarrow z_i, t_1 \leftarrow t_2, z_1 \leftarrow z_2, t_2 \to z_1$, as above. Since the tournament $T \setminus \{x, t_1, t_2, z_1, z_2\}$ is of even order $n - 5 \geq 41$ it is in fact of order at least 42, and so we may still conclude that it contains a y-BADHP. This yields the x-FADHP $x \to t_1 \leftarrow t_2 \to z_1 \leftarrow z_2 \to y \leftarrow \cdots$ in T.

It remains to argue that the algorithm has complexity $O(n)$. (Note that we must assume the tournament is already given, since to decribe a tournament with n vertices takes time proportional to n^2.) To find a vertex of indegree at least 21 we apply the result of Moon [5], which asserts that a tournament with N vertices must have a vertex with indegree at least $N/2$. We simply choose 42 arbitrary vertices of T and find, by exhaustive search, a vertex of indegree at least 21 in this subtournament, and thus in T as well. (If T has fewer than 42 vertices we may apply any technique without affecting the asymptotic behaviour of our algorithm.) Searching for vertices z, t with $z \to y, x \to t, z \to t$, we may restrict ourselves to 2 outneighbours t of x and 21 inneighbours of y. Then testing whether or not $z \to t$ for some z and t takes only constant time. The remaining operations are linear in view of our earlier remarks to the effect that Lemma 3.1 can be applied in time $O(n)$, and a TT_5 in a tournament on 16 vertices can be found in constant time. $\qquad \square$

4 Concluding remarks

We are hoping to extend our techniques to solve the following problem: Given a tournament T and two vertices x, y, find (if possible) an ADHP of T that begins in x and ends in y. In fact, it may be possible to use the above techniques to find (if possible) an x-FADHP, or x-BADHP, of T, that ends in y. It could be the case that such a path exists provided there exist 4 vertices x', x'', y', y'' such that the desired ADHP could begin $x, x', \ldots$ or $x, x'', \ldots$ and could end $\ldots, y', y$, or $\ldots, y'', y$.

As a last remark, note that from the second proof of our Theorem we can easily extract a parallel algorithm for finding an x-FADHP in T, provided x has outdegree at least 2. Its complexity is dominated by finding an ADHP, and thus is $O(\log n)$ time with $O(n^2 / \log n)$ processors. It is also easy to extend this to finding an x-FADHP from any vertex x: The sequential algorithm for this purpose explained earlier can be parallelized (within the same time and processor bound) by the standard 'pointer-jumping' tricks typified by the parallel prefix computation algorithms, [2], chapter 30.

References

1. E. Bampis, I. Milis, Y. Manoussakis, *NC algorithms for Antidirected Hamiltonian Paths and Cycles in tournaments*, in Springer Verlag LNCS 903, Graph Theoretic Concepts in Computer Science WG'94, (E. W. Mayr et al eds.) 387 - 394; journal version to appear in JCMCC.

2. T. H. Cormen, C. E. Leiserson, R. L. Rivest, **Introduction to Algorithms**, MIT Press, 1990.

3. B. Grünbaum, *Antidirected Hamiltonian paths in tournaments*, J. Combin. Theory (B) 11 (1971) 249-257.

4. P. Hell and M. Rosenfeld, *The complexity of finding generalized paths in tournaments*, Journal of Algorithms 4 (1983) 303-309.

5. J. W. Moon, **Topics on Tournaments**, Holt, Reinhart and Winston, New York 1969.

6. V. Petrovic, *Antidirected Hamiltonian circuits in tournaments*, In Proc. 4th Yugoslavian Seminar of Graph Theory, pp. 259-269, Novi Sad, 1983.

7. K. B. Reid and E. T. Parker, *Disproof of a conjecture of Erdös and Moser*, J. Combin. Theory (B) 9 (1970) 93-99.

8. M. Rosenfeld, *Antidirected Hamiltonian paths in tournaments*, J. Combin. Theory (B) 12 (1972) 93-99.

9. M. Rosenfeld, *Antidirected Hamiltonian circuits in tournaments*, J. Combin. Theory (B) 16 (1974) 234-242.

10. C. Thomassen, *Antidirected Hamiltonian circuits and paths in tournaments*, Math. Ann. 201 (1973) 231-238.

Complementary ℓ_1-Graphs and Related Combinatorial Structures

Michel Deza
LIENS-Ecole Normale Supérieure
45 rue d'Ulm
75230 Paris Cedex 05, France
and
Tayuan Huang*
Department of Applied Mathematics
National Chiao-Tung University
Hsinchu 30050, Taiwan R.O.C.

Abstract: The conditions under which both graph G and its complement $\overline{G}$ share some common properties, e.g., with diameter 2, with ℓ_1-addressings, with property of being strongly regular are studied. Many examples and counterexamples from different areas of graph theory regarding them are provided. In particular, a census of graphs with at most six vertices is given.

1. Introduction

The conditions under which a graph G and its complement $\overline{G}$ possess some specified properties were studied by Akiyama and Harary [2]. Following their spirit and the study of ℓ_1-embeddability of graphs by Deza, Grishukhin and Laurent [12,13], we shall study those ℓ_1-graphs with ℓ_1-complements. In particular, we pay special attention to those graphs which are self-complementary, to those graphs G such that both G and $\overline{G}$ are of diameter 2, or moreover, G and hence $\overline{G}$ are strongly regular.

A simple connected graph G is called an ℓ_1-graph if it admits a binary addressing such that, up to some scale, the Hamming distance between the binary addresses of two vertices concides with their distance in the graph, i.e., a result in [1], the metric space $(V(G), d)$ is isometrically embeddable in some ℓ_1-space. Terwilliger and Deza [26] proved that every ℓ_1-graph is an isometric subgraph of

---

*e-mail:thuang@cc.nctu.edu.tw

a product of half-cubes, Cocktail Party graphs and Gosset graphs. Shpectorov [25], and, later, Deza and Grishukhin [12] with a shorter proof, showed that a graph is an ℓ_1-graph if and only if it is an isometric subgraph of a direct product of Cocktail Party graphs and half-cubes.

For a connected graph $G = (V(G), E(G))$ with $V(G) = \{x_1, x_2, \cdots, x_n\}$, the matrix $d(G) = [d(x_i, x_j)]_{n \times n}$ is called the *distance matrix* of G, where $d(x_i, x_j)$ is the shortest path distance between vertices x_i and x_j in the graph G. The metric space $(V(G), d)$ is called *m-gonal* if for any $y_1, y_2, \cdots, y_m \in V(G)$ (not necessarily distinct),

$$\sum_{i,j=1}^{m} \lambda_i \lambda_j d(y_i, y_j) = \lambda \cdot d(G) \cdot \lambda^t \le 0$$

where $\lambda = (\lambda_1, \lambda_2, \cdots, \lambda_m) = (1, 1, \cdots, 1; -1, -1, \cdots, -1)$ with the first $\lceil \frac{m}{2} \rceil$ terms being 1 and λ^t the transpose of λ. G is called *hypermetric* if $(V(G), d)$ is $(2n + 1)$-gonal for all $n \in N$, and is called of *negative type* if it is $2n$-gonal for all $n \in N$. Remark that 3-gonal coincides with triangle inequality. It was proved in [10] that for any metric space, ℓ_1-embeddability implies hypermetricity and hypermetricity implies negative type, neither is reversable.

To prove that some graphs are not ℓ_1-graphs, we mainly show below that they are not 5-gonal, a little bit similar to the triangle inequality. The notion of ℓ_1-embeddability (or even weaker condition of 5-gonality) is a very strong condition, it does not hold for many highly symmetric graphs. For example, Grotzsch, Chvatal graphs, their relatives and small triangle-free graphs [19] are non-5-gonal because they contain a non-5-gonal $K_5 - K_{1,3}$ as an isometric subgraph. Many non-5-gonal graphs can also be found in [12].

An ℓ_1-*embedding* is a binary matrix such that its rows are binary addresses of vertices of a graph in some order. Two ℓ_1-embeddings are *equivalent* if one can be obtained from the other by a sequence of the following procedures:

    1. permutations of columns,

    2. adding (module 2) some binary vector to all rows, and

    3. adding or deleting columns consisting of either all ones or all zeros.

An ℓ_1-graph is called ℓ_1-*rigid* if its ℓ_1-embedding is unique, up to equivalence. The smallest quotient of the numbers of columns required for ℓ_1-embeddings over its scale is called the *size* of the graph G. Any ℓ_1-rigid graph is with scale either 1 or 2 [25]. Note that an ℓ_1-graph of scale 1 is exactly an isometric subgraph of a hypercube $H(m, 2)$ and the graph is ℓ_1-rigid [13]; if the scale is 2, then it is an isometric subgraph of a half-cube $\frac{1}{2}H(m, 2)$.

Notations used in this paper follow from [4,5]. The disjoint union of two graphs G and H is denoted by $G + H$, the induced subgraph of the graph G over a subset X of its vertex set is denoted by $G[X]$, $L(G)$ is the line graph of G, $\overline{G}$ is the complement of G, $G - x$ is the graph obtained from G by deleting a vertex x of G, $K_v - H$ is the complement of H in K_v, $K_{n_1, n_2, \cdots, n_m}$ is the complete m-partite graph with parts of cardinalities $n_1, n_2, \cdots$, and n_m respectively. The *suspension* $\nabla(G)$ of the graph G is obtained from G by adding a vertex to $V(G)$ which is adjacent to each vertex of G. C_n is the cycle of length n, $n \times n$-

grid is the line graph of $K_{n,n}$, O_3 the Petersen graph and its complement is the triangular graph $T(5)$, P_n is the path of length $n - 1$. The following graphs are mentioned often in the rest of this paper: $P_4 = \overline{P_4}$, P_5, $P_5 + e$ (the hat), $C_5 = \overline{C_5}$, $\overline{P_5} = C_5 + e$, $H_1 = K_{2,3}$, $H_2 = \overline{K_2 + P_3}$, and $H_3 = \nabla(K_{1,3}) = K_5 - K_3$ (the crown). Note that $H_1 = K_{2,3}$ is a partial subgraph of H_2, and H_3.

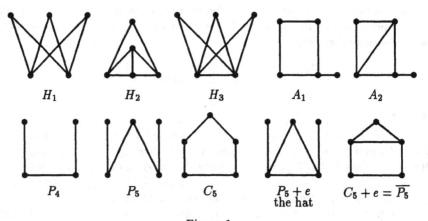

Figure 1.

This paper is organized as follows: ℓ_1-embeddability of graphs with small numbers of vertices is studied in Section 2; in particular, a census of graphs with at most six vertices is given. ℓ_1-embeddability of self-complementary graphs is studied in Section 3. The close relationship among strongly regular ℓ_1-graphs and designs is studied in Section 4.

2. ℓ_1-embeddability of graphs with small number of vertices

Simultaneous behavior of pairs of complementary graphs were heavily studied in the literature; for example, both are outerplanar, non-forests [2], both are almost self-complementary [9], and edge-integrity [20]. In this section, we shall study those complementary pairs such that both are ℓ_1-graphs with at most six vertices, both are of diameter 2, respectively.

First, we treat those graphs with small numbers of vertices. Following the list of connected graphs on six vertices, together with their numbering, given by Cvetković and Petric [8], we have:

Proposition 2.1 *If G is an ℓ_1-graph with ℓ_1-complement and with at most six vertices, then G is one of the following: P_4, $\{P_5, \overline{P_5}\}$, $P_5 + e$ (the hat), C_5, $\{A_1, \overline{A_1}\}$, $\{A_2, \overline{A_2}\}$ (all are ℓ_1-rigid), or one of the graphs with numbers*

$\{20, 108\}, \{23, 109\}, \{27, 110\}, \{34, 95\}, \{35, 101\}, \{39, 97\}, \{41, 103\}, \{42, 96\},$

$\{44, 102\}, \{47, 105\}, \{51, 106\}, \{54, 76\}, \{59, 77\}, \{60, 85\}, \{66, 80\}, \{67, 87\},$
$\{68, 86\}, \{70, 92\}$

(with 6 vertices) in [8], occurring in complementary pairs.

Those graphs on six vertices together with their numbering in the list of [8], are given in the Appendix for completeness. Some further observations for these 36 graphs are in order:

1. Eight of these graphs (namely graphs # 92, 101-103, 106, 108-110) are bipartite. In addition to these, 4 others are bipartite ℓ_1-graphs with 6 vertices, the complements of which are either disconnected ($K_{1,5}$, *i.e.*, graph #107) or not 5-gonal (the graphs #104, 111 and 112) and hence not ℓ_1-graphs.

2. All graphs, except #20,54, are ℓ_1-rigid. In addition to these, there are 8 more ℓ_1-graphs with 6 vertices which are not ℓ_1-rigid, complements of which are either disconnected (namely $K_6, K_6 - P_2, K_6 - 2P_2, K_5 + K_2 = K_6 - K_{1,5}, K_4 + K_3, K_4 + K_2 + K_2$ and graph #19) or not 5-gonal (the graph #55).

3. Among these 36 graphs, the graphs # 27, 41, 42, 47, 51, 66, 68 are of diameter 2, graphs #80, 103, 110 are of diameter 4, all others are of diameter 3. So there is no pair of complementary ℓ_1-graphs on six vertices of diameter 2 both. Also the cycle C_6, (*i.e.*, graph #106) is unique among these 36 graphs in having radius 3; all others have radius 2 [5 p. 37].

The following table provides some statistics for those graphs with small numbers of vertices.

properties number of graphs number of vertices	connectedness	ℓ_1-graphs		bipartite ℓ_1		non-separable ℓ_1-graphs	ℓ_1-graphs with ℓ_1-complements
		rigid	not rigid	trees	not trees		
≤ 4	9	8	1	4	1	5	1
5	21	15	3	3	1	8	4
6	112	68	10	6	6	46	18 pairs
7 with ≤ 8 edges	111	99	0	11	12	3	?

Table 1

Of these statistics, it is useful to note:

1. $H_1 = \overline{P_2 + C_3} = K_{2,3}, H_2 = \overline{P_2 + P_3},$ and $H_3 = \nabla K_{1,3}$ are exactly those connected but not ℓ_1-graphs on at most 5 vertices; all of them are not 5-gonal, though $H_3 = K_5 - K_3$ is of negative type.

2. The cycle C_7 and the following two ℓ_1-rigid graphs are exactly those non-separable (*i.e.*, blocks) ℓ_1-graphs on 7 vertices with at most 8 edges.

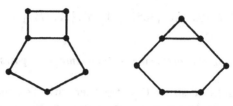

Figure 2.

3. $K_5, K_5 - P_2, K_4 + K_2 = K_5 - K_{1,3}$ are exactly those ℓ_1-graphs which are not ℓ_1-ridid on at most 5 vertices.

4. For all 253 graphs, the property of being 5-gonal implies ℓ_1-embeddability. Note that $K_7 - C_4$, which is 5-gonal, is not 7-gonal and hence not an ℓ_1-graph; also $K_7 - P_3, K_7 - P_4, K_7 - C_5$ are examples of hypermetric graphs which are not ℓ_1-graphs. (Exactly 12 such graphs exist on 7 vertices; they can be found in [12].)

5. Among the 204 ℓ_1-graphs, $K_6 - K_2$ is the only one of scale 4; exactly 44 of them are bipartite ℓ_1-graphs (*i.e.*, of scale 1), all others are of scale 2.

6. Among the 253 graphs, if both G and $\overline{G}$ are connected, then at least one of them is an ℓ_1-graph; this is not true in general, see counterexamples below (for 8 vertices).

Some other examples of small ℓ_1-graphs with ℓ_1-complements are the 3-cube $H(3, 2)$ with its complement 2×4-grid, the 3×3-grid (self-complementary), and O_3 with its complement $T(5)$. All these are distance-regular graphs which we shall treat in detail later.

The following four graphs provide examples of complementary pairs $\{G, \overline{G}\}$ such that both of them are of diameter 2, and have various ℓ_1-embeddability. The two graphs (#69, 93) in Figure 3 are complementary of diameter 2 each, moreover (i) is an ℓ_1-graph, but (ii) is not 5-gonal and hence not an ℓ_1-graph. The two complementary graphs of diameter 2 in Figure 3 iii) and iv) are not 5-gonal and hence not ℓ_1-embeddable; moreover, both are 4-partite, namely $\{1, 7\}, \{2, 8\}, \{3, 6\}, \{4, 5\}$ for (iii) and $\{1, 4\}, \{2, 6\}, \{3, 5\}, \{7, 8\}$ for (iv) and those 4-partitions give perfect matchings of the complete graph K_8.

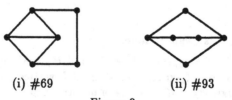

(i) #69   (ii) #93

Figure 3

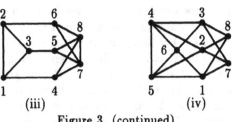

Figure 3. (continued)

Those graphs G such that both G and $\overline{G}$ are outerplanar but neither G nor $\overline{G}$ are forests are given in [2]. Among their list of 32 graphs, five of them are disconnected, the others are $P_5, C_5, P_5 + e$ (the hat), $K_4 + K_2$ (with 5 vertices) and the graphs # 77, 83, 85, 86, 87, 92, 96 and 102 (with 6 vertices), together with their complements. It is easy to check that the complement of the graph #83 is not 5-gonal and hence not an ℓ_1-graph. Based on this observation and Proposition 2.1, the following corollary is immediate.

Proposition 2.2 *If G is a connected graph such that both G and its complement $\overline{G}$ are outerplanar but not forests, then, except for graph #83 (Figure 4), both G and $\overline{G}$ are ℓ_1-graphs.*

Figure 4.

The complete list of all almost self-complementary graphs over 6 vertices are given in [9]. After inspecting this list of 6 graphs, the following corollary is also immediate.

Proposition 2.3 *Graph #54 and its complement #76 are the only almost self-complementary ℓ_1-graphs with ℓ_1-complements with at most 6 vertices.*

(i) #54   (ii) #76

Figure 5.

We now turn to the ℓ_1-embeddability of some graphs associated with Ramsey graph theory [17]. For a graph G, $R(G)$ is defined to be the smallest integer

r such that no matter how the edges of K_r are 2-colored, a mono-chromatic subgraph isomorphic to G is always formed. Some known Ramsey numbers are $R(K_2) = 2, R(K_3) = 6, R(K_4) = 18$ and it turns out that corresponding Ramsey graphs, *i.e.*, K_1, C_5 and the Greenwood-Gleason graph (Figure 6(i)) are unique and self complementary. Indeed, the Greenwood-Gleason graph (of diameter 2) has vertices from the field Z_{17} with i, j being adjacent if and only if $i - j$ is a square in Z_{17}, which is not 5-gonal since it contains $K_{5,6,10,14,15} - K_{5,10,15}$ as an induced subgraph. Furthermore, the Ramsey graph corresponding to $R(K_5, K_3) = 14$, given in Figure 6(ii) with diameter 2 on 13 vertices, is not 5-gonal either.

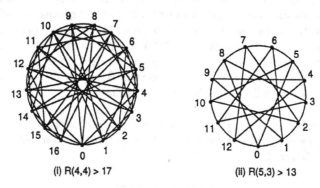

(i) R(4,4) > 17   (ii) R(5,3) > 13

Figure 6.

Any 5-universal graph G [22] contains $H_1 = K_{2,3}$ (also $H_2 = \overline{K_2 + P_3}$ and $H_3 = \nabla K_{1,3}$ as induced subgraphs, so G is not 5-gonal (because any induced subgraph of diameter 2 of a graph is an isometric subgraph). Hence orthogonality graphs [21] provide examples of complementary pair $\{G, \overline{G}\}$ such that both are strongly regular with 2^{10} vertices, and each of them contains all of H_1, H_2, H_3 as induced subgraphs. It is worthing mentioning here that for any graph G with at least $\max(R(H_2), R(H_3))$ vertices, either G or $\overline{G}$ contains each of the graph H_1, H_2, H_3 as partial subgraph; for example, $T(5)$ contains H_1, H_2, and H_3 but its complement O_3 does not contain any of H_1, H_2, H_3.

We now turn to the ℓ_1-embeddability of a class of extremal family of graphs in the sense of Entringer et al. [15]. For a connected graph G, let

$$\delta(G) = \sum_{x,y \in V(G)} d(x, y)/2.$$

Entringer et al. [15] showed that $\delta(G) + \delta(\overline{G}) \geq 3n(n-1)/2$. This bound is achieved by the graph $G(m_1, m_2, \cdots, m_5)$ consisting of n disjoint cliques $K_{m_1}, K_{m_2}, \cdots, K_{m_5}$, and each pair of vertices of K_{m_i} and of $K_{m_{i+1}}(i+1 \pmod{5})$ are adjacent, $i = 1, 2, 3, 4, 5$. The ℓ_1-embeddability of the graph $G(m_1, m_2, \cdots, m_5)$ is characterized in the following, its proof is straighforward and hence is ommited.

Proposition 2.4 *Let* $G = G(m_1, m_2, \cdots, m_5)$,
1. *if* $m_i \geq 2$ *for some* i, *then* $\overline{G}$ *is not 5-gonal,*
2. *if* $m_i \leq 2$ *for each* i, *then* G *is an* ℓ_1-*graph, and*
3. *both* G *and* $\overline{G}$ *are* ℓ_1-*graphs if and only if each* $m_i = 1$, *i.e.*, $G = C_5$.

While the number of complementary pairs $\{G, \overline{G}\}$ of graphs of diameter 2 which are both not 5-gonal is infinite (simply by taking some strongly regular graphs), it seems that the number of such pairs of complementary ℓ_1-graphs is finite, at least in the case of self-complementary pairs, or in the case that both G and $\overline{G}$ have diameter 2. For example, $K_n - K_m$ (for $n \geq m \geq 6$), $K_n - C_m$ (for $n \geq m \geq 7$) contains $H_3 = K_5 - (P_2 + P_3)$ as an isometric subgraph and hence they are not 5-gonal.

Let G be an ℓ_1-graph with v vertices and m edges with $m \leq \begin{pmatrix} v-1 \\ 2 \end{pmatrix}$, then $K_n - \overline{G} = \nabla^{n-v} G$ is not $(2v-1)$-gonal whenever $n \geq 2v - 1$, since it violates the condition that $\lambda d(G) \lambda^t \leq 0$ with v terms of 1 and $v-1$ terms of -1 in vector λ. The following Proposition follows easily.

Proposition 2.5 *Let* G *be a graph such that both* G *and its complement* $\overline{G}$ *are* ℓ_1-*graphs of diameter 2 each, on* $n \geq 3$ *vertices, then either* $K_{2n-1} - G$ *or* $K_{2n-1} - \overline{G}$ *is not* $(2n-1)$-*gonal and hence not an* ℓ_1-*graph.*

Proof: Since both G and $\overline{G}$ are connected, they have at least $n-1$ edges. The case $n-1$ (*i.e.*, a tree) is excluded since $n \geq 3$ and the diameter is 2. We may assume now that both have at least n edges and both have at most $\begin{pmatrix} n \\ 2 \end{pmatrix} - n = \begin{pmatrix} n-1 \\ 2 \end{pmatrix} - 1$ edges and the proposition follows easily. *Q.E.D.*

We finish this section with the following remark. Let G be an ℓ_1-graph; it has scale 1 if and only if it is bipartite, it has scale 2 if it is not a tree but with odd cycles only (since then each of its block is either K_2 or odd cycles. Though all bipartite ℓ_1-graphs are ℓ_1-rigid [13], and some 4-partite ℓ_1-graphs (for example $K_5 - P_3$ and graphs #23, 35, 39, 41 in the appendix) are examples (all for $n \leq 6$) of ℓ_1-rigid graphs, some other 4-partite ℓ_1-graphs (for example K_4, $K_{4\times2}$, and graph #54 from the appendix) are not ℓ_1-rigid. It seems quite reasonable to ask whether there exists any 3-partite ℓ_1-graph which is not ℓ_1-rigid; the answer is negative for $n \leq 6$.

3. ℓ_1-embeddability of self-complementary graphs

The attractiveness of self-complementarity for graphs led us to investigate ℓ_1-embeddability for them.

Two easy ways of building self-complementary graphs from P_4 and some other self-complementary graphs are given in the following [16]:

- $A(G)$: the graph obtained by joining the middle two vertices of P_4 with each vertex in G, and

- $B(G)$: the graph obtained by joining two end vertices of P_4 with each vertex in G.

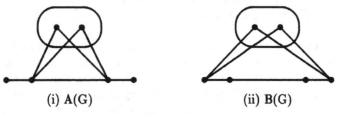

(i) A(G) (ii) B(G)

Figure 7.

For example, $A(K_0) = B(K_0) = P_4, A(K_1) = P_5 + e$, and $B(K_1) = C_5$ and the following four more from [3]:

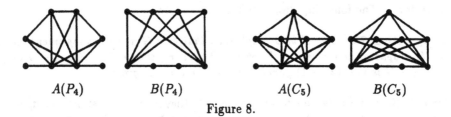

$A(P_4)$ $B(P_4)$ $A(C_5)$ $B(C_5)$

Figure 8.

These two operations preserve the property of self-complementarity, but reduce their diameters.

Lemma 3.1 *If G is self-complementary, then both $A(G)$ and $B(G)$ are self-complemntary with diameter 3 and 2 respectively.*

We now turn to the ℓ_1-embeddability of graphs of the form $A(G)$ or $B(G)$. Clearly, $A(K_0) = B(K_0) = P_4, A(K_1) = P_5 + e, B(K_1) = C_5$ and $A(P_4)$ are all examples of self-complementary ℓ_1-graphs. Since $K_7 - C_5$ is not an ℓ_1-graph, $A(C_5)$ is not an ℓ_1-graph either. Moreover, $B(P_4)$ is even not 5-gonal.

Lemma 3.2. *Suppose G is a self-complementary graph with v vertices, then*
1. *$A(G)$ is an ℓ_1-graph if $v = 4, 5, 7$,*
2. *$A(C_5)$ is hypermetric but not an ℓ_1-graph, and*
3. *$A(P_5 + e)$ and $A(G)$ for $|V(G)| \geq 10$ are not 5-gonal.*

Lemma 3.3 *Suppose G is self-complementary with v vertices, ·n $B(G)$ is 5-gonal if and only if $G = K_0$, or K_1.*

Lemma 3.4 *Let G be a self-complementary graph with v vertice, such that $\Gamma = \nabla^2 G = K_{v+2} - \overline{G}$ is 5-gonal, then Γ is one of the following: $\nabla^2(K_0) =$*

$P_2, \nabla^2(K_1) = C_3, \nabla^2(P_4) = K_6 - P_4,$ or $\nabla^2(C_5) = K_7 - C_5$.

Theorem 3.5 *Let G be a self-complementary graph on v vertices. Then $A(G)$ or $B(G)$ is an ℓ_1-graph if and only if they are $A(K_0) = B(K_0) = P_4, B(K_1) = C_5, A(K_1) = P_5 + e$ (the hat), or $A(P_4)$.*

Proof: ℓ_1-embeddability for $A(P_4)$ is clear. $G = K_v$ if and only if $v = 0, 1$. If $G \neq K_v$, then $B(G)$ contains $B(\overline{K}_2)$ as an isometric subgraph, which is not 5-gonal. If $G = \overline{G}$ contains a triangle, then $A(G)$ contains $A(\overline{K}_3)$ and is not 5-gonal. Self-complementary trees are K_1, P_4. Let t be the girth of G if G is not a tree. If $t \geq 6$ or $t = v = 5$, then $G = \overline{G}$ contains a triangle too. In the remaininng cases $A(C_5)$ contains $K_7 - C_5$, which is hypermetric, but not ℓ_1-graph. Remark that $K_7 - P_4, K_7 - P_3$ are hypermetric but not ℓ_1-graphs. Q.E.D.

A list of all 10 self-complementary graphs with 8 vertices is given in [3]. A direct inspection shows that exactly 4 of those 10 graphs, *i.e.*, 3×3-grid-v (of diameter 2, Figure 9 (i)), the graphs $A(P_4)$, and the graphs in Figure 9 (ii), (iii) (of diameter 3), are 5-gonal and, moreover, ℓ_1-graphs.

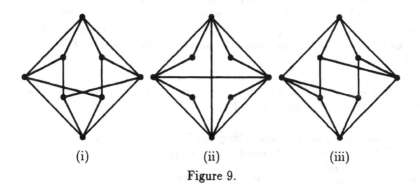

(i)   (ii)   (iii)

Figure 9.

4. Strongly regular ℓ_1-graphs and designs

The properties of self-complementarity, and ℓ_1-embeddability among strongly regular graphs are studied, for example, in [23,24] and in [12] respectively. A regular graph Γ is called *strongly regular* if the number of common neighbors of distinct vertices x and y depends only on whether x and y are adjacent in Γ or not, see [4] for more details.

Proposition 4.1 [12] $C_5, O_3, \frac{1}{2}H(5,2)$ (*i.e.*, the Clebsch graph), the Shrikhande graph, $K_{m \times 2}$ (*i.e.*, the Cocktail Party graph), $T(m)$ and $m \times m$-grids are the only strongly regular ℓ_1-graphs.

The following corollary follows easily through a direct checking of the ℓ_1-embeddability of their complements.

Corollary 4.2 $C_5, 3 \times 3$-*grid, the Petersen graph* O_3 *and its complement* $T(5)$ *are the only strongly regular* ℓ_1-*graphs with* ℓ_1-*complements.*

The strongly regular graphs mentioned above are ℓ_1-graphs of scale 2 and with sizes 5/2, 3, 3, 5/2 respectively. Both C_5 and O_3 are Moore graphs with diameter 2; however, the third known Moore graph with diameter 2, *i.e.*, the Hoffman-Singleton graph, is not an ℓ_1-graph.

Suppose that G is one of the above mentioned strongly regular graphs and $M(G), M(\overline{G})$ be their corresponding ℓ_1-embedding matrices (with respect to the same order of vertices along the rows). Let $N(G) = [M(G)|M(\overline{G})]$ be the concatenation of $M(G)$ with $M(\overline{G})$. It is worth noting that, since diameter is 2, the Hamming distance of the first half of any two rows is 2 if and only if the Hamming distance for the corresponding second half is 4 and vice versa. Regarding M as a point-block incidence matrix of a set system, some interesting combinatorial structures follow naturally.

If G is $\dot{C}_5$, self-complementary, then $N(C_5)$, given in Figure 10(i), is an incidence matrix of the trivial 2-(5,2,1) design. If G is the 3×3-grid, self-complementary too, then $N(3 \times 3$-grid), given in Figure 10(ii), is an incidence matrix of an affine plane 2-(9,3,1), which is the unique Steiner triple system on 9 points. The edges of 3×3-grid can be decomposed into three edge-disjoint 6-cycles, *i.e.*, $C_{1,2,5,6,9,7}$, $C_{1,3,9,8,5,4}$ and $C_{2,3,6,4,7,8}$ such that each pair of such cycles has three vertices in common. We summarize these observations in the following proposition.

Proposition 4.3 $N(C_5)$ *is the incidence matrix of the design* 2-(5,2,1), *and* $N(3 \times 3$-*grid) is an affine plane* 2-(9,3,1).

The case for O_3 is more interesting; $N(O_3)$ is given in Figure 10(iii). Let $N^* = N(O_3)^*$ be the matrix obtained from $N(O_3)$ by deleting the last row and then adjoining $\begin{bmatrix} 10101011000 \\ 01010111000 \end{bmatrix}$ to it, it follows that $N^*J = JN^* = 5J$, and $N^*N^{*t} = 2J + 3I$, the following proposition follows.

Proposition 4.4 N^*, *as defined above, is an incidence matrix of a* 2-(11,5,2) *design, which is symmetric.*

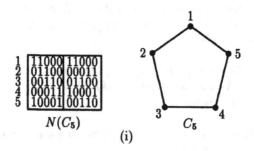

$$N(C_5)$$

(i)

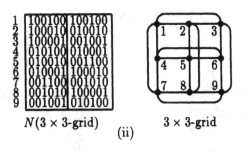

$N(3 \times 3\text{-grid})$ $3 \times 3\text{-grid}$

(ii)

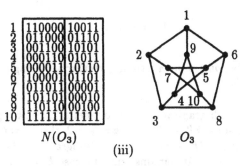

$N(O_3)$ O_3

(iii)

Figure 10.

The above three graphs and the self-complementary ℓ_1-graph $3 \times 3\text{-grid}-v$ provide all four known examples of decompositions of K_n into two complementary ℓ_1-graphs of diameter 2; namely
(i) $3d(K_5) = d(C_5) + d(\overline{C_5})$,
(ii) $3d(K_8) = d(3 \times 3\text{-grid} - v) + d(\overline{3 \times 3\text{-grid} - v})$,
(iii) $3d(K_9) = d(3 \times 3\text{-grid}) + d(\overline{3 \times 3\text{-grid}})$ and
(iv) $3d(K_{10}) = d(T(5)) + d(O_3)$.

The embeddings of K_n in (i), (ii) and (iii) can be extended to be ℓ_1-embeddings (with scale 6) of K_{2n+1}. Indeed, the cases (i) and (ii) are restrictions of the case (iii) on some 5 or some 8 points, respectively. Moreover, the case (iii) is a restriction of an ℓ_1-embedding, presented by an affine plane of order 3. But the case (iv) is not extendable, compare it with an extendable ℓ_1-embedding (with scale 6) of K_{11} given by teh biplane (11,5,2) [11]. Two other graphs with 13 vertices such that $d(G) + d(\overline{G}) = 3d(K_n)$ are given in the following.

There are exactly two vertex-transitive self-complementary graphs on 13 vertices, i.e., the Paley graph $P(13) = G(\{1,3,4\}, 13)$ and the graph $G(\{1,3,6\}, 13)$, Figure 11(i), as given in [24]. Both of them are 6-regular and of diameter 2, so $d(G) + d(\overline{G}) = 3d(K_{13})$ for them. Furthermore, the Paley graph is strongly regular, not 5-gonal (see Proposition 4.5 below); and the graph $G(\{1,3,6\}, 13)$, consisting of three edge-disjoint C_{13}, is not 5-gonal since its induced subgraph over $\{3,4,6,8,9\}$ forms a $\nabla K_{1,3}$. Moreover, the graph in [18], see Figure 11(ii), is another 6-regular self-complementary graph on 13 vertices which is not 5-gonal either.

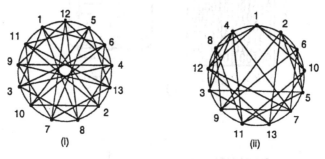

Figure 11.

Proposition 4.5 *The only 5-gonal strongly regular self-complementary graphs are* C_5 *and* 3×3*-grid.*

Proof: In fact, any such graph must have diameter 2, be $2k$-regular and with $n = 4k+1$ vertices for some k [23,p.225]. If $k = 1$ or 2, then $G = C_5$ or 3×3-grid. If $k \geq 4$, then G contains $\nabla K_{1,3}$ as an induced subgraph, by Lemma 7(ii) [23], and hence G is not 5-gonal. It remains only the case $k = 4$, *i.e.*, $n = 13$ and hence G must be the Paley graph $P(13)$ [24]. Since $P(13)$ is a circulant graph in which 0 is adjacent to 1, $4 = 2^2$, $9 = 3^2$, $12 = 5^2 - 13$, $10 = 6^2 - 2 \cdot 13$, and $3 = 4^2 - 13$. The induced subgraph of $P(13)$ over $\{0, 1, 2, 3, 12\}$ is $K_{2,3} + e$, and hence $P(13)$ is not 5-gonal. Q.E.D.

The decomposition $d(G) + d(\overline{G}) = 3d(K_n)$ can be seen as ℓ_1-embeddings of K_n with scale 6. The following lemma provides a sufficient condition for such decompositions.

Lemma 4.6 *If G is a graph such that both G and $\overline{G}$ are ℓ_1-graphs of diameter 2, then* $d(G) + d(\overline{G}) = 3d(K_n)$.

More specifically, $6d(K_5) = 2d(C_5) + 2d(\overline{C_5})$ gives the decomposition of K_5 into 2 disjoint hamiltonian cycles. In general, it is well-known (see, for example, [5, item 11.6] that for any odd n, K_n can be decomposed into $(n-1)/2$ hamiltonian cycles C^j obtained by successive rotations, (say $C_{1,2,3}$, $C_{1,3,5}$, $C_{1,4,7}$, ..., etc,) so

$$\sum_{j=1}^{(n-1)/2} d(C^j) = \sum_{j=1}^{(n-1)/2} j \cdot d(K_n) = (n^2 - 1)/8 \cdot d(K_n).$$

The decomposition $6d(K_{10}) = 2d(O_3) + 2d(\overline{O_3})$ can be seen as a case of general situation when a regular simplex of side t, (say, regular of side $\sqrt{6}$ simplex in R^{11}) is projected onto a simplex of smaller dimension n (say, onto 9-simplex in R^5, corresponding to ℓ_1-embedding of $\overline{O_3}$). In general, for any realization A of any addressable graph G, there exists some addressing B of K_n with scale $2t$, and with N columns such that A is a projection of B, as a projection of a regular $(n-1)$-simplex corresponding to B onto an $(n-1)$-simplex corresponding to A. It is interesting to find such t and N as small as possible. For more general cases, for any $(n-1)$-simplex A in R^n, Deza and Maehara [14] found that the smallest N is $2n - 2 - r$ where r is the multiplicity of the largest eigenvalue of -PDP, where D is the squared of distance (Euclidean) matrix of A, and $P = I_n - J_n$.

The minimum n such that K_n can be decomposed into k factors of diameter 2 each, denoted by $f_k(2)$, was studied by Bosak et. al. [6,7]. They showed, among many other things, that $f_2(2) = 5$, and the graph C_5 and its complement give such a decomposition of K_5, furthermore as we mentioned before, C_5 is a self-complementary ℓ_1-graph. The 3×3-grid-v with complement 2×4-grid, the 3×3-grid (self-complementary), and O_3 with its complement $T(5)$ provide three such decompositions of K_8, K_9, and K_{10} respectively. It is quite natural to raise the following question: can any complete graph K_n be decomposed into three factors such that each factor is an ℓ_1-graph of diameter 2? (Note that $12 \leq f_3(2) \leq 13$).

Acknowledgement This research was initiated while the first author visited the Institute of Mathematics, Academia Sinica, Taipei, Fall 1992. Both authors would like to thank Dr. Koh-Wei Lih for providing a multitude of stimulating contacts during the preparation of this research.

References:

1. P. Assouad & M. Deza, Espaces metrique plongeables dans un hypercube: aspects combinatoires, *Annals of Discrete Mathematics* 8(1980) 197-210.

2. J. Akiyama & F. Harary, A graph and its complement with specified properties 1: connectivity, *International Journal of Mathematics and Mathematical Sciences* 2(1979) 223-228.

3. R. Alter, A characterization of self-complementary graphs of order 8, *Portugaliae Mathematica* 34(1975) 157-161.

4. A. E. Brouwer, A. M. Cohen, and A. Neumaier, "Distance-Regular Graphs" Springer-Verlag, 1989.

5. J. Bosak, "Decompositions of Graphs", Kluwer Academic Publishers, 1990.

6. J. Bosak, P. Erdös and A. Rosa, Decompositions of complete graphs into factors with diameter 2, *Matematiickj Casopis* 21(1971) 14-28.

7. J. Bosak, A. Rosa and S. Znam, On decomposition of complete graphs into factors with given diameters, in "Theory of Graphs", P. Erdös, G. Katona (eds.), Academic Press, (1969) 37-56.

8. D. Cvetković & M. Peric, A table of connected graphs on six verticces, *Discrete Mathematics* 50(1984) 37-49.

9. P. K. Das, Almost self-complementary graphs, *Ars Combinatoria* 31(1991) 267-276.

10. M. Deza, On the Hamming geometry of unitary cubes, *Doklady Akademii Nauk SSR* (in Russian) 1037-1040, *Soviet Physics Doklady* 5(1961) 940-943.

11. M. Deza and T. Huang, ℓ_1-embeddability of some block graphs and cycloids, *Bulletin of the Institute of Mathematics*, Academia Sinica, to appear (1996).

12. M. Deza and V.P. Grishukhin, Hypermetric graphs, *Quarterly Journal of Oxford (2)* (1993), 399-433.

13. M. Deza and M. Laurent, ℓ_1-rigid graphs, *Journal of Algebraic Combinatorics* 3(1994), 153-175.

14. M. Deza and H. Maehara, Projecting a simplex onto another one, *European Journal of Combinatorics* 15(1994), 13-16.

15. R. C. Entringer, D. E. Jackson and P. A. Snyder, Distance in graphs, *Czechoslovak Mathematical Journal* 26(1976), 283-296.

16. R. A. Gibbs, Self-complementary graphs, *Journal of Combinatorial Theory, Ser. B*, 16(1974) 106-123.

17. R. L. Graham, B. L. Rothschild, and J. H. Spencer, "Ramsey Theory" 2nd. ed., John-Wiley & Sons (1990).

18. N. Hartsfield, On regular self-complementary graphs, *Journal of Graph Theory* 11(1987) 537-538.

19. D. Hanson, G. MacGillary, On small triangle-free graphs, *Ars combinatoria* 35(1993) 257-263.

20. R. Laskar, S. Stueckle and B. Piazza, On the edge-integrity of some graphs and their complements, *Discrete Mathematics* 122(1993) 245-253.

21. T. D. Parsons, Orthogonality graphs, *Ars Combinatoria* 3(1977), 165-208.

22. T. D. Parsons and T. Pisanki, Exotic n-universal graphs, *Journal of Graph Theory* 12(1988) 159-167.

23. I. G. Rosenberg, Regular and strongly regular self-complementary graphs, *Annals of Discrete Mathematics* 12(1982) 223-238.

24. S. Ruiz, On strongly regular self-complementary graphs, *Journal of Graph Theory* 5(1981), 213-215.

25. S. V. Shpectorov, On scale embeddings of graphs into hypercubes, *European Journal of Combinatorics* 14(1993) 117-130.

26. P. Terwilliger and M. Deza, Classification of finite connected hypermetric spaces, *Graph and Combinatorics* 3(1987), 293-298.

Appendix:

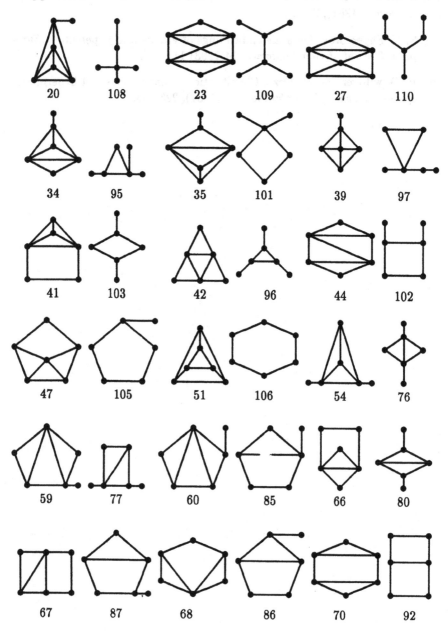

Double Description Method Revisited

Komei Fukuda[1] and Alain Prodon[2]

[1] Institute for Operations Research, ETHZ, CH-8092 Zürich, Switzerland
[2] Department of Mathematics, EPFL, CH-1015 Lausanne, Switzerland

Abstract. The double description method is a simple and useful algorithm for enumerating all extreme rays of a general polyhedral cone in $\mathbb{R}^d$, despite the fact that we can hardly state any interesting theorems on its time and space complexities. In this paper, we reinvestigate this method, introduce some new ideas for efficient implementations, and show some empirical results indicating its practicality in solving highly degenerate problems.

1 Introduction

A pair (A, R) of real matrices A and R is said to be a *double description pair* or simply a *DD pair* if the relationship

$$Ax \geq 0 \quad \text{if and only if} \quad x = R\lambda \text{ for some } \lambda \geq 0$$

holds. Clearly, for a pair (A, R) to be a DD pair, it is necessary that the column size of A is equal to the row size of R, say d. The term "double description" was introduced by Motzkin et al. [MRTT53], and it is quite natural in the sense that such a pair contains two different descriptions of the same object. Namely, the set $P(A)$ represented by A as

$$P(A) = \{x \in \mathbb{R}^d : Ax \geq 0\} \tag{1}$$

is simultaneously represented by R as

$$\{x \in \mathbb{R}^d : x = R\lambda \text{ for some } \lambda \geq 0\}. \tag{2}$$

A subset P of $\mathbb{R}^d$ is called *polyhedral cone* if $P = P(A)$ for some matrix A, and a matrix A is called a *representation matrix* of the polyhedral cone $P(A)$. We shall use the simpler term *cone* for polyhedral cone for the sequel since we only deal with such cones.

When a cone P is represented by (2), we say R is a *generating matrix* for P or a matrix R *generates* the polyhedron. Clearly, each column vector of a generating matrix R lies in the cone P and every vector in P is a nonnegative combination of some columns of R.

Minkowski's Theorem states that every polyhedral cone admits a generating matrix.

Theorem 1 (Minkowski's Theorem for Polyhedral Cones). *For any $m \times d$ real matrix A, there exists some $d \times n$ real matrix R such that (A, R) is a DD pair, or in other words, the cone $P(A)$ is generated by R.*

Here, the nontriviality is in that the row size n of R is a finite number. If we allow the size n to be infinite, there is a trivial generating matrix consisting of all vectors in the cone. In this respect, the essence of the theorem is the statement "every cone is finitely generated."

We have also the converse of the theorem, known as Weyl's Theorem.

Theorem 2 (Weyl's Theorem for Polyhedral Cones). *For any $d \times n$ real matrix R, there exists some $m \times d$ real matrix A such that (A, R) is a DD pair, or in other words, the set generated by R is the cone $P(A)$.*

These two theorems suggest two fundamental problems, one to construct a matrix R from a given matrix A and the converse. It is well known that these two problems are computationally equivalent, meaning they are linear-time reducible to each other. In fact, one can easily prove by using Farkas' Lemma that (A, R) is a DD pair if and only if (R^T, A^T) is a DD pair. Thus, we can treat the latter problem with a given R as the first problem with A being R^T. For this reason, we shall concentrate on the first problem to find a generating matrix R for a given A.

Clearly a more appropriate formulation of the problem is to require the minimality of R, meaning: find such a matrix R so that no proper submatrix is generating $P(A)$. Note that a minimal set of generators is unique up to positive scaling when we assume the regularity condition that the cone is *pointed*, i.e. the origin 0 is an extreme point of $P(A)$. Geometrically, the columns of a minimal generating matrix are in 1-to-1 correspondence with the extreme rays of P. Thus the problem is also known as the *extreme ray enumeration problem*. In this paper, this problem is the main concern since both theoretically and practically there is neither necessity nor importance to generate redundant information.

The extreme ray enumeration problem has been studied, directly or indirectly, by many researchers in mathematics, operations research and computational geometry etc. and considerable efforts have been devoted to finding better algorithms for the problem. Despite that, no efficient enumeration algorithm for the general problem is known. Here, we mean by an *efficient enumeration algorithm* one which runs in time polynomial in both the input size and the output size. Such efficient algorithms exist for certain special cases only.

The problem is said to be *degenerate* if there is a vector x in $P(A)$ satisfying more than d inequalities in $Ax \geq 0$ with equality, and *nondegenerate* otherwise. Avis and Fukuda [AF92] showed that when the problem is nondegenerate it is possible to solve the extreme ray enumeration problems in time $O(mdv)$ and in space $O(md)$ where v is the number of extreme rays. Note that the problem treated in [AF92] is the vertex enumeration problem, which is the non-homogeneous version of our problem and there are simple transformations between them. (A C-implementation of Avis-Fukuda algorithm is publicly available [Avi93].)

In this paper, we examine a classical algorithm, known as the double description (DD) method [MRTT53], which is an extremely simple algorithm. We are primarily interested in practicality of this algorithm applied to highly degenerate problems. Let us describe the algorithm briefly. Let A^k be the submatrix of

A consisting of the first k rows of A. It is easy to find a generating matrix R^k for $P(A^k)$ when k is small, say $k = 1$. In the general iteration k, the algorithm constructs a DD pair (A^{k+1}, R^{k+1}) from a DD pair (A^k, R^k) already computed.

This algorithm was rediscovered repeatedly and given different names. The method known as Chernikova's algorithm [Che65] is basically the same method. In the dual setting of computing A from the generating matrix R (known as the convex hull computation), the method is essentially equivalent to the beneath-and-beyond method, see [Ede87, Mul94]. On the other hand, the algorithm known as the Fourier-Motzkin elimination is more general than the DD method, but often considered as the same method partly because it can be used to solve the extreme ray enumeration problem. This algorithm is essentially different because it does not use the crucial information of double description of the associated convex polyhedra even when it is applied to the problem.

The importance of the DD method is not well understood theoretically at least for the moment, since the worst-case behavior of this algorithm is known to be exponential in the sizes of input and output. The algorithm is sensitive to the permutations of rows of A, and an exponential behavior is exhibited when the (column) size of R^k explodes with some bad ordering of rows for certain class of matrices A, see [Dye83]. Yet, it has been observed that once it is properly implemented with a certain heuristic strategy of selecting an ordering and further techniques, it becomes a very competitive algorithm, in particular for highly degenerate problems, see also [DDF95]. To present some useful implementation techniques for the DD algorithm is the main purpose of this paper.

In order to argue for the necessity of various implementation techniques, let us list serious pitfalls of a naive DD method which must be avoided whenever possible:

(a) the algorithm is terribly sensitive to the ordering of rows of A in practice as well and a random ordering often leads to prohibitive growths of intermediate generator matrices R^k (especially for highly degenerate problems);

(b) it generates enormous number of redundant generators and goes beyond any tractable limit of computation very quickly, even when some good ordering (a) is known;

(c) if the input data A is perturbed to resolve degeneracy, the size of perturbed output can grow exponentially in the sizes of original input and output;

(d) data structures and computational methods should be part of the algorithm, since they can significantly improve the efficiency of computation when appropriately chosen.

It should be noted the original paper [MRTT53] gives a satisfactory remedy to (b) and explains how to generate the minimum generating matrix R^k. Our implementations in fact follow their ideas. Also the point (c) has been discussed in a recent paper [AB95] where it is shown that the perturbation makes the DD method exponentially slower for a certain class of polyhedra. While this serves as the formal justification of not using perturbation techniques, we find that the DD method can exploit a very little from nondegeneracy assumptions. Thus our implementations do not apply any perturbations to input data. Our main contribution is thus to show some practical ideas to handle the difficulties (a) and (d).

With respect to (a), we tested several different orderings of rows. Orderings can be divided into two classes, the static ordering and the dynamic ordering. The former is an ordering prefixed at the beginning of the computation procedure and the latter selects the next row dynamically as the computation proceeds. A typical static ordering is the lexicographic ordering (lex-min) and a typical dynamic ordering is the "max-cut-off" ordering which amounts to select the row that removes as many generators (i.e. columns of R^k) as possible. Experimentally we found that two dynamic orderings, max-cut-off and min-cut-off, perform much worse than the lex-min ordering for relatively large set of degenerate test problems. Also, we study a static ordering which gradually increases the dimension of the intermediate cones. This ordering motivates us to introduce a decomposition technique for the DD method, called the column decomposition. This technique is especially useful for the cones arising from combinatorial optimization problems.

Concerning (d), we are proposing two different data structures. Both data structures are for efficient handling of the minimum generating matrix R consisting of extreme rays and of the adjacency among the extreme rays. Our implementations differ considerably from the data structure suggested by the books (e.g.) for the dual version (the beneath-and-beyond method) of the algorithm which stores the complete adjacency of extreme rays. The main idea in our implementations is to store only necessary adjacency relations. This makes the implementation less expensive in both storage and time. We report some experimental results showing the practicality of our implementations.

Since the DD method must store a DD pair for each intermediate cone during the computation, one can easily exhaust any memory space available in computers. For this reason, it is useful if we can decompose the given problem to smaller problems. We introduce the row decomposition technique for this purpose. This technique is useful not only for reducing the size of a given problem but for parallelizing the computation. We have not yet tested the practicality of this technique and this will be a subject of future research.

Finally we would like to emphasize that although we found our implementations quite competitive, it is not our interest to compare our implementations with the other existing implementations, e.g. [BDH93, CR95, Ver92, Wil93]. It is very difficult to make fair comparisons and we do not think such comparisons are very useful. Instead, we concentrate on showing what we learned from our experiences with the DD method, and present some techniques we found useful. One can reproduce most of our experimental findings since one of the implementations we use is publicly available by anonymous ftp [Fuk93].

2   Double Description Method: Primitive Form

In this section, we give the simplest form of the DD method [MRTT53]. Suppose that an $m \times d$ matrix A is given, and let $P(A) = \{x : Ax \geq 0\}$. The DD method is an incremental algorithm to construct a $d \times n$ matrix R such that (A, R) is a DD pair.

For the sequel of this paper, we assume for simplicity that the cone $P := P(A)$ is *pointed*, i.e., the origin 0 is an extreme point of P or equivalently A has rank d. To reduce a problem to the pointed case, one can simply restrict the cone to the orthogonal subspace of the lineality space $\{x : A\, x = 0\}$ of P. We describe in Section 4.2 how to handle directly this case. We also assume that the system $Ax \geq 0$ is irredundant.

Let K be a subset of the row indices $\{1, 2, \ldots, m\}$ of A and let A_K denote the submatrix of A consisting of rows indexed by K. Suppose we already found a generating matrix R for $P(A_K)$, or equivalently (A_K, R) is a DD pair. If $A = A_K$, clearly we are done. Otherwise we select any row index i not in K and try to construct a DD pair (A_{K+i}, R') using the information of the DD pair (A_K, R).

Once this basic procedure is described, we have an algorithm to construct a generating matrix R for $P(A)$. This procedure can be easily understood geometrically by looking at the cut-section C of the cone $P(A_K)$ with some appropriate hyperplane h in R^d which intersects with every extreme ray of $P(A_K)$ at a single point. Let us assume that the cone is pointed and thus C is bounded. Having a generating matrix R means that all extreme rays (i.e. extreme points of the cut-section) of the cone are represented by columns of R. Such a cut-section (of a cone lying in $\mathbb{R}^4$) is illustrated in Fig. 1. In this example, C is the cube $abcdefgh$.

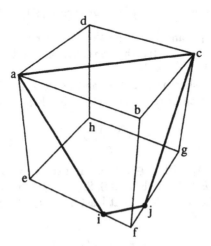

Fig. 1. Illustration of the mechanism of the DD method.

The newly introduced inequality $A_i\, x \geq 0$ partitions the space $\mathbb{R}^d$ into three parts:

$$H_i^+ = \{x \in \mathbb{R}^d : A_i\, x > 0\}$$
$$H_i^0 = \{x \in \mathbb{R}^d : A_i\, x = 0\} \tag{3}$$
$$H_i^- = \{x \in \mathbb{R}^d : A_i\, x < 0\}.$$

The intersection of H_i^0 with P and the new extreme points i and j in the cut-section C are shown in bold in Fig. 1.

Let J be the set of column indices of R. The rays r_j $(j \in J)$ are then partitioned into three parts accordingly:

$$J^+ = \{j \in J : r_j \in H_i^+\}$$
$$J^0 = \{j \in J : r_j \in H_i^0\} \tag{4}$$
$$J^- = \{j \in J : r_j \in H_i^-\}.$$

We call the rays indexed by J^+, J^0, J^- the *positive, zero, negative* rays with respect to i, respectively. To construct a matrix R' from R, we generate new $|J^+| \times |J^-|$ rays lying on the ith hyperplane H_i^0 by taking an appropriate positive combination of each positive ray r_j and each negative ray $r_{j'}$ and by discarding all negative rays.

The following lemma ensures that we have a DD pair (A_{K+i}, R'), and provides the key procedure for the most primitive version of the DD method.

Lemma 3 (Main Lemma for Double Description Method). *Let (A_K, R) be a DD pair and let i be a row index of A not in K. Then the pair (A_{K+i}, R') is a DD pair, where R' is the $d \times |J'|$ matrix with column vectors r_j $(j \in J')$ defined by*

$$J' = J^+ \cup J^0 \cup (J^+ \times J^-), \text{ and}$$
$$r_{jj'} = (A_i\, r_j)r_{j'} - (A_i\, r_{j'})r_j \text{ for each } (j, j') \in J^+ \times J^-$$

Proof. Let $P = P(A_{K+i})$ and let P' be the cone generated by the matrix R'. We must prove that $P = P'$. By the construction, we have $r'_{jj'} \in P$ for all $(j, j') \in J^+ \times J^-$ and $P' \subset P$ is clear.

Let $x \in P$. We shall show that $x \in P'$ and hence $P \subset P'$. Since $x \in P$, x is a nonnegative combination of r_j's over $j \in J$, i.e., there exist $\lambda_j \geq 0$ for $j \in J$ such that

$$x = \sum_{j \in J} \lambda_j r_j. \tag{5}$$

If there is no positive λ_j with $j \in J^-$ in the expression above then $x \in P'$. Suppose there is some $k \in J^-$ with $\lambda_k > 0$. Since $x \in P$, we have $A_i\, x \geq 0$. This together with (5) implies that there is at least one $h \in J^+$ with $\lambda_h > 0$. Now by construction, $hk \in J'$ and

$$r_{hk} = (A_i\, r_h)r_k - (A_i\, r_k)r_h. \tag{6}$$

By subtracting an appropriate positive multiple of (6) from (5), we obtain an expression of x as a positive combination of some vectors r_j $(j \in J')$ with new coefficients λ_j where the number of positive λ_j's with $j \in J^+ \cup J^-$ is strictly smaller than in the first expression. As long as there is $j \in J^-$ with positive λ_j, we can apply the same transformation. Thus we must find in a finite number of steps an expression of x without using r_j such that $j \in J^-$. This proves $x \in P'$, and hence $P \subset P'$. □

It is quite simple to find a DD pair (A_K, R) when $|K| = 1$, which can serve as the initial DD pair. Another simple (and perhaps the most efficient) way to obtain an initial DD form of P is by selecting a maximal submatrix A_K of A consisting of linearly independent rows of A. The vectors r_j's are obtained by solving the system of equations:

$$A_K R = I,$$

where I is the identity matrix of size $|K|$, R is a matrix of unknown column vectors r_j, $j \in J$. As we have assumed $rank(A) = d$, i.e. $R = A_K^{-1}$, the pair (A_K, R) is clearly a DD pair, since $A_K x \geq 0 \leftrightarrow x = A_K^{-1} \lambda, \lambda \geq 0$.

Here we write the DD method in procedural form:

```
procedure DoubleDescriptionMethod(A);
begin
    Obtain any initial DD pair (A_K, R);
    while K ≠ {1, 2, ..., m} do
    begin
        Select any index i from {1, 2, ..., m} \ K;
        Construct a DD pair (A_{K+i}, R') from (A_K, R);
            /* by using Lemma 3 */
        R := R';   K := K + i;
    end;
    Output R;
end.
```

The DD method given here is very primitive, and the straightforward implementation will be quite useless, because the size of J increases very fast and goes beyond any tractable limit. One reason for this is that many (perhaps, most) vectors $r_{jj'}$ the algorithm generates (defined in Lemma 3), are unnecessary. In the next section we will show how to avoid generating redundant vectors.

The DD method in dual form, for constructing the convex hull of a finite set of points, is known as the beneath-beyond method and studied extensively in the field of computational geometry. An implementation proposed in [Mul94], when interpreted as the DD method, is basically to use random inequality ordering and to store the complete adjacencies of the rays in the intermediate cones for efficient updates. One serious problem is that the analysis which yields its "optimality" depends on the assumption that the input is nondegenerate (or dually, the points are in general position). This assumption is rather nonrealistic because there is an efficient (linear-time) algorithm [AF92] for nondegenerate case and thus problems for which the DD method is superior must be degenerate. Furthermore, we show that it is not necessary to store the adjacencies of rays. In fact one of our new technique is essentially to store only those adjacencies which are needed for the computation.

3 Practical Implementations

In this section, we shall describe three practical implementations of the DD method. The first one is the simplest and might be called standard in the sense that anyone implementing the DD method is quite likely to come up with this implementation, and is in fact suggested in the original paper [MRTT53]. We shall give new implementations which are more efficient than the standard algorithm.

3.1 The standard implementation

We first introduce some definitions and properties which we will use in strengthening Lemma 3. Remember that we are assuming that P is pointed, that is $rank(A) = d$.

A vector r is said to be a *ray* of P if $r \neq 0$ and $\alpha r \in P$ for all positive α. We identify two rays r and r' if $r = \alpha r'$ for some positive number α. We shall denote this equivalence by $r \simeq r'$. For any vector x in P, we define the *zero set* or *active set* $Z(x)$ as the set of inequality indices i such that $A_i x = 0$. The word zero comes from "slack variables being zero" at the associated inequalities.

Proposition 4. *Let r be a ray of P, $\bar{F} := \{x : A_{Z(r)} x = 0\}$, $F := \bar{F} \cap P$ and $rank(A_{Z(r)}) = d - k$. Then*

(a) $rank(A_{Z(r) \cup \{i\}}) = d - k + 1$ *for all $i \notin Z(r)$;*
(b) *F contains k linearly independent rays;*
(c) *If $k \geq 2$ then r is a nonnegative combination of two distinct rays r_1 and r_2 with $rank(A_{Z(r_i)}) > d - k$, $i = 1, 2$.*

Proof. (a) If A_i is a linear combination of rows of A_K with $K \subseteq Z(r)$, then $A_i r = 0$.

(b) Clearly $\bar{F}$ contains k linearly independent vectors r, v_2, v_3, ..., v_k. Let $r_1 := r$, $r_i := r + \alpha_i v_i$, $i = 2, ..., k$. These vectors are linearly independent for $\alpha_i \neq 0$. When the coefficients α_i are chosen such that $\alpha_i > 0$ and $\alpha_i \leq min(-A_j r / A_j v_i$ for all j s.t. $A_j v_i < 0)$ for $v_i \notin P$, the vectors r_i, $i = 1, ..., k$, are k linearly independent vectors in F.

(c) Let $k \geq 2$. $\bar{F}$ contains a vector v s.t. neither v nor $-v$ is in P : since $rank(A_{Z(r)}) \leq rank(A) - 2$, there exist i and j such that $rank(A_{Z(r) \cup \{i,j\}}) = rank(A_{Z(r)}) + 2$, and thus there exist v_1 and v_2 in $\bar{F}$ with $A_i v_1 = 0$, $A_j v_1 > 0$, $A_i v_2 < 0$, $A_j v_2 = 0$. Then $v = v_1 + v_2$ satisfies $A_i v < 0$ and $A_j v > 0$. Let $r_1 = r + \alpha_1 v_1$, $r_2 = r - \alpha_2 v_2$. Then there exist positive maximal values for α_i s.t. $r_i \in F$.

For these α_i we get $Z(r_i) \supset Z(r)$, $i = 1,2$ and $r = 1/(\alpha_1 + \alpha_2)(\alpha_2 r_1 + \alpha_1 r_2)$. Thus (c) holds. □

A ray r is said to be *extreme* if it is not a nonnegative combination of two rays of P distinct from r.

Proposition 5. *Let r be a ray of P. Then*

(a) *r is an extreme ray of P if and only if the rank of the matrix $A_{Z(r)}$ is $d-1$;*
(b) *r is a nonnegative combination of extreme rays of P.*

Proof. Observe that if there exist some $\lambda_j > 0$ and $r_j \in P$ such that $r = \sum_j \lambda_j r_j$ then each of these r_j belongs to $\bar{F} := \{x : A_{Z(r)} x = 0\}$, since for $i \in Z(r)$, $A_i r = 0$ and $A_i r$ is a sum of nonnegative terms.

If $rank(A_{Z(r)}) = d - 1$, then $\bar{F} = \{kr : k \in \mathbb{R}\}$ and 0 and $-r$ are the only elements distinct of r in $\bar{F}$, hence r is an extreme ray.

If $rank(A_{Z(r)}) < d - 1$, then by Proposition 4 there exist rays r_1 and r_2 such that r is a nonnegative combination of these rays, and hence r is not an extreme ray. This proves (a). Moreover, since $rank(A_{Z(r_i)}) > rank(A_{Z(r)})$ for $i = 1, 2$, repeating the last argument for r_1 and r_2 until $rank(A_{Z(r_i)}) = d - 1$, $i = 1, 2$, proves (b). □

Since every extreme ray is certainly necessary to generate P we have the following corollary.

Corollary 6. *Let R be a minimal generating matrix of P. Then R is the set of extreme rays of P.*

Observe that the assumption that P be pointed is important here, because then computing the unique generating matrix is a well formulated problem. When P is not pointed, there are infinitely many minimal generating matrices.

We say that two distinct extreme rays r and r' of P are *adjacent* if the minimal face of P containing both contains no other extreme rays. This is equivalent to say that, if r'' is an extreme ray of P with $Z(r'') \supseteq Z(r) \cap Z(r')$, then either $r'' = r$ or $r'' = r'$.

Proposition 7. *Let r and r' be distinct rays of P. Then the following statements are equivalent:*

(a) *r and r' are adjacent extreme rays;*
(b) *r and r' are extreme rays and the rank of the matrix $A_{Z(r) \cap Z(r')}$ is $d - 2$.*
(c) *if r'' is a ray with $Z(r'') \supseteq Z(r) \cap Z(r')$ then either $r'' \simeq r$ or $r'' \simeq r'$;*

Proof. Let r and r' be distinct rays of P and $\bar{F} := \{x : A_{Z(r) \cap Z(r')} x = 0\}$. Then $F := \bar{F} \cap P$ is the minimal face of P containing r and r'.

To prove the equivalence of (a) and (b) let r and r' be extreme rays of P. Since P is pointed, we have $Z(r) \neq Z(r')$.

If (b) holds, i.e. $rank(A_{Z(r) \cap Z(r')}) = d - 2$, then $dim(\bar{F}) = 2$ and thus r and r' generate $\bar{F}$, that is each x in $\bar{F}$ can be written as $x = \alpha r + \alpha' r'$. Moreover, $x \in F$ implies α and α' are nonnegative, since there exists some $i \in Z(r') - Z(r)$ (respectively $Z(r) - Z(r')$), for which $A_i x \geq 0$ and $A_i x = \alpha A_i r$ (respectively $A_i x = \alpha' A_i r'$). Thus x is a nonnegative combination of r and r' and there is no other extreme ray in F. Thus (a) holds. Observe also that, since a two dimensional face contains at least two extreme rays by Proposition 4, it contains exactly two extreme rays.

If (b) does not hold, i.e. $rank(A_{Z(r) \cap Z(r')}) = d - k$ with $k \geq 3$, then by Proposition 4(b) $dim(F) = k$ and at least k extreme rays are necessary to generate F by Proposition 5(b). Thus r and r' are not adjacent.

We now use the former equivalence to prove (a) $\Leftrightarrow$ (c).

If (a) holds, then $rank(A_{Z(r) \cap Z(r')}) = d - 2$. It follows from Proposition 4(a) that $Z(r'') \supset Z(r) \cap Z(r')$ imply $rank(A_{Z(r'')}) = d - 1$, i.e. r'' is an extreme ray by Proposition 5. Thus, since r and r' are adjacent, (c) holds.

If (a) does not hold, that is, either r and r' are nonadjacent extreme rays or at least one is not an extreme ray. Then if $rank(A_{Z(r) \cap Z(r')}) < d - 2$ there exist in F at least one extreme ray different from r and r', which can serve as r''. If $rank(A_{Z(r) \cap Z(r')}) = d - 2$ then we know from (b) that r or r' is nonextreme and that $\bar{F}$ contains two extreme rays, thus at least one of these extreme rays is different from r and r' and can serve as r'', showing the invalidity of (c). $\square$

In the proposition above, the statement (c) can be called a combinatorial characterization of the adjacency, and (b) an algebraic characterization.

Now the main lemma 3 can be strengthened for practical purposes as follows:

Lemma 8 (Strengthened Main Lemma for the DD Method). *Let (A_K, R) be a DD pair such that $rank(A_K) = d$ and let i be a row index of A not in K. Then the pair (A_{K+i}, R') is a DD pair, where R' is the $d \times |J'|$ matrix with column vectors r_j ($j \in J'$) defined by*

$$J' = J^+ \cup J^0 \cup Adj,$$
$$Adj = \{(j, j') \in J^+ \times J^- : r_j \text{ and } r_{j'} \text{ are adjacent in } P(A_K)\}; \text{ and}$$
$$r_{jj'} = (A_i \, r_j) r_{j'} - (A_i \, r_{j'}) r_j \text{ for each } (j, j') \in Adj.$$

Furthermore, if R is a minimal generating matrix for $P(A_K)$ then R' is a minimal generating matrix for $P(A_{K+i})$

Proof. Let all the assumptions be satisfied. We know from Proposition 5 that each extreme ray of $P(A_K)$ must belong to R, and that only the extreme rays of $P(A_{K+i})$ are necessary in R'. Denote for abbreviation $W := Z(r_j) \cap Z(r_{j'}) \cap K$. Observe that $Z(r_{jj'}) \cap (K + i) = W \cup \{i\}$.

If r_j and $r_{j'}$ are adjacent extreme rays of $P(A_K)$, then $rank(A_W) = d - 2$. Then, by Proposition 4(a), $rank(A_{Z(r_{jj'}) \cap (K+i)}) = d - 1$. Thus $r_{jj'}$ is an extreme ray of $P(A_{K+i})$ and must belong to R'.

Suppose r_j and $r_{j'}$ are not adjacent extreme rays. Then if $rank(A_W) < d-2$ we get $rank(A_{Z(r_{jj'})\cap(K+i)}) < d-1$ and thus $r_{jj'}$ is not necessary. If $rank(A_W) = d-2$, then we know from Proposition 7 that r_j and $r_{j'}$ cannot be both extreme rays of $P(A_K)$. But they belong to a two dimensional face containing exactly two extreme rays of $P(A_K)$ which thus belong to R: This adjacent pair will then produce a new ray equal to $r_{jj'}$, so $r_{jj'}$ is not necessary.

Hence all new rays are extreme rays of $P(A_{K+i})$ and R' is minimal if R was minimal.   □

By using this lemma, we can write a straightforward variation of the DD method which produces a minimal generating set for P:

> **procedure** DDMethodStandard(A);
> **begin**
>   Obtain any initial DD pair (A_K, R) such that R is minimal;
>   **while** $K \neq \{1, 2, \ldots, m\}$ **do**
>   **begin**
>   Select any index i from $\{1, 2, \ldots, m\} \setminus K$;
>   Construct a DD pair (A_{K+i}, R') from (A_K, R);
>   /* by using Lemma 8 */
> $R := R';$ $K := K + i;$
>   **end;**
>   Output R;
> **end.**

To implement DDMethodStandard, we must check for each pair of extreme rays r and r' of $P(A_K)$ with $A_i\ r > 0$ and $A_i\ r' < 0$ whether they are adjacent in $P(A_K)$. As we state in Proposition 7, there are two ways to check adjacency, the combinatorial and the algebraic way. We do not know any theoretical reason to conclude which method is more efficient. Our computational experiments indicate that the combinatorial method is almost always faster than the algebraic method. See Section 5.1.

The asymptotic complexity of this algorithm depends strongly on the number of extreme rays of the intermediate cones, and thus on the ordering of the rows of A. However, if we assume that this number remains in $O(v)$, v denoting the size of the output and m the number of rows of A, then the whole complexity is in $O(m^2 v^3)$, the dominating part coming from the combinatorial test applied at each iteration : for each pair (there are at most $O(v^2)$) the test makes at most $O(v)$ comparisons of vectors of size m.

3.2   A new implementation with prefixed row ordering

From our computational experiences, the variation of the DD method described in the previous section is a reasonably efficient algorithm if it incorporates a good ordering of inequalities. However this method scans all pairs (r, r') of rays of $P(A_K)$ with $A_i\, r > 0$ and $A_i\, r' < 0$ in each iteration, and generates a new ray only when they are identified adjacent in $P(A_K)$. It is natural to imagine that most of these pairs are non-adjacent pairs and thus scanning all the pairs would be very inefficient.

One way to reduce the unnecessary scanning might be to store all the adjacencies of the rays, and to scan only the adjacent pairs. This modification might look efficient but it has several serious drawbacks. The first one is that the storage space can be quite large since the number of adjacencies can be quadratic in the number of rays. Secondly, it is not clear how to select only those pairs (r, r') of rays with $A_i\, r > 0$ and $A_i\, r' < 0$ without scanning through the large adjacency data. Furthermore updating the adjacency data is quite expensive.

These observations lead us to seek for an implementation of the DD method which stores only those adjacent pairs of rays in $P(A_K)$ that produce the new rays at later iterations. We shall show that this can be done if the ordering of inequalities is prefixed.

The idea is quite simple. Suppose we fix the ordering of inequalities as the natural ordering. For convenience, let

$$P^j = \{x : A_i\, x \geq 0 \text{ for } i = 1, 2, \ldots, j\},$$

for $j = 1, 2, \ldots, m$. Thus $P^m = P$. Suppose we are at ith iteration to add the ith inequality and the remaining inequality indices are $i+1, i+2, \ldots, m$. At this iteration, a minimal generating set R for the cone P^{i-1} is at hand and the DD method will produce a new set of rays satisfying the ith inequality with equality and produce a minimal generating set for the cone P^i. Let (r, r') be a pair of adjacent rays in P^i. Since the ordering of inequalities is fixed, we know exactly what will happen to this pair. Namely, there are three cases.

(a) Both r and r' are rays of P (and thus this adjacency won't create any new rays);

(b) Both r and r' stay rays of P^j for $j = i, i+1, \ldots, k-1$, and both becomes infeasible for P^k for some $k \leq m$ (and thus this adjacency won't create any new rays either);

(c) Both r and r' stay rays of P^j for $j = i, i+1 \ldots, k-1$, and only one of the rays, say r, becomes infeasible for P^k for some $k \leq m$. (A new ray is created only when the other ray r' satisfies the kth inequality with strict inequality, i.e. $k \notin Z(r')$.)

Furthermore it is easy to recognize which case holds for any pair of adjacent rays if the *minimum infeasible index* defined by

$$minf(r) = \begin{cases} m+1 & \text{if } r \text{ is feasible for } P \\ min\{k : A_k\, x < 0, 1 \leq k \leq m\} & \text{otherwise} \end{cases}$$

is stored for each ray r. Clearly, case (a) occurs if and only if $minf(r) = minf(r') = m+1$, case (b) occurs if and only if $minf(r) = minf(r') = k$ for some $k \leq m$, and case (c) occurs if and only if $k = minf(r) < minf(r')$ for some $k \leq m$. Moreover, a new ray will be created if and only if

(c$_k$) $k = minf(r) < minf(r')$ and $k \notin Z(r')$.

To incorporate the above observations, we prepare the linked list L_k of pairs of adjacent rays which willbe used to generate new rays at each iteration $k \leq m$. We must make sure that at the beginning of a general iteration i adding the ith inequality, the list L_i contains exactly those pairs of adjacent rays that create new rays. We know that each of these rays must satisfy the condition (c$_k$) in some earlier iteration. In order to construct the linked list L_k successfully and free of duplications, we simply check for each candidate of adjacent rays r and r' of P^i in any earlier iteration i whether they satisfy (c$_k$) and it is the last chance that the condition be checked. If it is the case, store the pair in L_k.

Any pair of adjacent rays r and r' of P^i which were not adjacent in the previous cone P^{i-1} must be examined as a possible candidate for membership of L_k. Such newly-born adjacencies are of two types:

(i) the adjacency of r and r' is inherited from the previous cone P^{i-1}, that is, one of the two rays, say r, is old and the other is newly created by r and another old ray $r'' \in P^{i-1} \setminus P^i$ adjacent to r in P^{i-1} (i.e. the pair (r, r'') is in the linked list L_i);

(ii) the adjacency of r and r' is new, that is, both rays lie on the hyperplane $H_i^0 = \{x : A_i\, x = 0\}$ (i.e. $i \in Z(r) \cap Z(r')$), and either

  (iia) at least one of the rays is newly created; or

  (iib) both are old but they were not adjacent in P^{i-1}.

This classification of newly-born adjacencies is illustrated in Fig. 1 : the pairs (i, e), (j, g) are of type (i), the pairs (i, j), (i, a), (j, c) of type (iia) and (a, c) of type (iib). The enumeration of the pairs of type (i) is very simple by using the linked list L_i, but that of type (ii) is not trivial. We must take all pairs of rays lying on H_i^0 and check their adjacency.

In the algorithm to be described below, we store not only each newly created ray but also its minimum infeasible index. A procedure ConditionalStoreEdge() simply checks whether a new pair of adjacent rays satisfies the condition (c$_k$) for the last time, and stores the adjacency if it is the case, so that the lists L_k's are free of duplicates.

```
procedure ConditionalStoreEdge(r,r',k,i);
begin
    if there is no index i' with i < i' < k with i' ∈ Z(r) ∩ Z(r') then
        if r and r' are adjacent then
            store the pair (r, r') in the linked list L_k
        endif
    endif
end.
```

```
procedure DDMethodVariation1(A);
begin
    Obtain any initial DD pair (A_K, R) with R being minimal;
    Permute row indices so that K = {1, 2, ... s};
    Initialize all linked lists L_i (i = s + 1, ..., m);
    for each pair (r, r') of adjacent rays in R satisfying (c_k) do
        ConditionalStoreEdge(r,r',k,s)
    endfor;
    for i = s + 1 to m do
```
(1)
```
        Obtain a DD pair (A_{K+i}, R');
            /* by generating the new rays for each pair in the linked list L_i */
```
(2)
```
        for each pair (r, r') of rays in R' lying in H_i^0 and satisfying (c_k) do
            ConditionalStoreEdge(r,r',k,i);
        endfor;
        R := R';    K := K + i;
    endfor;
    Output R
end.
```

This new variation is almost optimal in the sense that the generation of new rays in (1) is done as efficiently as possible; it is linear time. The only inefficiency, if it exists, comes from the line (2) to list all new pairs of adjacent rays. Our computational experiments indicate considerable speed-up of this implementation over the standard algorithm. See Section 5.3.

3.3 Another implementation without prefixed row ordering

What makes the combinatorial adjacency test rather expansive in DDMethod-Standard is that checking point (b) of Proposition 7 is linear in the number of extremal rays of the current cone P^i. One way to try to improve its efficiency is to first execute some faster tests to detect non adjacency. Such tests may be derived from the two following necessary conditions for adjacency and take advantage of the fact that computing the cardinality of a subset is relatively cheap.

Let $K(i)$ be the set of row indices considered in the first i iterations, and let $P^i = P(A_{K(i)})$. From point (b) of Proposition 7 we get

Proposition 9 (NC1). *A necessary condition for a pair (r, r') being adjacent in P^i is that*

$$|Z(r) \cap Z(r') \cap K(i)| \geq d - 2.$$

From the property described in the previous section of adjacency being either inherited or new, we get another, usually stronger necessary condition.

Let $b(r)$ be the birthindex of extremal ray r, that is the iteration i in which it was created. Let also $f(r)$ be the father of r, defined as the element of the pair (r', r'') from which r was created that was feasible with respect to the separating hyperplane.

Proposition 10 (NC2). *Let r and r' be distinct extreme rays of P^{i-1} separated by the hyperplane introduced at iteration i, $b(r)$ and $b(r')$ their birthindices, and $k = max\{b(r), b(r')\}$.*

Then, if the pair (r, r') did not inherit adjacency from P^{k-1}, a necessary condition for it to be adjacent in P^{i-1} is that at least one hyperplane introduced in some iteration l with $k \leq l < i$ contain both r and r', that is

$$Z(r) \cap Z(r') \cap (K(i-1) - K(k-1)) \neq \emptyset.$$

In order to implement this test we have to store for each ray its birthindex $b(r)$ and a pointer to its father $f(r)$ to recognize inherited adjacency: a pair of extreme rays of P^{i-1} has inherited adjacency if and only if one element is the father of the other.

```
procedure DDMethodVariation2(A);
begin
    Obtain any initial DD pair (A_K, R) with R being minimal;
    while K ≠ {1, 2, ..., m} do
        Select i in {1, 2, ..., m} \ K and partition J in J+ ∪ J⁰ ∪ J−;
        /*Obtain a DD pair (A_{K+i}, R') */
        R' := {r_j : j ∈ J+ ∪ J⁰};
        for each pair (r_j, r_{j'}) with j ∈ J+ and j' ∈ J− do
            if (r_j, r_{j'}) has inherited adjacency then
                add r_{jj'} to R'
            else
                if NC2((r_j, r_{j'})) then
                    if NC1((r_j, r_{j'})) then
                        if Adj((r_j, r_{j'})) then
                            add r_{jj'} to R'
                        endif
                    endif
                endif
            endif
            R := R';    K := K + i;
    endwhile;
    Output R
end
```

Observe that, for a fixed ordering of the rows of A, the two proposed variations differ in two aspects: the pairs that are scanned, and the iteration in which the combinatorial adjacency test is executed. Variation1 scans all the pairs lying in the current hyperplane while Variation2 scans all the pairs separated by that hyperplane. The iteration in which the combinatorial test is performed is that of the last hyperplane containing both rays in Variation1, that preceding the iteration of the separating hyperplane in the later. However, similar operations are executed in both Variations, namely testing if a small subset of the zero set of a pair is void for conditional storage in the first or for Proposition 10 in the second, and the combinatorial test is executed for exactly the same pairs when test NC1 is introduced in DDMethodVariation1.

4 Decomposition into Smaller Subproblems

For large problems, in which both the number of rows of A and the expected number of extreme rays are large, the computing time may become prohibitive (the running time for one iteration grows like the cube of the number of extreme rays in the actual cone). A natural approach to solve such a problem is to decompose it into smaller ones, and a nice feature of the Double Description method is its flexibility to handle such cases. The two following paragraphs describe simple modifications of the basic method to enumerate the extremal rays in each facet of the given cone and to preserve partial results for successive computations.

4.1 Row Decomposition

A natural way to decompose the ray enumeration problem for P is to solve the extreme ray enumeration problem for each cones $P(i)$ associated with P defined by

$$P(i) = \{x \in \mathbb{R}^d : A_i\, x = 0 \text{ and } A_k\, x \geq 0 \quad \forall k = i+1, i+2, \ldots, m\}.$$

Since the subproblem on $P(i)$ has fewer inequalities than the original problem and has one equality constraint, the extreme ray enumeration might be much easier for $P(i)$. Also, for each $P(i)$, if we output only those rays r that satisfy the ignored inequalities with strict inequality, that is, $A_1\, r > 0, \ldots, A_{i-1}\, r > 0$, then each extreme ray of the original cone will be output exactly once. Furthermore, each subproblem can be solved independently and perhaps in parallel.

The problem on $P(i)$ may be solved using the equation to eliminate one variable, but this might destroy some nice structure of A. A better way to treat equalities in the Double Description method is to modify the algorithm as follows: If i is an iteration corresponding to an equality, the iteration is performed as usual and at the end of the iteration all extremal rays which do not satisfy the equality (i.e. the rays which are strictly feasible with respect to this hyperplane) are eliminated. Then, by construction, all rays created during further iterations will still satisfy the equality.

Although in theory this row decomposition process can be applied recursively, a difficulty may occur, namely the rows of A which are redundant in $P(i)$ should be eliminated in order to make the enumeration problem really easier on $P(i)$ than on P. Redundancy of a row can be checked by solving a linear program, but repeating such a procedure will hardly be efficient. In practice, when P has a nice structure, faster heuristic or exact methods may exist to check redundancy, but the facets of P will hardly inherit this structure.

Another related technique is to enumerate extreme rays facet by facet. Unlike the row decomposition technique above, we do not eliminate already considered inequalities for each ith subproblem. Such a technique might become more efficient when P has many symmetries. In such a case, appropriately choosing one representative in each class of facets may significantly decrease the whole computing time. This occurs in many combinatorial problems, specially those with variables associated to the edges of a complete graph.

4.2 Block structure

Another decomposition, a kind of column decomposition, can be used when the matrix A has a special structure. This special structure, which we call block structure, appears frequently in combinatorial problems.

Let $i_1, \ldots, i_j, \ldots, i_d$ the indices of the first linearly independent rows of A, i.e. A_{i_j} is the first row which is not a linear combination of $A_{i_1}, \ldots, A_{i_{j-1}}$. We say that a matrix A has a block structure if

(i) for each j, all the rows A_i with $i_j \leq i < i_{j+1}$ are linear combination of $A_{i_1}, \ldots, A_{i_j}$ with nonzero coefficient of A_{i_j},
(ii) for each j and each i with $i_j \leq i < i_{j+1}$ the elements a_{ik} of the row vector A_i satisfy $a_{ik} = 0$, $\forall k > j$.

A block structure has the following interesting property. Let $\hat{P}^j$ be the cone defined by the first j columns and the first $i_{j+1} - 1$ rows of A. Then $\hat{P}^j \subseteq R^j$ and, as the system $A\,x \geq 0$ is irredundant, $\hat{P}^j$ is the projection of $\hat{P}^{j+1}$ on the hyperplane $x_{j+1} = 0$. Thus the number of extreme rays of $\hat{P}^j$ is not greater than the one of $\hat{P}^{j+1}$.

If an ordering of the rows and columns of the input matrix exists producing such a block structure, it should be maintained through the iterations of the algorithm, since it provides some control points on the number of extreme rays of the actual cone: for each successive dimension j, at the end of iteration $i_{j+1} - 1$ the number of extreme rays cannot exceed the number of extreme rays of P.

Observe that there always exists an ordering of the rows satisfying property (i) and having thus a projection property. The block structure is a special case of such an ordering, where the projection property holds for each successive dimension and where the cones $\hat{P}^j$ generally have a direct interpretation (typically the solution of the main problem on a subgraph).

We now describe how to modify the basic double description algorithm in order to maintain the block structure, that is to work in a general iteration i on a nonpointed cone.

Let $P^{i-1} = \{x \in \mathbb{R}^d : A_k\,x \geq 0 \text{ for } k = 1, \ldots, i-1\}$ be described by a minimal set of generators (R, B) such that $P^{i-1} = \{x \in \mathbb{R}^d : x = R\lambda + B\mu \text{ for some } \lambda \geq 0 \text{ and some } \mu\}$, where $B = \{b_1, \ldots, b_t\}$ is a basis of the lineality space and $R = \{r_1, \ldots, r_s\}$ a set of representatives of the minimal proper faces (faces of dimension t+1) of P^{i-1}.

Then a minimal set of generators (R', B') of $P^i = \{x \in \mathbb{R}^n : A_k\,x \geq 0 \quad \forall k = 1, \ldots, i\}$ is determined by the following rules:

a) if $A_i \perp B$, then $B' = B$ and R' is obtained as stated in Lemma 8 , two generators r_j and $r_{j'}$ in R being adjacent if they are contained in a $(t+2)$-dimensional face of P^{i-1}.

b) else choose $\hat{B} = \{b'_1, \ldots, b'_t\}$ such that $A_i \cdot b'_k = 0$ for $k = 1, \ldots, t-1$ and $A_i \cdot b'_t > 0$, and set $B' = \{b'_1, \ldots, b'_{t-1}\}$ and $R' = \{r'_1, \ldots, r'_s, b'_t\}$, with $r'_j = (A_i \cdot b'_t) r_j - (A_i \cdot r_j) b'_t$ for $j = 1, \ldots, s$.

Initializing the algorithm with $P^0 = \{x \in \mathbb{R}^d : x = B^0 \mu \text{ for some } \mu\}$ and $B^0 = A_{LS}^{-1}$ where $A_{LS} = \{A_{i_j}, j = 1, \ldots, d\}$ avoids the computation of $\hat{B}$ in case b) as the vectors of B^0 already satisfy the required property. When the matrix A has the block structure, B^0 may be chosen as the canonical basis of $\mathbb{R}^d$ and the iterations i_j are particularly easy to perform since only the sign of the jth unit vector has to be properly chosen.

Note also that after an iteration i with $A_i \not\perp B$ each pair (r'_j, b'_t) , $j = 1, \ldots, s$ is adjacent in P^i.

A block structure is not only part of a good ordering, it can also often be used in conjunction with row decomposition to reduce the computing time: if h is the last iteration completed on the main problem P and the generators of P^h have been saved, the computation of each subproblem corresponding to a facet $P \cap H_i^0$ may be started at iteration $h + 1$ with either P^h if $i > h$ or those generators of P^h lying in H_i^0 if $i \leq h$.

5 Computational Results

In this section we present some of the experiments we carried out with the different implementations of the DD Method. The test examples are part of the distribution of CDD and thus publically available by anonymous ftp [Fuk93]. The input matrices A come from various combinatorial, geometric and practical applications, namely the vertices of the hypercube (cube) or its dual (cross), the complete cut cone (ccc) or cut polytope (ccp), the arborescences - rooted directed trees - in small graphs (prodmT), the vertices of a regular polytope (reg), a ternary alloy ground state analysis (mit). All cones except the hypercube are degenerate.

5.1 Experiments on Adjacency Tests

Table 1 contains the running times of algebraic and combinatorial adjacency test on some examples. It shows that the combinatorial test is almost always faster. A partial explanation for that is the following: One can expect the algebraic test to perform better for proving adjacency (exhibit $d - 2$ linearly independent rows) and the combinatorial for proving nonadjacency (exhibit one extreme ray violating Proposition 7), but the number of tests executed on nonadjacent pairs is usually much larger.

Table 1. Comparison of Algebraic and Combinatorial Adjacency Tests

test problems	sizes			Algebraic Test (in seconds)	Combin. Test (in seconds)
	dim	m	#output		
reg24-5	4	24	24	7	4
reg600-5	4	600	120	695	523
cross8	8	256	16	189	45
ccc5	10	15	40	9	6
ccp5	10	16	56	17	7
cube10	10	20	1,024	87	25
ccc6	15	31	210	903	101
ccp6	15	32	368	1,663	179

(On Mac IIci running Mach OS; cdd-038 compiled by gcc-2.6.0)

5.2 Experiments on Row Ordering

Different static and dynamic strategies for row ordering are compared in Table 2. The dynamic strategies used consist in selecting an hyperplane that cuts a maximum (minimum) number of the actual extremal rays. Except for the hypercube, for which all strategies perform equally well, the lexicographic ordering is always better, dominating mincutoff, then random and maxcutoff. For problems with a large number of rows in higher dimension a lexicographic ordering or some variant (obtained for example by permuting the columns) is the only useful one.

Table 2. Comparison of Different Row Orderings

test problems	sizes			max intermediate size (average size)			
	dim	m	#output	random	lexmin	mincutoff	maxcutoff
reg600-5	4	600	120	494 (319)	168 (94)	187 (134)	444 (276)
cross8	8	256	16	813 (272)	22 (17)	293 (128)	948 (270)
cube10	10	20	1,024	1024 (210)	1024 (231)	1024 (209)	1024 (231)
ccc6	15	31	210	757 (372)	575 (186)	575 (248)	752 (391)
ccp6	15	32	368	1484 (752)	693 (257)	1092 (441)	1682 (739)

(Solved with Variation1 (cdd-055a))

5.3 Experiments on the New Edge Data Updates

Table 3 shows the speedup which can be obtained with the improvements on the standard method. Clearly these ratios depend also on the dimensions and the ordering used.

Table 3. Comparison of DD Standard and Variation1

test problems	sizes			DDStandard (in seconds)	DDVariation1 (in seconds)	speedup ratio (times faster)
	dim	m	#output			
reg600-5	4	600	120	13	5	2.60
mit729-9	8	729	4,862	42,587	5,959	7.15
cube12	12	24	4,096	15	5	3.00
cross12	12	4,096	24	111	89	1.25
cube14	14	28	16,384	238	69	3.45
ccc6	15	31	210	2	2	1.00
ccp6	15	32	368	5	3	1.67
cube16	16	32	65,536	4,812	1,512	3.18
prodmT5	19	711	76	39	25	1.56
ccc7	21	63	38,780	152,694	142,443	1.07
ccp7	21	64	116,764	621,558	469,681	1.32
prodmT62	24	3,461	168	15,083	10,047	1.50

(Standard(cdd-038) and Variation1(cdd-055a) compiled by gcc-2.6.2 on Spark-Server 1000)

5.4   Comparison of DD Variation1 and Variation2

Relative performances of Variation1 and 2 are presented in Table 4. None of these variations dominates the other, but Variation1 is usually slightly faster since there are less pairs lying in a common hyperplane than pairs separated.

Table 4. Comparison of DDVariation1 and DDVariation2

test problems		reg600-5	prodmT62	ccp6	cross14	cube14
dimensions(A)		600x5	3,461x25	32x16	16,384x15	28x15
#output		120	168	368	28	16,384
DDV1	#pairs scanned	24,386	25,021,758	563,086	2,728,111	89,494,867
	#pairs separated	12,443	5,041,661	125,643	372,827	16,382
	time(sec)	3	1,038	4	560	72
DDV2	#pairs scanned	512,257	19,723,402	168,038	196,610	16,382
	time(sec)	9	1,158	5	544	3
both	#comb.tests	1,304	2,091,514	4,691	17,327	0
	#rays (max)	524	1,146	693	40	16,384
	#rays (total)	1,889	65,239	1,768	16,504	16,384

(Variations 1 and 2 coded in pascal and run on Silicon Graphics)

References

[AB95]   D. Avis, D. Bremner and R. Seidel. How good are convex hull algorithms. *Computational Geometry: Theory and Applications* (to appear).

[AF92]   D. Avis and K. Fukuda. A pivoting algorithm for convex hulls and vertex enumeration of arrangements and polyhedra. *Discrete Comput. Geom.*, 8:295–313, 1992.

[Avi93]   D. Avis. *A C implementation of the reverse search vertex enumeration algorithm.* School of Computer Science, McGill University, Montreal, Canada, 1993. programs lrs and qrs available via anonymous ftp from mutt.cs.mcgill.ca (directory pub/C).

[BDH93]   C.B. Barber, D.P. Dobkin, and H. Huhdanpaa. The quickhull algorithm for convex hulls. Technical Report GCC53, The Geometry Center, Minnesota, U.S.A., 1993.

[CGAF94]   G. Ceder, G.D. Garbulsky, D. Avis, and K. Fukuda. Ground states of a ternary fcc lattice model with nearest and next-nearest neighbor interactions. *Physical Review B*, 49(1):1–7, 1994.

[Che65]   N.V. Chernikova. An algorithm for finding a general formula for non-negative solutions of system of linear inequalities. *U.S.S.R Computational Mathematics and Mathematical Physics*, 5:228–233, 1965.

[CR95]   T. Christof and G. Reinelt. Combinatorial Optimization and Small Polytopes. To appear in TOP 96.

[DDF95]   A. Deza, M. Deza and K. Fukuda. On Skeletons, Diameters and Volumes of Metric Polyhedra. Proceedings CCS'95 (this issue).

[DFL93]   M. Deza, K. Fukuda, and M. Laurent. The inequicut cone. *Discrete Mathematics*, 119:21–48, 1993.

[Dye83]   M.E. Dyer. The complexity of vertex enumeration methods. *Math. Oper. Res.*, 8:381–402, 1983.

[Ede87]   H. Edelsbrunner. *Algorithms in Combinatorial Geometry.* Springer-Verlag, 1987.

[Fuk93]   K. Fukuda. *cdd.c : C-implementation of the double description method for computing all vertices and extremal rays of a convex polyhedron given by a system of linear inequalities.* Department of Mathematics, Swiss Federal Institute of Technology, Lausanne, Switzerland, 1993. program available from ifor13.ethz.ch (129.132.154.13), directory /pub/fukuda/cdd.

[Grü67]   B. Grünbaum. *Convex Polytopes.* John Wiley and Sons, New York, 1967.

[MRTT53]   T.S. Motzkin, H. Raiffa, GL. Thompson, and R.M. Thrall. The double description method. In H.W. Kuhn and A.W.Tucker, editors, *Contributions to theory of games, Vol. 2.* Princeton University Press, Princeton. RI, 1953.

[MS71]   P. McMullen and G.C. Shephard. *Convex Polytopes and the Upperbound Conjecture.* Cambridge University Press, 1971.

[Mul94]   K. Mulmuley. *Computational Geometry, An Introduction Through Randamized Algorithms.* Prentice-Hall, 1994.

[Sch86]   A. Schrijver. *Theory of Linear and Integer Programming.* John Wiley & Sons, New York, 1986.

[Ver92]   H. Le Verge. A note on Chernikova's algorithm. Technical report internal publication 635, IRISA, Rennes, France, February 1992.

[Wil93]   D. Wilde. A library for doing polyhedral operations. Technical report internal publication 785, IRISA, Rennes, France, December 1993.

On Skeletons, Diameters and Volumes of Metric Polyhedra

Antoine Deza[1]*, Michel Deza[2] and Komei Fukuda[3]

[1] Tokyo Institute of Technology, Department of Mathematical and Computing Sciences, Tokyo, Japan
[2] CNRS, Ecole Normale Supérieure, Département de Mathématiques et d'Informatique, Paris, France
[3] ETH Zürich, Institute for Operations Research, CH-8092 Zürich, Switzerland and University of Tsukuba, Graduate School of Systems Management, Tokyo, Japan

Abstract. We survey and present new geometric and combinatorial properties of some polyhedra with application in combinatorial optimization, for example, the max-cut and multicommodity flow problems. Namely we consider the volume, symmetry group, facets, vertices, face lattice, diameter, adjacency and incidence relations and connectivity of the metric polytope and its relatives. In particular, using its large symmetry group, we completely describe all the 13 orbits which form the 275 840 vertices of the 21-dimensional metric polytope on 7 nodes and their incidence and adjacency relations. The edge connectivity, the i-skeletons and a lifting procedure valid for a large class of vertices of the metric polytope are also given. Finally, we present an ordering of the facets of a polytope, based on their adjacency relations, for the enumeration of its vertices by the double description method.

1   Introduction

We first recall the definition of the *metric polytope m_n* and some of its relatives and present some applications to well known optimization problems of those polyhedra. The general references are BAYER AND LEE [8] and ZIEGLER [31] for polytopes and BROUWER, COHEN AND NEUMAIER [9] for graphs. For a complete study of the applications and the combinatorial optimization aspects of those polyhedra, we refer, respectively, to the surveys DEZA AND LAURENT [17] and POLJAK AND TUZA [29].

For all 3-sets $\{i, j, k\} \subset N = \{1, \ldots, n\}$, we consider the following inequalities:

$$x_{ij} - x_{ik} - x_{jk} \leq 0 \ . \tag{1}$$

The inequalities (1) induce the $3\binom{n}{3}$ facets which define the *metric cone M_n*. Then, bounding the later by the following inequalities:

$$x_{ij} + x_{ik} + x_{jk} \leq 2 \tag{2}$$

---

* Research supported by Japanese Ministry of Education, Science and Culture for the first author.

we obtain the *metric polytope* m_n. The $3\binom{n}{3}$ facets defined by (1), which can be seen as triangle inequalities for distance x_{ij} on $\{1, 2, \ldots, n\}$, are called *homogeneous triangle facets*. The $\binom{n}{3}$ facets defined by the inequalities (2) are called *non-homogeneous triangle facets*, and by *triangle facet* we denote a facet of either type (1) or (2).

While the *cut cone* C_n is the conic hull of all, up to a multiple, $\{0, 1\}$-valued extreme rays of the metric cone, the *cut polytope* c_n is the convex hull of all $\{0, 1\}$-valued vertices of the metric polytope. Those two polyhedra can also be defined independently from the metric cone and polytope in the following ways.

Given a subset S of $N = \{1, 2, \ldots, n\}$, the *cut* defined by S consists of the pairs (i, j) of elements of N such that exactly one of i, j is in S. By $\delta(S)$ we denote both the cut and its incidence vector in $\mathbb{R}^{\binom{n}{2}}$, that is, $\delta(S)_{ij} = 1$ if exactly one of i, j is in S and 0 otherwise for $1 \le i < j \le n$. By abuse of language, we use the term cut for both the cut itself and its incidence vector, so $\delta(S)_{ij}$ are considered as coordinates of a point in $\mathbb{R}^{\binom{n}{2}}$. The cut polytope of the complete graph c_n, which is also called the complete bipartite subgraphs polytope, is the convex hull of all 2^{n-1} cuts, and the cut cone C_n is the conic hull of all $2^{n-1} - 1$ nonzero cuts. Those polyhedra were considered by many authors, see for instance [2, 7, 15, 16, 17, 18, 19, 21, 23, 24] and references therein. One of the motivations for the study of these polyhedra comes from their applications in combinatorial optimization, the most important being the max-cut and multicommodity flow problems.

Given a graph $G = (N, E)$ and nonnegative weights w_e, $e \in E$, assigned to its edges, the *max-cut* problem consists in finding a cut $\delta(S)$ whose weight $\sum_{e \in \delta(S)} w_e$ is as large as possible. It is a well-known NP-complete problem. By setting $w_e = 0$ if e is not an edge of G, we can consider without loss of generality the complete graph K_n. Then the max-cut problem can be stated as a linear programming problem over the cut polytope c_n as follows:

$$\max \quad w^T \cdot x$$
$$\text{subject to} \quad x \in c_n \ .$$

Since the metric polytope is a relaxation of the cut polytope, optimizing $w^T \cdot x$ over c_n instead of m_n provides an upper bound for the max-cut problem [7].

With E the set of edges of the complete graph K_n, an instance of the *multicommodity flow problem* is given by two nonnegative vectors indexed by E: a capacity $c(e)$ and a requirement $r(e)$ for each $e \in E$. Let $U = \{e \in E : r(e) > 0\}$. If T denotes the subset of N spanned by the edges in U, then we say that the graph $G = (T, U)$ denotes the *support* of r. For each edge $e = (s, t)$ in the support of r, we seek a flow of $r(e)$ units between s and t in the complete graph. The sum of all flows along any edge $e' \in E$ must not exceed $c(e')$. If such a set of flows exists, we call c, r *feasible*. A necessary and sufficient condition for feasibility is given by the Japanese theorem of IRI [22] and ONAGA AND KAKUSHO [26]: a pair c, r is feasible if and only if $(c - r)^T x \ge 0$ is valid over the metric cone. For example, the triangle facet induced by (1) can be seen as an elementary solvable flow problem with $c(ij) = r(ik) = r(jk) = 1$ and $c(e) = r(e) = 0$ otherwise, so the inequalities (1) correspond to $(c - r)^T x \geq 0$ for $x \in M_n$. In other words, the dual metric cone is the cone of all feasible multicommodity flow problems.

2   Skeletons and Diameters

2.1   Previous Results

The polytope c_n is a $\binom{n}{2}$ dimensional 0–1 polyhedron with 2^{n-1} vertices and m_n is a polytope of same dimension with $4\binom{n}{3}$ facets inscribed in the cube $[0, 1]^{\binom{n}{2}}$. We have $c_n \subseteq m_n$ with equality only for $n \leq 4$. It is easy to see that the point $\omega_n = (\frac{1}{2}, \frac{1}{2}, \ldots, \frac{1}{2})$ is the center of gravity of both c_n and m_n and is also the center of the sphere of radius $r = \frac{1}{2}\sqrt{n(n-1)}$ where all the cuts lie. Another two geometric characteristics of the cut polytope c_n are its *width* and *geometric diameter*. We recall that while the width of a polytope P is equal to the minimum distance between a pair of parallel hyperplanes containing P in the slice between them, the geometric diameter of P is the maximum distance between a pair of supporting hyperplanes. The width of c_n is 1 and its geometric diameter is $\frac{n}{2}$ for n even and $\frac{1}{2}\sqrt{n^2 - 1}$ for n odd. Any facet, respectively *subfacet* (that is, a face of codimension 2), of the metric polytope contains a facet, respectively a subfacet, of the cut polytope and the vertices of the cut polytope are vertices of the metric polytope, in fact the cuts are precisely the integral vertices of the metric polytope. Actually the metric polytope m_n wraps the cut polytope c_n very tightly since, in addition to the vertices, all edges and 2-faces of c_n are also faces of m_n, for 3-faces it is false for $n \geq 4$, see [14, 19]. In other words, c_n is a *segment of order* 2, but not 3, of m_n and its dual, m_n^*, is a segment of order 1 of c_n^* in terms of [25]: a polytope P is a segment of order s of a polytope Q if they have the same dimension and if every i-face of P is a face of Q for $0 \leq i \leq s$. The polytope c_n is 3-neighbourly, see [19]. Any two cuts are adjacent both on c_n and on m_n [7, 27]; in other words m_n is *quasi-integral* in terms of [30], that is, the skeleton of the convex hull of its integral vertices, i.e. the skeleton of c_n, is an induced subgraph of the skeleton of the metric polytope itself. While the diameter of m_n^* is 2, the diameters of c_n^* and m_n are respectively conjectured to be 4 and 3, see [13, 23]. We recall that the skeleton of a polytope is the graph formed by its vertices and edges.

The metric polytope and the cut polytope share the same symmetry group, that is, the group of isometries preserving a polytope. This group is isomorphic to the automorphism group of the *folded n-cube*: $Aut(\Box_n) \approx Is(m_p) = Is(c_p)$, see [15, 23]. We recall that the folded n-cube is the graph whose vertices are the partitions of $N = \{1, \ldots, n\}$ into two subsets, two partitions being adjacent when their common refinement contains a set of size one, see [9]. More precisely, for $n \geq 5$, $Is(m_n) = Is(c_n)$ is induced by permutations on $N = \{1, \ldots, n\}$ and *switching reflections by a cut*. Given a cut $\delta(S)$, the switching reflection $r_{\delta(S)}$ is defined by $y = r_{\delta(S)}(x)$ where $y_{ij} = 1 - x_{ij}$ if $(i, j) \in \delta(S)$ and $y_{ij} = x_{ij}$ otherwise. These symmetries preserve the adjacency relations and the linear independency. Using the partition of the faces of m_n and c_n into orbits of their symmetry group, the face lattice for small dimensions ($d = 3, 6$ and 10) was given in [14].

We finally mention the following link with metrics. There is an evident $1 - 1$ correspondence between the elements of the metric cone and all the semi-metrics on n points. Moreover the elements of th. cut cone correspond precisely to the semi-metrics on n points that are isometrically embeddable into some l_1^m, see [1], it is easy to check that such minimal m is smaller or equal to $\binom{n}{2}$.

Another relative of the metric cone is the *solitaire cone* S_B, that is, the cone generated by all the possibles moves of a Solitaire Peg game played on a board B. This cone shares a lot of similar properties with the metric cone, see [5]. In particular, for a game played on the line graph T_n of the complete graph K_n, the complete solitaire cone S_{T_n} equals the dual metric cone M_n^*, see [5].

2.2 New Results

The Metric Polytope on Seven Nodes. In Table 1 we present the 13 orbits under permutations and switching which form the 275 840 vertices of the metric polytope m_7. For each orbit O_i, we give a representative vertex v_i, the size of the orbit $|O_i|$, its size $|O_i \cap F|$ restricted to a facet and the incidence I_{v_i} and the adjacency A_{v_i} of any vertex belonging to the orbit O_i.

Table 1. The orbits of vertices of the metric polytope on seven nodes

Orbit O_i	Representative vertex v_i	$|O_i|$	$|O_i \cap F|$	I_{v_i}	A_{v_i}
O_1	$(0,0,0,0,0,0,0,0,0,0,0,0,0,0,0,0,0,0,0,0,0)$	64	48	105	55 226
O_2	$\frac{2}{3}(1,1,1,1,1,1,1,1,1,1,1,1,1,1,1,1,1,1,1,1,1)$	64	16	35	896
O_3	$\frac{2}{3}(1,1,1,1,1,0,1,1,1,1,1,1,1,1,1,1,1,1,1,1,1)$	1 344	384	40	763
O_4	$\frac{2}{3}(1,1,1,1,0,1,1,1,1,1,0,1,1,1,1,1,1,1,1,1,1)$	6 720	2 160	45	594
O_5	$\frac{2}{3}(1,1,1,1,0,0,1,1,1,1,1,1,1,1,1,1,1,1,1,1,0)$	2 240	784	49	496
O_6	$\frac{1}{4}(1,2,3,1,2,1,1,2,2,1,2,1,1,2,3,2,3,2,1,2,1)$	20 160	4 320	30	96
O_7	$\frac{1}{3}(1,1,1,1,1,1,2,2,1,1,1,2,1,1,1,1,1,2,2,2)$	4 480	832	26	76
O_8	$\frac{2}{5}(2,1,1,1,1,2,2,1,1,1,1,2,1,1,1,2,1,1,2,1,2)$	23 040	4 608	28	57
O_9	$\frac{1}{3}(2,2,1,1,1,2,2,1,1,1,1,2,1,1,1,2,1,1,2,1,2)$	40 320	6 336	22	46
O_{10}	$\frac{1}{3}(1,1,1,1,1,1,2,2,1,1,1,2,1,1,1,2,1,1,2,2,2)$	40 320	6 624	23	39
O_{11}	$\frac{2}{7}(1,2,3,2,1,2,1,2,1,2,1,1,2,1,1,1,2,2,1,1,1)$	40 320	7 200	25	30
O_{12}	$\frac{1}{5}(3,2,3,3,1,1,1,2,2,2,2,3,3,3,3,4,4,2,2,4,2)$	16 128	2 880	25	27
O_{13}	$\frac{1}{6}(1,2,4,2,2,2,1,3,3,3,3,2,2,2,4,2,2,2,4,4,4)$	80 640	13 248	23	24
Total		275 840	49 440		

Lemma 1. *For any vertex v_i of m_n, with $|O_i|$ denoting the size of the orbit of v_i, $|O_i \cap F|$ the size of its restriction to a facet a* *l I_{v_i} the incidence of v_i, we have:*

$$|O_i| \cdot I_{v_i} = |O_i \cap F| \cdot 4 \binom{n}{3} \tag{3}$$

Proof. Let $\{v_1, \ldots, v_K\}$ and $\{F_1, \ldots, F_L\}$ be respectively an ordering of the orbit O_i and of the triangle facets, and set $\chi_{kl} = 1$ if the vertex v_k belongs to the triangle facet F_l and 0 otherwise. We have:

$$\sum_{k,l} \chi_{kl} = \sum_l (\sum_k \chi_{kl}) = \sum_l (|O_i \cap F|) = |O_i \cap F| \cdot 4 \binom{n}{3}$$

and also,

$$\sum_{k,l} \chi_{kl} = \sum_k (\sum_l \chi_{kl}) = \sum_k (I_{v_i}) = |O_i| \cdot I_{v_i} \ . \qquad \Box$$

Table 2. Orbit-wise adjacencies relations of a cut in the skeleton of m_7

O_1	O_2	O_3	O_4	O_5	O_6	O_7	O_8	O_9	O_{10}	O_{11}	O_{12}	O_{13}
63	56	945	3 570	980	7 560	1 120	5 400	8 820	6 930	6 930	2 772	10 080

In Table 2 we present orbit-wise the 55 226 neighbours of a vertex belonging to the orbit O_1, that is a cut. For example, 945 in the third column means that a cut is adjacent to 945 vertices belonging to the orbit O_3, see Section 4 for details. Since all the facets incident to the origin $\delta(\emptyset)$ are precisely the $3\binom{n}{3}$ homogeneous triangle facets, to each vertex adjacent to $\delta(\emptyset)$ corresponds an extreme ray of the metric cone. In other words, the adjacency A_{v_1} of a cut equals the number of extreme rays of the metric cone M_n. We recall that the 41 orbits under permutations of the extreme rays of M_7 were previously found by GRISHUKHIN[21]. Table 2 also implies that the cuts form a dominating clique in the skeleton of m_7, that is, every vertex is adjacent to a cut, as conjectured by LAURENT AND POLJAK [24]. We have:

Corollary 2. *The metric cone on seven nodes has exactly 55 226 extreme rays.*

Corollary 3. *The diameter of the metric polytope on seven nodes is $\delta(m_7) = 3$.*

Proof. The cuts forming a dominating clique, we have $\delta(m_7) \leq 3$. Then, v_{13} and its switching by $\delta(3)$ having no common neighbour, see [12] , we have $\delta(m_7) \geq 3$.

Connectivity. A graph is said to be c edge connected provided it has at least $c + 1$ vertices and no two vertices can be separated by removing fewer that c edges. With C such maximal c, let $C(P)$ denote the edge connectivity of the skeleton of a polytope P. We have:

Theorem 4. *The edge connectivity of the metric and cut polytope is:*

1. $C(m_n^*) = 2\frac{(n-3)(n^2-7)}{3}$ *for* $n \geq 4$ *and* $C(m_3^*) = 3$.

2. $C(m_4) = 7$, $C(m_5) = 10$, $C(m_6) = 35$, $21 \leq C(m_7) \leq 24$.

3. $C(c_n^*) = \binom{n}{2}$.

4. $C(c_n) = 2^{n-1} - 1$.

Proof. We recall the following result of PLESNÍK [28]. The connectivity of a graph of diameter 2 equals its minimum degree. Then, the skeleton of m_n^* being of diameter 2 and with constant degree $k = 2\frac{(n-3)(n^2-7)}{3}$ for $n \geq 4$, it implies *1*. The diameter of m_4, m_5 and m_6 being 2, it also implies *2* for $n \leq 6$. The facet F_n of c_n induced by the following inequality:

$$\sum_{1 \leq i < j \leq n} b_i b_j x_{ij} \leq 2 \quad \text{where} \quad b = (-(n-4), 1, 1, \ldots, 1)$$

is a simplex facet which contains exactly the $\binom{n}{2}$ cuts $\delta(\{i\})$ for $2 \leq i \leq n$ and $\delta(\{i,j\})$ for $2 \leq i < j \leq n$. This implies that $C(c_n^*) \leq \binom{n}{2}$. Then, BALINSKI's theorem [6] stating that the connectivity of the skeleton of a polytope is at least its dimension, we obtain *3*. The skeleton of c_n being the complete graph, *4* is straightforward. $\square$

The i-Skeletons. We consider the following two families of graphs. while $G^i(P)$ denotes the graph which vertices are all the i-faces of a polytope P, two i-faces being adjacent if and only if $f_i^1 \cap f_i^2$ is a $(i-1)$-face of P, $G_i(P)$ is the graph which vertices are all the i-faces of P, two i-faces being adjacent if and only if f_i^1 and f_i^2 belong to the same $(i+1)$-face of P. We have:

Proposition 5.

1. $G_0(c_n) = K_{2^{n-1}}$.

2. $G_1(c_n) = L(K_{2^{n-1}})$.

3. $G_2(c_n)$ *has* $\binom{2^{n-1}}{3}$ *vertices and two vertices* f_2^1 *and* f_2^2 *are adjacent if and only if:*

   $$|f_2^1 \cap f_2^2| = 2 \quad \text{or} \quad |f_2^1 \cup f_2^2| = 4, \quad \text{and} \quad f_2^1 \cup f_2^2 \text{ is a face of } c_4.$$

4. *The complement of* $G^{\binom{n}{2}-1}(m_n)$ *is locally the bouquet of* $(n-3)$ (3×3)-*grids with common* K_3.

Proof. The cut polytope being 3-neighbourly, *1* and *2* are straightforward. The $\binom{2^{n-1}}{3}$ 2-faces of c_n are partitioned into the orbits respectively represented by

$f_2^{r,s,t} = \{\delta(\emptyset), \delta(1,\ldots,r+s), \delta(r+1,\ldots,r+s+t)\}$ for all triplets of integers $\{r,s,t\}$ such that $1 \leq r \leq \lfloor\frac{n}{3}\rfloor, 0 \leq s \leq r, r \leq t \leq min(\lfloor\frac{n-r}{2}\rfloor, \lfloor\frac{n}{2}\rfloor - s, n - 2r - s)$ and their incidence relations follows. For 4, that is the skeleton of the dual metric polytope, see [13]. □

Volumes. In Table 3 we give the volumes of m_n and c_n for $n \leq 6$. Both volumes seam to quickly vanish to 0 and their ratio, which can be consider as a measure of the tightness of the relaxation of c_n by m_n, seams to stay relatively close to 1. For $n \geq 5$, the volumes were computed using the reverse search method for vertex enumeration using lexicographic pivoting, implemented by AVIS. The code used was lrs Version 2.5i, an earlier version of the code is described in [3]. Since all facets of m_n are equivalent under permutation and switching, the volume of m_n equals $4\binom{n}{3}$ times the volume of the pyramid with basis one facet and apex the center of gravity ω_n of m_n. Comparing the volume of this pyramid and of c_n to the volume of the standard $\binom{n}{2}$-simplex of edge length 2, we have:

$$\frac{Vol(m_n) \cdot \binom{n}{2}!}{2^{\binom{n}{2}} 4\binom{n}{3}} = 2^{-4}, 2^{-5}, \frac{5 \cdot 2^{-3}}{3}, \frac{7 \cdot 281}{3^4} \qquad \text{for } n = 3,\ldots,6.$$

$$\frac{Vol(c_n) \cdot \binom{n}{2}!}{2^{\binom{n}{2}}} = 2^{-2}, 2^{-1}, 2^3, 11 \cdot 149 \qquad \text{for } n = 3,\ldots,6.$$

Table 3. Volumes of small metric and cut polytopes

#n nodes	Volume (m_n)	Volume (c_n)	Vol(c_n)/Vol(m_n)
3	1/3	1/3	100%
4	2/45	2/45	100%
5	4/1 701	32/14 175	≈ 96%
6	71 936/1 477 701 225	2 384/58 046 625	≈ 84%

2.3 Summary Tables

In Tables 4, 5 and 6 we sum up known and conjectured results concerning the skeletons and diameters of the metric and cut polytopes. In particular, we give the number of vertices #V and facets #F of those polytopes, the incidences I_v and I_f of their vertices and facets, the adjacencies A_v and A_f of their vertices and facets, and the diameter and connectivity of m_n and c_n and of their dual polytopes m_n^* and c_n^*. For example, the last value of the column I_f of Table 5 means that a facet of the cut polytope contains at least $\binom{n}{2}$ vertices, that is, is a simplex and at most $3 \cdot 2^{n-3}$ vertices, that is $\frac{3}{4}$ of the total number of vertices of c_n, this bound being reached only by the $4\binom{n}{3}$ triangle facets, see [13]. In the last row of Tables 4 and 5, $A_{\delta(S)}$, A_{Tr} and $\#F_{C_n}$ respectively denote the adjacency of a cut in m_n, the adjacency of a triangle facet in c_n and the number of facets of the cut cone.

Table 4. Skeletons and diameters of metric polytopes

#nodes	#V	I_v	A_v	#F	I_f	A_f	$\delta(m_n)$	$\delta(m_n^*)$
3	4	3	3	4	3	3	1	1
4	8	12	7	16	6	6	1	2
5	32	10~30	10~25	40	16	24	2	2
6	544	20~60	35~296	80	176	58	2	2
7	275 840	22~105	24~55 226	140	49 440	112	3	2
n		$\binom{n}{2}? \sim 3\binom{n}{3}$	$\binom{n}{2}? \sim A_{\delta(S)}?$	$4\binom{n}{3}$		$\frac{2(n-3)(n^2-7)}{3}$	3?	2

Table 5. Skeletons and diameters of cut polytopes

#nodes	#V	I_v	A_v	#F	I_f	A_f	$\delta(c_n)$	$\delta(c_n^*)$
3	4	3	3	4	3	3	1	1
4	8	12	7	16	6	6	1	2
5	16	40	15	56	10~12	10~28	1	2
6	32	210	31	368	15~24	15~142	1	3
7	64	38 780	63	116 764	21~48	21~11 432	1	$3 \leq \delta(c_7^*) \leq 4$
8	128	49 604 520	127	217 093 472	28~96	28~?	1	?
n	2^{n-1}	$\#F_{C_n}$	$2^{n-1}-1$		$\binom{n}{2} \sim 3 \cdot 2^{n-3}$	$\binom{n}{2} \sim A_{Tr}?$	1	4?

Table 6. Connectivity of the metric and cut polytopes

#nodes	$C(m_n)$	$C(m_n^*)$	$C(c_n)$	$C(c_n^*)$
3	3	3	3	3
4	7	6	7	6
5	10	24	15	10
6	35	58	31	15
7	$21 \leq C(m_7) \leq 24$	112	63	21
n	$\binom{n}{2}?$	$2\frac{(n-3)(n^2-7)}{3}$	$2^{n-1}-1$	$\binom{n}{2}$

Conjecture 6.

1. *The adjacency of a cut, that is, the number of extreme rays of the metric cone, is maximal in the skeleton of m_n. It holds for $n \leq 7$.*
2. *For n large enough, at least one vertex of m_n is simple, (that is, the incidence equals the dimension of the polytope). If true, it would imply that the edge connectivity, the minimal incidence and the minimal adjacency of the skeleton of m_n are equal to $\binom{n}{2}$. It holds for $n = 3$ and 5.*
3. *The adjacency of a triangle facet is maximal in the skeleton of c_n^*. It holds for $n \leq 7$.*

Table 7. Skeletons and diameters of metric cones

#nodes	#R	I_r	A_r	#F	I_f	A_f	$\delta(M_n)$	$\delta(M_n^*)$
3	3	2	2	3	2	2	1	1
4	7	8~9	6	12	5	5	1	2
5	25	9~24	9~20	30	14	19	2	2
6	296	16~50	23~190	60	113	45	2	2
7	55 226	20~90	20~18 502	105	12 821	86	3	2
n	$A_{\delta(S)}^{m_n}$	$\binom{n}{2}-1? \sim (n-1)\binom{n-1}{2}$	$\binom{n}{2}-1? \sim A_{\delta(\{1\})}?$	$3\binom{n}{3}$	$A_{\delta(S)/F}^{m_n}$	$\frac{(n-3)(n^2-6)}{2}$	3?	2

Table 8. Skeletons and diameters of cut cones

#nodes	#R	I_r	A_r	#F	I_f	A_f	$\delta(C_n)$	$\delta(C_n^*)$
3	3	2	2	3	2	2	1	1
4	7	8~9	6	12	5	5	1	2
5	15	27~30	14	40	9~11	9~22	1	2
6	31	114~130	30	210	14~23	14~98	1	3
7	63	11 343~16 460	62	38 780	20~47	20~4 928	1	$3 \le \delta(C_7^*) \le 4$
8	127	?	126	49 604 520	27~95	27~?	1	?
n	$2^{n-1}-1$	$I_{\delta(E)}? \sim I_{\delta(\{1\})}?$	$2^{n-1}-2$	$I_{\delta(S)}^{c_n}$	$\binom{n}{2}-1 \sim 3\cdot 2^{n-3}-1$	$\binom{n}{2}-1 \sim A_{Tr}?$	1	4?

Table 9. Connectivity of the metric and cut cones

#nodes	$C(M_n)$	$C(M_n^*)$	$C(C_n)$	$C(C_n^*)$
3	2	2	2	2
4	6	5	6	5
5	9	19	14	9
6	23	45	30	14
7	20	86	62	20
n	$\binom{n}{2}-1?$	$\frac{(n-3)(n^2-6)}{2}$	$2^{n-1}-2$	$\binom{n}{2}-1$

In Tables 7, 8 and 9 we give corresponding results concerning the skeletons and diameters of the metric and cut cones. Those results can be almost directly deduced from the ones given in Tables 4, 5 and 6. In the last row of Table 7, $A_{\delta(\{1\})}$, $A_{\delta(S)}^{m_n}$ and $A_{\delta(S)/F}^{m_n}$ respectively denote the adjacency of the cut $\delta(\{1\})$ in M_n, the adjacency of a cut in m_n and its restriction to a facet of m_n. In the last row of Table 8, $I_{\delta(\{1\})}$, $I_{\delta(E)}$, $I_{\delta(S)}^{c_n}$ and A_{Tr} respectively denote the incidence of the cut $\delta(S)$ with $|S| = 1$ and $|S| = \lfloor \frac{n}{2} \rfloor$ in C_n, the incidence of a cut in c_n and the adjacency of a triangle facet in C_n. For example, the column I_r of Table 7 gives that the maximal incidence of the extreme rays of M_n equals the one of a cut $\delta(S)$ with $|S| = 1$, that is, $I_{max} = I_{\delta(\{1\})} = (n-1)\binom{n-1}{2}$.

Remark. The values $\#F$ for $n = 8$ in Tables 5 and 8 are due to CHRISTOF AND REINELT who recently computed the facets of c_8 and C_8, see [10, 11]. The 217 093 472 facets of c_8 form 147 orbits under its symmetry group; for more information about those facets and the 49 604 520 on s of C_8 see the following WWW site: **http://www.iwr.uni-heidelberg.de/iwr/comopt/soft/SMAPO**.

Theorem 7. *The edge connectivity of the metric and cut cone is:*

1. $C(M_n^*) = \frac{(n-3)(n^2-6)}{2}$ *for $n \geq 4$ and $C(M_3^*) = 2$.*

2. $C(M_4) = 6$, $C(M_5) = 9$, $C(M_6) = 23$, $C(M_7) = 20$.

3. $C(C_n^*) = \binom{n}{2} - 1$.

4. $C(C_n) = 2^{n-1} - 2$.

Proof. The cuts forming a clique and the skeleton of M_n^* being of diameter 2 with constant degree $k = (n-3)(n^2-6)/2$ for $n \geq 4$, we have *1* and *4*. A switching of the facet F_n given in the proof of Theorem 4 is a simplex facet of C_n, this implies *3*. Applying BALINSKI's theorem [6] to a section of C_n by a bounding hyperplane, we have $C(C_n^*) = \binom{n}{2} - 1$. The same arguments as for the proof of Theorem 4 give item *2*. $\square$

Proposition 8.

1. *A facet of C_n contains at most $3 \cdot 2^{n-3} - 1$ extreme rays; this bound being reached only by the $3\binom{n}{3}$ triangle facets.*
2. *At least one facet of C_n is a simplex. This implies that the minimal incidence and the minimal adjacency of the skeleton of C_n^* are equal to $\binom{n}{2} - 1$.*
3. *An extreme ray of M_n belong to at most $(n-1)\binom{n-1}{2}$ facets; this bound being reached by only the n cuts $\delta(S)$ of size $|S| = 1$.*
4. *The cuts $\delta(S)$ and the extreme rays $\hat{\delta}(S)$ defined for $2 \leq |S| \leq n-2$ by $\hat{\delta}(S) = d(K_{S,\bar{S}})$ (that is $\hat{\delta}(S)_{st} = 1$ if s and t adjacent and 2 otherwise) form a subgraph of diameter 2 in the skeleton of M_n.*

Proof. Item *1* can be easily deduced form the corresponding result for c_n. A switching of the facet F_n given in the proof of Theorem 7 is a simplex facet of C_n stated in *2*. To prove item *3*, we first recall the following property of the vertices of m_n given in [13]. A vertex v of m_n belongs to at most $3\binom{n}{3}$ facets, that is $\frac{3}{4}$ of the total number of facets of m_n, this bound being reached only by the cuts. More precisely, for v a vertex of m_n and any 3-set $\sigma = \{i, j, k\} \subset N$, we have:

1. either v belongs to exactly 3 of the 4 facets supported by σ; and then $\{v_{ij}, v_{ik}, v_{jk}\} \subset \{0, 1\}$,

2. or v belongs to exactly 2 of the 4 facets supported by σ; and then, with $0 < \alpha < 1$, we have $\{v_{ij}, v_{ik}, v_{jk}\} = \{0, \alpha, \alpha\}$ o $\{1, \alpha, 1 - \alpha\}$,

3. or v belongs to at most 1 of the 4 facets supported by σ; and then we have $\{v_{ij}, v_{ik}, v_{jk}\} \cap \{0, 1\} = \emptyset$.

Then, one can easily check that, in M_n, a cut $\delta(S)$ of size $|S| = s$ belongs to exactly $3\binom{n}{3} - (n - s)\binom{s}{2} - s\binom{n-s}{2}$ triangle facets with the convention $\binom{i}{j} = 0$ for $i < j$. This, with above items 1 and 2, implies that the incidence in M_n of a cut is higher than the one of any other extreme rays. A cut of size $|S| = 1$ being of maximal incidence among the cuts, this completes the proof of item 3. Using the same notation for the extreme rays of M_n and the corresponding vertices of m_n, the relation in m_n: $\delta(\emptyset)$ not adjacent to $\hat{\delta}(S)$ if and only if $|S| \leq 1$ implies the following relation in M_n: $\delta(\{i\})$ not adjacent to $\hat{\delta}(S)$ if and only if $S = \{i\}$ or $\{i, j\}$. Then, for example, a common neighbour of $\hat{\delta}(\{i, j\})$ and $\hat{\delta}(\{k, l\})$ and of $\hat{\delta}(\{i, j\})$ and $\delta(\{i, j\})$ is $\delta(\{r\})$ for any 5-tuple $\{i, j, k, l, r\}$. This implies 4. $\square$

Conjecture 9.

1. The adjacency of a cut $\delta(S)$ with $|S| = 1$ is maximal in the skeleton of M_n. It holds for $n \leq 7$.

2. For n large enough, at least one extreme ray of M_n is simple, (that is, the incidence plus one equals the dimension of the cone). If true, it would imply that the edge connectivity, the minimal incidence and the minimal adjacency of the skeleton of M_n are equal to $\binom{n}{2} - 1$. It holds for $n = 3, 5$ and 7.

3. The incidence of a cut $\delta(S)$ in C_n is minim l, respectively maximal, for $|S| = \lfloor \frac{n}{2} \rfloor$, respectively for $|S| = 1$. It holds for $n \leq 7$.

4. The adjacency of a triangle facet is maximal in the skeleton of C_n^*. It holds for $n \leq 7$.

3 Lifting Construction

In this section we present a construction which, under given conditions on a vertex v of m_n, maps v to a vertex of a higher dimensional metric polytope. Let v be a point in $\mathbb{R}^{\binom{n}{2}}$, the *diameter* $\delta(v)$ and *radius* $r(v)$ of v are defined by:

$$\delta(v) = 2r(v) = \max_{1 \leq i < j \leq n} v_{ij} . \tag{4}$$

We consider the following mapping:

$$\Lambda_\alpha^m : \qquad \mathbb{R}^{\binom{n}{2}} \rightarrow \mathbb{R}^{\binom{n+m}{2}}$$

$$
\begin{aligned}
\Lambda_\alpha^m(v)_{ij} &= v_{ij} && \text{for } 1 \leq i < j \leq n \\
&= \alpha && \text{for } 1 \leq i \leq n < j \leq n + m \\
&= 2\alpha && \text{for } n < i < j \leq n + m
\end{aligned}
$$

Then, $\Lambda_\alpha^m(v)$ is a vertex of m_{n+m} if and only if $codim(T_{n+m}(\Lambda_\alpha^m(v))) = 0$ where $T_{n+m}(v)$ is the set of all triangle facets of m_{n+m} containing v.

Case $m = 1$. With $T_{ij,k}$ and P_{ijk} respectively denoting the facet induced by (1) and (2), we have by construction:

$$T_{n+1}(\Lambda_\alpha^1(v)) = T_n(v) \cup T . \tag{5}$$

Where

$$T = \bigcup_{v_{ij}=2\alpha} \{T_{ij,n+1}\} \bigcup_{v_{ij}=0} \{T_{i\,(n+1),j}\} \bigcup_{v_{ij}=0} \{T_{j\,(n+1),i}\} \bigcup_{v_{ij}=2-2\alpha} \{P_{ij\,(n+1)}\}.$$

The equality (5) clearly implies

$$\Lambda_\alpha^1(v) \in m_{n+1} \Longleftrightarrow r(v) \le \alpha \le 1 - r(v) \tag{6}$$

and

$$r(v) < \alpha < 1 - r(v) \Longrightarrow codim(T_{n+1}(\Lambda_\alpha^1(v))) \ge n . \tag{7}$$

This means that a necessary condition for $\Lambda_\alpha^1(v)$ to be a vertex of m_{n+1} is $\alpha = r(v)$ or $\alpha = 1 - r(v)$. Since we have $\Lambda_{1-\alpha}^1(v) = r_{\delta(\{n+1\})}(\Lambda_\alpha^1(v))$, we can consider only the case $\alpha = r(v)$ (we recall that $r_{\delta(\{n+1\})}$ is the switching by the cut $\delta(\{n + 1\})$, see Sect. 2.1.). We call $\Lambda_{r(v)}^1(v)$ the *radial extension* of v and denote it by $\Lambda^1(v)$.

Before stating the conditions on v to lift it to m_{n+1}, we need the following two definitions. Call a graph $G = (N, E)$ *good*, $N = \{1, 2 \ldots, n\}$, if it has a partial subgraph $G' = (N, E')$ with $|E'| = |N|$ which does not admit a non-zero edge-weighting $f: E' \to \mathbb{R}$ with $\sum_{v \in e \in E'} f_e = 0$ for each $v \in N$. The graph $\Gamma(v)$ on N is defined by: s and t adjacent if and only if $v_{st} = \delta(v)$. For example, if $v = \frac{1}{3}d(G)$ for a graph G of diameter 2 (that is $v_{st} = \frac{1}{3}$ if s and t adjacent and $\frac{2}{3}$ otherwise), then $\Gamma(v)$ is the complement of G and $\Lambda^1(v) = \frac{1}{3}d(\nabla G)$ where ∇G is the suspension of G, that is, G plus one vertex adjacent to all vertices of G.

Theorem 10. *For any vertex v of m_n such that $\Gamma(v)$ is good, the radial extension $\Lambda^1(v)$ is a vertex of m_{n+1}.*

Proof. Since $\Gamma(v)$ is good, it has a partial subgraph $\Gamma' = (N, E')$ with $|E'| = n$ which does not admit a non-zero edge-weighting. Clearly, any connected graph with n vertices and less than n edges is either a tree, or an *odd cycled tree* or an *even cycled tree*, where an odd cycled tree, respectively even cycled tree, is a tree plus one edge forming with it an odd, respectively even, cycle. Since a tree has $n - 1$ edges and an even cycled tree admits unwanted edge-weighting, they are both not good and therefore Γ' can only be a *odd cycled forest*, that is, contains for each connected components of Γ its spanning odd cycled tree. Now, since v is a vertex of m_n, $T_n(v)$ contains $\binom{n}{2}$ linearly independent triangle facets which form the set $T_n'(v)$. Then, the $\binom{n}{2} + n = \binom{n+1}{2}$ facets of the set $T_n'(v) \cup_{ij \in E'} T_{ij,n+1}$ are linearly independent facets containing $\Lambda^1(v)$, since if not, Γ' admits a non-zero weighting and therefore Γ is not good. This implies $codim(T_{n+1}(\Lambda^1(v))) = 0$ and completes the proof. $\square$

Case $m \geq 2$. As for the case $m = 1$, we need to consider only the case $\alpha = r(v)$. Similarly, $\Lambda^m_{r(v)}(v)$ is called the *radial m-extension* of v and denoted by $\Lambda^m(v)$. By construction, for $m \geq 2$ we have:

$$T_{n+m}(\Lambda^m(v)) = T_n(v) \cup T \ . \tag{8}$$

Where

$$T = \bigcup_{v_{ij}=\delta(v),\, 1\leq i<j\leq n<k\leq m} \{T_{ij,k}\} \quad \bigcup_{v_{ij}=0,\, 1\leq i<j\leq n<k\leq m} \{T_{ik,j}\} \quad \bigcup_{v_{ij}=0,\, 1\leq i<j\leq n<k\leq m} \{T_{jk,i}\}$$

$$\bigcup_{v_{ij}=1,\, 1\leq i<j\leq n<k\leq m} \{P_{ijk}\} \quad \bigcup_{1\leq k\leq n<i<j\leq n+m} \{T_{ij,k}\} \quad \bigcup_{\delta(v)=1,\, 1\leq k\leq n<i<j\leq n+m} \{P_{ijk}\}$$

$$\bigcup_{m\geq 3,\, n<i<j<k\leq n+m,\, \delta(v)=\frac{2}{3}} \{P_{ijk}\}.$$

The equality (8) implies:

$$\Lambda^2(v) \in m_{n+2} \text{ and, for } m \geq 3,\ \Lambda^m(v) \in m_{n+m} \iff \delta(v) \leq \frac{2}{3} \ . \tag{9}$$

Theorem 11. *For any vertex v of m_n such that $\Gamma(v)$ is good and, for $m \geq 3$, $\delta(v) \leq \frac{2}{3}$, the radial m-extension $\Lambda^m(v)$ is a vertex of m_{n+m}.*

Proof. The proof is similar to the one of Theorem 10. We consider the following set of $\binom{n}{2} + n \cdot m + \binom{m}{2} = \binom{n+m}{2}$ triangle facets containing $\Lambda^m(v)$: $T'_n(v) \cup (\cup_{ij \in E',\, n<k<n+m} T_{ij,k}) \cup_{1\leq n<i<j} T_{ij,k}$. The graph $\Gamma(v)$ being good, they are linearly independent and therefore we have $codim(T_{n+m}(\Lambda^1(v))) = 0$. □

Remark.

1. The condition that v is a vertex of m_n is not necessary. For example, $v = \frac{2}{3}d(K_4)$ is not a vertex of m_4 but $\Lambda^1(v) = \frac{2}{3}d(K_5)$ is a vertex of m_5.

2. We do not know any vertex of m_n with no good graph $\Gamma(v)$ such that $\Lambda^1(v)$ is a vertex of m_{n+1}.

3. Among the 13 representatives given in Table 1, for $i = 2,3,4,5,8,9$ the vertices v_i are both good and satisfy $\delta(v) \leq \frac{2}{3}$. We have $v_2 = \frac{2}{3}d(K_7)$, $v_7 = \frac{1}{3}d(K_7 - C_{2,3,4} - C_{5,6,7})$, $v_8 = \frac{2}{5}d(K_7 - C_7)$, $v_9 = \frac{1}{3}d(K_7 - C_7 - P_{1,3})$ and $v_{10} = \frac{1}{3}d(K_7 - C_{2,3,4} - C_{5,6,7} - P_{4,5})$ where C_s and P_s respectively denotes the cycle and the path on the subset $s \subset \{1,2,\ldots,7\}$, C_7 being the cycle on 7 nodes.

4. For $n \geq 5$, v a vertex of m_n and $\Gamma(v) = \bar{T}$ for a tree T which is not a star, LAURENT [23] proved that $\Lambda^1(v)$ is a vertex of m_{n+1}.

5. With G an almost complete t-partite graph, AVIS [2] proved that $\frac{1}{3}d(G)$ is a vertex of m_n, Theorem 11 implies that $\Lambda^1(\frac{1}{3}d(G))$ and $\Lambda^2(\frac{1}{3}d(G))$ are vertices of, respectively, m_{n+1} and m_{n+2} as well.

Proposition 12. *For G a complete t-partite graph on 8 nodes, $v = \frac{1}{3}d(G)$ is a vertex of m_8 only for $G = K_{4,3,1}$ and $K_{3,3,2}$. The point $v = \frac{1}{3}d(G_e)$ is also a vertex of m_8 for $G_e = K_{3,3,1,1} - e$, $K_{4,2,2} - e$ and $K_{6,1,1} - e$ where e is an edge of, respectively, the subgraph $K_{3,3}$, $K_{4,2}$ and $K_{1,1}$.*

Proof. Theorem 11 gives that $v = \frac{1}{3}d(G)$ is a vertex of m_8 for $G = K_{4,3,1}, K_{3,3,2}$ and $K_{3,3,1,1} - e$. To check if the others complete t-partite graphs induce a vertex of m_8, we built the set $T(v)$ of triangle facets containing the point $v = \frac{1}{3}d(G)$ and then check by computer if they intersect in a vertex. Considering some subsets of $T(v)$, we found that the graphs $K_{4,2,2} - e$ and $K_{6,1,1} - e$ induce a vertex of m_8.

4 Computational Aspects

All facets of the metric polytopes being equivalent under permutations and switching, it is enough to compute all the vertices belonging to one facet. In [21] GRISHUKHIN used this technique to compute the 41 orbits of extreme rays under permutations of the metric cone on 7 nodes. This *vertex enumeration* problem was solved using the double description method *cdd* implemented by FUKUDA [20]. The algorithm first constructs a simplex starting with a non-degenerate subset of $d + 1$ inequalities where d is the dimension, then at each step one inequality is inserted. The efficiency of this algorithm highly depends on the order in which the inequalities are inserted. It is observed that the results seem to be good when the size of the intermediate polytope produced at each step stay as small as possible. For this important ordering issues we refer to AVIS, BREMMER AND SEIDEL [4] where, in particular, worst case behavior polyhedra are constructed.

To obtain the 275 840 vertices of the 21-dimensional polytope m_7 we used the following ordering. The 140 facets were inserted such that $F_1 - F_4$, $F_5 - F_8, \ldots, F_{137} - F_{140}$ form the 35 maximal cocliques of the skeleton of m_7^*, that is, by set of 4 facets with the same support. Then to order those cocliques, we consider the following Hausdorff distance between cocliques of facets. With C and C' two cocliques, we have $d(C, C') = \max d(F, G)$ where F, respectively G, is a facet of C, respectively C' and $d(F, G) = 0$ if $codim(F \cap G) = 2$ and 1 otherwise. The cocliques are then ordered by the maximal cocliques (of cocliques) of the graph which nodes are the cocliques of facets and edges given by the previous Hausdorff distance. The same operation being repeated for cocliques of cocliques of facets and so on.

This ordering gave us much better results that the classical *lexico-graphic*, *min-cut off* and *max-cut off* ordering which respectively selects a facet which cuts off the minimum, respectively maximum, number of vertices of the intermediate polytope, see [20]. This ordering by maximal cocliques of the dual skeleton gave also excellent results for the computation of the Solitaire cone and its relatives, see [5]. In all those cases, including the metric polytope, the maximal size of the intermediate polyhedra was less than twice the size of the final one.

Computation of Table 2. For each representative vertex v_i we computed the cone C_i generated by the set $T(v_i)$ of all triangle facets containing v_i. Clearly, to each extreme ray of this cone pointed on v_i corresponds a neighbour of v_i, in other words, the size of C_i equals the adjacency A_{v_i} of v_i in m_7. Then, by a tedious one by one checking of all the extreme rays of C_i, we listed all rays pointing to a cut. Finally, using the relation $|O_i| \cdot a_{ij} = |O_j| \cdot a_{ji}$ where $|O_i|$ and a_{ij} respectively denotes the size of the orbit O_i and the number of vertices of O_j adjacent to v_i, we filled Table 2. For example, the 30 facets containing v_6 form the cone C_6 which have 96 extreme rays, that is, $A_{v_6} = 96$. Out of those 96 rays, exactly 24 point to a cut. Then, $64 \times a_{1,6} = 20160 \times 24$ implies $a_{1,6} = 7560$.

Remark. Clearly we have $a_{1,1} = 2^{n-1}-1$; the values $a_{2,1} = 2^{n-1}-n-1$ and $a_{3,1} = 2^{n-1}-3n+2$ were given in [13]. So we have $a_{i,1} = 63, 56, 45, 34, 28, 24, 16, 15, 14, 11, 11, 11, 8$ for $i = 1, 2, \ldots 13$. The complete list of cuts adjacent to v_i for $i = 4, \ldots, 13$ is:

- v_4 adjacent to $\delta(S)$ for $S = \{i, j\}$ with $3 \le i < j \le 5$ and for $S = \{i, j, k\}$ with $\{i, j, k\} \cap \{3, 4, 5\} \ne \emptyset$,
- $v_5 \sim \delta(S)$ for $S = \{i, j\}$ with $2 \le i < j \le 5$, $S = \{1, i, j\}$ with $2 \le i < j \le 5$ and for $S = \{i, j, k\}$ with $2 \le i < j < k \le 7$ and $j \ne 6$.
- $v_6 \sim \delta(S)$ for $S = \emptyset, \{1\}, \{4\}, \{6\}, \{1, 2\}, \{1, 5\}, \{1, 7\}, \{2, 6\}, \{3, 4\}, \{4, 7\}, \{5, 6\}, \{6, 7\}, \{1, 2, 3\}, \{1, 2, 7\}, \{1, 3, 5\}, \{1, 4, 7\}, \{1, 5, 7\}, \{2, 3, 4\}, \{2, 3, 6\}, \{2, 6, 7\}, \{3, 4, 5\}, \{3, 5, 6\}, \{4, 6, 7\}, \{5, 6, 7\}$,
- $v_7 \sim \delta(S)$ for $S = \emptyset$, $S = \{i\}$ with $i \ne 1$ and for $S = \{i, j\}$ with $i = 2, 3, 4$ and $j = 5, 6, 7$,
- $v_8 \sim \delta(S)$ for $S = \emptyset, \{1, 3\}, \{1, 4\}, \{1, 5\}, \{2, 5\}, \{2, 6\}, \{3, 6\}, \{4, 7\}, \{1, 3, 5\}, \{1, 3, 6\}, \{1, 4, 6\}, \{2, 4, 6\}, \{2, 4, 7\}, \{2, 5, 7\}, \{3, 5, 7\}$,
- $v_9 \sim \delta(S)$ for $S = \emptyset, \{1\}, \{3\}, \{1, 4\}, \{1, 5\}, \{3, 6\}, \{3, 7\}, \{1, 3, 5\}, \{1, 3, 6\}, \{1, 4, 6\}, \{2, 4, 6\}, \{2, 4, 7\}, \{2, 5, 7\}, \{3, 5, 7\}$,
- $v_{10} \sim \delta(S)$ for $S = \emptyset, \{4\}, \{5\}, \{2, 5\}, \{2, 6\}, \{2, 7\}, \{3, 5\}, \{3, 6\}, \{3, 7\}, \{4, 6\}, \{4, 7\}$,
- $v_{11} \sim \delta(S)$ for $S = \emptyset, \{1\}, \{3\}, \{1, 2\}, \{1, 6\}, \{3, 4\}, \{4, 5\}, \{2, 3, 7\}, \{2, 5, 7\}, \{3, 6, 7\}, \{5, 6, 7\}$,
- $v_{12} \sim \delta(S)$ for $S = \{3\}, \{5\}, \{1, 3\}, \{4, 5\}, \{4, 7\}, \{5, 6\}, \{1, 3, 4\}, \{1, 4, 7\}, \{1, 5, 6\}, \{1, 6, 7\}, \{2, 3, 5\}$,
- $v_{13} \sim \delta(S)$ for $S = \emptyset, \{5\}, \{6\}, \{7\}, \{4, 7\}, \{1, 2, 7\}, \{4, 5, 7\}, \{4, 6, 7\}$.

Acknowledgments The authors thank DIMITRII PASECHNIK who helped to complete Table 1.

References

1. Assouad P. and Deza M.: Metric subspaces of L^1. Publications mathématiques d'Orsay **3** (1982)
2. Avis D.: On the extreme rays of the metric cone. Canadian Journal of Mathematics **XXXII 1** (1980) 126-144
3. Avis D.: In H. Imai ed. RIMS Kokyuroku A C Implementation of the Reverse Search Vertex Enumeration Algorithm. **872** (1994)
4. Avis D., Bremmer D. and Seidel R.: How ood are convex hull algorithms. Computational Geometry: Theory and Applications (to appear)
5. Avis D. and Deza A.: Solitaire Cones. (in preparation)
6. Balinski M.: On the graph structure of convex polyhedra in n-space. Pacific Journal of Mathematics **11** (1961) 431-434
7. Barahona F. and Mahjoub R.: On the cut polytope. Mathematical Programming **36** (1986) 157-173
8. Bayer M. and Lee C.: Combinatorial aspects of convex polytopes. In P. Gruber and J. Wills eds. Handbook on Convex Geometry North Holland (1994) 485-534
9. Brouwer A., Cohen A. and Neumaier A.: Distance-Regular Graphs. Springer-Verlag, Berlin (1989)
10. Christof T. and Reinelt G.: Combinatorial optimization and small polytopes. To appear in Spanish Statistical and Operations Research Society **3** (1996)
11. Christof T. and Reinelt G.: Computing linear descriptions of combinatorial polytopes. (in preparation)
12. Deza A.: Metric polyhedra combinatorial structure and optimization. (in preparation)
13. Deza A. and Deza M.: The ridge graph of the metric polytope and some relatives. In T. Bisztriczky, P. McMullen, R. Schneider and A. Ivic Weiss eds. Polytopes: Abstract, Convex and Computational (1994) 359-372
14. Deza A. and Deza M.: The combinatorial structure of small cut and metric polytopes. In T. H. Ku ed. Combinatorics and Graph Theory, World Scientific Singapore (1995) 70-88
15. Deza M., Grishukhin V. and Laurent M.: The symmetries of the cut polytope and of some relatives. In P. Gritzmann and P Sturmfels eds. Applied Geometry and Discrete Mathematics, the "Victor Klee Festschrift" DIMACS Series in Discrete Mathematics and Theoretical Computer Science **4** (1991) 205-220
16. Deza M. and Laurent M.: Facets for the cut cone I. Mathematical Programming **56 (2)** (1992) 121-160
17. Deza M. and Laurent M.: Applications of cut polyhedra. Journal of Computational and Applied Mathematics **55** (1994) 121-160 and 217-247
18. Deza M. and Laurent M.: New results on facets of the cut cone. R.C. Bose memorial issue of Journal of Combinatorics, Information and System Sciences **17 (1-2)** (1992) 19-38
19. Deza M., Laurent M. and Poljak S.: The cut cone III: on the role of triangle facets. Graphs and Combinatorics **8** (1992) 125-142
20. Fukuda K.: cdd reference manual, version 0.56. ETH Zentrum, Zürich, Switzerland (1995)
21. Grishukhin V. P.: Computing extreme rays of the metric cone for seven points. European Journal of Combinatorics **13** (1992) 153-165
22. Iri M.: On an extension of maximum-flow minimum-cut theorem to multicommodity flows. Journal of the Operational Society of Japan **13** (1970-1971) 129-135

23. Laurent M.: Graphic vertices of the metric polytope. Discrete Mathematics **145** (1995) (to appear)

24. Laurent M. and Poljak S.: The metric polytope. In E. Balas, G. Cornuejols and R. Kannan eds. Integer Programming and Combinatorial Optimization Carnegie Mellon University, GSIA, Pittsburgh (1992) 274-28⁻

25. Murty K. G. and Chung S. J.: Segments in enumerating faces. Mathematical Programming **70** (1995) 27-45

26. Onaga K. and Kakusho O.: On feasibility conditions of multicommodity flows in networks. IEEE Trans. Circuit Theory **18** (1971) 425-429

27. Padberg M.: The boolean quadric polytope: some characteristics, facets and relatives. Mathematical Programming **45** (1989) 139-172

28. Plesník J.: Critical graphs of given diameter. Acta Math. Univ. Comenian **30** (1975) 71-93

29. Poljak S. and Tuza Z.: Maximum Cuts and Large Bipartite Subgraphs. In W. Cook, L. Lovasz and P. D. Seymour eds. DIMACS **20** (1995) 181-244

30. Trubin V.: On a method of solution of integer linear problems of a special kind. Soviet Mathematics Doklady **10** (1969) 1544-1546

31. Ziegler G. M.: Lectures on Polytopes. Graduate Texts in Mathematics **152** Springer-Verlag, New York, Berlin, Heidelberg (1995)

Improving Branch and Bound
for Jobshop Scheduling
with Constraint Propagation

Yves Caseau

Bouygues - Direction Scientifique

1 avenue E. Freyssinet

78 061 St Quentin en Yvelines, France

caseau@dmi.ens.fr

François Laburthe

Ecole Normale Supérieure

D.M.I.

45, rue d'Ulm, 75005 PARIS, France

laburthe@dmi.ens.fr

Abstract. Task intervals were defined in [CL94] for disjunctive scheduling so that, in a scheduling problem, one could derive much information by focusing on some key subsets of tasks. The advantage of this approach was to shorten the size of search trees for branch&bound algorithms because more propagation was performed at each node.

In this paper, we refine the propagation scheme and describe in detail the branch&bound algorithm with its heuristics and we compare constraint programming to integer programming. This algorithm is tested on the standard benchmarks from Muth & Thompson, Lawrence, Adams et al, Applegate & Cook and Nakano & Yamada. The achievements are the following:

• Window reduction by propagation : for 23 of the 40 problems of Lawrence, the proof of optimality is found with no search, by sole propagation; for typically hard 10×10 problems, the search tree has less than a thousand nodes; hard problems with up to 400 tasks can be solved to optimality and among these, the open problem LA21 is solved within a day.

• Lower bounds very quick to compute and which outperform by far lower bounds given by cutting planes. The lower bound to the open 20×20 problem YAM1 is improved from 812 to 826

keywords: Jobshop scheduling, branch and bound, heuristics, propagation, constraints

1. Introduction

Disjunctive scheduling problems are combinatorial problems defined as follows : a set of uninterruptible tasks with fixed durations have to be performed on a set of machines. The problem is constrained by precedence relation between tasks. Moreover, the problem is said to be disjunctive when a resource can handle only one task at a time (as opposed to cumulative scheduling problems). The problem is to order the tasks on the different machines so as to minimize the total makespan of the schedule. These problems have been extensively studied in the past twenty years and many algorithmic approaches have been proposed, including branch & bound ([CP 89], [AC 91], [CP 94]), mixed integer programming with cutting planes ([AC 91]), simulated annealing ([VLA92]), tabu search ([Ta 89], [DT 93]), genetic algorithm ([NY 92], [DP 95]). In this paper, we describe a branch and bound algorithm which requires very small search trees to give good lower bounds, to find optimal solutions and to prove their optimality.

The paper is organized as follows : Section 2 defines scheduling problems and recalls how it can be modelled with time windows and as a mixed integer program, Section 3 explains the difference between propagation rules and cutting planes, Section 4 describes in details our model with task intervals and the associated propagation scheme, Section 5 shows computational results and compares them with those of Applegate and Cook given in [AC91].

2. Disjunctive scheduling

2.1. Jobshop scheduling

A scheduling problem is defined by a set of tasks T and a set of resources R. Tasks are constrained by precedence relationships, which bind some tasks to wait for other ones to complete before they can start. Tasks are not interruptible (non-preemptive scheduling) and mutually exclusive: a resource can perform only on task at a time (disjunctive versus cumulative scheduling). The goal is to find a schedule that performs all tasks in the minimum amount of time.

Formally, to each task t, a non-negative duration $d(t)$ and a resource $use(t)$ are associated. For precedence relations, $precede(t_1, t_2)$ denotes that t_2 cannot be performed before t_1 is completed. The problem is then to find a set of starting times $\{time(t)\}$, that minimizes the total makespan of the schedule defined as $Makespan := \max\{time(t) + d(t)\}$ under the following constraints:

$$\forall t_1, t_2 \in T, \quad precede(t_1, t_2) \Rightarrow time(t_2) \geq time(t_1) + d(t_1)$$

$$\forall t_1, t_2 \in T, \quad use(t_1) = use(t_2) \Rightarrow time(t_2) \geq time(t_1) + d(t_1) \vee$$
$$time(t_1) \geq time(t_2) + d(t_2)$$

In the general case of disjunctive scheduling, precedence relationships can link a task to several other ones. Job-shop scheduling is a special case where the tasks are grouped into jobs $j^1, ..., j^n$. A job j^i is a sequence of tasks $j^i_1, ..., j^i_m$ that must be performed in this order, i.e. for all $k \in \{1, ..., m-1\}$, one has $precede(j^i_k, j^i_{k+1})$. Such problems are called $n \times m$ problems, where n is the number of jobs and m the number of resources. The precedence network is thus very simple: it consists of n "chains". The simplification does not come from the matrix structure (one could always add empty tasks to a scheduling problem) but rather from the fact that precedence is a functional relation. It is also assumed that each task in a job needs a different machine. For a task j^i_k, the head will be defined as the sum of the durations of all its predecessors on its job and similarly the tail as the sum of the durations of all its successors on its job, e.g.:

$$head(j^i_k) = \sum_{l=1}^{k-1} d(j^i_l) \quad \text{and} \quad tail(j^i_k) = \sum_{l=k+1}^{m} d(j^i_l).$$

Although general disjunctive scheduling problems are often more appropriate for modelling real-life situations, little work concerning them has been done (they have been studied more by the Artificial Intelligence community than by Operations

Researchers and most of the published work concerns small instances - like a famous bridge construction problem with 42 tasks [VH89]-)

The interest of n × m scheduling problems is the attention they have received in the last 30 years. The most famous instance is a 10 × 10 problem of Fisher & Thompson [MT63] that was left unsolved until 1989 when it was solved by Carlier & Pinson [CP89]. Classical benchmarks include problems randomly generated by Adams, Balas & Zawak in 1988 [ABZ88], Applegate & Cook in 1991 [AC91] and by Lawrence in 1984 [La84]. Out of the 40 problems of Lawrence, one is still unsolved (a 20 × 10 referred to as LA29). The size of these benchmarks ranges from 10 × 5 to 30 × 10.

2.2 Branch and Bound with time windows

Branch and bound algorithms have, however, undergone much study, and the method effectively used in [CP89] to solve MT10 is a branch & bound scheme called "edge-finding". Since a schedule is a set of orderings of tasks on the machines, a natural way to compute them step after step is to order a pair of tasks that share the same resource at each node of the search tree (which corresponds to getting rid of a disjunction in the constraint formulation). There are many variations depending on which pair to pick, how to exploit the disjunctive constraint before the pair is actually ordered, etc., but the general strategy is almost always to order pairs of tasks [AC91].

The domain associated with $time(t_i)$ is represented as an interval : to each task t_i, a window $\left[\underline{t_i}, \overline{t_i} - d(t_i)\right]$ is associated, where $\underline{t_i}$ is the minimal starting date and $\overline{t_i}$ is the maximal completion date (thus the starting date $time(t_i)$ must be between $\underline{t_i}$ and $\overline{t_i} - d(t_i)$). During the search, a partial ordering ($<<$) of tasks is built, with the following meaning :

$$t_1 << t_2 \Leftrightarrow time(t_1) + d(t_1) \leq time(t_2)$$

In order to prune efficiently the search space, one needs to be able to propagate the decisions taken at each node of the search tree. Thus, whenever an ordering is selected, say $t_1 << t_2$, the bounds of the domains can be updated as follows: $\underline{t_2} \geq \underline{t_1} + d(t_1)$ and $\overline{t_1} \leq \overline{t_2} - d(t_2)$. With this model, inconsistency can be detected when one has $\overline{t} - \underline{t} < d(t)$ for some task t (t can no longer fit in its window).

2.3 Mixed integer programming

As reported in [AC91], the problem can be described as a mixed integer program. In this case, the variables are $\{time(t), for\ t \in T\}$, *Makespan* and extra 0-1 variables $\{Y_{t,t'}, for\ t,t' \in T\ such\ that\ use(t) = use(t')\}$; these variables are used to represent the disjunctive constraints: $Y_{t,t'} = 1$ if t is scheduled before t' on their common resource and $Y_{t,t'} = 0$ otherwise. The problem is then a linear problem, where *Makespan* is the objective to minimize, under the following constraints :

- precedence constraints
- constraints defining the makespan : $Makespan \geq time(t) + d(t)$ for all tasks
- constraints linking the 0-1 variables : $Y_{t,t'} = 1 - Y_{t',t}$

- constraints representing the disjunctions : $time(t') \geq time(t) + d(t) - K.Y_{t',t}$ for all pairs of tasks (t,t') sharing the same resource. (K is a large constant)
- $Y_{t,t'} \in \{0,1\}$

To solve the problem, the last set of constraints is relaxed into $0 \leq Y_{t,t'} \leq 1$ and the linear program is solved (for which only integral solutions correspond to actual schedules). Inequalities that are valid for any integral solution but not for the current best fractional solution are added one by one. In the end, if the solution is still fractional, branching is performed on the values of $Y_{t,t'}$. In this process of making the solution more and more integral, the makespan increases, so, at any time, it provides a lower bound. Another approach consists in using the solutions to the fractional problem for Lagrangean relaxation, as described in [VDV91].

Several classes of such cutting inequalities are known, such as the "two job cuts" [Ba 85], the "late job cuts" [DW90] or "half cuts" [AC91] (for more details, see [DW90] or [AC91]). In order to avoid too large a branching phase, many such smart cuts must be added to the system. However, these cutting planes soon become complex, and moreover, it is not always easy to find which of the available cuts separates the current solution from all integral solution (as mentioned in [AC91], finding a violated "clique cut" requires knowing the solution to a linear subprogram).

3. A geometric comparison between constraint propagation and cutting planes

A constraint propagation system is a programming environment where the programmer can specify the behaviour of an algorithm in two different ways : with traditional imperative code and with logic formulaes called propagation rules. More precisely, a propagation rule has the following syntax :

cond ⇒ exp

where cond is a logical assertion and exp is any expression. The meaning of the propagation rule is that whenever cond becomes true, exp is evaluated (this corresponds to "propagation" of information : once a situation has been detected, a decision is taken). The condition of the rule may be almost any first-order logical formula (with universal and existential quantifiers) over the problem variables, the conclusion may be anything (usually updating the problem variables).

What the system does is to formally differentiate the logical assertion cond. From this, an algebraic engine produces guards that watch over updates to certain variables, in order to detect when cond may become true. These guards can be produced very efficiently in order to detect when cond becomes true with as few computations as possible.

Geometrically, propagation rules are complex objects : a rule associates a hyperplane H to a region of space P, with the following meaning :

$x \in P \Rightarrow H.x \geq 0$ (as soon as x is in P, H can be used as a cutting plane). If the logical assertion remain simple, P can be thought of as a convex body, say a polytope.

This geometrical object is rather different from a cutting plane. With an integer program :

max $c'x$, under the constraints
$Ax \leq b$ and $x \in Z$,

the algorithm starts with a polytope defined by $Ax \leq b$ (except for the case of a totally unimodular matrix A, this polytope may have fractional vertices) and finds cutting planes that separate integer and non-integer solutions. Step after step, the shape of the polytope is refined with cutting planes and the polytope is eventually reduced to the convex hull of its integral interior points.

These two cutting techniques are extremely different: in both cases, the search starts with a region of space, but cutting planes can globally refine this region to make it a more integral polytope (up to the point when the region is the convex hull of all schedules) whereas propagation rules are local tools to guide the search of the starting region (it fragments the initial region into smaller ones where more precise information can be drawn). In fact, if one thinks about constraints in their negative form $\neg(x \in P \wedge H.x < 0)$, these reduction rules rather correspond to forbidding certain areas in the initial polytope (making holes). This difference of approach (conditional cuts versus global cuts) accounts for the complexity of advanced cutting planes in integer programming : it is indeed very hard to say many things about all schedules, whereas propagation rules can have very precise conclusions (sharp planes) because they distinguish disconnected subregions of the starting area.

In fact, the main difference between the two approaches is that linear and integer programming model combinatorial problems with convex bodies whereas propagation rules do not. As will be shown in the next section, propagation rules require almost no encoding of the problem. Moreover, scheduling is not convex by essence, what stands half-way between two schedules is not a schedule (it is a major issue of genetic algorithms for scheduling to find a reproduction procedure that takes two schedules and produces a child schedule that ressembles both parents [DP 95]). In order to turn around this drawback and to force the disjunctive problem to become convex, integer programming needs the addition of many new variables (like the $Y_{t,t'}$) : the dimension of the objects grows up to the point where it can be convex, while the projection on the space defined by the initial variables remains frankly unconvex. For an $m \times n$ scheduling problem, the initial dimension is $mn + 1$ to which the integer programming approach adds $m\dfrac{n(n-1)}{2}$ more 0-1 variables.

Finally, it should be stressed that writing logical conditions with group operations and quantifiers allows to describe very precise situations with natural mathematical notations, when the associated inequation with 0-1 variables would be very complex. Moreover, quantifiers allow for a very dense description of the cuts, by permitting to group similar equations on different variables into one single quantified assertion.

4. Task intervals and their application to scheduling

This part describes a redundant model, called *task intervals*, -introduced in [CL94]- that gives better insight about the feasibility of the scheduling problem than solely information from time windows. The idea is to focus not only on tasks but on sets of tasks sharing the same resource in order to reflect the disjunctive sharing constraints. This additional model has several interests: first, it allows us to propagate more information from an ordering decision between two tasks and so reduces the size of search trees, second, it detects inconsistencies early and third, it is particularly well-suited for a branching scheme using edge-finding.

4.1 Intervals as Sets of Tasks

A trick that is often used with algorithms that shrink domains is to add redundant constraints to improve pruning. The most obvious redundant constraint that is used by all constraint-based schedulers is the resource interval constraint. If we denote by $T(r)$ the set of tasks that use the resource r, $\underline{T(r)} = \min\{\underline{t}, t \in T(r)\}$ the earliest starting time of all tasks in $T(r)$ and $\overline{T(r)} = \max\{\overline{t}, t \in T(r)\}$ the latest completion time of all tasks in $T(r)$, the task constraint

$$\overline{t} - \underline{t} \geq d(t)$$

also applies to $T(r)$, with $d(T(r)) = \sum\{d(t), t \in T(r)\}$:

$$\overline{T(r)} - \underline{T(r)} \geq d(T(r)) \qquad (1)$$

If the time window $\overline{T(r)} - \underline{T(r)}$ is not sufficient for all tasks in $T(r)$, this additional constraint detects an inconsistency, whereas without this constraint, some ordering on r would have been necessary before an inconsistency for a $t \in T(r)$ could have been detected. The novel idea of task intervals is to apply this constraint (1) to all subsets of tasks that use a common resource. For an n × m problem, this generates m × $(2^n -1)$ constraints. Fortunately, checking them on all subsets of tasks is equivalent to checking it only on task intervals, which are at most m × n^2.

Definition : If t_1 and t_2 are two tasks (possibly the same) satisfying :

$$\text{use}(t_1) = \text{use}(t_2) = m \text{ (the tasks share the same resource)}$$

$$\underline{t_1} \leq \underline{t_2} \text{ and } \overline{t_1} \leq \overline{t_2}$$

then the *task interval* $[t_1, t_2]$ is the set of tasks t such that use(t)=m, $\underline{t_1} \leq \underline{t}$ and $\overline{t} \leq \overline{t_2}$. By convention, if t_1 and t_2 do not verify the order condition $(\underline{t_1} \leq \underline{t_2} \wedge \overline{t_1} \leq \overline{t_2})$, then $[t_1, t_2]$ will denote the empty set. Task intervals that are not empty are called *active*.

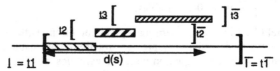

Figure 1: Task Intervals :

$I = \left[t_1, t_1 \right]$ *represents the set* $\{t_1, t_2, t_3\}$

Note that to any set of tasks $S = \{t_1,...,t_i\}$, we can associate t_p and t_q such that $\underline{t_p} = \underline{S}$ and $\overline{t_q} = \overline{S}$. Since $d([t_p,t_q]) \geq d(S)$ by construction, the equation (1) for $[t_p,t_q]$ subsumes that for S. On all following figures, the time will be represented on a horizontal axis, each task t on a horizontal line with two brackets to denote its time window[1] $\left[\underline{t}, \overline{t} \right]$ and a box of length $d(t)$ between both brackets. Figure 1 shows an example of a task interval $I = \left[t_1, t_1 \right]$. The corresponding window is represented by larger brackets.

---

[1] To avoid confusion, we will reserve the word "interval" for task intervals and the expression "time window" for the laps of time during which the task may be performed.

By construction, there are at most $m \times n^2$ task intervals to consider, which is only n times more than the number of tasks. It is possible to further reduce the number of task intervals that must be considered if we notice that the same time window may be covered by several pairs of tasks (when two tasks have a bound of their time windows in common). We can use a total ordering on tasks to select a unique task interval to represent each time window. However, the maintenance of such "critical" task intervals has shown to be too computationally expensive to gain any benefits from a reduced number of task intervals. In the rest of the paper, we shall call the *slack* of an interval I, written $\Delta(I)$, the value $\bar{I} - \underline{I} - d(I)$.

Implementation note: Each task interval can be seen as the representation of a dynamic redundant constraint. In addition to the two bounds t_1 and t_2 that do not change, we need to store for each interval $[t_1, t_2]$ its set extension (for propagation [Section 4.2] and maintenance [Section 4.3]). To see if an interval is active, we simply check if its extension is not empty. The set extension is a dynamic value that will change throughout the search and that needs to be backtracked. To avoid useless memory allocation, it is convenient to code it with a bit vector mechanism. Task intervals are stored in a matrix with cross-access, so that we have direct access to the set of intervals with t as a left bound, denoted $[t, \_]$ and to the set of intervals with t as a right bound, denoted $[\_, t]$.

4.2 Reduction with Intervals

Our approach for scheduling differs from traditional constraint programs (as described in [VH89] in the sense that we have added a redundant structure (task intervals) that needs to be modified and backtracked throughout propagation and search: our algorithm can thus be described as *constraints* (the reduction rules below) and *control* (triggering of the rules, maintenance of task intervals and branching heuristics). For a thorough comparison with other constraint approaches, see [CL94].

In addition to the constraints from the equation (1), we use three sets of reduction rules, corresponding respectively to ordering (a rule between two tasks), edge finding (a rule between an interval and one of its tasks) and exclusion (a rule between an interval and a task not in it).

Ordering rules use the precedence relation among tasks and a dynamic ordering relation $\ll$ that will be built during the search (cf. Section 4.5). The rules are as follows.

$$\forall t_1, t_2, \left(precede(t_1, t_2) \wedge \underline{t_2} < \underline{t_1} + d(t_1) \right) \Rightarrow \underline{t_2} := \underline{t_1} + d(t_1)$$

$$\forall t_1, t_2, \left(t_1 \ll t_2 \wedge \underline{t_2} < \underline{t_1} + d(t_1) \right) \Rightarrow \underline{t_2} := \underline{t_1} + d(t_1)$$

Symmetrical rules apply to upper bounds (the relations *precede* and $\ll$ have the same meaning: the only difference is that *precede* is known before hand as data of the problem and $\ll$ is built dynamically and backtracked).

The second set of rules, *edge finding*, determines if a task can be the first or the last to be performed in a given task interval. For a task t belonging to a task interval S, the value of $\bar{S} - \underline{t} - d(S)$ is considered. If it is strictly negative, we know that t

cannot be first in S, therefore t cannot start before the earliest ending time over all other tasks in S (cf. Figure 2). Thus the rule that we apply is the following :

$$\forall t, S, (t \in S \wedge \overline{S} - \underline{t} - d(S) < 0) \Rightarrow \underline{t} \geq \min\left\{\underline{t_i} + d(t_i), t_i \in S - \{t\}\right\}$$

Here also, a symmetrical rule applies to see if a task cannot be the last member of a given interval. Finding the first and last members of task intervals is known as "edge finding" [CP89] [AC91] and is a proven way to improve the search. We will complete these two rules in Section 4.5 with a search strategy that also focuses on edge finding.

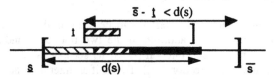

Figure 2: Edge Finding

The last set of rules, *exclusion*, tries to order tasks and intervals. More precisely, we check to see if a task can be performed before an interval to which it does not belong (but which uses the same resource). This is done by computing the value of $\overline{S} - \underline{t} - d(S) - d(t)$ (same as previously but t no longer belongs to S). If it is negative, then t cannot be performed before S, thus t must be performed after some tasks in S. In all cases, t must be performed after the first task in S, but in two special cases, it can be deduced that t must be performed after all tasks in S. Either because the interval S is too tight to allow t to be performed between two tasks of S, or because $\underline{t}$ is already greater than the latest start of all possible latest task in S. The functions packed? and is_after? respectively detect these two situations :

$$\text{packed?}(S,t) := \left(\overline{S} - \underline{S} < d(S) + d(t)\right)$$

$$\text{is\_after?}(t,S) := \left(\underline{t} + d(t) > \max\left\{\overline{t_i} - d(t_i), t_i \in S\right\}\right)$$

Figure 3: Exclusion - case packed
(packed?(S,t) = (d(t) > Δ))

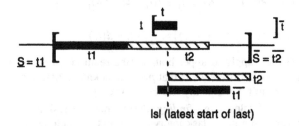

lsl (latest start of last)

Figure 4: Exclusion - case where t is after S
(is_after?(S,t) = ($\underline{t}$ + d(t) > lsl))

Finally, we use the following rule (and its symmetrical counterpart):

$$\forall t, S = [t_1, t_2], \ (t \notin S \wedge \bar{S} - \underline{t} - d(S) - d(t) < 0) \Rightarrow$$

if packed?(S,t) or is_after?(t,S)

then $\quad \underline{t} \geq \underline{S} + d(S) \wedge \forall t_i \in S, \bar{t_i} \leq \bar{t} - d(t)$

else $\quad \underline{t} \geq \min\{\underline{t_i} + d(t_i), t_i \in S\}$

Finer tuning of propagation: Some refinements can be made for the edge-finding and the exclusion rules (case unpacked). Indeed, more information can be propagated when we reach the conclusion that a task t is performed after some other task in a task interval S. When we come to this conclusion, the quantity $D = d(S \cup \{t\}) - (\bar{S} - \underline{t})$ is strictly positive. And so, there exists a subset V of tasks of S, which durations account for more than D, such that t is performed after V and before $S - V$. The candidates to be this subset V are $\{V \subseteq S \text{ such that } \underline{V} + d(S \cup \{t\}) \leq \bar{S} \text{ and } d(V) \geq D\}$ and S itself. If S is the only candidate, $\underline{t}$ is increased to $\underline{S} + d(S)$, otherwise, $\underline{t}$ could be increased to a value x defined by :

$$x = \min\{earliest\_end(V) \quad \text{for } V \subseteq S \text{ such that } \underline{V} + d(S \cup \{t\}) \leq \bar{S} \text{ and } d(V) \geq D\}$$

with $earliest\_end(V) = \max\{\underline{W} + d(W) \quad \text{for } W \subseteq V\}$

Computing such a value x exactly is too long (it amounts to a knapsack problem). However, the first lower bound used (i.e. $\min\{\underline{t} + d(t) \text{ for } t \in S\}$) is too poor an estimate of x. We refined it with the value computed by the following procedure:

```
[after first(S:Task_Interval, t:Task, D:integer) : integer
 -> let v := ∞ in
    (for t' in S - {t}
       let d' := d(t'), v' := t' + d' in
          (if (v' < v) (if (d' ≥ D) v := v'
                        else D := D - d')),
     if (v = ∞) v := S + d(S),
     v)]
```

which considers sequentially all tasks t' that may be performed before t and takes the minimum of their earliest ending time either if their duration is larger than D (the set {t'} is a candidate to be performed before t) or if other tasks {t1, ..., tk} with duration less than D have already been examined and if the total duration of {t1, ..., tk,t'} accounts for more than D. This algorithm runs in linear time (all tasks of S are examined only once) and approximates the function $earliest\_end$ by the following estimate :

$estimate(V) = \underline{t'} + d(t') \quad \text{for some } t' \in V$

Note also that the function is_after? can be sharpened to take the dynamic ordering into account. Indeed the computation of the latest start of the last task in S can ignore tasks t' for which the ordering $t' \ll t$ has already been selected (at a node of the search tree).

Triggering of these rules: If all these equations were checked each time some new information is drawn, the system would be terribly slow. So, deciding when to trigger the evaluation of these equations is a key point in the algorithm. There is fine tradeoff between too much triggering which brings redundant checks and is too slow, and not enough triggering which is faster but misses consequences that could have been drawn.

• ordering rules for a task t are triggered upon changes of $\underline{t}$, $\bar{t}$, or $<<$. This corresponds to a classical dynamic reevaluation of the PERT.

• Concerning edge finding, the rule to decide whether t could be scheduled first in S is triggered upon changes of $\underline{t}$, $\bar{S}$ but also $\underline{S}$. Indeed, changes to $\underline{S}$ will not change the triggering condition of the rule, but may change its conclusion, by changing the value of the bound after_first(S,t,D). The rule is not triggered upon changes to the set extension of S, because it prunes too little for its cost in time.

• The exclusion rule that decides whether a task t can be scheduled before all tasks of S or not is triggered upon changes to $\underline{t}$, $\bar{S}$ and to the set extension of S, but also upon changes to $\underline{S}$ (which might change the boolean value of packed?(S,t))

For rapidity purposes, we also preferred to group the triggering upon time window bounds for the two edge finding and the two exclusion rules into a single one. It avoided redundant checks. The code also contains guards to avoid propagating the consequences of a change concerning a task t as soon as one of the consequences is to further reduce the window of t.

Note that unlike many problems solved with propagation rules, these rules may not be confluent. With the simplest version of the exclusion rule -case unpacked- (i.e. without the computations of after_first), these rules are not commutative (since applying them to a task interval I no longer subsumes it for all other subsets of I [BL95]). Hence, it is not clear whether propagation leads to a fix point or not. It remains an open problem to find an appropriate set of reduction rules (subsuming this one) that can be proved to be confluent.

4.3 Interval Maintenance

Taking task intervals into account is a powerful technique that supports focusing very quickly on bottlenecks. However, the real issue is the incremental maintenance of intervals (indeed, recomputing all intervals upon each update on a time window would be much too slow). Resources are interdependent because of the precedence relationship (as displayed in Figure 5, where the arrows symbolize the precedence constraints). While a resource is being scheduled, the changes to the windows of the tasks are propagated to other resources. We need to be able to compute the changes on task intervals very quickly (new active intervals, intervals that are no longer active and changes to the set extensions of the intervals). More precisely, there are two types of events that we need to be reacted to: the increase of a $\underline{t}$ and the decrease of a $\bar{t}$.

The correct algorithm for updating $\underline{t}$ can be derived from the definition of the set extension of a task interval:

$$[t_1, t_2] = \left\{ t, \ \underline{t_1} \le \underline{t} \wedge \bar{t} \le \bar{t_2} \right\}$$

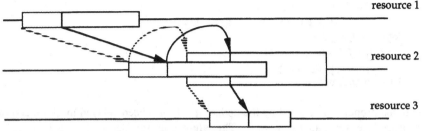

Figure 5 Resource Interdependence

From this definition, we see that, when increasing the values of t from a to b,
- we must deactivate intervals $[t, t_2]$ if $b > t_2$
- we must remove tasks t' from active intervals $[t, t_2]$ when $t' < b$
- we must create new active intervals $[t_1, t]$ if $a < t_1 \le b$ and $\overline{t_1} < t$
- we must add t to active intervals $[t_1, t_2]$ if $a < t_1 \le b$ and $t < \overline{t_2}$

The interesting issues are the order in which we need to perform these operations, and the detection that we are in a state stable enough to propagate the changes and trigger the rules. It turns out that negative changes (removing tasks from set extension) do not need to be propagated, because all the rules we use always apply to subsets of intervals (i.e., if a rule can be applied to an interval, it could also be applied to any subset and would not yield more changes). Thus we perform the two "negative" actions first. Then, we need to swell the other intervals, which may trigger rules (as the window or the set extension of the interval changes). This requires that we have set t to its new value b (but not propagated this change yet). After the negative updates, rules can legally be triggered, because one can be sure that all intervals have a set extension *that can only be smaller than what it should be*. Therefore, since all rules that are used are monotonic with respect to set extension, we know that any conclusion that may be drawn will be valid.

The last action is then to propagate the change to t (triggering the rules which condition depends on the value of t). To avoid duplicate work, we also need to check that t was not subsequently changed to a higher value by the propagation of another rule. Therefore, we are using two invariants H0 and H1 (cf. the following algorithm) to make sure that we stop all propagation work if b is no longer the new value for t.

The algorithm that we use to increase t from a to the new value b is therefore as follows :

```
increase(t:Task, a:integer, b:integer) :              ;; b > a
    for I = [t,t₂] in [t,_]
        if ( a < t₂ ≤ b )∧(t₂ ≠ t)  set(I) := ∅       ;; I is no longer active
        else for t' in I
                if ( a < t' ≤ b ) ∧ ( t' < t₂ )
                    ( I := I - {t'},   d(I) := d(I) - d(t') ),
    t := b,                                            ;; H0 ⟺ (t = b)
    for all t₁ ≠ t such that use(t) = use(t₁)
        if ( a < t₁ ≤ b )                              ;; H1 ⟺ (t₁ ≤ b)
            for I' = [t₁, t₂] in [t₁, _]
```

$$\text{if } (\overline{t} \le \overline{t_2}) \wedge H0 \wedge H1$$
$$(I' := I' \cup \{t'\}, \quad d(I') := d(I') + d(t')),$$
$$\text{for } I = [t_1, t] \text{ in } [\_,t]$$
$$\text{if } (\underline{t_1} \le b) \wedge (\overline{t_1} \le \overline{t}) \wedge H0$$
$$(I := \{...\}, d(I) := ...),$$
$$\text{if } H0 \quad \text{propagate}(\underline{t} = b)$$

So, the propagation caused by the update ($\underline{t} := b$) has been delayed until the end of the procedure (propagate($\underline{t}$ = b)). The algorithm for decreasing $\overline{t}$ is exactly symmetrical. We have tried two different variations of this algorithm. First, as mentioned earlier, we tried to restrict ourselves to "critical intervals", using a total ordering on tasks to eliminate task intervals that represented the same time window (and thus the same set). It turns out that the additional complexity does not pay off. Moreover, the maintenance algorithm is so complex that it becomes very hard to prove. The other idea that we tried is to only maintain the extension of task intervals (represented by a bit vector) and to use m pre-computed duration matrices of size 2^n representing the durations of all possible subsets of tasks. It turns out that the duration is used very heavily during the computation and that caching its value improves performance substantially.

4.4 Comparison with related work

The structure of task intervals together with the reduction rules has two highlights : it is conceptually simple, but gives a very sharp insight of the tightness of the window situation. It provides an elegant unified frame for interpreting many former techniques used by Operational Researchers.

Operationally, the reduction rules presented here are very similar to those presented, in a different terminology, in [CP94]. Carlier and Pinson also do some window reduction, but call it adjusting the heads and tails of the tasks. There are however some real differences due to the fact that they do not maintain an extra structure such as the task intervals : As far as complexity is concerned, the procedure increase which is called every time that one of the bounds of a task has been changed is in $O(n^3)$, whereas theirs is in $O(n \log(n))$, but their triggering is less efficient (since they do not reason about intervals, they have to consider more subsets after each modification to the window bounds of the tasks). As far as the expressiveness is concerned, Carlier and Pinson also include some lookahead in their propagation scheme (trying to replace a time window [a,b] by its left half or its right half and hoping to come to an impossibility in one of the cases). This operation amounts to a one-step breadth exploration of the search tree and explains the relatively smaller number of nodes for their search trees since each node encapsulates a fair amount of search.

The other contribution of task intervals is to give a unified framework that allows the expression complex reduction rules in a simple way. For example, all the sophisticated cutting planes described in [AC91] for semi-definite integer programming are subsumed by the three reduction rules.

4.5 Branching strategies

As we mentioned previously, a classical branching scheme for the job-shop is to order pairs of tasks that share the same resource [AC91]. The search algorithm, therefore, proceeds as follows. It picks a pair of tasks $\{t_1, t_2\}$ and a preferred ordering $t_1 \ll t_2$ The algorithm then explores sequentially the two branches ($t_1 \ll t_2$ and $t_2 \ll t_1$) recursively. The key point is the selection of the pair and of the preferred order. This algorithm produces a feasible schedule within the given makespan. When the makespan is not large enough, the algorithm explores the whole tree without finding any solution. The classical scheme with reduction rules is to iterate the algorithm many times with decreasing makespans to obtain an optimal solution, up to the point when the makespan is one unit too short and the algorithm comes to a dead-end in all branches of the tree, which proves optimality. The search strategy presented here is specially designed for proofs of optimality, therefore we only describe the selection of the pair to order since both branches are visited. This heuristic is well-suited for tight situations and therefore also works correctly to find an optimal (or near-optimal) solution. However, it performs very poorly for finding an initial feasible schedule.

The choice of the task pairs is directly inspired from the edge-finding method described in [AC91], which is itself inspired from the work of Carlier & Pinson [CP89]. This principle consists in considering the set of unscheduled tasks for a given resource, and picking a pair of tasks that could be both first (resp. last) in this set. The choice between first and last is based on cardinality. Our adaptation of this idea is to focus on the most constrained subset of tasks for the resource instead of the set of tasks that are currently unscheduled. This allows faster focusing on bottlenecks and takes advantage of the task intervals that are being carefully maintained.

For all task intervals I, let $\{t_1, \dots, t_p\}$ be the set of tasks that could be scheduled first in I, let $\{t_1', \dots, t_q'\}$ be the set of tasks that could be last and let $NC(I) := \min(p, q)$ be the number of choices associated to I. To each resource r, we associate the most critical task interval $Crit(r)$ as the one minimizing (over all tasks intervals using the considered resource) the quantity $\Delta(S) \times NC(S)$. Minimizing the slack forces to concentrate on bottlenecks and minimizing the number of choices insures that there will be much propagation. Indeed, the faster the first task of an interval is known, the faster the exclusion constraints between it and the other tasks in the interval can be propagated. Over all resources, we select the one (and the associated most critical task interval) that minimizes the quantity

$$ff(r) = \Delta(Crit(r)) \times \Delta(r) \times \min(par, NC(Crit(r))).$$

where par is a fixed parameter, empirically set around 3. This heuristic combines several criteria into a single numerical objective function :

- first, the slack of the most critical interval of the resource forces to concentrate on bottlenecks,
- second, the slack of r, which denotes the smallest slack over all intervals using r, forces the algorithm to schedule the tightest machines first (if most of the ordering has been done on one machine, it may be worth finishing it before considering other even tighter machines),

- the third quantity is taken into account to force the algorithm to take into account the "first-fail principle" (concentrating first on choices with the smallest number of possibilities), but only for choices offering less than *par* possibilities. As mentioned above, driving the search by first-fail makes sense only when few alternatives need to be considered, therefore *par* is given a relatively small value,

- the final quantity is the product of these three, rather than a linear combination. Indeed, scaling problems arise with linear combinations (linear coefficients seem to be quiete sensitive to the size of the problem instance). Taking the product showed to be more robust.

Finally, once the task interval I has been selected, suppose that the set of possibly first tasks, $\{t_1, \ldots, t_p\}$ has been picked (this is the case when $p \leq q$); among this set, there remains to pick two tasks t_a and t_b such that both choices $(t_a \ll t_b)$ and $(t_b \ll t_a)$ will have the maximal impact (we try to reduce the entropy of the scheduling system, in a manner similar to what is described in [CGL93]). The ordering $(t_a \ll t_b)$ is evaluated by predicting the consequent changes on the bounds of certain windows. If Δ is the slack of a window and Δ-δ is the slack after the ordering decision is taken, we want to minimize the resulting slack (Δ-δ) and to maximize the change δ. After many attempts, our best evaluation function (to be minimized) is the following:

$f(\Delta, \delta) =$ if $(\delta = 0)$ M else if $(\Delta < \delta)$ 0 else $(\Delta - \delta)^2 / \Delta$,

where M is the current allowed makespan.

We evaluate the consequences of the ordering $t_a \ll t_b$:

$$t_a \ll t_b \Rightarrow \underline{t_b} := \max(\underline{t_b}, \underline{t_a} + d(t_a)) \quad \wedge \quad \overline{t_a} := \min(\overline{t_a}, \overline{t_b} - d(t_b))$$

$$\delta(t_b) = \max(0, \underline{t_a} + d(t_a) - \underline{t_b}) \quad and \quad \delta(t_a) = \max(0, \overline{t_a} - (\overline{t_b} - d(t_b)))$$

We assess the impact of an ordering by its heaviest consequences :

$$g(t_a \ll t_b) = \min(f(\Delta(t_a), \delta(t_a)), f(\Delta(t_b), \delta(t_b)))$$

We always take $t_a = t_1$ (the left bound of the interval) and select t_b by minimizing the following function :

$$h(t_1, t_b) = \max\big(g(t_1 \ll t_b), \ \min\big(g(t_b \ll t_1), f(\Delta(S), \delta(S))\big)\big)$$

The function h is a maximum over both branches because one wants both possibilities to perform much propagation. Notice that the change to $\Delta(S)$ is taken into account for the branch $(t_b \ll t_1)$. We can now summarize how to select the next pair of tasks that will be ordered.

```
next_pair()
    find r such that ff(r) is minimal,
    let S = Crit(r), S= [t1, t2]
```
$S1 := \{t \mid use(t) = r \wedge t \neq t_1 \wedge not(t_1 \ll t) \wedge \underline{t} \leq \underline{t_1} + \Delta(S)\}$;; could be first

$S2 := \{t \mid use(t) = r \wedge t \neq t_2 \wedge not(t \ll t_2) \wedge \overline{t} \geq \overline{t_2} - \Delta(S)\}$;; could be last

```
    if IS1I ≤ IS2I
```
$\delta(S) := \min(\underline{t}, \ t \in set(S) - \{t_1\}) - \underline{t_1}$
```
                find t in S1 such that h(t1, t) is minimal
                return (t1,t) if g(t1 << t) ≤ g(t << t1) and (t,t1) otherwise
        else
```
$\delta(S) := \overline{t_2} - \max(\overline{t}, \ t \in set(S) - \{t_2\})$
```
                find t in S2 such that h(t, t2) is minimal
                return (t,t2) if g(t << t2) ≤ g(t2 << t) and (t2,t) otherwise
```

5. Computational results

Up to now, we have described a propagation mechanism that can be efficiently used, within a branching scheme, to explore a solution space for a given makespan. However, for large scheduling problems, this exact approach takes too much time and the scheduler is expected to give solutions quickly (even if not optimal) and some guarantees about their quality (by means of lower bounds) within seconds. We have built a full algorithm that addresses both the problem of finding a feasible solution as well as that of detecting unfeasible situations: on the one hand, it finds an approximate solution to start with, makes local changes and repairs on it to quickly decrease the upper bound and finally, when the upper bound is close to the optimum, performs a branching algorithm for decreasing makespans; on the other hand, it can give good lower bounds (to estimate the distance of a solution to the optimal) and perform proofs of optimality. This algorithm is beyond the scope of this paper, but it is fully described in [CL95]. (An interesting fact is that the propagation mechanism takes part in this whole process, not only for the exhaustive search part of the algorithm, but also for local optimization.), we will describe here only our experiments with the branching algorithm.

5.1. Lower bounds

The first experiment was to look for lower bounds. We used two techniques that we compared with the methods described in [AC91]. Several methods are presented in [AC91] : the first one "Preempt" gives instantaneously a lower bound by constructing the optimum of the relaxed preemptive scheduling problem (for details, see [Ca 82]) and the other ones ("Cuts 1", "Cuts 2" and "Cuts 3") are cutting plane heuristics with increasingly complex cuts. Our first bound (reported in the column "Task Intervals") is to determine what maximum allowed total time window would cause a contradiction when the reduction rules are applied. The second method is also a common technique, that consists in computing the minimal schedule of one machine, leaving the propagation rules active. That is to say that only one machine is scheduled, but decisions made on this machine are propagated to the other machines, where contradictions can be raised. This second lower bound (reported in the column "Task Intervals - 1 machine") is more expensive to obtain since some search is involved. We indicated the number of backtracks (b.) and running times (measured on an IBM 3081D for those given by [AC91] and on a Sun Sparc 10 for ours).

A few remarks need to be made. The preemptive bound is at an average distance of more than 13% to the optimum, bounds obtained with cutting planes are at an average distance of around 11%, and our two bounds are respectively within 8% and 2% of the optimum. Taking running times into account, the first bound (almost instantaneous) should be related to the preemptive bound. The distance to optimum is almost divided by a factor 2 for the same running times. The second bound is also obtained within reasonable times (less than 25 seconds for 8/10 problems) and is extremely precise. Sophisticated cutting planes are not competitive in terms of quality of the bound (by a factor 5 or 6) and in terms of running times (becoming unreasonable).

	Preempt [AC 91]	Cuts 1 [AC 91]	Cuts 2 [AC 91]	Cuts 3 [AC 91]	Task Intervals	Task Intervals one machine
MT 10 opt = 930	808 (0,1 s.)	823 (5,23 s.)	824 (305 s.)	827 (7552 s.)	868 (0,3 s.)	915 (25 s - 541b.)
ABZ 5 opt = 1234	1029 (0,1 s.)	1074 (5,61 s.)	1076 (611 s.)	1077 (4971 s.)	1127 (0,3 s.)	1208 (18 s. - 426 b.)
ABZ 6 opt = 943	835 (0,1 s.)	835 (4,87 s.)	837 (335 s.)	840 (5257 s.)	890 (0,3 s.)	936 (3,5 s. - 59 b.)
La 19 opt = 842	709 (0,1 s.)	709 (5,57 s.)	716 (917 s.)	not available	763 (0,3 s.)	813 (7,8 s. - 202 b.)
La 20 opt = 902	807 (0,1 s.)	807 (5,13 s.)	807 (806 s.)	not available	851 (0,3 s.)	874 (1,6 s. - 29 b.)
ORB1 opt = 1059	929 (0,1 s.)	930 (7,16 s.)	931 (358 s.)	not available	975 (0,3 s.)	1045 (100 s. - 2k b.)
ORB 2 opt = 888	766 (0,1 s.)	768 (10 s.)	769 (327 s.)	not available	815 (0,3 s.)	867 (9,3 s. - 189 b.)
ORB 3 opt = 1005	865 (0,1 s.)	869 (5,95 s.)	870 (449 s.)	not available	907 (0,3 s.)	971 (19 s. - 436 b.)
ORB 4 opt = 1005	833 (0,1 s.)	891 (5,58 s.)	895 (555 s.)	not available	898 (0,3 s.)	1004 (473 s. 10k b.)
ORB 5 opt = 887	801 (0,1 s.)	801 (6,90 s.)	801 (323 s.)	not available	822 (0,3 s.)	873 (14 s. - 303 b.)

Figure 6: Lower Bounds for ten 10 × 10 problems

These results clearly show the power of task intervals and their reduction rules as a cutting mechanism.

5.2. Proofs of optimality

The next tables contains the results for proofs of optimality on the same ten problems. Our times are given on a Sparc 10, and those from [AC91] on a Sparc 1. A good performance is achieved on MT10 with ten times less backtracks than [AC 91] and on La19, La20 and ORB2 which were supposed to be particularly hard 10 × 10 problems. In fact, performances are stable : 8 out of 10 problems are solved under 1600 backtracks. A few thousands of backtracks or less seems to be the typical measure for 10 × 10 problems

	[AC 91]	Task Intervals
MT 10	372 s. - 16 kb.	106 s. - 1575 b.
ABZ 5	951 s. - 58 kb.	85 s. - 1350 b.
ABZ 6	91 s. - 1,3 kb.	9,9 s. - 157 b.
La 19	1460 s. - 94 kb.	63 s. - 1109 b.
La 20	1402 s. - 82 kb.	52 s. - 901 b.
ORB 1	1482 s. - 72 kb.	550 s. - 7265 b.
ORB 2	2484 s. - 153 kb.	36 s. - 456 b.
ORB 3	2297 s. - 130 kb.	340 s. - 4323 b.
ORB 4	1013 s. - 44 kb.	82 s. - 1060 b.
ORB 5	526 s. - 23 kb.	61 s. - 799 b.

Figure 7: Proof of optimality for ten 10 × 10 problems

The table below reports experiments with other classical benchmarks. 8 more 10 × 10 problems confirm the previous conclusion. The interesting fact is that among the 40 problems published by Lawrence ([La 84]), 23 can be solved with no backtrack and no search. For a time window one unit smaller than the optimal, reduction rules come to a contradiction. Among these problems, large ones are solved (all five 30 × 10). This indicates that the large problems of Lawrence are not exceptionally hard, as are the 15 × 10 (such as La21) or 20 × 10 (such as La29), but are just large. On such normal instances, task intervals perform especially well to detect bottlenecks.

problem	size	time/backtracks	problem	size	time/backtracks
La1,2,3,5	10 × 5	0,1 s. - 0 b.	La31-35	30 × 10	1,7 s. - 0 b.
La4	10 × 5	0,6 s. 16 b.			
La6-10	15 × 5	0,2 s. - 0 b.	MT20	5 × 20	0,4 s. - 0 b.
La11-15	20 × 5	0,3 s. - 0 b.			
La16	10 × 10	3,7 s. - 54 b.	ORB 6	10 × 10	202 s. - 2770 b.
La17	10 × 10	0,4 s. - 5 b.	ORB 7	10 × 10	44 s. - 631 b.
La18	10 × 10	4,6 s. - 80 b.	ORB 8	10 × 10	0,2 s. - 4 b.
La23	15 × 10	0,6 s. - 0 b.	ORB 9	10 × 10	8,3 s. - 142 b.
La26,28,30	20 × 10	0,8 s. - 0 b	ORB 10	10 × 10	21 s. - 255 b.

Figure 8: Proof of optimality for other classical benchmarks

5.3. Finding a solution given the optimal makespan

We use branch and bound for two different purposes: for finding optimal (or near-optimal, in the case of open problems) solutions and for proofs of optimality. In both cases, branching is done by edge-finding, and the pair on which to branch is selected with an entropic function, as described in section 4.5. There are however a few shallow differences between these cases. Indeed, for proofs of optimality one needs to visit the whole search tree while for finding a solution at a given makespan, one just needs to come to a leaf of the tree. Therefore, in the first case, the branching pair is selected to maximize the minimum entropic change over both branches (since both of them have to be explored), with the function h (cf paragraph 4.5) :

$$h(t_1, t_b) = \max\left(g(t_1 \ll t_b), \ \min\left(g(t_b \ll t_1), f(\Delta(S), \delta(S))\right)\right)$$

(recall that small values of g account for important changes, therefore the minimal change over both branches is a max), whereas in the second case, the function is modified (to reflect the change over the best of the two branches since one hopes to visit only this branch):

$$h'(t_1, t_b) = \min\left(g(t_1 \ll t_b), \ \max\left(g(t_b \ll t_1), f(\Delta(S), \delta(S))\right)\right)$$

	backtracks	time
MT 10	363 b.	20 s.
ABZ 5	1 164 b.	61 s.
ABZ 6	67 b.	4,5 s.
La 19	1 008 b.	50 s.
La 20	267 b.	10,6 s.
ORB 1	1 539 b.	89 s.
ORB 2	54 b.	4,7 s.
ORB 3	6 690 b.	347 s.
ORB 4	1 156 b.	60 s.
ORB 5	658 b.	40 s.

Figure 9: Search trees for finding an optimal solution

5.4. Branch and bound

Finding an optimal solution and giving the proof of optimality can also be integrated within a single search tree. The proof of optimality is then performed "on the fly", by dynamically reducing the makespan after the first optimal solution has been found and continuing the exploration of the same search tree. This technique is the classical technique of branch and bound procedures. For MT10, starting from 930 (the optimum), it takes a total of 6281 backtracks with h' (and even more with h). This figure may seem surprisingly high since part of the tree has already been discarded when the proof of optimality starts. But this part is located deep in the tree and therefore, the search for optimality is performed in a tree which top nodes have been selected for a larger (feasible) makespan. Since the entropic analysis is very fine, the pairs selected for a makespan of 930 are less relevant for 929 and therefore, the relevant alternatives are reconsidered many times, below the less relevant nodes inherited from the first part of the algorithm (930). The price to pay for the acuteness of the entropic analysis is the inability to reuse parts of a tree. When the value of the optimum is known beforehand, it is always worth restarting a search from scratch for the proof of optimality, rather than performing it in the same tree as the one for the optimal solution. However, when the optimal value is still unknown, such trees may be of interest. For example, when this algorithm is run on MT10, starting with upper bound 940 (recall that the optimum is 930), it takes 779 backtracks to find a first solution at 938; this solution is successively improved to 937, 936, 935, 934 and 930, requiring between each solution another 30 to 80 backtracks. The system then takes 5492 backtracks for the proof of optimality. The solution to avoid this dramatic increase of the number of backtracks for the proof of optimality, is to systematically stop the search and start it over when the number of backtracks became too large compared to the number of backtracks necessary for finding solutions.

5.5 The full algorithm on famous instances

MT10 is a very famous 10 × 10 job shop instance that remained unsolved for 25 years. It was first solved by Carlier & Pinson, using branch & bound and edge-finding. The shortest proof of optimality reported today is the one in [CP94], which take 35 nodes (but this figure does not account for some lookahead exploration) in a couple of minutes. The figure below describes the behavior of our full algorithm (described in [CL95])

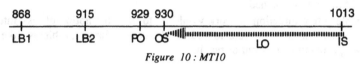

Figure 10 : MT10

An initial solution (IS) is found at 1013 in 1s., this solution is repaired and local optimization (LO) goes down to 930 in 70 s. The proof of optimality is obtained in 1575 backtracks and 80 seconds. So, the total procedure takes 150 s. Aside, the first lower bound (LB1) obtained by simple propagation comes in 0,3 seconds, and the second (LB2 : trying to schedule only the tightest resource) is obtained in 22 s. and 542 bk; If the optimum makespan is given, the branch and bound algorithm takes 363 backtracks and 20 s. to find an optimal solution (OS).

The problem LA21 is a hard 15 × 10 instance published by Lawrence in 1984 (for which Applegate & Cook report the 1040 as best lower bound and 1053 as best upper bound). This problem was solved early 1995 for the first time by three separate teams, one using mixed integer programming and the other two (including ours) using constraint propagation. To our knowledge, our approach yields the best performance.

1033	1045	1046		1220
LB1= LB2	PO	OS	·····ıılll¡ıııııı.....LO	IS

Figure 11: LA21

The two lower bounds coincide at 1033 (and take 1 s.), the proof of optimality is obtained in 2 millions backtracks and 24 hours. The initial solution is found at 1220 in 10 s. and the local optimisation takes 200 s. to go from 1220 down to an optimal solution.

LA29 is the last of the problems by Lawrence to remain open. It is a hard 20 × 10 instance. Our first lower bound gives the value of 1119 in two seconds (1120 is the best lower bound known), and our local optimization procedure takes 1200 s. to go down from 1428 to 1168 (1160 is the best reported upper bound).

Finally, we have been able to improve the lower bound on one of the four 20×20 problems published by Nakano and Yamada. The lower bound for YAM1 has now gone from 812 to 826 (it required 50 000 bactracks and 33 000s. to prove this bound)

6. Conclusion

In this paper, we have presented reduction rules integrated in a propagation scheme for jobshop scheduling. Propagation is based on the notion of task intervals which give an elegant framework for complex cuts. Propagation has been compared to integer programming, and we have explained why propagation works better than integer programming on disjunctive problems.

Our branch and bound algorithm has been tested on many standard problems on which it performs very well. Among these, the 15×10 problem LA21, unsolved since 1984, was solved for the first time in a day of computations, and the lower bound to YAM1 was sharpened.

A possible future direction of work would be to limit the notion of task intervals to smaller subsets of tasks in order to address the case of larger problems (like 50×20), where optimality is now out of reach.

Acknowledgments

This paper and the work on job-shop scheduling has been strongly influenced by numerous people. We are especially grateful to Bill Cook and Claude Le Pape for their insights and their comments throughout our work. We also want to thank François Fages, Clyde Monma, Marie Claude Portmann, Jean-François Puget and Pascal Van Hentenryck for their kind encouragement and their valuable comments. Last, we are grateful to Jacques Carlier and Eric Pinson for an enlightning conversation about disjunctive scheduling.

References

[ABZ88]   J.Adams, E. Balas & D. Zawak. *The Shifting Bottleneck Procedure for Job Shop Scheduling.* Management Science 34, p391-401. 1988

[AC91]   D. Applegate & B. Cook. *A Computational Study of the Job Shop Scheduling Problem.* Operations Research Society of America 3, 1991

[Ba 69]   E. Balas. *Machine Sequencing via Disjunctive Programming: an Implicit Enumeration Algorithm.* Operations Research 17, p 941-957. 1969

[Ba 85]   E. Balas *On the facial structure of Scheduling Polyhedra,* Mathematical Programming studies 24, p. 179-218.

[BL 95]   P. Baptiste, C. Le Pape *A theoretical and experimental comparison of constraint propagation techniques for disjunctive scheduling.* Proc. of the 14th International Joint Conference on Artificila Intellignece, 1995.

[Ca 82]   J. Carlier. *The one machine sequencing problem* European Journal of Operations Research 11, p. 42-47, 1982.

[CP89]   J. Carlier & E. Pinson. *An Algorithm for Solving the Job Shop Problem.* Management science, vol 35, no 2, february 1989

[CP94]   J. Carlier & E. Pinson. *Adjustments of heads and tails for the job-shop problem* , European Journal of Operations Research, vol 78, 1994, p. 146-161.

[Ca 91]   Y. Caseau. *A Deductive Object-Oriented Language.* Annals of Maths and Artificial Intelligence, Special Issue on Deductive Databases, March 1991.

[CGL93]   Y. Caseau, P.-Y. Guillo & E. Levenez. *A Deductive and Object-Oriented Approach to a Complex Scheduling Problem.* Proc. of DOOD'93, 1993

[CK92]   Y. Caseau & P. Koppstein. *A Cooperative-Architecture Expert System for Solving Large Time/Travel Assignment Problems.* International Conference on Databases and Expert Systems Applications, Spain, 1992.

[CL94]   Y. Caseau & F. Laburthe. *Improved CLP Scheduling with Task Intervals.* Proc. of the Eleventh International Conference on Logic Programming, ed: P. van Hentenryck, The MIT Press, 1994.

[CL95]   Y. Caseau & F. Laburthe. *Disjunctive Scheduling with Task Intervals.* LIENS report 95-25, École Normale Supérieure, 1995.

[DP95]   L. Djerid & M.-C. Portmann. *Comment entrecroiser des procédures par séparation et évaluation et des algorithmes génétiques: application à des problèmes d'ordonnancement à contraintes disjonctives.* FRANCORO, Mons, 1995, p. 84-85

[DT93]   M. Dell'Amico & M. Trubian. *Applying Tabu-Search to the Job-Shop Scheduling Problem.* Annals of Operations Research, vol 41, 1993, p. 231-252

[DW90]   M. Dyer & L.A. Wolsey. *Formulating the Single Machine Sequencing Problem with Release Dates as a Mixed Integer Program.* Discrete Applied Mathematics 26, p.255-270. 1990.

[La84]   S. Lawrence. *Resource Constrained Project Scheduling: an Experimental Investigation of Heuristic Scheduling Techniques.* GSIA, C.M.U. 1984

[MT63]   J.F. Muth & G.L. Thompson *Industrial scheduling.* Prentice Hall, 1963.

[NY92]   Nakano & Yamada *A Genetic Algorithm applicable to Large Scale Job-Shop Problems.* Parallel Problem solving from Nature 2, Elsevier, 1992.

[Ta89]   E. Taillard. *Parallel Taboo Search Technique for the Jobshop Scheduling Problem.* Internal Report ORPWP 89/11, École Polytechnique Fédérale de Lausanne, 1989

[VH89]   P. Van Hentenryck. *Constraint Satisfaction in Logic Programming.* The MIT press, Cambridge, 1989.

[VDV91]   S.L. Van De Velde. *Machine scheduling and Lagrangian relaxation.* Doctoral Thesis, CWI, Amsterdam, 1991

[VLA92]   P van Laarhoven, E.Aarts & J.K. Lenstra. *Job Shop Scheduling by Simulated Annealing.* Operations Research vol 40, no 1, 1992

A New Efficiently Solvable Special Case of the Three-Dimensional Axial Bottleneck Assignment Problem *

Bettina Klinz   Gerhard J. Woeginger

Institut für Mathematik B, TU Graz
Steyrergasse 30, A-8010 Graz, Austria

Abstract. Given an $n \times n \times n$ array $C = (c_{ijk})$ of real numbers, the three-dimensional axial bottleneck assignment problem (3-BAP) is to find two permutations ϕ and ψ of $\{1, \ldots, n\}$ such that $\max_{i=1,\ldots,n} c_{i\phi(i)\psi(i)}$ is minimized.

We first present two closely related conditions on the cost array C, the wedge property and the weak wedge property, which guarantee that an optimal solution of 3-BAP is obtained by setting ϕ and ψ to the identity permutation. In order to enlarge this class of efficiently solvable special cases of the 3-BAP, we then propose an $O(n^3 \log n)$ time algorithm which, given an $n \times n \times n$ array C, either finds three permutations ρ, σ and τ such that the permuted array $C_{\rho,\sigma,\tau} = (c_{\rho(i)\sigma(j)\tau(k)})$ satisfies the wedge property, or proves that no such permutations exist.

1   Introduction

Let an $n \times n \times n$ cost array $C = (c_{ijk})$ of real numbers be given and let $\mathcal{P}_n$ denote the set of permutations of $\{1, \ldots, n\}$. The *three-dimensional axial bottleneck assignment problem*, or 3-BAP for short, is to find two permutations $\phi, \psi \in \mathcal{P}_n$ such that the maximum cost entry given by

$$\max_{i=1,\ldots,n} c_{i\phi(i)\psi(i)} \tag{1}$$

is minimized. The 3-BAP is obtained from the well-known *three-dimensional axial assignment problem*, or 3-AP, by replacing the usual sum objective function $\sum_{i=1}^{n} c_{i\phi(i)\psi(i)}$ by its bottleneck analogue (1). Most of the papers in the literature on three-dimensional assignment problems deal with the sum case only. For references on solution methods for the 3-AP, see e.g. Balas and Saltzman [1] or Burkard and Rudolf [4]. For applications of three-dimensional axial assignment problems, see Pierskalla [11, 12].

---

* This research has been partially supported by the Spezialforschungsbereich F 003 "Optimierung und Kontrolle", Projektbereich Diskrete Optimierung. The second author acknowledges financial support by a research fellowship of the Euler Institute for Discrete Mathematics and its Applications.

By introducing 0-1 variables x_{ijk} where $x_{ijk} = 1$ if and only if $\phi(i) = j$ and $\psi(i) = k$, the 3-BAP can be formulated as the following 0-1 integer program:

$$\min \quad \max_{i,j,k=1,\ldots,n} \quad c_{ijk} x_{ijk}$$

$$\text{s.t.} \quad \sum_{i=1}^{n} \sum_{j=1}^{n} x_{ijk} = 1 \quad \text{for all } k = 1, \ldots, n$$

$$\sum_{i=1}^{n} \sum_{k=1}^{n} x_{ijk} = 1 \quad \text{for all } j = 1, \ldots, n$$

$$\sum_{j=1}^{n} \sum_{k=1}^{n} x_{ijk} = 1 \quad \text{for all } i = 1, \ldots, n$$

$$x_{ijk} \in \{0,1\} \quad \text{for all } i, j, k = 1, \ldots, n.$$

It is easily seen that 3-AP resp. 3-BAP generalize the classical 2-dimensional assignment problem with sum resp. bottleneck objective. The term *axial* in the name of the 3-AP and the 3-BAP is used to distinguish this type of three-dimensional assignment problem from a different one which is known as *planar* three-dimensional assignment problem. In the planar case the set of feasible solutions is described by

$$\sum_{i=1}^{n} x_{ijk} = 1 \quad \text{for all } j, k = 1, \ldots, n$$

$$\sum_{j=1}^{n} x_{ijk} = 1 \quad \text{for all } i, k = 1, \ldots, n$$

$$\sum_{k=1}^{n} x_{ijk} = 1 \quad \text{for all } i, j = 1, \ldots, n$$

$$x_{ijk} \in \{0,1\} \quad \text{for all } i, j, k = 1, \ldots, n.$$

Hence a feasible solution in the planar case contains exactly n^2 variables with value 1, while in the axial case exactly n variables have value 1. In the following we will only deal with axial three-dimensional assignments. From the point of view of special cases which is adopted in this paper, the axial version seems to be easier to attack due to the relatively small number of 1s in a feasible 0-1 solution.

Previous results. Both 3-AP and 3-BAP are known to be NP-hard even when the costs c_{ijk} are restricted to be from $\{0,1\}$ (this follows from the NP-completeness of the three-dimensional matching problem proved in the seminal paper by Karp [10]). For this reason several researchers tried to identify special cases which can be solved in polynomial time.

Most of the results known in this area, however, are of negative type. Burkard, Rudolf and Woeginger [5] show that 3-AP and 3-BAP remain NP-hard even for decomposable costs of the form $c_{ijk} = \alpha_i \beta_j \gamma_k$ for nonnegative integers α_i, β_j and γ_k. Spieksma and Woeginger [13] consider two geometric special cases of the 3-AP where three sets of n gridpoints, say N_1, N_2 and N_3, are given in the Euclidean plane and the entry c_{ijk} of the cost array C is obtained either from the circumference or the area of the triangle with points $i \in N_1$, $j \in N_2$ and $k \in N_3$. Both versions are proven to be NP-hard in [13] and it is not difficult to see that analogous results hold for the corresponding 3-BAP.

Not much is known on the positive side. One of the few efficiently solvable special cases of the 3-BAP results from restricting the cost array C to the class of 3-dimensional bottleneck Monge arrays. An $n \times n \times n$ array C is called a *(3-dimensional) bottleneck Monge array* if it fulfills the following property for all $i_p = 1, \ldots, n$ and $j_p = 1, \ldots, n$, $p = 1, 2, 3$

$$\max\{ c_{k_1 k_2 k_3}, c_{\ell_1 \ell_2 \ell_3} \} \le \max\{ c_{i_1 i_2 i_3}, c_{j_1 j_2 j_3} \}, \tag{2}$$

where $k_p := \min\{i_p, j_p\}$ and $\ell_p := \max\{i_p, j_p\}$ for $p = 1, 2, 3$. The 3-dimensional bottleneck Monge property results from the better known 3-dimensional Monge property (or submodularity property) by replacing the operation "+" in the 3-dimensional Monge property by "max" (for more details on multidimensional Monge properties see e.g. the survey paper by Burkard, Klinz and Rudolf [3]).

The following result on the 3-BAP follows from a more general result on algebraic multi-dimensional assignment problems described in Burkard, Klinz and Rudolf [3].

Proposition 1 *Let C be an $n \times n \times n$ bottleneck Monge array. Then an optimal solution of the 3-BAP with cost array C is obtained by choosing ϕ and ψ to be the identity permutation ε_n, i.e. $\varepsilon_n(i) := i$ for all $i = 1, \ldots, n$.*

An analogous theorem for the 3-AP restricted to Monge arrays was earlier obtained by Bein, Brucker, Park and Pathak [2] as a corollary to a result on multi-dimensional transportation problems. Recently, Fortin and Rudolf [9] showed that a weaker version of the 3-dimensional (bottleneck) Monge property is sufficient for guaranteeing that the identity solution $\phi = \psi = \varepsilon_n$ is an optimal solution of 3-AP resp. 3-BAP.

Our results. In this paper we identify a new special case of the 3-BAP which can be solved in polynomial time. An $n \times n \times n$ cost array $C = (c_{ijk})$ is said to have the *wedge property* if it satisfies

$$c_{u\ell\ell} < \min\{ c_{ijk} \colon 1 \le i \le \ell; \; i \le j, k \le n; \; j + k \ne 2i \} \qquad \text{for } \ell = 1, \ldots, n, \tag{3}$$

and to have the *weak wedge property*, if the inequalities in (3) may hold with equality, i.e. if

$$c_{u\ell\ell} \le \min\{ c_{ijk} \colon 1 \le i \le \ell; \; i \le j, k \le n; \; j + k \ne 2i \} \qquad \text{for } \ell = 1, \ldots, n. \tag{4}$$

We first show that the identity solution $\phi = \psi = \varepsilon_n$ constitutes an optimal solution of the 3-BAP restricted to the class of cost arrays with the (weak) wedge property. Then we present an $O(n^3 \log n)$ time algorithm which, given an $n \times n \times n$ array $C = (c_{ijk})$, either finds three permutations ρ, σ and τ such that the permuted array $C_{\rho,\sigma,\tau} = (c_{\rho(i)\sigma(j)\tau(k)})$ satisfies the wedge property, or prove that no such permutations exist.

Organization of the paper. In Section 2 it is shown that setting $\phi = \psi = \varepsilon_n$ yields an optimal solution of the 3-BAP restricted to cost arrays C satisfying property (4). Section 3 presents an $O(n^3 \log n)$ time algorithm for deciding whether a given $n \times n \times n$ array C can be permuted into an array $C_{\rho,\sigma,\tau}$ satisfying the wedge property. Section 4 briefly discusses the difficulties that arise in the more general case when $C_{\rho,\sigma,\tau}$ is only required to satisfy the weak wedge property. In Section 5 the results of this paper on the 3-BAP are generalized to d-dimensional bottleneck assignment problems with $d > 3$. The paper is concluded with a summary in Section 6.

2 The Wedge Property: A New Solvable Case of the 3-BAP

Recall that an $n \times n \times n$ array C is said to have the *wedge property* if the following condition is fulfilled for all $\ell = 1, \ldots, n$:

$$c_{\ell\ell\ell} < \min\{c_{ijk}: 1 \le i \le \ell,\ i \le j, k \le n \text{ and } j + k \ne 2i\}. \tag{5}$$

Similarly, C is said to have the *weak wedge property* if for all $\ell = 1, \ldots, n$

$$c_{\ell\ell\ell} \le \min\{c_{ijk}: 1 \le i \le \ell,\ i \le j, k \le n \text{ and } j + k \ne 2i\}. \tag{6}$$

Let $\mathcal{W}^<$ denote the class of arrays with the wedge property and $\mathcal{W}^\le$ denote the class of arrays with the weak wedge property. Note that obviously $\mathcal{W}^< \subset \mathcal{W}^\le$.

The definition of the wedge property is motivated by the class of trapezoidal matrices introduced in Butkovič, Cechlárová and Szabó [6] (see also Cechlárová [8]). In [6] an $n \times n$ matrix A is termed *trapezoidal* if it satisfies the following property for all $\ell = 1, \ldots, n$:

$$a_{\ell\ell} > \max\{a_{ij}: 1 \le i \le \ell,\ i + 1 \le j \le n\}. \tag{7}$$

In [6] it is shown that, given a trapezoidal matrix A, the identity permutation ε_n is an optimal solution of the following variant of the bottleneck assignment problem:

$$\max_{\phi \in \mathcal{P}_n} \min_{i=1,\ldots,n} a_{i\phi(i)}.$$

In order to translate the results of [6] for the type of bottleneck assignment problem treated in this paper, the order has to be reversed, i.e. $>$ has to be replaced by $<$ and max by min. After this transformation, condition (7) reads as

$$a_{\ell\ell} < \min\{a_{ij}: 1 \le i \le \ell,\ i+1 \le j \le n\}. \tag{8}$$

Hence, the wedge property can be viewed as generalization of the trapezoidal property to three dimensions.

The following theorem and the subsequent corollary show that the weak wedge property and thus also the wedge property lead to a polynomially solvable special case of the 3-BAP.

Theorem 1 *Given an $n \times n \times n$ cost array $C \in \mathcal{W}^{\le}$, setting $\phi = \psi = \varepsilon_n$ yields an optimal solution of the 3-BAP.*

Proof. Suppose that there exists an optimal solution (ϕ^*, ψ^*) of the 3-BAP such that $(\phi^*, \psi^*) \ne (\varepsilon_n, \varepsilon_n)$. Let $p := \min\{q : 1 \le q \le n, \phi^*(q) \ne q \text{ or } \psi^*(q) \ne q\}$, i.e. p denotes the minimum index where (ϕ^*, ψ^*) deviates from the identity solution $(\varepsilon_n, \varepsilon_n)$. Hence we have $p < n$, $\phi^*(p) \ge p$, $\psi^*(p) \ge p$ and $\phi^*(p)+\psi^*(p) > 2p$. Since $C \in \mathcal{W}^{\le}$, it follows that

$$c_{rrr} \le c_{p\phi^*(p)\psi^*(p)} \qquad \text{for all } r = p, \ldots, n. \tag{9}$$

Since $\phi^*(q) = q$ and $\psi^*(q) = q$ holds for all $q = 1, \ldots, p-1$ by the definition of p, the inequalities (9) imply that

$$\max_{r=1,\ldots,n} c_{rrr} \le \max_{r=1,\ldots,n} c_{r\phi^*(r)\psi^*(r)}.$$

Thus the identity solution $\phi = \psi = \varepsilon_n$ is another optimal solution of the 3-BAP.
□

Corollary 2 *The 3-BAP restricted to $n \times n \times n$ cost arrays C with the weak wedge property can be solved in $O(n^3)$ time, or even in $O(n)$ time if it is a-priori known that $C \in \mathcal{W}^{\le}$.*

Comparing Theorem 1 to Proposition 1, one is tempted to conjecture that the bottleneck Monge property and the (weak) wedge property are related in some sense, since in both cases the identity solution yields an optimal solution of the 3-BAP. The following two examples demonstrate, however, that neither of the two properties is a special case of the other.

It is easy to check that the $3 \times 3 \times 3$ array $C = (c_{ijk})$ with

$$(c_{1jk}) = \begin{pmatrix} 0\ 1\ 1 \\ 1\ 1\ 1 \\ 1\ 1\ 1 \end{pmatrix} \qquad (c_{2jk}) = \begin{pmatrix} 1\ 0\ 0 \\ 0\ 0\ 1 \\ 1\ 1\ 1 \end{pmatrix} \qquad (c_{3jk}) = \begin{pmatrix} 0\ 0\ 0 \\ 0\ 0\ 0 \\ 0\ 0\ 0 \end{pmatrix}$$

fulfills the wedge property, but does not satisfy the bottleneck Monge property since e.g. $\max\{c_{211}, c_{222}\} = 1 > 0 = \max\{c_{212}, c_{221}\}$. (If the entry c_{233} is changed to 2, then the modified matrix does not even fulfill the weak bottleneck Monge property introduced in Fortin and Rudolf [9]). On the other hand, the $3 \times 3 \times 3$ array $B = (b_{ijk})$ with

$$(b_{1jk}) = \begin{pmatrix} 0\ 1\ 1 \\ 1\ 1\ 1 \\ 1\ 1\ 1 \end{pmatrix} \qquad (b_{2jk}) = \begin{pmatrix} 1\ 1\ 1 \\ 1\ 1\ 1 \\ 1\ 1\ 0 \end{pmatrix} \qquad (b_{3jk}) = \begin{pmatrix} 1\ 1\ 1 \\ 1\ 1\ 1 \\ 1\ 1\ 1 \end{pmatrix}$$

fulfills the bottleneck Monge property, but not the (weak).wedge property since $b_{222} = 1 > 0 = \min\{ b_{223}, b_{232}, b_{233} \}$.

In Theorem 1 we identified a new class of cost arrays for which the 3-BAP can be solved in polynomial time. This class can be enlarged by observing that, given a 3-BAP with cost array $C = (c_{ijk})$, it obviously leads to an equivalent problem if arbitrary permutations $\rho, \sigma, \tau \in \mathcal{P}_n$ are applied to C to yield a new permuted cost array $C_{\rho,\sigma,\tau} = (c_{\rho(i)\sigma(j)\tau(k)})$. This motivates the following definition:

An $n \times n \times n$ array C is said to have the *permuted (weak) wedge property* if there exist three permutations ρ, σ and τ such that the permuted array $C_{\rho,\sigma,\tau}$ satisfies the (weak) wedge property. For notational convenience, let us refer by $\mathcal{PW}^<$ to the class of arrays with the permuted wedge property and by $\mathcal{PW}^{\leq}$ to the class of arrays with the permuted weak wedge property.

In order to be able to exploit the presence of the permuted wedge property in connection with the 3-BAP, it is, however, necessary to have an algorithm available which solves the following recognition problem:

(RP) Given an $n \times n \times n$ array C, decide whether $C \in \mathcal{PW}^<$, and if so, find three permutations ρ, σ and τ such that $C_{\rho,\sigma,\tau} \in \mathcal{W}^<$.

Another recognition problem of interest arises if we require that $C_{\rho,\sigma,\tau} \in \mathcal{W}^{\leq}$. This variation will be referred to as (RP$^{\leq}$).

3 Recognizing the Permuted Wedge Property

In this section we will present an efficient algorithm for recognizing the permuted wedge property, i.e. for solving the recognition problem (RP) defined above. For (RP) an approach turns out to be successful which is similar to the method proposed by Butkovič, Cechlárová and Szabó [6] for deciding whether there exist permutations σ and τ which transform a given matrix A into a trapezoidal matrix $A_{\sigma,\tau}$.

In order to describe the algorithm for solving (RP), we need the following definitions. Given an $n \times n \times n$ array C and three sets $I, J, K \subseteq \{1, \ldots, n\}$, we denote by $C[I, J, K]$ the array that results from C by keeping only entries c_{ijk} for which $i \in I$, $j \in J$ and $k \in K$ and removing all other entries. For the ease of notation, we will still refer to the entries in $C' = C[I, J, K]$ with their original indices they had in the array C, i.e. entry c'_{ijk} corresponds to entry c_{ijk} in C. If

$I = \{i^*\}$ and $J = K = \{1,\ldots,n\}$, the subarray $C[I,J,K]$ will be referred to as *1-plane* i^*, since it contains all entries c_{ijk} with first coordinate $i = i^*$.

For simplicity, we will sometimes refer to an entry of C with smallest resp. second smallest value as *smallest entry* resp. *second smallest* entry of C. For a given array C, let $v_i^{(1)}$ resp. $v_i^{(2)}$ denote the value of the smallest resp. the second smallest entry in 1-plane i of C. (C is omitted from this notation for simplicity, since it will always be clear from the context to which array we refer.) In case, 1-plane i consists of only one entry, i.e. $n = 1$, we define $v_i^{(2)} = \infty$.

Algorithm 1: Recognition of arrays $C \in \mathcal{PW}^<$.

1. Set $h := 1$, $u_{min} := \infty$, $I = J = K := \{1,\ldots,n\}$ and $\tilde{C} := C$.
2. Let i^* be a 1-plane of $\tilde{C}$ such that
   (a) $v_{i^*}^{(1)} < v_{i^*}^{(2)}$, i.e. 1-plane i^* has a unique smallest entry, say $\tilde{c}_{i^* y(i^*) z(i^*)}$,
   (b) $\tilde{c}_{i^* y(i^*) z(i^*)} < u_{min}$, i.e. $\tilde{c}_{i^* y(i^*) z(i^*)}$ is smaller than the minimum of the values of the upperdiagonal entries defined so far, and
   (c) the value of the second smallest entry of 1-plane i^* is as large as possible among all 1-planes satisfying (2a) and (2b).
   If such a 1-plane does not exist, we may stop, since then $C \notin \mathcal{PW}^<$.
3. Set $\rho(h) := i^*$, $\sigma(h) := y(i^*)$ and $\tau(h) := z(i^*)$ and update u_{min} according to $u_{min} := \min\{u_{min}, v_{i^*}^{(2)}\}$.
4. Redefine $I := I \setminus \{i^*\}$, $J := J \setminus \{y(i^*)\}$ and $K := K \setminus \{z(i^*)\}$ and let $\tilde{C} := \tilde{C}[I,J,K]$ (i.e. all entries $\tilde{c}_{ijk}$ with $i = i^*$ or $j = y(i^*)$ or $k = z(i^*)$ are removed from $\tilde{C}$).
5. If $h < n$, increase h by one and goto Step 2.

In proving the correctness of Algorithm 1, the following three lemmata will be essential.

Lemma 1 *For any array $C \in \mathcal{PW}^<$ there exists a 1-plane i with a unique smallest entry.*

Proof. Since $C \in \mathcal{PW}^<$, there exist permutations $\rho, \sigma, \tau \in \mathcal{P}_n$ such that $C_{\rho,\sigma,\tau} \in \mathcal{W}^<$. Setting $i = \rho(1)$ yields the 1-plane with the desired property since in any array $B \in \mathcal{W}^<$, the entry b_{111} has to be strictly smaller than all other entries b_{1jk}, $(j,k) \neq (1,1)$. $\square$

Lemma 2 *Let C be an $n \times n \times n$ array with $C \in \mathcal{W}^<$ and let i^* be a 1-plane of C such that the value of its second smallest entry is maximum among the values of the second smallest entries of the 1-planes $1,\ldots,n$. Furthermore, let $c_{i^* y(i^*) z(i^*)}$ be an entry of 1-plane i^* with minimum value. Then the following holds:*

1. *$y(i^*) = i^*$ and $z(i^*) = i^*$, i.e. the smallest entry in 1-plane i^* is unique and is located on the main diagonal.*
2. *There exist three permutations $\rho, \sigma, \tau \in \mathcal{P}_n$ such that*
   (i) $\rho(1) = i^*$, $\sigma(1) = i^*$ and $\tau(1) = i^*$.

(ii) $C_{\rho,\sigma,\tau} \in \mathcal{W}^<$.

(iii) $c_{\rho(\ell)\sigma(\ell)\tau(\ell)} < \min\{c_{i\bullet jk}\colon j,k = 1,\ldots,n \text{ and } (j,k) \neq (i^*,i^*)\}$ holds for $\ell = 2,\ldots,n$.

Proof. Assume that the first claim does not hold, i.e. $(y^*(i), z^*(i)) \neq (i^*, i^*)$. Thus $c_{i\bullet i\bullet i\bullet}$ and $c_{i\bullet y(i^*)z(i^*)}$ are two distinct entries of the 1-plane i^* and by the choice of $c_{i\bullet y(i^*)z(i^*)}$ we have $c_{i\bullet y(i^*)z(i^*)} \leq c_{i\bullet i\bullet i\bullet}$. Hence, it follows that $v_{i\bullet}^{(2)} \leq c_{i\bullet i\bullet i\bullet}$ (recall that $v_{i\bullet}^{(2)}$ is the value of the second smallest entry of 1-plane i^*). Since by assumption $v_{i\bullet}^{(2)} \geq v_p^{(2)}$ for all $p = 1,\ldots,n$ this in turn implies that

$$c_{i\bullet i\bullet i\bullet} \geq v_1^{(2)}. \tag{10}$$

On the other hand, by using the wedge property of C for $\ell = i^*$ we obtain

$$c_{i\bullet i\bullet i\bullet} < \min\{c_{1jk}\colon j,k = 1,\ldots,n \text{ and } (j,k) \neq (1,1)\},$$

i.e. $c_{i\bullet i\bullet i\bullet} < v_1^{(2)}$ which contradicts (10). Hence $y(i^*) = i^*$ and $z(i^*) = i^*$.

The second claim of the lemma is proved by constructing appropriate permutations ρ, σ and τ which do the required job. In our construction the same permutation π will be used for all three permutations ρ, σ and τ. We set $\pi = (i^*, i^* - 1, \ldots, 1)(i^* + 1), \ldots, (n)$, i.e. $\pi(1) = i^*$, $\pi(i) = i - 1$ for $i = 2,\ldots,i^*$ and $\pi(i) = i$ for $i = i^* + 1,\ldots,n$. Thus (i) is satisfied.

To prove (ii), we need to show that the permuted array $B := C_{\pi,\pi,\pi}$ satisfies the wedge property. This is, however, easy to check. First, note that $C \in \mathcal{W}^<$, the choice of i^* and the first claim of the lemma imply

$$c_{\ell\ell\ell} < \min\{c_{1jk}\colon j,k = 1,\ldots,n \text{ and } (j,k) \neq (1,1)\} = v_1^{(2)}$$
$$\leq v_{i\bullet}^{(2)} = \min\{c_{i\bullet jk}\colon j,k = 1,\ldots,n \text{ and } (j,k) \neq (i^*,i^*)\}$$

for all $\ell = 1,\ldots,n$. By the choice of π this leads to

$$b_{\ell\ell\ell} < \min\{b_{1jk}\colon j,k = 1,\ldots,n \text{ and } (j,k) \neq (1,1)\} \tag{11}$$

for all $\ell = 1,\ldots,n$. The validness of the remaining inequalities

$$b_{\ell\ell\ell} < \min\{b_{ijk}\colon i = 2,\ldots,\ell, \ j,k = i,\ldots,n \text{ and } (j,k) \neq (i,i)\} \tag{12}$$

for $\ell = 2,\ldots,n$ follows from observing that due to the choice of π all these inequalities are already required to hold because $C \in \mathcal{W}^<$.

Finally, property (iii) follows directly from the inequalities (11) by noting that $\pi(1) = i^*$. $\square$

Lemma 3 *Let C be an $n \times n \times n$ array and let $s \in \{1,\ldots,n-1\}$. Furthermore, suppose that for all $\ell = 1,\ldots,s$*

$$c_{\ell\ell\ell} < \min\{c_{ijk}\colon i = 1,\ldots,\ell, \ j,k = i,\ldots,n \text{ and } j + k \neq 2i\}. \tag{13}$$

Then the following two statements are equivalent:

(i) There exist permutations $\rho, \sigma, \tau \in \mathcal{P}_n$ such that $C_{\rho,\sigma,\tau} \in \mathcal{W}^<$, and $\rho(\ell) = \tau(\ell) = \sigma(\ell) = \ell$ holds for all $\ell = 1, \ldots, s$, i.e. the position of the entries c_{ijk} with $i, j, k \in \{1, \ldots, s\}$ is not changed when transforming C into $C_{\rho,\sigma,\tau}$.

(ii) There exist permutations $\rho', \sigma', \tau' \in \mathcal{P}_{n-s}$ such that applying ρ', σ' and τ' to the $(n - s) \times (n - s) \times (n - s)$ array $B = (b_{ijk})$ with $b_{ijk} = c_{i+s,j+s,k+s}$ results in the array $F = B_{\rho',\sigma',\tau'}$ which has the wedge property and satisfies additionally

$$f_{\ell\ell\ell} < \min\{c_{ijk} \colon i = 1, \ldots, s, \ j, k = i, \ldots, n \ \text{and} \ j + k \neq 2i\}$$

for all $\ell = 1, \ldots, n - s$.

The proof of Lemma 3 is straightforward and is therefore omitted. Summarizing we have obtained the following theorem and the subsequent corollary.

Theorem 3 *Given an $n \times n \times n$ array C, Algorithm 1 finds in $O(n^3 \log n)$ time either three permutations ρ, σ and τ such that $C_{\rho,\sigma,\tau} \in \mathcal{W}^<$, or proves that such permutations do not exist.*

Proof. The correctness of Algorithm 1 follows from Lemmata 1 through 3. Lemma 1 guarantees that in the first execution of Step 2 there exists at least one 1-plane satisfying properties (2a) and (2b) (provided that $C \in \mathcal{PW}^<$). It then follows from Lemma 2 that we can set $\rho(1) = i^*$, $\sigma(1) = y(i^*)$ and $\tau(1) := z(i^*)$. Lemma 3 justifies Step 4 in which the $n \times n \times n$ array C is reduced to an $(n - 1) \times (n - 1) \times (n - 1)$ array containing only entries c_{ijk} with $i \neq i^*$, $j \neq y(i^*)$ and $k \neq z(i^*)$. Then property (iii) of Lemma 2 and Lemma 1 imply that in the next iteration of Step 2 there again exists at least one 1-plane satisfying properties (2a) and (2b) (provided that $C \in \mathcal{PW}^<$). Hence the procedure can be continued in this manner until either $\tilde{C}$ becomes empty or we have to stop in Step 2 since $C \notin \mathcal{PW}^<$

In order to prove that Algorithm 1 can be implemented to run in $O(n^3 \log n)$ time, note that there are at most n iterations of Steps 2–5. By sorting the entries of each 1-plane in nondecreasing order in a preprocessing step, and maintaining pointers to the smallest and second smallest entry of each 1-plane, at most $O(n^3)$ steps are needed over all executions of Step 2. The preprocessing step takes $O(n^2 \log n)$ time per 1-plane, thus $O(n^3 \log n)$ time overall. □

Corollary 4 *The 3-BAP restricted to $n \times n \times n$ cost arrays C with the permuted wedge property is solvable in $O(n^3 \log n)$ time.*

4 Some Comments on Recognizing the Permuted Weak Wedge Property

At first sight, Algorithm 1 seems to be easily adapted to recognize the class $\mathcal{PW}^\leq$ of arrays with the permuted weak wedge property in polynomial time. Unfortunately, this is not the case: Since the existence of a 1-plane with a unique smallest entry is not necessary for an array to belong to $\mathcal{PW}^{\leq}$, Step 2 of Algorithm 1 obviously has to be modified such that condition (2a) is removed. Furthermore, in condition (2b) we have to check for $\tilde{c}_{i \cdot y(i^*)z(i^*)} \leq u_{min}$. However, the following main problems arise with this modified algorithm:

1. If in Step 2 there does not exist a unique 1-plane i^* with largest value $v_{i \cdot}^{(2)}$ among all 1-planes fulfilling the modified condition (2b) and an arbitrary choice is made among the available candidates, then the algorithm might stop prematurely even in cases where $C \in \mathcal{PW}^{\leq}$.
2. Since the smallest entry in 1-plane i^* is not necessarily unique, the choice of $\tilde{c}_{i \cdot y(i^*)z(i^*)}$ is also not unique. Again an arbitrary choice might lead to wrong conclusions about the given array C.

To illustrate these problems, consider the following example: Let C be the following $3 \times 3 \times 3$ array with

$$(c_{1jk}) = \begin{pmatrix} 1\,1\,2 \\ 1\,2\,2 \\ 2\,2\,2 \end{pmatrix} \quad (c_{2jk}) = \begin{pmatrix} 0\,0\,2 \\ 0\,1\,2 \\ 2\,2\,2 \end{pmatrix} \quad (c_{3jk}) = \begin{pmatrix} 2\,2\,2 \\ 2\,2\,2 \\ 0\,2\,1 \end{pmatrix}$$

Obviously, we have $C \in \mathcal{W}^{\leq}$. Since $v_1^{(2)} = v_3^{(2)} = 1 > v_2^{(2)} = 0$, we either have $i^* = 1$ or $i^* = 3$. Suppose we choose $i^* = 3$. Thus we have $y(i^*) = 3$ and $z(i^*) = 1$. Consequently, the following reduced $2 \times 2 \times 2$ array $\tilde{C} = (\tilde{c}_{ijk})$ remains after Step 4 of Algorithm 1:

$$(\tilde{c}_{1jk}) = \begin{pmatrix} 1\,2 \\ 2\,2 \end{pmatrix} \quad (\tilde{c}_{2jk}) = \begin{pmatrix} 0\,2 \\ 1\,2 \end{pmatrix}$$

Thus in the next iteration we get $i^* = 1$ because $v_1^{(2)} = 2 > v_2^{(2)} = 1$. Since only $\tilde{c}_{111}, \tilde{c}_{211}$ and $\tilde{c}_{221}$ are $\leq u_{min} = 1$ and all these three entries are removed from $\tilde{C}$ in the second iteration, Algorithm 1 stops in the third iteration with the wrong conclusion $C \notin \mathcal{PW}^{\leq}$.

The above problems are due to the fact that the Lemmata 1 and 2 cannot be carried over to arrays $C \in \mathcal{W}^{\leq}$. (Lemma 3 holds in a modified version which is obtained by replacing $\mathcal{W}^{<}$ by $\mathcal{W}^{\leq}$ and all occurrences of $<$ by $\leq$.)

Recall that Lemma 2 enabled us to reduce $\tilde{C}$ in Step 4 of Algorithm 1 and still to be sure that if $C \in \mathcal{PW}^{<}$, the reduced array $\tilde{C}$ fulfills the following condition after the h-th iteration of Steps 2–4:

There exist permutations $\rho', \sigma', \tau' \in \mathcal{P}_{n-h}$ such that

$$\max_{q=1,\ldots,n-h} \tilde{c}_{\rho'(q)\sigma'(q)\tau'(q)} \leq u_{min}. \tag{14}$$

(Actually, for $C \in \mathcal{PW}^{<}$ we even had $\max_{q=1,\ldots,n-h} \tilde{c}_{\rho'(q)\sigma'(q)\tau'(q)} < u_{min}$.) Obviously, for a successful termination of Algorithm 1 for arrays $C \in \mathcal{PW}^{\leq}$, it is necessary that condition (14) holds in each iteration of Algorithm 1. We conjecture that this is also sufficient. Note, however, that (14) is equivalent to requiring that the set M consisting of exactly those triples $(i, j, k) \in I \times J \times K$ for which $c_{ijk} \leq u_{min}$ contains a *three-dimensional matching*, i.e. a subset $M' \subseteq M$ with cardinality $|M'| = n - h$ (recall that $|I| = |J| = |K| = n - h$) such that no two elements of M' agree in any coordinate. Since deciding the existence of a three-dimensional matching is NP-complete (see Karp [10]), it is rather unlikely that an approach along the lines of Algorithm 1 will be successful in recognizing the permuted weak wedge property.

5 Generalization to Higher Dimensions

Let $d > 3$. The *d-dimensional axial bottleneck assignment problem*, *d-BAP* for short, is a natural generalization of the 3-BAP and NP-hard as well. Given a d-dimensional array C of order $n \times n \times \cdots \times n$, the d-BAP is to find $d - 1$ permutations $\phi_1, \ldots, \phi_{d-1}$ such that the objective function

$$\max_{i=1,\ldots,n} c_{i\phi_1(i)\ldots\phi_{d-1}(i)}$$

is minimized.

The following two properties play an analogous role for the d-BAP as the weak wedge property and the wedge property for the 3-BAP. A d-dimensional array C of order $n \times n \times \cdots \times n$ is said to have the *d-dimensional weak wedge property* if for all $\ell = 1, \ldots, n$

$$c_{\ell \ldots \ell} \leq \min \left\{ c_{i_1 i_2 \ldots i_d} : 1 \leq i_1 \leq \ell; i_1 \leq i_2, \ldots, i_d \leq n; \sum_{r=2}^{d} i_r \neq (d-1)i_1 \right\} \quad (15)$$

If all inequalities in (15) are strict, then C is said to have the *d-dimensional wedge property*. Similarly, as in the case $d = 3$, we say that C has the *permuted (weak) d-dimensional wedge property* if there exist d permutations $\pi_1, \ldots, \pi_d$ such that the permuted array $C_{\pi_1, \ldots, \pi_d} = (c_{\pi_1(i_1) \ldots \pi_d(i_d)})$ fulfills the d-dimensional (weak) wedge property.

By the same proof technique as used for Theorem 1, the following result on the d-BAP can be proved.

Theorem 5 *The d-BAP restricted to the class of d-dimensional cost arrays C of order $n \times n \times \cdots \times n$ with the weak d-dimensional wedge property is solved optimally by the identity solution $\phi_1 = \phi_2 = \ldots = \phi_{d-1} = \varepsilon_n$.*

Corollary 6 *The d-BAP restricted to the class of d-dimensional cost arrays C of order $n \times n \times \cdots \times n$ with the weak d-dimensional wedge property can be solved in $O(n^d)$ time, or even in $O(n)$ time if it is known a-priori that the given cost array satisfies the d-dimensional weak wedge property.*

A larger class of efficiently solvable special cases of the d-BAP is obtained by generalizing Algorithm 1.

Theorem 7 *Given a d-dimensional array C of order $n \times n \times \cdots \times n$, in $O(n^d \log n)$ time we can either find d permutations $\pi_1, \ldots, \pi_d$ such that the permuted array $C_{\pi_1, \ldots, \pi_d}$ fulfills the d-dimensional wedge property or prove that no such permutations exist.*

Corollary 8 *The d-BAP restricted to d-dimensional cost arrays of order $n \times n \times \cdots \times n$ which satisfy the permuted d-dimensional wedge property is solvable in $O(n^d \log n)$ time.*

6 Summary and Concluding Remarks

In this paper we have identified a new efficiently solvable special case of the three-dimensional axial bottleneck assignment problem 3-BAP.

The main open problem in connection with the topic of this paper is whether there exists an efficient algorithm for recognizing arrays $C \in \mathcal{PW}^{\leq}$. To the authors' knowledge even the following (seemingly easier) problem for matrices is open:

(P) Given an $n \times n$ matrix A, either find two permutations $\sigma, \tau \in \mathcal{P}_n$ such that the permuted matrix $\tilde{A} = A_{\sigma, \tau}$ satisfies the following weak trapezoidal property

$$\tilde{a}_{\ell\ell} \geq \max\{\tilde{a}_{ij} \colon 1 \leq i \leq \ell, \ i + 1 \leq j \leq n\},$$

for all $\ell = 1, \ldots, n$ or prove that no such permutations exist.

Butkovič, Cechlárová and Szabó [6] erroneously claimed ([8]) that a minor modification of their algorithm for deciding whether an $n \times n$ matrix A can be permuted into a trapezoidal matrix also works for the more general problem (P) above.

Another challenging research question is to search for further classes of efficiently solvable special cases of three-dimensional axial and planar assignment problems. For the axial case, it would be of particular interest to find special cases where the optimal solution is not fixed (as the identity solution in the case dealt with in this paper), but depends on the given input data. For the planar case, nothing is known on special cases. Here the main difficulty is that unlike the axial case where we used the identity solution $\phi = \psi = \varepsilon_n$, there is no obvious simple-structured candidate optimal solution.

Acknowledgement: We would like to thank Katarína Cechlárová for a discussion on the recognition problem (P) mentioned in Section 6.

References

1. E. Balas and M.J. Saltzman, Facets of the three-index assignment polytope, *Discrete Applied Mathematics* **23**, 1989, 201–229.
2. W.W. Bein, P. Brucker, J.K. Park and P.K. Pathak, A Monge property for the *d*-dimensional transportation problem, *Discrete Applied Mathematics* **58**, 1995, 97–109.
3. R.E. Burkard, B. Klinz and R. Rudolf, Perspectives of Monge properties in combinatorial optimization, SFB-Report 2, Spezialforschungsbereich "Optimierung und Kontrolle", Institut für Mathematik, TU Graz, Austria, July 1994, to appear in *Discrete Applied Mathematics*.
4. R.E. Burkard and R. Rudolf, Computational investigations on 3-dimensional axial assignment problems, *Belgian Journal of Operations Research, Statistics and Computer Science* **32**, 1993, 85–98.
5. R.E. Burkard, R. Rudolf and G.J. Woeginger, Three-dimensional axial assignment problems with decomposable cost coefficients, Report 238-92, Institut für Mathematik, TU Graz, Austria, October 1992, to appear in *Discrete Applied Mathematics*.
6. P. Butkovič, K. Cechlárová and P. Szabó, Strong linear independence in bottleneck algebra, *Linear Algebra and its Applications* **94**, 1987, 133-155.
7. K. Cechlárová, Trapezoidal matrices and the bottleneck assignment problem, *Discrete Applied Mathematics* **58**, 1995, 111–116.
8. K. Cechlárová, Personal communication, 1995.
9. D. Fortin and R. Rudolf, Weak algebraic Monge arrays, SFB-Report 33, Spezialforschungsbereich "Optimierung und Kontrolle", Institut für Mathematik, TU Graz, Austria, June 1995, submitted for publication.
10. R.M. Karp, Reducibility among combinatorial problems, in: *Complexity of Computer Computations*, R.E. Miller and J.W. Thatcher (eds.), Plenum Press, New York, 1972, pp. 85-103.
11. W.P. Pierskalla, The tri-substitution method for the three-dimensional assignment problem, *CORS J.* **5**, 1967, 71–81.
12. W.P. Pierskalla, The multidimensional assignment problem, *Operations Research* **16**, 1968, 422–431.
13. F.C.R. Spieksma and G.J. Woeginger, Geometric three-dimensional assignment problems, SFB-Report 3, Spezialforschungsbereich "Optimierung und Kontrolle", Institut für Mathematik, TU Graz, Austria, July 1994, to appear in *European Journal of Operational Research*.

Ramsey Numbers by Stochastic Algorithms with New Heuristics

Jihad Jaam

Laboratoire d'Informatique de Marseille - LIM
Centre National de la Recherche Scientifique - CNRS URA 1787,
Faculté des Sciences de Luminy Case 901, 163, avenue de Luminy
13288 Marseille Cedex 9, France

Abstract. In this paper, we are interested in combinatorial problems of graph and hypergraph colouring linked to Ramsey's theorem. We construct correct colourings for the edges of these graphs and hypergraphs, by stochastic optimization algorithms in which the criterion of minimization is the number of monochrome cliques. To avoid local optima, we propose a technique consisting of an enumeration of edge colourings involved in monochrome cliques, as well as a method of simulated annealing. In this way, we are able to improve some of the bounds for the Ramsey numbers. We also introduce cyclic colourings for the hypergraphs to improve the lower bounds of classical ternary Ramsey numbers and we show that cyclic colourings of graphs, introduced by Kalbfleisch in 1966, are equivalent to symmetric Schur partitions.

Key Words: stochastic optimization, cyclic colouring, Ramsey number, Schur number, graph, hypergraph.

Introduction

The success in the resolution of some combinatorial NP–Hard problems by stochastic optimization methods such as the "N–Queens" problem for $N > 3 \times 10^6$ [43, 44] and the salesman problem [29], has spurred research on the use of this type of algorithm for solving other combinatorial problems. In particular, the problem of graph colouring as linked to the Ramsey's theorem. The use of these methods gives promising experimental results for improving some bounds on Ramsey numbers ([9, 10, 11, 12, 13],[24, 25, 26]). Other applications on NP–Hard problems have also been proposed by Selman *et al.* [41, 42], Adorf and Johnson [2], Minton *et al.* [35] and by Gu [20].

In this paper, we are interested in the NP–Hard problem of the evaluation of Ramsey numbers. These numbers are particularly used for calculating the complexity of some algorithms on parallel machines [45]. It has been confirmed in [18] that Ramsey numbers serve to construct the "best" communication network.

Ramsey theory deals with the distribution of subsets of elements of sets. The theorem of Ramsey is a generalization of the "pigeon–hole" principle: suppose that a flock of pigeons flies into a set of pigeonholes to roost. The *pigeonhole principle* (also called the *Dirichlet drawer principle*) states that if there are more pigeons than pigeonholes, then there must be at least one pigeonhole with at least two pigeons in it. This principle can be applied to other objects besides pigeons and pigeonholes. In general:

if $(k-1)T+1$ objects are distributed in T drawers then at least one of the drawers contains k objects.

We can use Ramsey numbers to solve the following problems:

Problem 1 What is the minimum number of students required in a class to be sure that at least six of them will receive the same grade, if there are five possible grades, A, B, C, D, and F?

Problem 2 What is the least number of area codes needed to guarantee that the 25 million phones in a state have distinct 10-digit telephone numbers? (Assume that telephone numbers are of the form $NXX–NXX–XXXX$, where the first three digits form the area code, N represents a digit from 2 to 9 inclusive, and X represents any digit.)

Problem 3 Assume that in a group of eighteen people, each pair of individuals consists of two friends or two enemies. Show that there are either four mutual friends or there are four mutual enemies in the group.

Problem 4 A computer network consists of six computers. Each computer is directly connected to at least one of the other computers. Show that there are at least two computers in the network that are directly connected to the same number of other computers.

Problem 5 During a month with 30 days a baseball team plays at least 1 game a day, but no more than 45 games. Show that there must be a period of some number of consecutive days during which the team must play exactly 14 games.

Problem 6 Show that, if there are 101 people of different heights standing in a line, it is possible to find 11 people in the order they are standing in the line with heights that are either increasing or decreasing and in general, every sequence of n^2+1 distinct real numbers contains a subsequence of length $n+1$ that is either strictly increasing or strictly decreasing.

The evaluation of each Ramsey number is a combinatorial NP–Hard problem (see [7] Appendix NP–Complete Problems (Monochromatic Triangle)). So, it is costly to explore the whole space or the search-tree needed to find a solution. Indeed, the search for a solution through the enumeration of all cases leads to procedures of a disastrous efficiency. Thus, to determine these numbers, or to improve their bounds, we propose three stochastic optimization algorithms with colouring based criterion. These algorithms, which are incomplete constraint satisfaction search, are generally methods of rapid descent, completed by enumeration or "backtracking" algorithms when we are "close" to a solution, and methods of simulated annealing to avoid local optima.

To improve the lower bounds of binary Ramsey numbers we use the properties of cyclic colouring of graphs introduced by Kalbfleisch [28] in 1966. We generalize the cyclic colouring for hypergraphs associated with Ramsey numbers.

1 Ramsey's General Theorem

Definition 1. Let X be a set of n elements, and let $\mathcal{P}_h(X)$ be the set of parts with h elements of X, $(|\mathcal{P}_h(X)| = \binom{n}{h})$. The couple $H = (X, \mathcal{P}_h(X))$ constitutes a *complete hypergraph of rank h and order n.* Elements of X and $\mathcal{P}_h(X)$ denote respectively vertices and edges of H. Each edge of H is therefore a set of cardinal h. Such hypergraph is called a *n–clique of rank h* and denoted by K_n^h. We also denote by K_n (or K_n^2), *the simple complete graph of order n.*

Definition 2. Let H be a complete hypergraph of n vertices and rank h, and let $k_1, k_2, \ldots, k_m$ be the positive integers greater than h. A $(k_1, k_2, \ldots, k_m; h)$–*colouring* is an application that attributes to each edge of H a "colour" among a set of m colours. This $(k_1, k_2, \ldots, k_m; h)$–colouring is said to be *correct* if none of the k_i–cliques (i.e., a complete sub–hypergraph of rank h and order k_i) of H has all edges of colour i. In other words, there is no monochrome k_i–clique of colour i and rank h, for $i = 1, 2, \ldots, m$.

Definition 3. We say that the complete hypergraph K_n^h, is $(k_1, k_2, \ldots, k_m; h)$–*colourable,* if we can associate a correct $(k_1, k_2, \ldots, k_m; h)$–colouring with its edges.

To formulate Ramsey's general theorem as a colouring of hypergraph edges we express this theorem in the following manner:

Theorem 4. *Consider m colours $1, 2, \ldots, m$ and $m + 1$ positive integers $k_1, k_2, \ldots, k_m, h$ with $k_1, k_2, \ldots, k_m \geq h$. There exists a finite integer n $(n < \infty)$ such that any complete hypergraph of rank h and order n has no correct $(k_1, k_2, \ldots, k_m; h)$–colouring for its edges.*

The smallest integer n is called the *classical Ramsey number of rank h in* $k_1, k_2, \ldots, k_m$ which is denoted by $R(k_1, k_2, \ldots, k_m; h)$. The existence of this number has been proven by Ramsey [38] and by Ryser [39].

According to this theorem, all $(k_1, k_2, \ldots, k_m; h)$–colourings of the edges of K_R^h, where R is the number $R(k_1, k_2, \ldots, k_m; h)$, contains by definition, a certain number of monochrome k_i–cliques of colour i (for $i = 1, 2, \ldots, m$) and rank h. The minimum number of these monochrome k_i–cliques is called *multiple Ramsey number of rank h in* $k_1, k_2, \ldots, k_m$ which is denoted by $r(k_1, k_2, \ldots, k_m; h)$.

Obviously, we are able to determine the value of the multiple Ramsey number, $r(k_1, k_2, \ldots, k_m; h)$, only when the classical Ramsey number, $R(k_1, k_2, \ldots, k_m; h)$, is known.

The *multiplicity function* $M_{(n)}(k_1, k_2, \ldots, k_m; h)$ *of rank h in* $k_1, k_2, \ldots, k_m$ is defined as the minimum number of monochrome k_i–cliques of colour i (for $i = 1, 2, \ldots, m$) in the colouring of the edges of K_n^h (i.e., the complete hypergraph of rank h and order $n > R(k_1, k_2, \ldots, k_m; h) + 1$). Obviously, the multiple Ramsey number $r(k_1, k_2, \ldots, k_m; h)$ is equal to $M_{(R)}(k_1, k_2, \ldots, k_m; h)$ where R is the classical Ramsey number $R(k_1, k_2, \ldots, k_m; h)$.

2 Evaluation of Ramsey Numbers

The evaluation of each Ramsey number, in itself, is a combinatorial NP–Hard problem [19][1]. It is not surprising that, classical "backtracking" methods fail to find "good" lower and upper bounds for Ramsey numbers $R(k_1, k_2, \ldots, k_m; h)$ even when $h = 2$ (i.e., the hypergraph $H = (X, \mathcal{P}_2(X))$ becomes a simple complete graph $G = (X, E)$ of n vertices K_n, with $n = |X|$), as the complexity of these methods is exponential and the number of trials to get a solution (i.e., the number of possible colourings), with exhaustive search, is very large. So, we have to be satisfied with the improvement of lower and upper bounds of these numbers (see [6], [14, 15, 16], [21, 22], [23], [27], [32, 33] etc.).

In general, to improve lower and upper bounds associated with Ramsey numbers, we look for the greatest integer $n < R(k_1, k_2, \ldots, k_m; h)$, where n is the number of vertices of the complete hypergraph of rank h, $H = (X, \mathcal{P}_h(X))$, for which there exists a correct $(k_1, k_2, \ldots, k_m; h)$–colouring for its edges. Similarly, to improve upper bounds of classical Ramsey numbers, we look for the smallest integer $n \geq R(k_1, k_2, \ldots, k_m; h)$, for which no cor--ct $(k_1, k_2, \ldots, k_m; h)$–colouring exists for K_n^h edges.

2.1 Binary Ramsey Numbers

To find the exact value of *the classical binary Ramsey number* $R(k_1, k_2, \ldots, k_m; 2)$, it is necessary to construct a correct $(k_1, k_2, \ldots, k_m; 2)$–colouring for the K_{R-1} edges, where R is the Ramsey number $R(k_1, \ldots, k_m; 2)$, and to prove that no such correct $(k_1, k_2, \ldots, k_m; 2)$–colouring exists for the edges of K_R. This is a very difficult task. Indeed, the value of the number $R(k_1, k_2, \ldots, k_m; 2)$ increases very "rapidly" when the values of k_i ($i \in \{1, \ldots, m\}$) increase. The number of possible colourings of the graph edges associated with $R(k_1, k_2, \ldots, k_m; 2)$ is about $m^{(R(\ldots)-1)(R(\ldots)-2)/2}$. Thus, for $|X| = 16$, $m = 3$ (i.e., three colours) with $k_1 = k_2 = k_3 = 3$ (i.e., a simple case), the number of possible colourings to get a solution is about $3^{120} \simeq 10^{57}$.

In 1947, Erdös [8] proposed a probabilistic method that associates lower bounds with binary Ramsey numbers. In particular, he proved that the Ramsey number $R(34, 34; 2)$ is greater than 10^6. This method gives information about the value of Ramsey numbers but does not indicate how to construct the correct colourings for their associated graphs. Greenwood and Gleason [17], have also given some properties which make it possible to associate upper bounds with

---

[1] To illustrate the difficulty of the evaluation of Ramsey numbers, the Hungarian mathematician, Paul Erdös, (see [18]) proposed the following anecdote:

the extra-terrestrials invade the Earth and threaten to annihilate it within a year if humanity does not find the classical Ramsey number $R(5, 5; 2)$. Under the threat, the best mathematicians and the most powerful computers are mobilized, and the catastrophe is avoided. On the other hand, if the extra-terrestrials ask for the classical Ramsey number $R(6, 6; 2)$, the only possibility would be war.

every binary Ramsey number. Abott [1] also proved some inequalities that give lower bounds for binary Ramsey numbers. However, all the values obtained by these inequalities are very far from the exact values of Ramsey numbers. Therefore, it was necessary to seek other methods for improving the bounds associated with Ramsey numbers.

Much work has been done during the last fifty years to improve the bounds associated with binary Ramsey numbers (see Giraud [14, 15, 16], McKay and Radziszowski [32, 33, 34] Exoo [9, 10, 11, 12, 13]). Only eleven numbers have been exactly evaluated, ten of them for $m = h = 2$ (i.e., two colours) and only one for $h = 2, m = 3$ (i.e., three colours) ($R(3, 3, 3; 2) = 17$). Most of these numbers were evaluated by Greenwood-Gleason (1955) and by Kalbfleisch (1966).

2.2 Ternary Ramsey Numbers

The evaluation of *the classical ternary Ramsey numbers*, $R(k_1, k_2, \ldots, k_m; 3)$, is more difficult than that of the binary Ramsey numbers $R(k_1, k_2, \ldots, k_m; 2)$, due to the fact that the number of possible colourings of edges of their associated hypergraphs is very large (i.e., this number is about $m^{\binom{R(\ldots)}{3}}$). In 1966, Kalbfleisch [28] showed that ternary and binary Ramsey numbers are closely linked by the following inequality:

$$R(k_1, k_2; 3) \le R(x, y; 2) + 1, \text{ with } x = R(k_1 - 1, k_2; 3), y = R(k_1, k_2 - 1; 3).$$

This shows that the first non–trivial ternary Ramsey number, $R(4, 4; 3)$, is lower than 20. This means that any correct $(4, 4; 3)$-colouring does not exist for the edges of K_{20}^3. Kalbfleisch also proposed two methods to construct correct $(4,4;3)$–colourings for the edges of K_{11}^3. However, these two methods have failed in their attempt to construct a correct $(4,4;3)$–colouring for the edges of K_{12}^3. Therefore, we have $12 \le R(4, 4; 3) \le 19$. In 1969, Giraud [16] proved that the ternary Ramsey number $R(4, 4; 3)$ is smaller than 15. Also in 1969, Isbell [23] improved the lower bound of this number by 1. Recently (1991), by using Turán numbers, McKay and Radziszowski [34] have shown that any $(4, 4; 3)$–colouring of the edges of K_{13}^3 always contains a monochrome 4–clique of rank 3. They have also found a correct $(4, 4; 3)$–colouring of the edges of K_{12}^3. Consequently, we have: $R(4, 4; 3) = 13$.

3 Theoretical Results

The theoretical results given in this section are of Greenwood-Gleason [17], Erdös [8] and Abott [1]. These results permit to approximate "roughly" the Ramsey numbers.

Lemma 5. *The Ramsey number $R(k_1, k_2, \ldots, k_m; 2)$ is invariant for any permutation of k_i.*

Lemma 6. *The Ramsey number $R(2, k_2, \ldots, k_m; 2)$ is equal to $R(k_2, \ldots, k_m; 2)$ and, we also have: $R(2, k_2; 2) = R(k_2; 2) = k_2$.*

Lemma 7. *Let G be a $(k_1, k_2, \ldots, k_m; 2)$–colourable graph, for each vertex of G there are at most $R(k_1, k_2, \ldots, k_i - 1, \ldots, k_m; 2) - 1$ edges of colour i leading to the other vertices.*

Lemma 8. $R(k_1, k_2; 2) \leq R(k_1 - 1, k_2; 2) + R(k_1, k_2 - 1; 2)$; *moreover, this inequality is strict if these last two terms are both even integers.*

Lemma 9. *The following inequality gives an upper bound for the classical binary Ramsey number $R(k_1, k_2, \ldots, k_m; 2)$:*

$$R(k_1, k_2, \ldots, k_m; 2) \leq 2 - m + \sum_{i=1}^{m} R(k_1, k_2, \ldots, k_i - 1, \ldots, k_m; 2) .$$

Theorem 10 (Abott, 1966). *This theorem gives a lower bound for binary Ramsey numbers.*

$$R(k_1, k_2, \ldots, k_m; 2) \geq 1 + ([R(k_1, \ldots, k_s; 2) - 1] \times [R(k_{s+1}, \ldots, k_m; 2) - 1]) .$$

4 Ramsey Numbers and Chromatic Number

To study the binary Ramsey numbers turns out to be equivalent to study the chromatic number of a hypergraph, $\mathcal{X}(H)$, whose concept was introduced in 1966 by Erdös and Hajnal (see [4]). Indeed, let $H = (X, \mathcal{E})$ be a hypergraph of order q with $\mathcal{E} \subseteq \mathcal{P}_q(X)$ where $q = 2, \ldots, h$ (i.e., H is a complete hypergraph or not). A set $S \subset X$ is called *stable* if it contains no edge $E_i \in \mathcal{E}$ of cardinal greater than 1. The *chromatic number* $\mathcal{X}(H)$ is the smallest number of the necessary colours that we can use to colour vertices of H, in such a way that no edge E_i (with $|E_i| > 1$) has its vertices coloured with the same colour. An m–colouring of vertices of H (with $m \geq \mathcal{X}(H)$) is a partition of the set of vertices X into m stable sets $S_1, S_2, \ldots, S_m$, while a correct $(k_1, k_2, \ldots, k_m; 2)$–colouring of the edges of the simple complete graph $G = (X, \mathcal{P}_2(X))$ is a partition of $\mathcal{P}_2(X)$ into m sets (or m colours) $\mathcal{B}_1, \mathcal{B}_2, \ldots, \mathcal{B}_m$, such that for each part A_i of X of cardinal k_i, we have $\mathcal{P}_2(A_i) \not\subset \mathcal{B}_i$. Therefore we can consider a hypergraph of rank k, whose vertices are the edges of the simple complete graph G, and whose edges are the k–cliques of rank 2 of G. This hypergraph will be denoted by G^k. The proposed problem of the evaluation of classical binary Ramsey numbers consists in studying the chromatic number $\mathcal{X}(G^k)$. We then have the following inequality:

$$\mathcal{X}(G^k) \leq R(k_1, k_2, \ldots, k_m; 2) - 1 \text{ (with } k_1 = k_2 = \cdots = k_m) .$$

5 Stochastic Optimization Algorithms

In this section, we give the stochastic optimization algorithms that we have used to improve the bounds associated with Ramsey numbers.

The principle of these algorithms consists in minimizing an objective function depending on the problem being considered. For example, to evaluate the Ramsey number $R(k_1, k_2, k_3; 2)$, the objective function is the number of monochrome k_i–cliques of colour i (for $i = 1, 2$ and 3) that are found in the coloured graph associated with this number. The optimization criterion is therefore the number of monochrome k_i–cliques of colour i.

To increase the efficiency of these algorithms, we use a strategy similar to the *"divide and conquer"* technique [30]. The general idea is to decompose the problem in sub–problems, which are less difficult to solve (if this is possible) and then solve each sub–problem independently. The obtained solutions of the sub–problems are "pasted" together to give a solution of the initial problem. Thus, to evaluate the classical Ramsey number $R(k_1, k_2, \ldots, k_m; 2)$ (with $k_m \geq k_{m-1} \geq \cdots \geq k_1$), we eliminate first, in the corresponding coloured graph, all monochrome k_m–cliques of colour m, then all monochrome k_{m-1}–cliques of colour $m-1$ and so on until the elimination of all the monochrome k_1–cliques of colour 1.

We will describe the stochastic algorithms, that we have used, to construct correct colourings of the hypergraphs associated with Ramsey numbers. To simplify the description, we assume that $k_1 = k_2 = \cdots = k_m$.

5.1 Rapid Descent Algorithm

The resolution process of this algorithm (denoted by algorithm 1) starts with the creation of a random colouring C of the edges of the simple complete graph G of order n which produces a minimum number of monochrome k_i–cliques (for $i = 1, 2, \ldots, m$). In order to accomplish this, we proceed in the following manner: we have $\binom{n}{2}$ edges to colour with m colours. We iteratively select, randomly, an edge and we colour it with the colour that produces a minimum number of monochrome k_i–cliques of colour i (for $i = 1, 2, \ldots, m$). In the resulting colouring, we have, generally, a certain number of monochrome k_i–cliques of colour i, to be calculated, that we will try to delete.

Local Minima. This algorithm cannot find the solution for all random colourings of the edges of the graph G. There are some cases in which whatever the number of undertaken permutations of colour edges is, the algorithm can no longer reduce the number of monochrome k_i–cliques. If, for a given colouring C, we undertake ten thousand changes in the colour of the edges (i.e., empirical value), without being able to decrease the value of $N(C)$, we estimate that we have reached a *local minimum*. In this case, we choose another random colouring C' and we iterate the process.

In a simplified manner, this algorithm starts from an initial state $(C_1, \mathcal{F}_1)$ (i.e., that is the random colouring, C_1, of the edges of G and the number of monochrome k_i–cliques $\mathcal{F}_1 \geq 0$ in C_1) and tries to reach (i.e., by changing of the colour of the edges) a final state $(C_n, \mathcal{F}_n)$, with:

C_n is the correct colouring of the edges of G and

$\mathcal{F}_n = 0$ is the number of monochrome k_i–cliques in $\mathcal{C}_n$, for all $i \in \{1, 2, \ldots, m\}$.

The local minimum results from the fact that, during the resolution, we reach an intermediate state $(\mathcal{C}_i, \mathcal{F}_i > 0)$, in which, the number of monochrome k_i–cliques remains stable whatever the number of changes of the colour of the edges is. In other words, for every colouring $\mathcal{C}_j$ obtained by changing the colour of the G edges, we will always have $\mathcal{F}_j \geq \mathcal{F}_i > 0$.

Algorithm 1.

This algorithm tries to construct a correct $(k_1, k_2, \ldots, k_m; 2)$–colouring for the edges of the complete graph G of order n. With this colouring we show that $R(k_1, k_2, \ldots, k_m; 2) > n$. At the beginning, we fix the number of the initial colourings (i.e., the number of different colourings generated to stop the algorithm) as well as the number of consecutive colourings that we accept; this number denoted by NCC corresponds, in fact, to a *local minimum* (i.e., empirical value).

1- *Choose a random colouring C for G edges which produces a minimum number of monochrome k_i–cliques of colour i (for $i = 1, \ldots m$).*
   Introduce a counter K and initialize it with zero.
   Initialize the successive number of colourings NCC.

2- *If the number $N(C)$ of monochrome k–cliques is null then stop (i.e., a solution is found). Otherwise, set K to zero.*

3- *Choose an edge e that is included in the maximum of monochrome cliques. Introduce a colouring C' that coincides with C except for $C'(e) \neq C(e)$.*
   - *If $N(C') < N(C)$ then replace C by C' and go to 2.*
   - *If $K = NCC$ then go to 1.*
   - *If $N(C') = N(C)$ then replace C by C', increase K by 1 and go to 3.*
   - *If $N(C') > N(C)$ then increase K by 1 and go to 3.*

Fig.1. The algorithm of rapid descent.

5.2 Random Drawing and the Arborescent Method

In this section, we propose another stochastic optimization algorithm for "avoiding" local minima. It consists in combining the rapid descent method with the arborescent method (i.e., a backtracking or a branch and bound method[30]).

This algorithm operates as follows: first, we use a rapid descent algorithm to reduce the size of the search area. As soon as we reach a small enough number of monochrome k_i–cliques, we enumerate all possible colourings of the edges contained in the monochrome k_i–cliques. The enumeration is made by the arborescent or "backtracking" method [3]. At this level, the number of possible colourings to get a solution is relatively small. Indeed, each monochrome k_i–clique is formed by $\binom{k_i}{2}$ edges (i.e., a complete sub–graph of order k_i), then if there remain T monochrome k_i–cliques in the coloured graph, the number of possible colourings is about $\binom{k_i}{2}^T$. Thus, we reduce the size of the search area or paths to cover from $\binom{k_i}{2}^{\binom{n}{2}}$ to $\binom{k_i}{2}^T$ (with $3 \leq T \leq 18$ (i.e., empirical value) and n the order of the complete graph G).

In this algorithm, we can say that the rapid descent method is a pre-treatment method [46]. We use it to get a partial solution of the problem.

Algorithm 2.

1- *Choose a random colouring C for the edges of G which produces the minimum number of monochrome k_i–cliques of colour i (for $i = 1, \ldots m$).*

2- *Minimize the number of monochrome k_i–cliques in G by changing the colour of the edges.*

3- *If a pseudo–local–minimum is reached (i.e. empirical value) then apply the* **enumeration procedure** *after constructing a list U of the edges implied in all the monochrome k_i–cliques.*
Otherwise, stop (i.e., a solution is found).

> **PROCEDURE enumeration** (G, U);
> $\{C$ is a variable that takes values $1, 2, \cdots, m$ representing the colours $\}$
> **BEGIN**
> **IF** $U = \emptyset$ **THEN** *stop (i.e., a solution is found)*; **OTHERWISE**
> $\{$*Choose an edge $e \in U$*;
> **FOR** $C \leftarrow 1$ **TO** m **DO**
> $\{$*Allocate the colour C to e;*
> **IF** *(number–monochrome–cliques (G))* $= 0$ **THEN**
> **enumeration** $(G, U - \{e\})$;$\}$
> $\}$
> **END.**

Fig.2. The algorithm of stochastic descent and enumeration.

After the construction of the U list, the graph G contains $|U|$ uncoloured edges and zero monochrome k_i–cliques of colour i for all $i \in \{1, 2, \ldots, m\}$.

The edges of the list U are ordered (i.e., the first edge is implied in the maximum of monochrome k_i–cliques).

5.3 Simulated Annealing Method

The Simulated Annealing method is a widespread technique used to solve combinatorial problems where the search space is vast. It has been proposed in 1953, by statistical physics specialists (Metropolis et al. [31]). An application of this method for the salesman problem has been proposed by Kirkpatrick and Gelatt [29]. This method is an extension of the Monte Carlo method [30].

The purpose of this method is to escape from the local minima. There is always an objective function $\mathcal{F}$ to minimize. In our case, it is the number of monochrome k_i–cliques (for $i = 1, 2, \ldots, m$).

Schematically, the principle of this method is the following: from an initial value of T (i.e., a control parameter) and an initial configuration (i.e., a colouring of the edges of G, in our case), S, of the considered problem, we choose arbitrarily another configuration S' (i.e., after changing an edge colour, for example). If the value of the objective function in S', $\mathcal{F}(S')$, is lower than its value in S (i.e., $\mathcal{F}(S') < \mathcal{F}(S)$), this means that if the drawn configuration is better than the current one, S' replaces S and we repeat the process. If this is not the case (i.e., if S' is worse than S) we can however decide to replace S by S'. This decision is made with a certain probability given by:

$$\text{Prob} = 1 - \exp^{-k \times (T - T_0)} \text{ or } \exp^{\frac{-\Delta}{k \times T}} \text{ with } \Delta = \mathcal{F}(S') - \mathcal{F}(S) \; .$$

k being a constant and T being a control parameter that goes down to T_0. For $T = T_0$, only "beneficial" changes, for which we have $\mathcal{F}(S') < \mathcal{F}(S)$, are admitted, while for $T > T_0$, it is permitted to increase the value of the objective function $\mathcal{F}$. These techniques will probably make it possible to avoid a local minimum which is not global.

Algorithm 3.

1- *Initialize T, $T_{\min}$ (i.e., the minimum value of T), T_{red} (i.e., the reduction factor of T), k and the number of iterations N_{itr}.*

2- *Choose a random colouring C for G edges which produces a minimum number of monochrome k_i–cliques of colour i (for $i = 1, \ldots m$).*

3- *Calculate the number $N(C)$ of monochrome k_i–cliques in C.*

REPEAT
    $i \leftarrow 0$;
    WHILE $(i < N_{\text{itr}})$ **AND** $(N(C) > 0)$ **DO**
        • *Choose randomly an edge e;*
        • *Introduce a colouring C' that coincides with C except for $C'(e) \neq C(e)$;*
        • $\Delta \leftarrow N(C') - N(C)$, $i \leftarrow i + 1$;
        • **IF** $(\Delta \leq 0)$ **OR** (**RAND** $<$ **EXP**$(\frac{-\Delta}{k \times T})$) **THEN** *replace C by C';*
    END
    $T \leftarrow T \times T_{\text{red}}$;

UNTIL$(N(C) = 0)$ **OR** $(T < T_{\min})$

Fig.3. The algorithm of Simulated Annealing.

The **RAND** function is a generator of random numbers between 0.0 and 1.0. In this type of algorithm, several questions may be asked. For example, how can we determine the value of T and $T_{\min}$? How to reduce T ? etc. The values of all these parameters are empirical. However, T is initialized with a value relatively large, which makes it possible to accept all the configurations obtained by "penalizing" permutations of colours of edges. Practically, the value of T is chosen in the following manner: we generate n random colourings (i.e., a thousand, for example), and we calculate the average value of Δ for these colourings. The value of T will be equal to $\frac{-\Delta}{\ln \text{Const}}$ (with $0.12 \leq$ Const ≤ 0.95 (empirical values)).

6 New Heuristics to Colour the Hypergraph Edges

We will introduce the cyclic colourings for the hypergraphs to improve the lower bounds associated with classical ternary Ramsey numbers and we show that the cyclic colourings of graphs are equivalent to symmetric Schur partitions [40].

The properties of cyclic colourings of graphs (see [10]) introduced in 1966 by Kalbfleisch [28] are very useful for constructing correct colourings for the edges of the graphs associated with binary Ramsey numbers.

6.1 Cyclic Colouring of Graphs

Definition 11. Let G be a simple complete graph of order n with its vertices labelled by the integers $1, 2, \ldots, n$, and let e be an edge of G. We define the *length* of e, $l(e)$, by the difference between its two extremities. The length $l(e)$ (with $e = \{i, j\}$ and $i < j$) is equal to:

$$\begin{cases} |i - j|, & \text{if } |i - j| < \lfloor \frac{n}{2} \rfloor, \\ i - j + n & \text{otherwise .} \end{cases}$$

this length varies between 1 and $\lfloor \frac{n}{2} \rfloor$.

The lengths of the edges of the simple complete graph of order 5, K_5, are:

$$l(\{1,2\}) = 1, \; l(\{3,4\}) = 1, \; l(\{4,5\}) = 1, \; l(\{1,5\}) = 1, \; l(\{2,3\}) = 1,$$
$$l(\{2,4\}) = 2, \; l(\{2,5\}) = 2, \; l(\{1,3\}) = 2, \; l(\{3,5\}) = 2, \; l(\{1,4\}) = 2.$$

Definition 12. Let G be a $(k_1, k_2, \ldots, k_m; 2)$–colourable graph of order n, We say that G is *Ramsey–regular*, if there exist a correct $(k_1, k_2, \ldots, k_m; 2)$–colouring of the edges of G, in which all edges with the same length are coloured with the same colour. The correct $(k_1, k_2, \ldots, k_m; 2)$–colouring that associates only one colour with each edge of G is called *a correct cyclic $(k_1, k_2, \ldots, k_m; 2)$–colouring*.

According to this definition, the $(3, 3; 2)$–colourable graph of order 5, K_5, is Ramsey–regular. Indeed, to have a correct cyclic $(3, 3; 2)$–colouring for the edges of K_5 it suffices to assign the colour Red to all the edges of length 1 and the colour Blue to those of length 2.

Property of non Ramsey–Regular Graphs. The non Ramsey–regular graphs have an interesting property: if there exists a $(k_1, \ldots, k_m ; 2)$–colourable graph K_n, then there exists a $(k_1, k_2, \ldots, k_m; 2)$–colourable graph K_{n-1}. To have a correct $(k_1, k_2, \ldots, k_m; 2)$–colouring for the edges of K_{n-1}, it suffices to delete a vertex of K_n with all the edges that connect it to the other vertices. However, this is not always true for Ramsey–regular graphs. That is to say, it is possible that there exists a correct cyclic $(k_1, k_2, \ldots, k_m; 2)$–colouring for the edges of K_n, which is not the case for the edges of K_{n-1}; for example, the case of K_{10} that has a correct cyclic $(3, 3, 3; 2)$–colouring for its edges (i.e., easy to find) while K_9 has no such cyclic $(3, 3, 3; 2)$–colouring. This leads to the following theorem:

Theorem 13. *Given m colours and an integer $i > 2$, the simple complete graph of order $3i$ has no correct cyclic $(\underbrace{3, 3, \ldots, 3}_{m}; 2)$–colouring for its edges.*

Proof of theorem. The simple complete graph K_n (with $n = 3i$) contains always a simple complete sub–graph $G = (X, \mathcal{P}_2(X))$ of order 3, with $X = \{x_1, x_2, x_3\}$ and $l(e_1 = \{x_1, x_2\}) = l(e_2 = \{x_2, x_3\}) = l(e_3 = \{x_1, x_3\}) = i$. Indeed, we can take for example, $x_1 = 1$, $x_2 = 1 + i$ and $x_3 = 1 + 2i$. It is clear that $l(e_1) = l(e_2) = i$. The length of e_3 cannot be equal to $|1 - 2i - 1|$ because $2i$ is strictly greater than $\lfloor \frac{n}{2} \rfloor$ and consequently, $l(e_3) = n + 1 - 1 - 2i = i$. The edges of G have therefore the same length and thus they can be coloured with the same colour. So, G forms a monochrome 3–clique of rank 2. Consequently, K_n has no correct cyclic $(\underbrace{3, 3, \ldots, 3}_{m}; 2)$–colouring for its edges. $\qquad\square$

According to this theorem, the graphs $K_9, K_{12}, K_{15}, K_{21}, \ldots, K_{63}$ etc. have no correct cyclic $(3, \ldots, 3; 2)$–colouring for their edges whatever the number of colours m is.

6.2 The Cyclic Colouring and the Symmetric Schur Partition

In this section we show the relation between the cyclic colouring of graphs associated with binary Ramsey numbers of the form $R(3, 3, \ldots, 3; 2)$ and the symmetric Schur partition. We can generalize this relation for all Ramsey numbers of rank h, $R(k_1, k_2, \ldots, k_m; h)$.

Definition 14. Let $E = \{1, \ldots, n\}$, a *Schur $(H_1, H_2, \ldots, H_k)$–partition*, is a partition of E into k classes $H_1, \ldots, H_k$, in such a way that if two elements i and j of E belong to the class H_l, then their sum m (if $m \leq n$) does not belong to this class. Schur's lemma [40] asserts that there exists a finite set E of $S(k)$ positive elements that has a $(H_1, H_2, \ldots, H_k)$–partition and for any set of n' elements with $n' > S(k)$, this $(H_1, H_2, \ldots, H_k)$–partition does not exist. The integer $S(k)$ is called *the Schur number of k classes*.

In other words, if we colour the elements of E ($|E| > S(k)$) with k colours ($k > 0$), then there exist always three elements, x, y and z, coloured with the same colour such that: $x + y = z$.

Definition 15. The Schur $(H_1, H_2, \ldots, H_k)$–partition is called *symmetric* iff for all pairs (i, j) of E of sum $n + 1$ (with $|E| = n$) we have i and j belong to the same class.

The symmetric Schur $(H_1, H_2, H_3, \ldots, H_k)$–partition and the correct cyclic $\underbrace{(3, 3, 3, \ldots, 3}_{m}; 2)$–colouring of the edges of the simple complete graphs are equivalent. In other words, finding a correct $(3, 3, \ldots, 3; 2)$–cyclic colouring for the edges of K_n, is the same problem as finding a symmetric Schur $(H_1, H_2, \ldots, H_k)$–partition of the set E of $n - 1$ elements.

Example 1. Let E be a set of 13 elements, $E = \{1, 2, \ldots, 13\}$. A symmetric Schur (H_1, H_2, H_3)–partition of E is the following:

$$H_1 = \{1, 4, 7, 10, 13\}, \ H_2 = \{5, 6, 8, 9\} \text{ and } H_3 = \{2, 3, 11, 12\}.$$

It is known that there is no symmetric Schur (H_1, H_2, H_3)–partition if $|E| = 14$ because $S(3) = 14$ (see [19, 9]). So, there is no correct cyclic $(3, 3, 3; 2)$–colouring for all the edges of the simple complete graph of order $n > 14$.

The correct cyclic $(3, 3, 3; 2)$–colouring of the edges of K_{14} is obtained by colouring the edges of length 1, 4 and 7 ($\{1, 4, 7\} \in H_1$) with the colour Red, those of length 5 and 6 ($\{5, 6\} \in H_2$) with the colour Blue and finally those of length 2 and 3 ($\{2, 3\} \in H_3$) with the colour Green. In the same way, we can easily construct the symmetric Schur $(H_1, H_2, \ldots, H_m)$–partition from classes of Ramsey–regular graphs.

To improve the lower bound associated with classical binary Ramsey numbers of the form $R(3, 3, \ldots, 3; 2)$, we can use the corresponding symmetric Schur $(H_1, H_2, \ldots, H_k)$–partition. This leads to the following theorem:

Theorem 16. *All simple complete graphs of order $n > 45$ have no correct cyclic $(3, 3, 3, 3; 2)$–colouring for their edges.*

Proof of theorem. We have shown above the equivalence between the cyclic colouring and the symmetric Schur partition. The Schur number $S(4)$ is exactly equal to 44 (see [19]). This implies that there exists no symmetric Schur (H_1, H_2, H_3, H_4)–partition for any set of n elements with $n > 44$. Consequently, there is no correct cyclic $(3, 3, 3, 3; 2)$–colouring for the edges of K_n with $n > 45$. $\square$

The improvement of the lower bound associated with the Ramsey number $R(3, 3, 3, 3; 2)$ becomes then a very difficult problem due to the fact that its associated graph has no correct cyclic $(3, 3, 3, 3; 2)$–colouring. However, in 1972, Chung [5] constructed a correct $(3, 3, 3, 3; 2)$–colouring (not cyclic) for the edges of K_{50}. So, we have $R(3, 3, 3, 3; 2) > 50$. We mention that this bound, 50, has not been improved yet. Exoo [9] has used a Schur $(H_1, H_2, H_3, H_4, H_5)$–partition to prove that $R(3, 3, 3, 3, 3; 2) > 162$.

6.3   Cyclic Colouring of Hypergraphs

We introduce in this section, the cyclic colouring of the hypergraphs associated with the classical ternary Ramsey numbers.

Definition 17. Let H be a complete hypergraph of rank 3 and order n, $H = (X, \mathcal{P}_3(X))$ and let E be an edge of H defined as a set of 3 vertices x_i, x_j, x_k of X. We define the *length* of E, $L(E)$, by:

$$L(E) = l(\{x_i, x_j\}) + l(\{x_i, x_k\}) + l(\{x_j, x_k\}) \ .$$

The lengths of the edges of the complete hypergraph K_5^3 are:

$L(\{1,2,3\}) = 4$, $L(\{1,2,5\}) = 4$, $L(\{3,4,5\}) = 4$, $L(\{2,3,4\}) = 4$, $L(\{1,4,5\}) = 4$, $L(\{1,3,5\}) = 5$, $L(\{1,2,4\}) = 5$, $L(\{2,4,5\}) = 5$, $L(\{2,3,5\}) = 5$, $L(\{1,3,4\}) = 5$.

Definition 18. Let H be a $(k_1, k_2, \ldots, k_m; 3)$–colourable hypergraph of rank 3 and order n. We say that H is *Ramsey–regular* if all the edges with the same length can be coloured with the same colour. The correct colouring that associates only one colour with each edge of a Ramsey–regular hypergraph, is called *a correct cyclic $(k_1, k_2, \ldots, k_m; 3)$–colouring.*

Proposition 19. *A complete hypergraph of order 5 and rank 3, K_5^3, has a correct cyclic $(4, 4; 3)$–colouring for its edges.*

Proof of proposition. A complete hypergraph K_5^3 contains five complete sub–hypergraphs of rank 3 and order 4. These sub–hypergraphs are exactly:

$H_1 = (X_1, \mathcal{P}_3(X_1))$ with $X_1 = \{1, 2, 3, 4\}$, $H_2 = (X_2, \mathcal{P}_3(X_2))$ with $X_2 = \{1, 2, 3, 5\}$, $H_3 = (X_3, \mathcal{P}_3(X_3))$ with $X_3 = \{1, 2, 4, 5\}$, $H_4 = (X_4, \mathcal{P}_3(X_4))$ with $X_4 = \{1, 3, 4, 5\}$, $H_5 = (X_5, \mathcal{P}_3(X_5))$ with $X_5 = \{2, 3, 4, 5\}$.

Each H_i contains two edges of different lengths. Indeed, H_1 contains the two edges $e_i = \{1, 2, 3\}$ and $e_j = \{1, 2, 4\}$ with $L\{e_i\} = 4$ and $L\{e_j\} = 5$; and in the same way, H_2 contains $e_i = \{1, 2, 3\}$ and $e_j = \{1, 3, 5\}$, H_3 contains $e_i = \{1, 2, 5\}$ and $e_j = \{1, 2, 4\}$, H_4 contains $e_i = \{1, 2, 3\}$ and $e_j = \{1, 2, 4\}$, H_5 contains $e_i = \{1, 2, 3\}$ and $e_j = \{1, 2, 4\}$ with $L\{e_i\} = 4$ and $L\{e_j\} = 5$.

To associate a correct cyclic $(4, 4; 3)$–colouring with the edges of K_5^3, it suffices to assign to the edges of length 4 the Red colour, and to those of length 5 the Blue colour. □

The following is a list of Ramsey–regular hypergraphs $(4, 6; 3)$–colourable and $(5, 6; 3)$–colourable.

Table 1. Correct cyclic colourings of the edges of certain Ramsey hypergraphs.

$(4, 6; 3)$–colourable	K_{17}^3	K_{19}^3	
colour 1	4, 8, 17	4, 10, 19	
colour 2	6, 10, 12, 14, 16	6, 8, 12, 14, 16, 18	

$(5, 6; 3)$–colourable	K_{23}^3	K_{25}^3	K_{28}^3
colour 1	6, 8, 14, 20, 22	6, 8, 14, 20, 22, 24	8, 14, 20, 22, 24, 26
colour 2	4, 10, 12, 16, 18, 23	4, 10, 12, 16, 18, 25	4, 6, 10, 12, 16, 18, 28

The above tables show that $R(4,6;3) > 19$ and $R(5,6;3) > 28$. We can define, in the same way, the length of the edges of all complete hypergraphs of any order h. We conjecture that:

Conjecture 20. *Given $m+1$ positive integers, $k_1, k_2, \ldots, k_m, h$ with $k_i > h$ for all $i \in \{1, 2, \ldots, m\}$, there exists a positive integer n, such that all complete hypergraphs of rank h and order n have a correct cyclic $(k_1, k_2, \ldots, k_m; h)$-colouring for their edges.*

Property of non Ramsey–Regular Hypergraphs. The non Ramsey–regular hypergraphs have the same characteristics as the non Ramsey–regular graphs. That is to say, if there exist a $(k_1, k_2, \ldots, k_m; 3)$-colourable hypergraph K_n^3, then there exists a $(k_1, k_2, \ldots, k_m; 3)$-colourable hypergraph K_{n-1}^3. However, this is not true for the Ramsey–regular hypergraphs. Indeed, the hypergraph K_5^3 has a correct cyclic $(4, 4; 3)$-colouring, while K_6^3 has no cyclic colouring. In fact, K_6^3 contains a complete sub–hypergraph H' of rank 3, constituted by vertices 2, 3, 5 and 6. We can easily verify that all the edges of H' have the same length 6 and, by definition, they will be coloured with the same colour. This leads to the following theorem:

Theorem 21. *Given m colours and an integer i greater than 3, the complete hypergraph of rank 3 and order $n = 4i$, K_n^3, has no correct cyclic $(\underbrace{4, \ldots, 4}_{m}; 3)$-colouring for its edges.*

Proof of theorem. Let H be a complete hypergraph of rank 3 and order n, and let i be a positive integer greater than 3, with $n = 4i$. The hypergraph H contains always a complete sub–hypergraph of rank 3 and order 4, $H' = (X, \mathcal{P}_3(X))$ with $X = \{x_1, x_2, x_3, x_4\}$ and $l(e_1 = \{x_1, x_2\}) = l(e_2 = \{x_2, x_3\}) = l(e_3 = \{x_3, x_4\}) = l(e_4 = \{x_1, x_4\}) = i$. Indeed, we can take for example, $x_1 = 1$, $x_2 = 1+i$, $x_3 = 1 + 2i$ and $x_4 = 1 + 3i$. This gives: $l(e_1) = l(e_2) = l(e_3) = i$. The length l of e_4 cannot be equal to $|1 - 1 - 3i|$ because $3i$ is strictly greater than $\lfloor \frac{n}{2} \rfloor$. Consequently, $l(e_4) = 1 + 4i - 1 - 3i = i$. We can verify that the lengths of $e_5 = \{x_1, x_3\}$ and $e_6 = \{x_2, x_4\}$ are both equal to $2i$. All the edges of H' have the same length $(4i)$ and by definition, they will be coloured with the same colour. So, H' constitutes a monochrome 4–clique of rank 3. Consequently, there is no correct cyclic $(\underbrace{4, 4, \ldots, 4}_{m}; 3)$-colouring for the edges of the hypergraph H. $\square$

According to this theorem, the hypergraphs K_4^3, K_8^3, K_{12}^3, K_{16}^3, K_{20}^3, K_{24}^3, K_{28}^3, K_{32}^3, K_{36}^3 and K_{40}^3 etc. have no correct cyclic $(4, \ldots, 4; 3)$-colouring for their edges.

Finally, we conjecture that:

Conjecture 22. *Given m colours and two positive integers i and k with $i > k - 1$; then all complete hypergraphs of order $k \times i$ and rank $k - 1$ have no correct cyclic $(\underbrace{k, k, \ldots, k}_{m}; k - 1)$-colouring for their edges.*

Our previous two theorems (i.e., Theorem 13 and Theorem 21), assert that this conjecture is true for $h = 2$ and 3. We can also verify it for $h = 4$ by following exactly the same reasoning. We can, for example, take $x_1 = 1$, $x_2 = 1 + i$, $x_3 = 1 + 2i$, $x_4 = 1 + 3i$ and $x_5 = 1 + 4i$, as being vertices of the complete sub–hypergraph of order 5 and rank 4, H', and we prove that all the edges of H' have the same length, they will therefore be coloured with the same colour.

Results on Classical and Multiple Ramsey Numbers

To evaluate or to approximate the numbers of Ramsey, we apply our stochastic optimization algorithms (i.e., the rapid descent algorithm, the rapid descent and enumeration algorithm and the simulated annealing algorithm) and we use the cyclic colouring heuristics. We compare our results with those given in [1], [10], [18] and [36, 37]. The following tables illustrate our results. The symbol '*' is used to indicate new bounds on Ramsey numbers.

Table 2. Some new lower bounds for Ramsey numbers.

$R(k_1, k_2, k_3; 2)$	R(3,3,4;2)	R(3,3,5;2)	R(3,3,6;2)	R(3,3,7;2)	R(3,3,8;2)	R(3,3,9;2)
Theoretical bounds of Abott	≥ 17	≥ 27	≥ 35	≥ 45	≥ 55	≥ 71
Stochastic algorithms	≥ 30	≥ 45	$\geq 54^*$	$\geq 65^*$	$\geq 75^*$	$\geq 85^*$

Table 3. Two new upper bounds for $M_{(n)}(k_1, k_2; 2)$.

$M_{(n)}(k_1, k_2; 2)$	$M_{(16)}(3, 5; 2)$	$M_{(17)}(3, 5; 2)$	$M_{(18)}(3, 6; 2)$	$M_{(19)}(3, 6; 2)$	$M_{(18)}(4, 4; 2)$
Lower bounds of Exoo	≤ 13	≤ 18	≤ 2	≤ 5	≤ 9
Stochastic algorithms	$\leq 12^*$	$\leq 16^*$	≤ 2	≤ 5	≤ 9

Table 4. Some new upper bounds for $M_{(n)}(k_1, k_2; 2)$.

$M_{(28)}(3, 8; 2) \leq 2$	$M_{(17)}(3, 3, 3; 2) \leq 13^*$
$M_{(29)}(3, 8; 2) \leq 9^*$	$M_{(18)}(3, 3, 3; 2) \leq 19^*$
$M_{(30)}(3, 8; 2) \leq 20^*$	$M_{(19)}(3, 3, 3; 2) \leq 24^*$
$M_{(31)}(3, 8; 2) \leq 29^*$	$M_{(20)}(3, 3, 3; 2) \leq 32^*$
$M_{(14)}(4, 4; 3) \leq 14^*$	$M_{(15)}(4, 4; 3) \quad \leq 27^*$

Table 5. New upper bounds for multiple Ramsey numbers.

$r(3, 3, 3; 2) \leq 13^*$	$r(4, 4; 3) \leq 4^*$	$r(3, 8; 2) \leq 2^*$

Conclusion

These algorithms not only make it possible to find almost all known Ramsey numbers but also to improve considerably some bounds for those which remain unknown. Generally, they are capable to associate better bounds with Ramsey numbers. However, to construct a correct colourings, it is often necessary to produce a large number of initial colourings. This procedure requires a non negligible calculation time (i.e., about 15 CPU hours on a SUN SPARC machine for each Ramsey number, in averge), therefore, we are thinking about running them on a parallel machine (i.e., a Connection Machine: CM2 or CM5) hoping that this will improve their performance and make it possible to obtain new bounds on binary Ramsey numbers of two colours $R(k_1, k_2; 2)$ that are not yet evaluated.

We have also shown the existence of the cyclic colouring of hypergraphs associated with ternary Ramsey numbers. Thus, we can associate with each Ramsey number a complete hypergraph and construct a correct colouring by using the properties of the cyclic colouring that we have introduced. These new heuristics are very useful for an improvement of the bounds on ternary Ramsey numbers.

The advantage of our methods, despite their incompleteness, resides in their capacity to solve some difficult combinatorial problems of Ramsey theory, which cannot be solved by classical resolution methods.

We mention that, in the case of the associated graphs possessing many correct colourings, the algorithm of rapid descent (i.e., Algorithm 1) seems to be the best adapted one. The other two algorithms (i.e., Algorithm 2, Algorithm 3), and especially the last one (i.e., Simulated Annealing algorithm) are used in the opposite case.

At the present, methods of stochastic optimization seem to be those that give best results, especially in problems where the search space is particularly large.

Acknowledgment. I would like to thank Alain Colmerauer, Dhiaddin Husain, Vincent Vialard and the anonymous reviewers for many helpful comments and suggestions to improve the presentation of this paper.

References

1. Abott, H.L.: Some problems in combinatorial analysis. Ph.D. University of Alberta Edmonton (1965)
2. Adorf, H.M., Johnson, M.D.: A discrete stochastic neural network algorithm for constraint satisfaction problems. Proc. of the Int. Joint. Conf. on Neural Networks (1990)
3. Golomb, S.W., Baumert, L.O.: Backtrack Programming. Journal of the ACM, **12 4** (1965) 516–524
4. Berge, C.: Graphes et Hypergraphes. Dunod, Paris (1970)
5. Chung, F.R.K.: On the Ramsey numbers $N(3, 3, \ldots, 3; 2)$. J. of Disc. Math. **5** (1973) 317–321
6. Chvátal, V., Harary, F.: Generalized Ramsey theory for Graphs. Proc. of the Amer. Math. Soc. **32** (1972) 389–394
7. Garey, R.M., Johnson, D.S.: Computers and Intractability: A Guide to the Theory of NP–Completeness. Freeman and Company, New York, (1979).
8. Erdös, P.: Some remarks on the theory of Graphs. Bull. of the Amer. Math. Soc. **53** (1947) 292–294
9. Exoo, G.: A lower bound for Schur numbers and multicolor Ramsey numbers of K_3. Elec. J. of Combin. **1** (1994)
10. Exoo, G.: Applying optimization algorithms to Ramsey problems. J. of Graph Theory (1989) 175–197
11. Exoo, G.: A lower bound for $R(5, 5)$. J. of Graph Theory **13** (1989) 97–98
12. Exoo G.: On two classical Ramsey numbers of the form $R(3, n)$. J. of Disc. Math. **4** (1989) 488–490
13. Exoo, G.: Some constructions related to Ramsey multiplicity. Ars Combin. **26** (1988) 233–242

14. Giraud, G.: Sur le problème de Goodman pour les quadrangles et la majoration des nombres de Ramsey. J. of Combin. Theory **B 27** (1979) 237–253

15. Giraud, G.: Sur les proportions respectives de triangles uni, bi ou tricolores dans un tricoloriage des arêtes du n–emble. J. of Disc. Math. **16** (1976) 13–38

16. Giraud, G.: Majoration du nombre de Ramsey ternaire-bicolore en $(4, 4)$. Note CRAS Paris **A 269** (1969) 1173–1175

17. Greenwood, R.E., and Gleason, A.M.: Combinatorial relations and chromatic Graphs. Canad. J. of Math. **7** (1955) 1–7

18. Graham, R., Spencer, J.: La théorie de Ramsey. Pour la Science **155** (1990) 58–63

19. Graham, R.L., Rothschild, B.L., Spencer, J.H.: Ramsey Theory (second edition) John Wiley, New York (1990)

20. Gu, J.: Efficient local search for very large-scale satisfiability problems. Sigart Bull. **3 1** (1992) 8–12

21. Harary, F., Prins, G.: The Ramsey multiplicity of a Graph. Generalized Ramsey Theory for Graphs IV, Networks **4** (1974) 163–173

22. Harary, F.: A survey of generalized Ramsey theory. Graphs and Combinatorics, Springer-Verlag (1974) 10–17

23. Isbell, J.R.: $N(4, 4; 3) \geq 13$. J. of Combin. Theory **6** (1969) 210

24. Jaam, J.: Coloring of Hypergraphs associated with Ramsey numbers by stochastic algorithms. 4th Twente Workshop on Graphs and Combinatorial Optimization, (1995), 153–157

25. Jaam, J.: Coloriage cyclique pour les hypergraphes complets associés aux nombres de Ramsey classiques ternaires. Bull. of Symbolic Logic (extended abstract) **2** (1995) 241–242

26. Jaam J., Fliti T., Hussain D.: New bounds of Ramsey numbers via a top–down algorithm. CP'95-Workshop SSRHP, (1995), 110–118.

27. Jacobson, M.S.: A note on Ramsey multiplicity. J. of Disc. Math. **29** (1980) 201–203

28. Kalbfleisch, J.G.: Chromatic Graphs and Ramsey theorem. Ph.D. University of Waterloo (1966)

29. Kirkpatrick S., Gelatt, C.D., Vecchi, M.P.: Optimization by simulated annealing. Science **220** (1983) 671–680

30. Manber, U.: Introduction to Algorithms (A Creative Approch). Addison Wesley, USA (1989)

31. Metropolis, N., Rosenbluth, A., Rosenbluth, M., Teller, A.: Equation of state calculations by fast computation machines. J. of Chimical Physics **22** (1953) 1087–1092

32. McKay, D.B., Radziszowski, S.P.: The Death of Proof. Scientific Amer. (1993)

33. McKay, D.B., Radziszowski, S.P.: A new upper bound for the Ramsey number $R(5, 5)$. Aust. J. Combin. **5** (1992) 13–20

34. McKay, D.B, Radziszowski, S.P.: The first classical Ramsey number for Hypergraphs is computed. Proc. of the Second Annual ACM-SIAM Symp. on Discrete Algorithms. (1991) 304–308

35. Minton, S., Johnson, M.D., Philips, A.B., Laird, P.: Solving large-scale constraint satisfaction and scheduling problems using a heuristic repair method. Proc. of the AAAI-90 (1990) 17–24

36. Radziszowski, S.P.: Small Ramsey numbers. Elec. J. of Combin. **1** (1994)

37. Radziszowski, S.P., Kreher, D.L.: Search algorithm for Ramsey Graphs by union of group orbits. J. of Graph Theory **12** (1988) 59–72

38. Ramsey, F.P.: On a problem of formal logic. Proc. of the London Math. Soc. **30** 264–286 (1930)

39. Ryser, H.J.: Combinatorial mathematics. Carus Math. Monograph **14** (1963) 38–43

40. Schur, I.: Über die Kongruenz $x^m + y^m = z^m mod(p)$. Jahresber Deutsch Verein **25** (1916) 114–116

41. Selman, B., Kautz, H.A., and N. Cohen, N.: Noise strategies for improving local search. Proc. of the 12th Nat. Conf. on Artificial Intelligence **1** (1994) 337–343

42. Selman, B., Kautz, H.A.: Domain independent extensions to G-SAT: solving large structured satisfiability problems. Proc. of the IJCAI-93 (1993)

43. Sosic, R., Gu, J.: 3,000,000 Queens in less than one minute. Sigart Bull. **2 2** (1991) 22–24

44. Sosic, R., J. Gu, J.: A polynomial time algorithm for the N-Queens problem. Sigart Bull. **1 3** (1990) 7–11

45. Snir, M.: On parallel searching. SIAM J. of Comput. **14** (1985) 688–708

46. Waltz, D.: Understanding line drawings of scenes with shadows. The Psychology of Computer Vision, Mc-Graw-Hill, New York (1975) 19–91

On the Hybrid Neural Network Model for Solving Optimization Problems

Fouad B. Chedid

Faculty of Computer Science, Temple University Japan
2-2 Minami-Osawa, Hachioji-shi, Tokyo 192-03, Japan

Abstract: A recent model of neural networks, named the Hybrid Neural Network Model (HN), for solving optimization problems appeared in [3]. In [3], the main algorithm called the Hybrid Network Updating Algorithm (HNUA) is used to drive the HN model. The best thing about the HNUA is that it reaches a feasible solution very quickly. Our argument here is that while the HNUA is very quick to satisfy the constraints, it guarantees very little in terms of the quality of the generated solution. In this paper we rewrite one of the steps in the HNUA so that the goal function is better served. we demonstrate our work using the traveling salesman problem as an example.

1 Introduction

Ever since the work of Hopfield and Tank on solving optimization problems by neural networks appeared in [2], there has been a tremendous interest in the subject with results varying in their degrees of approval or disapproval of the method. The one thing that is sure; however, is that the well-researched theory of algorithms has produced a long list of NP-hard problems [1], most of which we would like to solve immediately, that do not have and may never have effective algorithms (the problem of whether P=NP is as much philosophical as it is mathematical). To this extent, any new approach to tackle NP-hard problems becomes important, and this is exactly what makes Hopfield Networks attractive in this context. We begin with a brief review of Hopfield Networks, and Hopfield and Tank's work on solving the Traveling Salesman Problem (TSP) [1].

Following [4], a Hopfield network is a single layer of neurons with total interconnectivity (every neuron is connected to every other neuron). Hopfield networks are massive feedback loops. Also, the output of each neuron depends upon the previous values of its own activation, so the individual neurons have time-dependent behavior. Figure 1 shows the general configuration of Hopfield networks with 5 neurons.

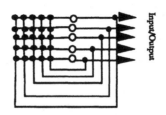

Fig. 1. Hopfield Network

In Fig. 1, black circles represent connections and white circles represents neurons. To enable the network to compute a solution to a problem, an energy function must be defined. In the case of the traveling salesman problem with n cities, Hopfield and Tank use n^2 neurons named v_{xi} $(1 \le x, i \le n)$. The neuron v_{xi} is set to 1 if the city x is visited on the $i\,th$ step during the tour. A feasible solution for the TSP is an instance of v_{xi} in which each row and each column has exactly one 1; all other entries should be 0. An optimum solution for the TSP is an instance of v_{xi} for which the following energy function E is minimum [2]:

$$E = E_s + Cost$$

$$E_s = \frac{A}{2}\sum_x\sum_i\sum_{j\neq i} v_{xi}v_{xj} + \frac{B}{2}\sum_i\sum_x\sum_{y\neq x} v_{xi}v_{yi} + \frac{C}{2}\left(\sum_x\sum_i v_{xi} - n\right)^2$$

$$Cost = \sum_x\sum_{y\neq x}\sum_i d_{xy}v_{xi}(v_{y,i+1} + v_{y,i-1})$$

where d_{xy} is the distance between cities x and y, and A, B, C, and D are constant values. It is clear that E_s is zero if there is no more than one entry per row or column of v_{xi}, and if $\sum_x\sum_i v_{xi} = n$. This is to say that no city can be visited more than once, no two cities can be visited at the same time, and the tour is a valid tour. During convergence, the network moves from states of "roughly defined tours" where cities are being considered for several positions simultaneously) to states of higher refinement (where some cities positions are still unfixed), until a single tour is left. Hopfield used his network to find a tour of 10 cities. He started from random states and in a trial of 20 convergences, 16 converged to legitimate tours. Half of the 16 tours produced optimal tours [4].

On the negative side, one can summon the major obstacles of doing optimization problems by neural networks in three issues: 1) the lack of any systematic approach to formulate an optimization problem as a neural network instance, 2) the issue of avoiding local minimums (valleys) on the energy surface, and 3) the uncertainty about whether a global minimum can really be achieved; that is, will the system converge to a globally stable state?, and how long will it take for it to do so? In a recent publication [3], Sun and Fu introduced a new model of neural networks, called the Hybrid Neural Network Model (HN), for solving optimization problems. The HN model contains two sub-networks; the constraints-network to satisfy the constraints, and the goal-network to optimize the goal function. These two sub-networks are put to work together to guarantee fast convergence of the constraints sub-network. In [3], the main algorithm called the Hybrid Network Updating Algorithm (HNUA) is used to drive the HN model. The best thing about the HNUA is that it reaches a feasible solution very quickly (no hill climbing, nothing of this sort). The worst thing about the HNUA is that it guarantees very little in terms of the quality of the generated solution (goal optimization). The HNUA, as described in [3], uses a basic greedy-like technique in its drive to maximize the goal function. In this paper, we take that greedy technique a step further so that the goal function is better served; we demonstrate our work using the TSP as an example. The rest of the paper is organized as follows: Section 2 reviews the HN model. Section 3 reviews the HNUA for the TSP. Our Modified HNUA (MHNUA) is described in section 4, and a detailed example is included in section 5.

2 The HN Model

The Hybrid Neural Network Model (HN) is designed with optimization problems in mind. As such, the HN model has two sub-networks: the constraints-network and the goal-network. These two sub-networks are put to work together to quickly satisfy the constraints while maintaining a feasible solution. Fig. 2 below depicts the HN model.

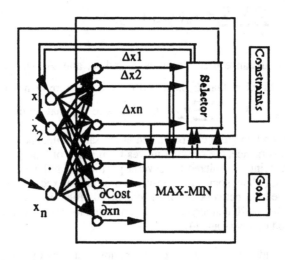

Fig. 2. The Hybrid Neural Network Model

In Fig. 2, the x_i are the problem's variables. These x_i assume values 0 and 1 only; that is to say false and true, respectively. The constraints are equations of the form $a_i = t_i$, where a_i involve the variables x_i, and t_i are some target values to be reached. The sum of squared errors is defined as $E_s = \sum (a_i - t_i)^2$. The term *Cost* is used to represent the goal function to be optimized, and the energy of the system is defined as $E = E_s + Cost$. For each neuron, the constraints sub-network computes the updating values Δx_i as:

$$\Delta x_i = -\frac{\frac{\partial E_s}{\partial x_i}}{\frac{\partial^2 E_s}{\partial x_i^2}}$$

which assumes that the second derivative of E_s is non-zero. These Δx_i are passed as well to the goal sub-network where the MAX-MIN portion determines which x_i corresponds to maximum Δx_i. The goal network computes $\frac{\partial Cost}{\partial x_i}$ and the MAX-MIN portion determines which one among the x_i (with maximum Δx_i) corresponds to optimum $\frac{\partial Cost}{\partial x_i}$. One of these x_i, say x_0, is chosen (at random) and passed to the selector. The selector passes the corresponding Δx_0 to the network input to update the value of x_0. During the next iteration, x_0 assumes the value $x_0 + \Delta x_0$. This process is repeatedly executed until the gradient of E_s is equal to zero; that is the constraints are satisfied.

3 The Hybrid Network Updating Algorithm (HNUA)

W.l.o.g, we write the algorithm to suit the traveling salesman problem (TSP). In [3], the TSP is formulated in a way similar to the formulation of Hopfield and Tank [2]; except Sun and Fu force the entries on each row and each column of matrix v_{xi} to be at most 1. Following [3], the energy function E is defined as

$$E = E_s + Cost$$

$$E_s = \sum_x \sum_i \sum_{j \neq i} (v_{xi}v_{xj})^2 + \sum_i \sum_x \sum_{y \neq x} (v_{xi}v_{yi})^2 +$$
$$\sum_x \sum_i \sum_{j \neq i} ((1-v_{xi})(1-v_{xj}))^2 + \sum_i \sum_x \sum_{y \neq x} ((1-v_{xi})(1-v_{yi}))^2.$$

$$Cost = \sum_x \sum_{y \neq x} \sum_i d_{xy} v_{xi} (v_{y,i+1} + v_{y,i-1} - v_{y,i+1}v_{y,i-1}).$$

Notice the absence of any constants in the expression of E_s. It is shown in [3] that the function E_s monotonically converges to a stable state; because E_s has non-zero

second partial derivatives over a convex set in which each variable is of order two --
For this and many other properties of E_s, the reader is referred to [3].

Procedure HNUA:
Let n be the number of cities.

[1] For each v_{xi} $(1 \leq x, i \leq n)$, compute

$$\Delta x_i = -\frac{\frac{\partial E_s}{\partial x_i}}{\frac{\partial^2 E_s}{\partial x_i^2}}$$

[2] For each v_{xi} $(1 \leq x, i \leq n)$, compute

$$\frac{\partial Cost}{\partial v_{xi}}$$

[3] Determine the maximum values Δv^* of $|\Delta v_{xi}|$'s,

$$\Delta v^* = \max((|\Delta v_{xi}|, 1 \leq x, i \leq n)),$$

then form a set Γ of variables v_{yj} that corresponds to the maximum updating value
Δv^*, i.e., $\Gamma = \{v_{yj} | |\Delta v_{yj}| = \Delta v^*, \forall y, j\}$.

[4] Among the v_{yj}'s in the set Γ, select a variable v_{wk} which corresponds to the
minimum value of the partial derivative of the *Cost* function.

$$v_{wk} = \min\left(\frac{\partial Cost}{\partial v_{yj}}, \forall v_{yj} \in \Gamma\right).$$

Update v_{wk} by adding the updating value obtained in step [1].

$$v_{wk} = v_{wk} + \Delta v_{wk}.$$

[5] Test the gradient of function E_s. If the gradient of E_s is equal to 0
$\left(\frac{\partial E_s}{\partial v_{xi}} = 0 \ (1 \leq x, i \leq n)\right)$, i.e. the E_s function converges to a stable state, then
stop. Otherwise return to step [1] for the next iteration.

4 The Modified HNUA (MHNUA)

We observe that while the HNUA satisfies the constraints very rapidly (the number of iterations is equal to $n-1$), the algorithm guarantees very little in terms of the quality of the generated solution. Notice that in step [4], the algorithm will almost always result in more than one choice for v_{wk}. To accept one of those v_{wk} at random is a very costly operation. That v_{wk} will determine once and for all that city w is to be visited on the kth step during the tour. Subsequently, this will have a very random effect on the length of the generated tour. Our argument here is that while the HNUA is very quick to satisfy the constraints, the major issue in the TSP is the length of the generated tour. In the next section, we rewrite step [4] in the HNUA so that the goal function (tour length) is better served. The basic idea in the MHNUA is to try to select the best candidate for v_{wk} at each iteration rather than a mere random selection.

We notice that a projection of the cost that a particular choice for v_{wk} may contribute to the length of the entire tour is

$$Costv_{wk} = \sum_{x} \max\left(\frac{\partial Cost}{\partial v_{xi}}, (1 \le i \le n)\right).$$

where the sum runs over all the cities x whose order has not been decided; i.e., $\sum_{i=1}^{n} v_{xi} = 0$. The term $\dfrac{\partial Cost}{\partial v_{xi}}$ measures the length of that part of the tour (two edges at most) associated with visiting city x at time i.

Our MHNUA computes $Costv_{wk}$ for all v_{wk} found at each iteration, and selects that v_{wk} which provides the smallest additional cost. Obviously it takes longer to run the MHNUA than running the HNUA. The judgment as to which algorithm to use lies in what is more important for a particular application; a mere tour or a good tour.

In the following we rewrite step [4] as it should appear in the MHNUA:

/* beginning of step [4] */

Let $\Delta Cost^* = \min\left(\dfrac{\partial Cost}{\partial v_{yj}}, \forall v_{yj} \in \Gamma\right)$.

Let Γ^* be the set $\left\{v_{wk} \in \Gamma \mid \dfrac{\partial Cost}{\partial v_{wk}} = \Delta Cost^*\right\}$.

For each v_{wk} **in** Γ^* **do**

    Work on a seperate copy of current v_{xi} $(1 \leq x, i \leq n)$

    Update v_{wk} by adding the updated value obtained in step [1]: $v_{wk} = v_{wk} + \Delta v_{wk}$.

    Compute $Costv_{wk} = \sum_x \max\left(\dfrac{\partial Cost}{\partial v_{xi}}, (1 \leq i \leq n) \right)$, where $\sum_{i=1}^{n} v_{xi} = 0$.

EndFor

Choose and update that v_{wk} which corresponds to minimum $Costv_{wk}$.

/* end of step [4] */

We have included a detailed example in Appendix A below; additional comments are included as well.

5 Appendix A

Suppose n=4 and the following distance matrix:

$$d = \begin{array}{c} \\ A \\ B \\ C \\ D \end{array} \begin{array}{c} \begin{array}{cccc} A & B & C & D \end{array} \\ \left(\begin{array}{cccc} 0 & 1 & 4 & 10 \\ 1 & 0 & 9 & 2 \\ 4 & 9 & 0 & 2 \\ 10 & 2 & 2 & 0 \end{array} \right) \end{array}$$

The variables for this instance are:

$$v_{xi} = \begin{pmatrix} v_{11} & v_{12} & v_{13} & v_{14} \\ v_{21} & v_{22} & v_{23} & v_{24} \\ v_{31} & v_{32} & v_{33} & v_{34} \\ v_{41} & v_{42} & v_{43} & v_{44} \end{pmatrix}$$

Iteration #1:

$$V_{xi} = \begin{pmatrix} 1 & 0 & 0 & 0 \\ 0 & 0 & 0 & 0 \\ 0 & 0 & 0 & 0 \\ 0 & 0 & 0 & 0 \end{pmatrix} \quad \Delta v_{xi} = \begin{pmatrix} 0 & .5 & .5 & .5 \\ .5 & 1 & 1 & 1 \\ .5 & 1 & 1 & 1 \\ .5 & 1 & 1 & 1 \end{pmatrix} \quad \frac{\partial Cost}{\partial v_{xi}} = \begin{pmatrix} 0 & 0 & 0 & 0 \\ 0 & 1 & 0 & 1 \\ 0 & 4 & 0 & 4 \\ 0 & 10 & 0 & 10 \end{pmatrix}$$

$$\Gamma = \{v_{22}, v_{23}, v_{24}, v_{32}, v_{33}, v_{34}, v_{42}, v_{43}, v_{44}\}$$
$$\Gamma^* = \{v_{23}, v_{33}, v_{43}\}$$

We are about to select the city to visit at step 3 during the tour.

1) Let $v_{wk} = v_{23}$ (try city B). We update v_{23}; $v_{23} = v_{23} + \Delta v_{23} = 0 + 1 = 1$. Next we see how does this choice influence the other cities (only those whose order has not been decided). We compute $\dfrac{\partial Cost}{\partial v_{xi}}$ for x = 3,4.

$$\frac{\partial Cost}{\partial v_{xi}} = \begin{pmatrix} . & . & . & . \\ . & . & . & . \\ 0 & 13 & 0 & 13 \\ 0 & 12 & 0 & 12 \end{pmatrix}$$

This says that now city C has an associated cost of length 13 if visited at step 2 or 4. City D has an associated cost of length 12 if visited at step 2 or 4. To verify the above data, examine the corresponding tours:

A->C->B->D->A (total length = 25)
A->D->B->C->A (total length = 25)

Finally , we compute $Costv_{23}$, the projected cost that this choice may contribute to the length of the entire tour:

$$Costv_{23} = \sum_{x=3,4} \max\left(\frac{\partial Cost}{\partial v_{xi}}, 1 \le i \le 4\right) = 25.$$

2) Let $v_{wk} = v_{33}$ (try city C). We update $v_{33} = v_{33} + \Delta v_{33} = 0 + 1 = 1$, and compute $\dfrac{\partial Cost}{\partial v_{xi}}$ for x = 2,4.

$$\frac{\partial Cost}{\partial v_{xi}} = \begin{pmatrix} \cdot & \cdot & \cdot & \cdot \\ 0 & 10 & 0 & 10 \\ \cdot & \cdot & \cdot & \cdot \\ 0 & 12 & 0 & 12 \end{pmatrix}$$

$Costv_{33} = 22.$

3) Let $v_{wk} = v_{43}$ (try city D). We update $v_{43} = v_{43} + \Delta v_{43} = 0 + 1 = 1$, and compute $\dfrac{\partial Cost}{\partial v_{xi}}$ for x = 2,3.

$$\frac{\partial Cost}{\partial v_{xi}} = \begin{pmatrix} \cdot & \cdot & \cdot & \cdot \\ 0 & 3 & 0 & 3 \\ 0 & 6 & 0 & 6 \\ \cdot & \cdot & \cdot & \cdot \end{pmatrix}$$

$Costv_{43} = 9.$

Since it is v_{43} which provides the minimum cost, we choose to visit city D at step 3 during the tour. The gradient of E_s

$$\nabla E_s = \begin{pmatrix} 0 & -2 & 0 & -2 \\ -2 & -4 & -2 & -4 \\ -2 & -4 & -2 & -4 \\ 0 & -2 & 0 & -2 \end{pmatrix}$$

Iteration #2:

$$v_{xi} = \begin{pmatrix} 1 & 0 & 0 & 0 \\ 0 & 0 & 0 & 0 \\ 0 & 0 & 0 & 0 \\ 0 & 0 & 1 & 0 \end{pmatrix} \quad \Delta v_{xi} = \begin{pmatrix} 0 & .5 & 0 & .5 \\ .5 & 1 & .5 & 1 \\ .5 & 1 & .5 & 1 \\ 0 & .5 & 0 & .5 \end{pmatrix} \quad \frac{\partial Cost}{\partial v_{xi}} = \begin{pmatrix} 0 & 10 & 0 & 10 \\ 0 & 3 & 0 & 3 \\ 0 & 6 & 0 & 6 \\ 0 & 10 & 0 & 10 \end{pmatrix}$$

$$\Gamma = \{v_{22}, v_{24}, v_{32}, v_{34}\}$$
$$\Gamma^* = \{v_{22}, v_{24}\}$$

This decides whether to visit city B during the second step or the fourth step.

1) Let $v_{wk} = v_{24}$ (try step 4). We update $v_{24} = v_{24} + \Delta v_{24} = 0 + 1 = 1$, and compute $\dfrac{\partial Cost}{\partial v_{xi}}$ for x = 3.

$$\frac{\partial Cost}{\partial v_{xi}} = \begin{pmatrix} \cdot & \cdot & \cdot & \cdot \\ \cdot & \cdot & \cdot & \cdot \\ 9 & 6 & 9 & 6 \\ \cdot & \cdot & \cdot & \cdot \end{pmatrix}$$

$Costv_{24} = 9$.

2) Let $v_{wk} = v_{22}$ (try step 2). We update $v_{22} = v_{22} + \Delta v_{22} = 0 + 1 = 1$, and compute

$$\frac{\partial Cost}{\partial v_{xi}} = \begin{pmatrix} \cdot & \cdot & \cdot & \cdot \\ \cdot & \cdot & \cdot & \cdot \\ 9 & 6 & 9 & 6 \\ \cdot & \cdot & \cdot & \cdot \end{pmatrix}$$

$Costv_{22} = 9$.

Since both choices project similar cost, we choose either one (at random); say v_{24}. The gradient of E_s

$$\nabla E_s = \begin{pmatrix} 0 & -2 & 0 & 0 \\ 0 & -2 & 0 & 0 \\ -2 & -4 & -2 & -2 \\ 0 & -2 & 0 & 0 \end{pmatrix}$$

Iteration #3:

$$v_{xi} = \begin{pmatrix} 1 & 0 & 0 & 0 \\ 0 & 0 & 0 & 1 \\ 0 & 0 & 0 & 0 \\ 0 & 0 & 1 & 0 \end{pmatrix} \quad \Delta v_{xi} = \begin{pmatrix} 0 & .5 & 0 & 0 \\ 0 & .5 & 0 & 0 \\ .5 & 1 & .5 & .5 \\ 0 & .5 & 0 & 0 \end{pmatrix} \quad \frac{\partial Cost}{\partial v_{xi}} = \begin{pmatrix} 1 & 10 & 1 & 10 \\ 0 & 3 & 0 & 3 \\ 9 & 6 & 9 & 6 \\ 2 & 10 & 2 & 10 \end{pmatrix}$$

$\Gamma = \{v_{32}\}, \quad \Gamma^* = \{v_{32}\}$

This decides that city C is to be visited during the 2nd step of the tour. We set $v_{32} = v_{32} + \Delta v_{32} = 0 + 1 = 1$, and compute the gradient of E_s

$$\nabla E_s = \begin{pmatrix} 0 & 0 & 0 & 0 \\ 0 & 0 & 0 & 0 \\ 0 & 0 & 0 & 0 \\ 0 & 0 & 0 & 0 \end{pmatrix}$$

Since $\nabla E_s = 0$ then the constraints are now satisfied. The algorithm ends, and the solution is stored in

$$v_{xi} = \begin{pmatrix} 1 & 0 & 0 & 0 \\ 0 & 0 & 0 & 1 \\ 0 & 1 & 0 & 0 \\ 0 & 0 & 1 & 0 \end{pmatrix}$$

which corresponds to the tour A->C->D->B->A of length 9 which is optimal in this case.

To clarify further the difference between HNUA and MHNUA, we compare the solutions that would otherwise be generated by HNUA on the above instance, together with the number of iterations made by each algorithm. Algorithm HNUA would have made 3 iterations and generated any of the following tours (following random values of v_{wk} at step [4]):

A->C->B->D->A	of length 25
A->D->B->C->A	of length 25
A->B->C->D->A	of length 22
A->D->C->B->A	of length 22
A->B->D->C->A	of length 9
A->C->D->B->A	of length 9.

On the other hand, MHNUA makes 6 iterations ($= \sum |\Gamma^*|$) exploring all possible values of v_{wk} and generating an optimal solution: A->C->D->B->A.

References

[1] M.R.Garey and D.S. Johnson, *Computers and Intractability: A Guide to the Theory of NP-completeness*. San Francisco, CA: Freeman, 1979.

[2] J.J. Hopfield and D.W. Tank, "Neural Computation of Decisions in Optimization Problems," *Biological Cybernetics*, vol. 52, pp. 1-25, 1985.

[3] K.T. Sun and H.C. Fu, "A Hybrid Neural Network Model for Solving Optimization Problems," *IEEE Trans. on Computers*, vol. 42, no. 2, pp. 218-227, Feb. 1993.

[4] *"Introduction to Neural Networks, Computer Simulations of Biological Intelligence,"* California Scientific Software, Pasadena, CA, 1988.

Constructive – non-Constructive Approximation and Maximum Independent Set Problem

Marc Demange[1,2]   Vangelis Th. Paschos[1]

[1] LAMSADE, Université Paris-Dauphine
Place du Maréchal De Lattre de Tassigny, 75775 Paris Cedex 16, France
e_mails: {demange,paschos}@lamsade.dauphine.fr
[2] CERMSEM, Université Paris I, 90, rue de Tolbiac, 75634 Paris Cedex 13

Abstract. We apply in the case of the maximum independent set, a general thought process consisting in integrating an information on the optimal objective value in its instance. This thought process for the study of the relative hardness between determining solutions of combinatorial optimization problems and computing (approximately or exactly) their optimal values, allows us to define classes of independent set problems the approximability of which is particularly interesting.

1   Introduction

In theoretical computer science, problems (a fortiori the NP-complete ones) are originally defined as decisions about the existence of solutions verifying some properties. So in the *decision framework* (denoted by **D** in what follows) a problem is a question whose objective is to answer by *yes* or *no*.

However, the polynomial approximation theory needs a different framework, called *optimization framework*, defined below, simultaneously establishing and using the notion of the *feasible solution* and the one of the *objective value*; both notions are crucial for defining the concept of the *approximate solution* and for characterizing its quality.

So, in the optimization framework (denoted by **O**), the instance of a problem Π is expressed in terms of an optimization program of the form

$$\begin{cases} \text{opt } f(\mathbf{x}) \\ \quad \mathbf{x} \in \mathcal{C} \end{cases}$$

where opt $\in \{\min, \max\}$, f is the objective function and $\mathcal{C}$ the set of feasible solutions. Then, Π consists of **determining an optimal solution $\mathbf{x}^*$** and an algorithm determining it is called *constructive*. An approximation algorithm dealing with this framework (i.e., **determining a feasible solution and guaranteeing a nearness of its value to the value of an optimal one**) will be called *constructive approximation algorithm*. The approximation point of view of **O** will be denoted by **O-ap**.

There exists also another framework, the one of the *optimal value* (denoted by **V**), where the instances are the same as in **O** but the goal is now to **determine the optimal value** and an algorithm doing this will be called *non-constructive*. In this context, one can also define a concept of approximation

(**V-ap**) and an algorithm determining a feasible **value** and guaranteeing a nearness of this value to the optimal one is called *non-constructive approximation algorithm.*

A broad discussion about these several computation frameworks is performed in [2].

In this paper we consider the relative hardness between the constructive and the non-constructive approximations in the framework of maximum independent set problem.

We propose in what follows a general thought process consisting in integrating in the instances of the problems an information carrying over their optimal value, in such a way that the new (modified) problem is a priori **V**-easier (easier when dealing with **V**) than the original one. Let us note that in this case, the recognition of the instances of the derived problems is not necessarily polynomial[3]. So, via this transformation, work in framework **O** allows to better understand the links between constructiveness and non-constructiveness and a pertinent question is: *in what an information about the optimal value of a problem helps in deteminining either the optimal or a good approximate solution for it?*

We study the above question in the case of one of the most famous NP-complete problems, the maximum independent set.

Let $G = (V, E)$ be a graph of order n; an *independent set* is a subset of $V' \subseteq V$ such that whenever $\{v_i, v_j\} \subseteq V'$, $v_i v_j \notin E$, and the *maximum independent set problem (IS)* is to find an independent set of maximum size. A natural generalization of IS is the one where positive integer weights are associated with the vertices of G; then, the objective becomes to maximize the sum of weights of the vertices of an independent set; we denote this problem by WIS; we suppose also that the (integer) weights are universal constants (independent of n).

We consider a type of information addressed to (restricted) classes of instances of WIS, classes defined using this information. This thought process can be rich enough, since it allows us to apprehend the boundaries between constructiveness and non-constructiveness. We consider graphs defined by means of an information on the optimal value of the stability problem to be solved, and we give results concerning the constructive solution (exact or approximate) of IS on such graphs.

Let us remark that these results can be presented alternatively as autonomous independent set results on restricted classes of graphs, or as results in a "constructive – non-constructive framework".

Let $G = (V, E)$ be a graph of order n. We denote by $\alpha(G)$ the stability number (cardinality of a maximum independent set) of G and (for WIS) by w_i the weight of the vertex v_i, $i = 1, \ldots, n$ (by **w**, we denote the vector of the weights); as usual, $\Gamma(v)$, $v \in V$, denotes the neighbourhood of v. For a set $V' \subseteq V$, we denote by $G[V']$ the subgraph of G induced by V'; for a set A of edges, we denote by $T[A]$ the set of vertices, endpoints of the edges of A.

Moreover, given an instance I of a problem Π, we denote by $v(I)$ its optimal (objective) value. Finally, whenever vector **1** is indexed by a set of vertices,

---

[3] This type of problems are, eventually, not NPO in the sense of [2].

it denotes the characteristic vector of this set. Especially for the four linear programs of section 2.2, since the dimensions of the vectors 1 and 0 are not always the same ones, these vectors will be indexed by their dimension.

In the sequel, in section 2 we give some constructive results for IS and WIS, while in section 3 we investigate the boundaries between constructive and non-constructive framworks for IS and their impacts on the approximability of IS and of other related problems.

2 Constructive algorithms for classes of independent set problems defined with respect to the optimal value of their instances

2.1 The König-Egervary graphs

The class of König-Egervary graphs (KE-graphs) is composed of graphs G for which the cardinality of a maximum independent set is equal to the cardinality of a minimum edge covering of G. So, we really have here a class of IS instances defined starting from an a priori information concerning the optimal value. In this particular case, the knowledge of this information allows to polynomially V-solve (solve in the framework V) IS, since we only have to solve minimum edge covering, and this can be done in polynomial time ([3]). So, IS in KE-graphs is polynomial for the non-constructive framework. Moreover, given that Deming ([8]) has proved that the recognition of a KE-graph can be done in polynomial time and, furthermore, IS is O-polynomial in KE-graphs, the constructive version of the problem is also polynomial.

More recently, Bourjolly et al. ([4]) have extended the class KE to the one b-KE, where positive integer weights were considered on the vertices of G, and proved that WIS remains polynomially O-solvable (solvable in the framework O) for this class.

Here, we propose another generalization of the KE-graphs in both weighted and non-weighted cases. For the non-weighted case generalizations are issued from a combinatorial interpretation of the relation maximum independent set – minimum edge covering. For the weighted one, this generalization contains a less restrictive information than for the non-weighted case and is based upon linear programming arguments. As we show, in both weighted and non-weighted cases these generalization preserve the polynomiality of WIS.

2.2 Some generalizations of König – Egervary graphs

The non-weighted case. A general instance of IS defined by a graph $G = (V, E)$ can be written as a 0-1 linear problem as follows:

$$IS(G) = \begin{cases} \max 1_n \cdot \mathbf{x} \\ A \cdot \mathbf{x} \leq 1_{|E|} \\ \mathbf{x} \in \{0,1\}^n \end{cases}$$

where A is the edge-vertex incidence matrix of G.

Let us denote by IS_r the following relaxed version of IS:

$$IS_r(G) = \begin{cases} \max \mathbf{1}_n \cdot \mathbf{x} \\ \quad A \cdot \mathbf{x} \leq \mathbf{1}_{|E|} \\ \quad \mathbf{x} \geq \mathbf{0}_n \end{cases}$$

The dual of IS_r denoted by EC_r is

$$EC_r(G) = \begin{cases} \min \mathbf{1}_{|E|} \cdot \mathbf{x} \\ \quad A^T \cdot \mathbf{x} \geq \mathbf{1}_n \\ \quad \mathbf{x} \geq \mathbf{0}_{|E|} \end{cases}$$

where this problem is denoted by EC_r in order to indicate that it is the relaxed version of the following problem EC known as the minimum edge covering:

$$EC(G) = \begin{cases} \min \mathbf{1}_{|E|} \cdot \mathbf{x} \\ \quad A^T \cdot \mathbf{x} \geq \mathbf{1}_n \\ \quad \mathbf{x} \in \{0,1\}^{|E|} \end{cases}$$

Remark that $\mathbf{0}_n$ and $\mathbf{1}_{|E|}$ are feasible for IS_r and EC_r, respectively. As these dual instances have their respective constraint sets non empty, they have the same optimal value. Then, the following inequalities hold[4]: $\alpha(G) \leq v(IS_r(G)) = v(EC_r(G)) \leq v(EC(G))$. Let us refer to the difference $v(EC(G)) - \alpha(G)$ as the *discrete duality gap*.

A very interesting idea seems to be the generalization of the class KE (we will see later how this generalization applies also to the class b-KE) by allowing a certain "freedom" to the discrete duality gap. To do that, we first propose a combinatorial interpretation of the difference $v(EC(G)) - \alpha(G)$.

Given a graph G, let us consider a minimum vertex cover C^* ($|C^*| = \tau(G)$) and the corresponding maximum independent set S^*. Let us also suppose that, given a maximum matching M ($|M| = m$), there are f matching edges such that both their endpoints belong to C^*, for the remaining ones, one of their endpoints belonging to C^* and the other one to S^*. Let us call this edges "dissymmetric" and denote by F the set of these "dissymmetric" edges ($|F| = f$). For M (whenever it is not perfect), let us denote by X the set of the exposed (non-saturated) vertices of G with respect to M, and by X_C ($|X_C| = g$) and X_S the subsets of X belonging to C^* and S^*, respectively (of course, $X = X_C \cup X_S$). The numbers f and g, consequently the sets F and X_C, depend not only on M but also, for a fixed matching, on the sets C^* (and S^*) considered. However, the sum $f + g$ is a quantity depending only on G ($f + g$ is the discrete duality gap). In fact, for every graph G, $\tau(G) = m + (f + g)$ and $|S^*| = \alpha(G) = n - m - (f + g)$. Consequently, $v(EC(G)) - v(IS(G)) = f + g$ and *a KE-graph is exactly a graph where* $f + g = 0$.

If we relax the constraint $f + g = 0$ by allowing a positive discrete duality gap bounded above, we get the following proposition.

---

[4] Recall that, following our notations, $v(IS(G)) = \alpha(G)$.

Proposition 1. *Consider a graph $G = (V, E)$ such that $0 \leq \tau(G) - m = f + g \leq \kappa$. Then, (i) if κ is a fixed positive integer constant, there exists an exact polynomial algorithm for maximum independent set problem in G; (ii) otherwise, there exists a polynomial time approximation algorithm (having κ among its input parameters) providing an independent set of cardinality, at least equal to $\lceil n/[2(\kappa + 1)] - 2 \rceil$.*

Proof. (i) The condition $f + g \leq \kappa$ implies that both f and g are bounded above by κ. So, for all integer $h \leq \min\{m, \kappa\}$ and for all integer $k \leq \min\{n - 2m, \kappa - h\}\}$, for all h-tuple H of matching edges and for all k-tuple K of exposed vertices of V with respect to M, we form the sub-graphs of G induced by the vertex set $V \setminus (K \cup T[H])$. We next apply the algorithm of [8] on all of the so-obtained sub-graphs of G. Then, for one of the pairs (H, K), the induced subgraph is KE verifying that a maximum independent set is identical to the maximum independent set of G (by definition of the quantities f and g). Hence, the maximum cardinality set between the so-obtained independent sets is a maximum independent set of G. Given that κ is a universal constant, there is a polynomial number of pairs (H, K) and, consequently, the whole of the described process remains also polynomial.

(ii) Obviously, $f \leq \kappa$ and $g \leq \kappa$. We arbitrarily partition the edges of M into $\kappa + 1$ sets $M_1, \ldots, M_{\kappa+1}$, where $|M_i| = \lfloor m/(\kappa + 1) \rfloor$, $i = 1, \ldots, \kappa$ and $M_{\kappa+1} = m - \sum_{i=1}^{\kappa} \lfloor m/(\kappa + 1) \rfloor$. We also arbitrarily partition the set X of the exposed vertices of G into $\kappa + 1$ sets $X_1, \ldots, X_{\kappa+1}$ sets, each of size at least $\lfloor (n - 2m)/(\kappa + 1) \rfloor$. We so obtain $(\kappa + 1)^2$ sub-graphs of G, each one induced by the vertex-set $X_i \cup T[M_j]$, $(i, j) \in \{1, \ldots, \kappa + 1\}^2$ (Cartesian square).

So, we can apply the algorithm of [8] on all of the so-obtained $(\kappa + 1)^2$ graphs. Since at least one of these graphs is KE, one of the obtained solutions will be of size at least $(n - 2m)/(\kappa + 1) + m/(\kappa + 1) - 2 = (n - m)/(\kappa + 1) - 2$. The minimum of this quantity, for $m \in [0, \frac{n}{2}]$, is obtained for $m = n/2$ and its value is, in this case, equal to $[n/[2(\kappa + 1)]] - 2$. $\blacksquare$

Corollary 2. *Given a fixed positive constant κ, deciding if a graph G verifies $0 \leq f + g \leq \kappa$ is polynomial.*

The proof of the above corallary results from an immediate application of the part (i) of proposition 1.

We have seen that we can allow some freedom on the hypothesis imposed on the elements of the class of KE-graphs; moreover, in the case of "bounded freedom", the relation between constructive and non-constructive versions remains invariant and polynomial. In the next paragraph we generalize this result in the case of WIS.

The weighted case. We have already seen that the KE-graphs are the ones for which the discrete duality gap $v(\text{EC}(G)) - \alpha(G)$ is equal to 0. Moreover, in the previous paragraph we have relaxed the condition $v(\text{EC}(G)) - \alpha(G) = 0$, by allowing to this gap to be bounded above by a fixed positive constant. On the other hand, $v(\text{IS}_r(G)) - \alpha(G) \leq v(\text{EC}(G)) - \alpha(G)$; consequently, KE-graphs are the ones for which $v(\text{IS}_r(G)) - \alpha(G) = 0$.

Of course, the question on the difference between the value of IS and the one of its linear relaxation can be also posed for WIS[5]. In [4], the authors introduce the class of graphs where $v(\text{WIS}(G)) = v(\text{WIS}_r(G))$ and call it the class of b-KE-graphs, for which they conceive an $O(n^{2.5})$ exact WIS-algorithm.

Here, we relax the condition $v(\text{WIS}(G)) = v(\text{WIS}_r(G))$ by a less restrictive one and we prove that, in the class of graphs defined by this relaxed information, WIS remains polynomial. Although the proof of the similar result in the non-weighted case resulted from a combinatorial interpretation leading to the consideration of an upper bound of the quantity $\alpha(G) - v(\text{IS}_r(G))$, such a combinatorial interpretation for the weighted case, or, more generally, for the primal – dual approach, is much less natural. So, we give here a straightforward proof based upon linear programming arguments.

The purpose of this section is to prove the following result.

Theorem 3. *Consider the class of graphs $G = (V, E)$ satisfying $v(\text{WIS}(G)) \geq v(\text{WIS}_r(G)) - \kappa$, where κ is a fixed constant. Then, the problems: (i) decide if a graph G belongs to this class and (ii) solve WIS in this class, are polynomial.*

It is well-known that the LP-relaxation WIS_r of WIS has the semi-integral property, i.e., each basic feasible optimal WIS_r-solution B assigns to the variables values drawn from the set $\{0, 1/2, 1\}$ (see [10], for a very interesting discussion about this fact). Starting from this property, it can be proved that if V_1 is the set of vertices corresponding to variables valued by 1 in B, then there exists a maximum independent set in G that contains V_1 (in what follows, we denote by V_1, $V_{1/2}$ and V_0 the subsets of V corresponding to LP-variables assigned by values 1, 1/2 and 0, respectively; of course, these three sets form a partition of V). Moreover, in [10], it is shown that WIS_r can be seen as the problem of computing the minimum edge covering in a bipartite graph. The key-operation of such a computation is the computation of a maximum matching, performed in $O(n^{2.5})$.

Always in [10] it is shown how, given an instance $G = (V, E)$ of WIS, one can (in $O(n^{4.5})$) either determine a partition of V into sets V_1, $V_{1/2}$ and V_0, such that $V_{1/2} = \emptyset$ and so $\text{WIS}(G)$ is solved to optimality, or one can reduce G to a subgraph where the unique optimal solution for WIS_r is formed by assigning to all of its vertices the value 1/2. This can be done by procedure **DECOMPOSITION**, strongly inspired from [10], which we give here for purposes of self-completeness of the paper.

In what follows, when speaking of WIS_r-solution, we refer to Nemhauser – Trotter's method ([10]).

---

[5] Let us note that the respective linear programs for $\text{WIS}(G)$ and $\text{WIS}_r(G)$ are identical to the ones for $\text{IS}(G)$ and $\text{IS}_r(G)$, up to the replacement of the unit-vector by a positive cost-vector with components w_i, $i = 1, \ldots, n$.

Procedure DECOMPOSITION
begin
    solve $\text{WIS}_r(G)$ to obtain sets V_0, V_1, $V_{1/2}$;
    stop $\leftarrow$ **false**;
    while stop = **false** and $V_{1/2} \neq \emptyset$ **do**
        $N' \leftarrow V_{1/2}$;
        repeat
            choose $v_0 \in N'$;
            $N' \leftarrow N' \setminus \{v_0\}$;
            $N \leftarrow V_{1/2} \setminus (\{v_0\} \cup \Gamma(v_0))$;
            solve $\text{WIS}_r(G[N])$;
$\{* \ V_0', V_1', V_{1/2}'$ are the basic feasible sets of $V_{1/2} \setminus (\{v_0\} \cup \Gamma(v_0)) \ *\}$
        until $\mathbf{w} \cdot (1_{V_1'} + (1/2)1_{V_{1/2}'} + 1_{\{v_0\}}) = (1/2)\mathbf{w} \cdot 1_{V_{1/2}}$ or $N' = \emptyset$;
        if $\mathbf{w} \cdot (1_{V_1'} + (1/2)1_{V_{1/2}'} + 1_{\{v_0\}}) = (1/2)\mathbf{w} \cdot 1_{V_{1/2}}$ **then**
            $V_1 \leftarrow V_1 \cup V_1' \cup \{v_0\}$;
            $V_0 \leftarrow V_0' \cup V_0$;
            $V_{1/2} \leftarrow V_{1/2}'$;
        else stop $\leftarrow$ **true**;
        fi od
end;

Lemma 4. ([10]). *Given a graph G, the $O(n^{4.5})$ procedure* **DECOMPOSITION** *determines a partition of V into sets V_1, $V_{1/2}$ and V_0, and a solution for $\text{WIS}_r(G)$ corresponding to this partition, such that either (i) $V_{1/2} = \emptyset$ and so $\text{WIS}(G)$ is solved to optimality (by considering $v(\text{WIS}(G)) = \sum_{v_i \in V_1} w_i$), or (ii) the unique optimal solution for $\text{WIS}_r(G[V_{1/2}])$ is formed by assigning, to all of the vertices of $V_{1/2}$, the value 1/2.*

Definition 5. Let κ be a fixed positive constant. The class κ-KE is defined as the class of graphs $G = (V, E)$, of order n, such that: (i) the solution of $\text{WIS}_r(G)$ consisting of assigning value $1/2$ on all of the vertices of V is unique for $\text{WIS}_r(G)$, and (ii) $(1/2)\mathbf{w} \cdot 1_n - \kappa \leq v(\text{WIS}(G)) \leq (1/2)\mathbf{w} \cdot 1_n$.

We are now going to prove lemma 6 (the proof of which constitutes the main part of the proof of theorem 3), i.e., that WIS is polynomial for the class of κ-KE-graphs.

Lemma 6. *Let κ be a fixed positive constant and let $G = (V, E)$ be a κ-KE-graph of order n. Algorithm* **STABLE** *determines a maximum-weight independent set for G in $O(n^{4.5})$. Also,* **STABLE** *polynomially decides if a graph is κ-KE.*

Proof. Let us first prove the following emphasized proposition: *if there exists a maximum-weight independent set of G not containing v_x, then G_x', constructed by algorithm* **STABLE**, *is $[\kappa - (1/2)]$-KE.*

Algorithm STABLE
begin
       if $\kappa \leq 0$ **then call** the procedure of [4] to optimally solve WIS(G);
       else
             choose $v_i v_j \in E$;
             for $x \in \{i, j\}$ **do**
                  $V_x \leftarrow V \setminus \{v_x\}$;
                  $G_x \leftarrow G[V_x]$;
                  call procedure **DECOMPOSITION** on G_x;
$\{* \ V_{x_1}, V_{x_{1/2}}, V_{x_0} \text{ the obtained sets } *\}$
                  if $V_{x_{1/2}} = \emptyset$ **then** $S_x \leftarrow \emptyset$;
                    else
                        $V'_x \leftarrow V_{x_{1/2}}$;
                        $G'_x \leftarrow G[V'_x]$;
                        $\kappa \leftarrow \kappa - (1/2)$;
                        $S_x \leftarrow \mathbf{STABLE}(\kappa, G'_x)$;
                fi
             od
             $\hat{x} \leftarrow \mathrm{argmax}_{x \in \{i,j\}} \{ \mathbf{w} \cdot (\mathbf{1}_{V_{x_1}} + \mathbf{1}_{S_x}) \}$;
             $S \leftarrow V_{\hat{x}_1} \cup S_{\hat{x}}$;
       fi
end.

In fact, let us evaluate $v(\text{WIS}(G'_x))$. Let us first note that, from the hypothesis on v_x, $v(\text{WIS}(G_x)) = v(\text{WIS}(G))$. On the other hand, procedure **DECOMPOSITION**, called by algorithm **STABLE**, works in such a way that a maximum-weight independent set of G_x is obtained by adding to a maximum-weight independent set of G'_x, the set V_{x_1} (the vertices of V_x valued by one in the basic feasible solution of $\text{WIS}_r(G_x)$); so,

$$
\begin{aligned}
v(\text{WIS}(G'_x)) &= v(\text{WIS}(G_x)) - \mathbf{1}_{V_{x_1}} \cdot \mathbf{w} \\
&= v(\text{WIS}(G)) - \mathbf{1}_{V_{x_1}} \cdot \mathbf{w} \\
&\geq (1/2)\mathbf{w} \cdot \mathbf{1}_n - \kappa - \mathbf{1}_{V_{x_1}} \cdot \mathbf{w} \\
&\geq (1/2)\mathbf{w} \cdot (\mathbf{1}_{\{v_x\}} + \mathbf{1}_{V_{x_1}} + \mathbf{1}_{V_{x_{1/2}}} + \mathbf{1}_{V_{x_0}}) - \kappa - \mathbf{1}_{V_{x_1}} \cdot \mathbf{w} \\
&\geq (1/2)\mathbf{1}_{V'_x} \cdot \mathbf{w} - (\kappa - (1/2)(\mathbf{1}_{\{v_x\}} + \mathbf{1}_{V_{x_0}} - \mathbf{1}_{V_{x_1}}) \cdot \mathbf{w}) \qquad (1)
\end{aligned}
$$

where the last inequality holds because $V'_x = V_{x_{1/2}}$.

By definition of the class κ-KE, the solution of $\text{WIS}_r(G)$, consisting in assigning value $1/2$ on all of the vertices of G, is the unique optimal solution of the problem; so, since the solution determined by procedure **DECOMPOSITION** (called by **STABLE**) is feasible for $\text{WIS}_r(G)$, we get

$$
(\mathbf{1}_{V_{x_1}} + (1/2)\mathbf{1}_{V_{x_{1/2}}}) \cdot \mathbf{w} < (1/2)\mathbf{w} \cdot (\mathbf{1}_{\{v_x\}} + \mathbf{1}_{V_{x_1}} + \mathbf{1}_{V_{x_{1/2}}} + \mathbf{1}_{V_{x_0}})
$$

or

$$(1_{V_{x_1}} + (1/2)1_{V_{x_{1/2}}}) \cdot \mathbf{w} \le (1/2)\mathbf{w} \cdot (1_{\{v_x\}} + 1_{V_{x_1}} + 1_{V_{x_{1/2}}} + 1_{V_{x_0}}) - 1/2$$

because the (first) strict inequality takes place between two semi-integers; we so deduce that

$$(1/2)(1_{\{v_x\}} + 1_{V_{x_0}} - 1_{V_{x_1}}) \cdot \mathbf{w} \ge 1/2. \tag{2}$$

From expressions (1) and (2), we get $v(\text{WIS}(G'_x)) \ge (1/2)1_{V'_x} \cdot \mathbf{w} - [\kappa - (1/2)]$.

Finally, let us recall that procedure **DECOMPOSITION** is conceived in such a way that the solution assigning $1/2$ on all of the vertices of G'_x is the unique optimal solution of $\text{WIS}_r(G'_x)$; this concludes the proof of the emphasized proposition.

The recursive algorithm **STABLE** explores a binary tree of depth 2κ, the nodes of which correspond to the edges $v_i v_j$ chosen at the first line of the else clause of the outer **if** condition. If there exists a branch along which v_x verifies the hypothesis of the emphasized proposition, then from its conclusion, and by an immediate induction, the graph, obtained by algorithm **STABLE**, corresponding to the leaf of this branch is b-KE, so, WIS is polynomial for this graph ([4]). Let us finally observe that, for all k, if we denote by $G^{(k)}$ the graph treated during the kth recursive call of algorithm **STABLE**, the sum of a maximum-weight independent set of $G'_x{}^{(k)}$ with the weight of the set $V_{x_1}^{(k)}$, constitutes a maximum-weight independent set for $G'_x{}^{(k)}$, hence for $G^{(k)}$ (because of the hypothesis that there exists at least a maximum-weight independent set of $G^{(k)}$ not containing v_x).

On the other hand, for every edge $v_i v_j$ of $G^{(k)}$, both of its endpoints cannot simultaneously belong to every independent set (a fortiori optimal), this fact guaranteeing that such a branch (leading to a b-KE-graph) always exists. So, algorithm **STABLE**, which, entirely explores this binary tree by choosing among the independent sets corresponding to the branches, the one of maximum weight, it (algorithm **STABLE**) determines a maximum-weight independent set of the initial graph (recall that the algorithm of [4], called here on the leaves of the binary tree, returns an empty solution if an input graph is not b-KE).

So, algorithm **STABLE** is called, at worst, on all of the nodes of a fictive binary tree of depth 2κ and order 4^κ (a constant since κ is a fixed constant), the dominating operation of each call being procedure **DECOMPOSITION** of complexity $O(n^{4.5})$; consequently, the overall algorithm's complexity is of $O(n^{4.5})$.

Finally, remark that **STABLE** can be trivially used to decide if, for a given (fixed constant) κ, a graph is κ-KE.

Let us remark that, in the light of the previous result, the complexity of the algorithms of proposition 1 is also of $O(n^{4.5})$.

We are now well-prepared to conclude the proof of theorem 3.

Consider a constant κ and a graph $G = (V, E)$ such that $v(\text{WIS}(G)) \ge v(\text{WIS}_r(G)) - \kappa$. Procedure **DECOMPOSITION** allows to partition V into three sets V_0, V_1 and $V_{1/2}$ in such a way that $v(\text{WIS}(G)) = v(\text{WIS}(G[V_{1/2}])) +$

$\mathbf{w} \cdot \mathbf{1}_{V_1}$ and, furthermore, $v(\text{WIS}_r(G)) = v(\text{WIS}_r(G[V_{1/2}])) + \mathbf{w} \cdot \mathbf{1}_{V_1}$. Moreover, in order to construct an optimal solution of $\text{WIS}(G)$, one has simply to complete an optimal solution of $\text{WIS}(G[V_{1/2}])$ with the vertices of V_1. The above expressions allow to establish that $v(\text{WIS}(G[V_{1/2}])) \geq v(\text{WIS}_r(G[V_{1/2}])) - \kappa$; furthermore, using lemma 4 one can be sure that the graph $G[V_{1/2}]$ is κ-KE. Then, lemma 6 concludes the proof of theorem 3.

3 Boundaries between constructive and non-constructive frameworks for independent set: problems S_κ

The classes considered in sections 2.1 and 2.2 were restrictive enough to allow facility for the constructive framework, when starting from an information about the optimal value; so, for these classes, the constructive and non-constructive points of view are of identical facilities and cannot be dissociated.

In order to capture the boundary between the two frameworks, let us now consider, starting from IS, the transformation consisting in integrating in the instance a vaguer information on the optimal value, i.e.,we consider only the order of the optimal value with respect to input-size n. This thought process leads us to the following problems introduced in [6].

Definition 7. Problems S_κ.
For every constant $\kappa > 1$, we define the stability problem S_κ corresponding to the restriction of IS to graphs (of order n) admitting stability number greater than or equal to n/κ; we also define the problem $S_{\kappa\infty}$, an instance of which is the pair (G, κ) where $\kappa \geq 1$, and G is a graph of order n with $\alpha(G) \geq n/\kappa$. Here also, the objective is to determine a maximum independent set of G.

The first remark concerning these problems is that the recognition of their instances is NP-complete for $\kappa \geq 2$ (contrarily to the problems considered in sections 2.1 and 2.2).

Proposition 8. *The problem of deciding if a graph G of order n verifies $\alpha(G) \geq n/\kappa$ is NP-complete* $\forall \kappa \geq 2$, $\kappa \in \mathbb{N}$.

Proof. The reduction is from IS, where we have to decide if, for a graph G, $\alpha(G) \geq \ell$, $\ell \in \mathbb{N}$.

Remark. We can consider that $\ell \leq n/\kappa$.

If not ($\ell > n/\kappa$), we add a clique K_c in G and we insert all edges between vertices of G and vertices of K_c. The new graph G' is of order $n + c$ vertices and, moreover, $\alpha(G) = \alpha(G')$. If $c \geq \kappa\ell - n$, then $\ell \leq (n + c)/\kappa$.

Remark. The quantities n and ℓ are multiples of $\kappa - 1$.

If not, we consider graph G' consisting of $\kappa - 1$ disjoint copies of G. Its order is now $(\kappa - 1)n$ and the question becomes if $\alpha(G') \geq \ell(\kappa - 1)$.

Remark. Finally, $\ell = n/\kappa$.

If $\ell \neq n/\kappa$, we add σ independent vertices in G. The resulting graph G' has $n+\sigma$ vertices and $\alpha(G') = \alpha(G) + \sigma$. So, we have to decide if $\alpha(G) + \sigma \geq \ell + \sigma$. The fact $\ell + \sigma = (n+\sigma)/\kappa$ can be assured if $\sigma = [n/(\kappa-1)] - [\kappa\ell/(\kappa-1)]$. The two first remarks assure $\sigma \in \mathbb{N}^+$.

The combination of the three above remarks conclude the NP-completeness claimed.

The above result indicates that the information integrated in the instance of the problems S_κ is quite strong. Consequently, it is reasonable to wonder if one can exploit this information for approximating these problems.

We have then the following result.

Theorem 9. *For every constant ρ, there is no polynomial time approximation algorithm for $S_{\kappa\infty}$ which guarantees approximation ratio greater than or equal to ρ, unless $P = NP$.*

In other words, $S_{\kappa\infty}$ does not admit a constant-ratio polynomial time approximation algorithm. For the problems S_κ, this means that, *if there exists a polynomial time approximation algorithm guaranteeing, for every problem S_κ, an approximation ratio $\rho(\kappa)$, then the mapping $\kappa \mapsto \rho(\kappa)$ tends to 0 whenever $\kappa \to \infty$.*

In fact, we can even precise the above remark by specifying the convergence velocity of ρ to 0.

Theorem 10. *If $P \neq NP$, then $\exists \varepsilon > 0$ and $\exists \kappa_0$ such that $\forall \kappa \geq \kappa_0$, there does not exist algorithm polynomial in n (but, eventually, exponential in κ) for S_κ guaranteeing approximation ratio $(1/\kappa)^\varepsilon$.*

Proof. In [1], it is proved that unless $P = NP$, IS, even for above-bounded-maximum-degree graphs (let us denote the class of IS defined on such graphs by $B(\Delta)S$), cannot be approximated by a polynomial time approximation schema[6] (PTAS).

Using the result of [1], we prove the following intermediate result (lemma 11).

Lemma 11. $\exists \kappa_0$ *such that S_{κ_0} does not admit a PTAS.*

Proof. (**Lemma 11.**) Let us prove that if, $\forall \kappa$, S_κ admits a PTAS, then $B(\Delta)S$ also admits a PTAS.

Recall that in a graph G of order n, every maximal independent set is greater than or equal to $n/(\Delta + 1)$; consequently, $\alpha(G) \geq n/(\Delta + 1)$ and, for each Δ, problem $B(\Delta)S$ is a sub-problem of $S_{\Delta+1}$, solved by a PTAS on the hypothesis that, $\forall \kappa$, S_κ can be solved by a PTAS, and this completes the proof of lemma 11.

---

[6] A sequence of polynomial time approximation algorithms, indexed by ε, guaranteeing, for every ε, approximation ratio of $1 - \varepsilon$.

Of course, lemma 11 implies that S_κ does not admit a PTAS for all $\kappa \geq \kappa_0$ (because S_κ is a sub-problem of $S_{\kappa+1}$, for all κ).

Let now G be an instance of S_{κ_0} of size n. For all integers m, consider the graph G^m defined as

$$G^1 = G$$
$$G^i = G \times G^{i-1} \quad i = 2, \ldots, m$$

where $G_1 \times G_2$ denotes the composition of the graphs G_1 and G_2 ([3,9]). It can be easily proved (see [3,9]) that every independent set of cardinality $\tilde{\alpha}(G)$ in G can be polynomially transformed into an independent set of cardinality $\tilde{\alpha}(G^m) = \tilde{\alpha}(G)^m$ in G^m and vice-versa. Consequently, $\alpha(G^m) = \alpha(G)^m$. Let n_m be its order; then, $n_m = n^m$. The properties of this construction allow to establish that, if G is an instance of S_κ, then G^m is an instance of S_{κ^m}.

Suppose now that, $\forall \epsilon > 0$, $\exists \kappa > \kappa_0$ such that S_κ is $(1/\kappa)^\epsilon$-approximable. For $\eta > 0$, choose $\epsilon > 0$ such that $(1/\kappa_0^{3/2})^\epsilon > 1 - \eta$ and $\kappa > \kappa_0$ such that S_κ is $(1/\kappa)^\epsilon$-approximable. We then determine integer m such that $\kappa_0^m \leq \kappa < \kappa_0^{m+1}$; G^m is an instance of S_κ, and so, the algorithm supposed to solve it allows to determine in polynomial time an independent set of cardinality $\alpha'(G^m) \geq (1/\kappa)^\epsilon)\alpha(G^m) \geq (1/\kappa_0^{1+m})^\epsilon)\alpha(G^m)$.

Following the above discussion, we can deduce an independent set of cardinality $\alpha'(G^m)^{1/m} \geq (1/\kappa_0^{(1+m)/m})^\epsilon)(\alpha(G^m))^{1/m} \geq ((1/\kappa_0^{3/2})^\epsilon)\alpha(G) \geq (1 - \eta)\alpha(G)$ for G. This holding, $\forall \eta \in]0, 1[$, we have determined a PTAS for S_κ, a contradiction. This completes the proof of theorem 10

As we have seen, lemma 11 rules out, for allmost all κ, the existence of PTAS for S_κ. Unfortunately, we cannot yet characterize more presisely the hardness of approximating problems S_κ for a fixed κ. The only remark we can make is that, for $\kappa < 2$, S_κ is polynomially constant-approximable.

In fact, let us consider an instance G of $S_{2-\epsilon}$ for a positive constant ϵ. Then, $\alpha(G) \geq n/(2 - \epsilon) \geq [(1/2) + \epsilon']n$ for $\epsilon' > 0$. Let us denote by e the number of vertices exposed[7] with respect to a maximum matching M of cardinality m. We have $n = 2m + e$ and $\alpha(G) \leq m + \epsilon$. If we choose as approximate solution for IS the independent set formed by these vertices, the previous expressions lead to $e \geq 2\epsilon'n$ and the constant ratio $2\epsilon'$ is guaranteed for $S_{2-\epsilon}$.

The approximability of problems S_κ is particularly interesting. If one could precise (see theorem 10) a threshold κ_0 such that, $\forall \kappa \geq \kappa_0$, S_κ is not constant-approximable, then one would bring to the fore a problem constant-approximable in the non-constructive framework within a ratio $1/\kappa$ which, at the same time, is not constant-approximable in the constructive framework.

Furthermore, the approximability of problems S_κ is strongly linked to the approximability of other combinatorial problems, for example convexe programming problems including quadratic programming as sub-case, or, even, to the open problem of improving the approximation ratio 2 for minimum vertex covering.

---

[7] The set of these vertices is non-empty because of the hypothesis $\alpha(G) > n/2$.

More precisely, the following conditional results are proved in [6].

Theorem 12. ([6]). *Let $\rho < 1$ be a fixed positive constant. Under the hypothesis $P \neq NP$, (i) the non-existence of a ρ-approximation polynomial time algorithm for S_3 implies that no polynomial time approximation algorithm for VC can guarantee an approximation ratio strictly smaller than $3/2$; (ii) if, on the contrary, such an algorithm exists for S_3, then there exists an algorithm[8] for VC guaranteeing an approximation ratio smaller than, or equal to, $2 - (\rho/6) < 2$.*

In the case $\kappa = 2$, we can obtain the following result.

Proposition 13. ([6]). *On the hypothesis $P \neq NP$, if S_2 is non-constant approximable, then no polynomial time approximation algorithm for VC can guarantee an approximation ratio $\rho < 2 - \epsilon$, for a fixed positive constant ϵ.*

The (maximization) convex programming problem considered in [6] is defined as follows:

$$CPM(\kappa) \ = \ \begin{cases} \max \sum\limits_{i \in \{1,\dots,n\}} f(x_i) \\ x \in \mathcal{P} \end{cases}$$

where $\mathcal{P}$ is a polytope defined by a finite number of constraints, and f belongs to a family $\mathcal{F}$ of functions increasing in $[0, 1]$, with $f(0) = 0$ for every $f \in \mathcal{F}$, verifying the property

$$\inf_{f \in \mathcal{F}} \left\{ \frac{f(\frac{1}{2})}{f(1)} \right\} \in \left[0, \frac{1}{\kappa}\right[, \quad \kappa \in \mathbb{R} \setminus [0, 2[.$$

We have then the following result.

Theorem 14. ([6]). *If there exists $\kappa \in \mathbb{R} \setminus [0, 2[$ such that S_κ does not admit a polynomial time algorithm guaranteeing a maximal independent set greater than ρn for a fixed positive constant $\rho < 1$, then there does not exist a polynomial time approximation algorithm for $CPM(\kappa)$ guaranteeing an approximation ratio greater than $\kappa \inf_{f \in \mathcal{F}} \{f(1/2)/f(1)\}$.*

By applying the result of theorem 14 on particular families $\mathcal{F}$ ([6]), we can deduce the following corollaries.

Corollary 15. ([6]). *If S_3 does not admit a polynomial time approximation algorithm of (universally) constant ratio, then the quadratic programming problem does not admit a polynomial time approximation algorithm guaranteeing an approximation ratio greater than or equal to $3/4$, unless $P = NP$.*

Corollary 16. ([6]). *If there exists κ such that S_κ does not admit a polynomial time algorithm guaranteeing a maximal independent set greater than ρn for a fixed positive constant $\rho < 1$, then the problem of maximizing a convex function in a polytope does not admit a constant-ratio polynomial time approximation algorithm.*

---

[8] This algorithm is constructive, since the proof of part (ii) of theorem 12 is entirely constructive.

Analogous results are given in [6] for concave programming problems.

A slightly different problem, with respect to the approximability of S_κ, is defined as follows: *given an instance G of S_κ, determine an independent set of size n/κ.* This is a purely constructive problem, harder than the approximability of S_κ with ratio $1/\kappa$.

For this new problem, starting from proposition 8, we can immediately deduce the following result.

Proposition 17. (Constructive version of proposition 8.)

If $P \neq NP$, then $\forall \kappa \geq 3$, $\kappa \in \mathbb{N}$, there cannot exist a pair of (A, f), where A is an algorithm and f a function, both (A and f) polynomial in n (but, eventually, exponential in κ) such that, for all graphs G of order n instances of S_κ, A determines an independent set greater than, or equal to n/κ with complexity less than, or equal to $f(n)$.

Acknowledgement. The proof of proposition 8 in its current version is due to an anonymous referee; we thank him also for many other pertinent suggestions and useful remarks contributing to improve the legibility of this paper.

References

1. Arora, S., Lund, C., Motwani, R., Sudan, M., Szegedy, M.: Proof verification and intractability of approximation problems. Proc. FOCS (1992) 14-23
2. Ausiello, G., Crescenzi, P., Protasi, M.: Approximate solution of NP Optimization Problems. Theoretical Computer Science (to appear)
3. Berge, C.: Graphs and hypergraphs. North Holland (1973)
4. Bourjolly, J. - M., Hammer, P. L., Simeone, B.: Node-weighted graphs having the König-Egervary Property. Math. Prog. Study **22** (1984) 44-63
5. Demange, M., Paschos, V. Th.: On an approximation measure founded on the links between optimization and polynomial approximation theory. Theoretical Computer Science (to appear)
6. Demange, M., Paschos, V. Th.: The approximability behaviour of some combinatorial problems with respect to the approximability of a class of maximum independent set problems. Computational Optimization and Applications (to appear)
7. Demange, M., Paschos, V. Th.: Relative hardness of constructive - non-constructive approximation: the case of maximum independent set problem, manuscript, 1995.
8. Deming, R. W.: Independence numbers of graphs – an extension of the König-Egervary theorem. Disc. Math. **27** (1979) 23-33
9. Garey, M. R., Johnson, D. S.: Computers and intractability. A guide to the theory of NP-completeness. Freeman (1979)
10. Nemhauser, G. L. Trotter, Jr, L. E.: Vertex packings: structural properties and algorithms. Math. Prog. **8** (1975) 232-248

Weakly Greedy Algorithm and Pair-Delta-Matroids

Takashi TAKABATAKE *

Graduate Division of International and Interdisciplinary Studies,
University of Tokyo
Komaba, Meguro-ku, Tokyo 153, Japan

Abstract. Weakly greedy algorithm is an extension of the greedy algorithm. It gives a solution of a combinatorial optimization problem on discrete systems in a properly wider class than the class of Δ-matroids. Discrete systems with a certain 2 to 2 exchangeability belong to this class. We characterize these systems in terms of their rank function. Excluded minors of Δ-matroids and these systems are also described.

1 Introduction

The concept of matroid explains various kinds of combinatorial structures in a consistent way [4]. On matroids and their generalizations, such as Δ-matroids, jump systems, etc. , greedy algorithms always give a solution of a certain combinatorial optimization problem [1, 2, 3]. These combinatorial structures are, in a sense, characterized by greediness.

In this paper we develop an extension of the greedy algorithm called *weakly greedy algorithm*. It is applicable to a combinatorial optimization problem on the class of *weak-Δ-matroids*, which properly contains the class of Δ-matroids. *Pair-Δ-matroids* compose a subclass of weak-Δ-matroids, where a certain 2 to 2 exchangeability is guaranteed. We describe the property of the rank function that characterizes pair-Δ-matroids. Some examples of pair-Δ-matroids are shown. We also prove that Δ-matroids and pair-Δ-matroids are characterized by some excluded minors.

2 Preliminaries

A *discrete system* is a pair $(S, \mathcal{F})$, where S is a finite set and $\mathcal{F}$ is a non-empty family of subsets of S. S is called its *support* and members of $\mathcal{F}$ are called its *independent sets*.

We consider the following optimization problem on a discrete system $(S, \mathcal{F})$ under a weight function $w : S \to \mathbb{R}$.

Find $X \in \mathcal{F}$ which maximizes $\sum_{i \in X} w(i)$.

---

* e-mail:takashi@klee.c.u-tokyo.ac.jp

In this paper we call this problem the *combinatorial optimization problem* (COP in short), and we use $(S, \mathcal{F}, w)$ to describe this COP. In what follows, $w(X)$ denotes $\sum_{i \in X} w(i)$.

Given a discrete system $(S, \mathcal{F})$, *twisting* [2] this discrete system at $N \subseteq S$ results in $(S, \mathcal{F}\Delta N)$, where $\mathcal{F}\Delta N = \{F\Delta N | F \in \mathcal{F}\}$.

The following greedy algorithm [1, 3] gives an answer of a COP. (In this paper, "answer" does not imply its correctness, while "solution" does.)

Greedy Algorithm
Input COP $(S, \mathcal{F}, w)$.
Output A subset $X \in \mathcal{F}$ which maximizes $w(X)$.
Step1 Consider an optimization problem $(S, \mathcal{F}\Delta N, w')$, where $w'(i) = |w(i)|$ $(i \in S)$ and $N = \{i \in S \mid w(i) < 0\}$. Sort the support $S = \{1, 2, \dots, n\}$ so that $w'(i) \geq w'(i+1)$ $(1 \leq i \leq n-1)$, where n denotes $|S|$.
Step2 $X := \phi$.
Step3 For $i := 1$ to n repeat,
$$X := X \cup \{i\} \text{ if there exists } F \in \mathcal{F}\Delta N \text{ such that } X \cup \{i\} \subseteq F.$$
End Return $X\Delta N$ as the output.

The greedy algorithm does not always give a solution of COP $(S, \mathcal{F}, w)$. But if the pair $(S, \mathcal{F})$ is a discrete system called Δ-matroid, the greedy algorithm always gives a solution of the COP [1, 3].

Definition 1. A discrete system $(S, \mathcal{F})$ is a Δ-*matroid*, if it satisfies (1)

$$\text{For all } X, Y \in \mathcal{F}, \text{ and for all } x \in X\Delta Y, \text{ there}$$
$$\text{exists } y \in X\Delta Y \text{ such that } X\Delta\{x, y\} \in \mathcal{F}. \tag{1}$$

Note that x and y may be identical. The operator Δ is symmetric difference and $X\Delta Y$ denotes $(X \cup Y) \setminus (X \cap Y)$. Obviously twisted Δ-matroids are also Δ-matroids.

The greedy algorithm requires a procedure which determines whether there exists an independent set that includes A and is disjoint to B. Bouchet gives the name *separation oracle* to a procedure which solve this question in a single step [1]. This algorithm uses the separation oracle n times, where n denotes the cardinality of the support.

Chandrasekaran and Kabadi define the rank function of a Δ-matroid [3].

Definition 2. Let $(S, \mathcal{F})$ be a Δ-matroid. Define a function f from an ordered pair of disjoint subsets of S as (2).

$$f(A, B) = \max_{X \in \mathcal{F}} (|A \cap X| - |B \cap X|) \tag{2}$$

Function f is called the *rank function* of $(S, \mathcal{F})$.

They show that the rank function f satisfies the following property called *bi-submodularity*.

$$f(A, B) + f(C, D) \geq f(A \cap C, B \cap D) + f((A \cup C) \setminus (B \cup D), (B \cup D) \setminus (A \cup C)) \tag{3}$$

Equivalence between bi-submodularity and Δ-matroid is also shown [3].

3 Weakly Greedy Algorithm

The weakly greedy algorithm is an improved greedy algorithm so that it can also deal with some discrete systems which are not Δ-matroids. The essence of the improvement is as follows.

The greedy algorithm returns, as its output, the first independent set under a certain order. This order is lexicographical with respect to the weight function.

Let us see in a little more detail. Suppose that the weight function is non-negative. For two independent sets X, Y, the order between X and Y is defined by the comparison between the heaviest element of $X \backslash Y$ and that of $Y \backslash X$.

The weakly greedy algorithm depends on another order. This order is defined by the comparison between the heaviest and the second heaviest elements of $X \backslash Y$ and those of $Y \backslash X$. The weakly greedy algorithm is an algorithm which returns, as its output, one of the first independent sets under this new order.

We now describe the weakly greedy algorithm.

Let $(S, \mathcal{F}, w)$ be a COP. In this algorithm, we use a list which consists of subsets of S linked together in a linear fashion. Let L be a list with $X_1, \ldots, X_l$ in this order. We use the expression $L = (X_1, \ldots, X_l)$ to describe L. The *addition* of Y to this list means $L := (X_1, \ldots, X_l, Y)$. The *deletion* of X_i from this list means $L := (X_1, \ldots, X_{i-1}, X_{i+1}, \ldots, X_l)$. $L' = ()$ means that the list L' is empty. The number of subsets in a list L is represented by $|L|$.

We also use the separation oracle. If there exists an independent set that includes A but contains no elements of B, we say that the pair (A, B) is *possibly independent*.

Weakly greedy algorithm
Input COP $(S, \mathcal{F}, w)$.
Output A subset $X \in \mathcal{F}$ which maximizes $w(X)$.
Step1 Consider an optimization problem $(S, \mathcal{F} \Delta N, w')$, where $w'(i) = |w(i)|$
   $(i \in S)$ and $N = \{i \in S \mid w(i) < 0\}$. Sort the support $S = \{1, 2, \ldots, n\}$ so
   that $w'(i) \geq w'(i+1)$ $(1 \leq i \leq n-1)$, where n denotes $|S|$.
Step2 $L(0) := (\phi)$.
Step3 For $i := 1, \ldots, n$, repeat;
   Sub-Step1 $L(i) := ()$.
   Sub-Step2 For $j := 1, \ldots, |L(i-1)|$, repeat;
      Let A_j be the j'th subset of $L(i-1)$.
      $A_j^+ := A_j \cup \{i\}$, $B_j^+ := \{1, \ldots, i\} \backslash A_j^+$,
      $A_j^- := A_j$, $B_j^- := \{1, \ldots, i\} \backslash A_j^-$.
      If (A_j^+, B_j^+) is possibly independent, then;
         Add A_j^+ to $L(i)$.
      end if.
      If (A_j^-, B_j^-) is possibly independent, then;
         flag := TRUE.
         While $L(i)$ contains a subset except A_j^+ that contains i
         and **flag** = TRUE, repeat;

Let X be the first subset in $L(i)$ that contains i.

If $w'(A_j^-) > w'(X)$, then;

Delete X from $L(i)$.

else;

flag := FALSE.

end if.

end while.

If no subsets in $L(i)$ except A_j^+ contain i, then add A_j^- to $L(i)$.

end if.

end for. (Sub-Step2)

end for. (Step3)

End Find an independent subset $A \in L(n)$ which maximizes $w'(A)$. Return $A \Delta N$ as the output.

We describe discrete systems where the weakly greedy algorithm always succeeds.

Definition 3. A discrete system $(S, \mathcal{F})$ is called a *weak-Δ-matroid* if it satisfies (4).

For all $X, Y \in \mathcal{F}$ and $x, y \in X \Delta Y$ ($x \neq y$), there exist $s, t \in X \Delta Y$ ($s \neq t$) which satisfy either $X \Delta \{x, y, s, t\} \in \mathcal{F}$, or $X \Delta \{x, s\}, X \Delta \{y, t\} \in \mathcal{F}$. (4)

It should be noted that s, t in (4) may be identical to x or y. Thus the condition $s \neq t$ makes sense only for $X \Delta \{x, s\}, X \Delta \{y, t\} \in \mathcal{F}$. It is easy to see that Δ-matroids are weak-Δ-matroids. (In the case of Δ-matroids, $X \Delta \{x, s\} \in \mathcal{F}$. If $s = y$, then (4) is satisfied with $t = x$. Otherwise $y \in Y \Delta (X \Delta \{x, s\})$. Thus $(X \Delta \{x, s\}) \Delta \{y, t\} \in \mathcal{F}$ for some $t \in Y \Delta (X \Delta \{x, s\})$.)

Theorem 4. *If a discrete system $(S, \mathcal{F})$ is a weak-Δ-matroid, the weakly greedy algorithm gives a solution of COP $(S, \mathcal{F}, w)$ under an arbitrary weight function $w : S \to \mathbb{R}$.*

Before proving Theorem 4, we make several definitions and an observation.

Given COP $(S, \mathcal{F}, w)$ with a nonnegative weight function w, sort the support $S = \{1, 2, \ldots, n\}$ so that $w(i) \geq w(i+1)$ ($1 \leq i \leq n-1$), where n denotes $|S|$.

For $i, j \in S$, we call i *smaller* than j if $i < j$.

Define $w(i)_\epsilon = w(i) + \epsilon^i$, where $0 < \epsilon \ll 1$.

For $X, Y \in \mathcal{F}$, let x_1, x_2 (y_1, y_2) be the smallest element and the second smallest element of $X \backslash Y$ ($Y \backslash X$), respectively. If there exist no such elements, we use dummy elements with zero weight.

We define $X \prec_1 Y$, if $X \neq Y$ and $w_\epsilon(\{x_1\}) \geq w_\epsilon(\{y_1\})$.

We define $X \prec_2 Y$, if $|X \backslash Y| \geq 2$ and $w_\epsilon(\{x_1, x_2\}) \geq w_\epsilon(\{y_1, y_2\})$.

Observation 5. If the list $L(i)$ contains subsets X, Y in this order, then $X \prec_1 Y$. Or, equivalently, the smallest element of $X \Delta Y$ is contained in X. (Let k be the smallest element of $X \Delta Y$. $(X \cap \{1, \ldots, k-1\}) \cup \{k\}$ is stored in $L(k)$ just before $Y \cap \{1, \ldots, k-1\}$.) And the other elements of $X \Delta Y$ are contained in Y. (Suppose not. Let m be the second smallest element of $X \backslash Y$. Then either $Y \cap \{1, \ldots, m-1\}$ is not added to $L(m)$ or $(X \cap \{1, \ldots, m-1\}) \cup \{m\}$ is deleted from $L(m)$.)

Now we are ready to prove Theorem 4.

Proof of Theorem 4. From a given COP $(S, \mathcal{F}, w)$, the weakly greedy algorithm twists the discrete system $(S, \mathcal{F})$ and consider a new COP $(S, \mathcal{F} \Delta N, w')$. If X is an independent set in $\mathcal{F} \Delta N$ which have the maximum weight under w', $X \Delta N$ is a solution of COP $(S, \mathcal{F}, w)$. It is obvious that twisted weak-Δ-matroids are weak-Δ-matroids. Thus we may assume that w is nonnegative. With this assumption, $w' = w$ and $\mathcal{F} \Delta N = \mathcal{F}$.

First we show there exists an independent set with the maximum weight which is not preceded, with respect to $\prec_2$, by any other independent set. Suppose not. Let X be the first independent set with respect to $\prec_1$ among all independent sets with the maximum weight. There exists an independent set Y which satisfies $Y \prec_2 X$. Let $x_1, x_2, (y_1, y_2)$ be the smallest and the second smallest elements of $X \backslash Y (Y \backslash X)$, respectively.

By (4), either $X \Delta \{y_1, s\}, X \Delta \{y_2, t\} \in \mathcal{F}$, or $X \Delta \{y_1, y_2, s, t\} \in \mathcal{F}$ for some $s, t \in X \Delta Y$ ($s \neq t$). If y_1 is smaller than x_1, then $X \Delta \{y_1, s\}$ and $X \Delta \{y_1, y_2, s, t\}$ precede X, with respect to $\prec_1$ and these two independent sets have greater weight than X. A contradiction. If x_1 is smaller than y_1, then $w(\{x_1, x_2\}) < w(\{y_1, y_2\})$. Thus $w(X \Delta \{y_1, y_2, s, t\})$ and either $w(X \Delta \{y_1, s\})$ or $w(X \Delta \{y_2, t\})$ are greater than $w(X)$, which is a contradiction.

Now we prove that all independent sets that are not preceded, with respect to $\prec_2$, belong to $L(n)$. Let $\mathcal{F}_i$ denote $\{F \cap \{1, \ldots, i\} | F \in \mathcal{F}\}$ for $i = 1, \ldots, n$. We show that $L(i)$ contains all subsets in $\mathcal{F}_i$ that are not preceded, with respect to $\prec_2$, by any other subset in $\mathcal{F}_i$.

$L(1)$ contains all subsets in $\mathcal{F}_1$. We assume that the $L(i-1)$ contains all subsets that are not preceded and then we see that $L(i)$ also does as follows.

Suppose not. Let X be the first subset with respect to $\prec_1$ among the independent sets X' such that $X' \cap \{1, \ldots, i\}$ is not preceded and does not belong to $L(i)$. Let $X_i (X_{i-1})$ denote $X_i \cap \{1, \ldots, i\}$ $(X_i \cap \{1, \ldots, i-1\})$, respectively. Then $X_{i-1} \in L(i-1)$ by the assumption of induction. Let Y be such an independent set that $Y \cap \{1, \ldots, i\}$ causes the disqualification of X_i in Sub-Step2. Let Y_i (Y_{i-1}) denote $Y \cap \{1, \ldots, i\}$ $(Y \cap \{1, \ldots, i-1\})$, respectively.

If $i \notin X_i$, then $i \in Y_i$, $w(Y_i) \geq w(X_i)$, and $Y_{i-1} \prec_1 X_{i-1}$ by Observation 5. Let x be the smallest element in $X \Delta Y$. Obviously $x \neq i$ and Y contains x, i. Since $w(Y_i) \geq w(X_i)$, thus $Y_i \prec_2 X_i$. This contradicts the choice of X.

If $i \in X_i$, then $i \notin Y_i$, $w(Y_i) > w(X_i)$, and $X_{i-1} \prec_1 Y_{i-1}$ by Observation 5. Let x be the smallest element in $X \Delta Y$. Let p, q be the smallest and the second smallest elements in $Y \backslash X$. Note that x and i are the smallest and the second smallest in $X \backslash Y$.

If $w(\{p, q\}) > w(\{x, i\})$, then $Y_i \prec_2 X_i$. A contradiction.

If $w(\{p, q\}) \leq w(\{x, i\})$, then $|Y_i \backslash X_i| \geq 3$. By (4), either $Y \Delta \{x, s\}$ or $Y \Delta \{x, i, s, t\}$ is independent for some $s, t \in X \Delta Y$.

In the case that $Y \Delta \{x, s\}$ is independent, the smallest and the second smallest elements in $((Y \Delta \{x, s\}) \cap \{1, \ldots, i\}) \Delta X_i$ belong to $(Y \Delta \{x, s\}) \cap \{1, \ldots, i\}$. Since $|((Y \Delta \{x, s\}) \cap \{1, \ldots, i\}) \backslash X_i| \geq 2$, it implies $(Y \Delta \{x, s\}) \cap \{1, \ldots, i\} \prec_2 X_i$, which is a contradiction.

In the case that $Y \Delta \{x, i, s, t\}$ is independent. Since $w(\{p, q\}) \leq w(\{x, i\})$, thus $w(Y \Delta \{x, i, s, t\} \cap \{1, \ldots, i\}) \geq w(Y_i)$. And $(Y \Delta \{x, i, s, t\}) \cap \{1, \ldots, i - 1\} \prec_1 Y_{i-1}$. Thus $(Y \Delta \{x, i, s, t\}) \cap \{1, \ldots, i\}$ is not in $L(i)$. (Suppose not. $(Y \Delta \{x, i, s, t\}) \backslash Y$ contains x and i. Then Y_i is not added to $L(i)$ in Sub-Step2, since $w(\{x, i\}) \geq w(\{p, q\}) \geq w(\{s, t\})$. A contradiction.) If $(Y \Delta \{x, i, s, t\}) \cap \{1, \ldots, i\}$ is not preceded, it contradicts the choice of X, for $(Y \Delta \{x, i, s, t\}) \cap \{1, \ldots, i\} \prec_1 X_i$. If it is preceded by some independent set in $\mathcal{F}_i$, X_i is also preceded by the same set. A contradiction.

Thus all independent sets that are not preceded belong to $L(n)$ and the algorithm find an independent set with the maximum weight from $L(n)$. $\square$

We examine the order of time required for this algorithm. We suppose that the separation oracle is given. Let n denote $|S|$.

In Step1 it takes $O(n \log(n))$ time.

In Step2 it takes a constant amount of time.

In Step3 two sub-steps are iterated n times. In Sub-Step1 it takes a constant amount of time. In Sub-Step2 one procedure is iterated $|L(i - 1)|$ times. This procedure determines whether two pairs of subsets are possibly independent or not. It takes a constant amount of time with the separation oracle. And at most two additions of subsets and several comparisons of weights and deletions are executed. But at most $2|L(i - 1)|$ comparisons and deletions happen in one iteration of Sub-Step2, since one comparison causes one subset to be deleted from or not to be added to $L(i)$. (Subsets in $L(i)$ are deleted from the first to the last.) Thus Sub-Step2 requires $O(|L(i - 1)|)$. A little consideration of the algorithm shows that $|L(i)|$ increases at most by one in each iteration of Step3. (Only one subset in $L(i - 1)$ is common to two subsets in $L(i)$.) Thus $|L(i)| \leq i + 1$. Therefore it requires $O(n^2)$ time in Step3.

To find a subset which maximizes its weight, $O(|L(n)|) \approx O(n)$ time is required.

Therefore, the time complexity of the algorithm is $O(n^2)$, if we use the separation oracle.

4 Pair-Δ-matroids

Pair-Δ-matroids compose an interesting subclass of weak-Δ-matroids. We characterize pair-Δ-matroids by some property of their rank function.

Definition 6. A discrete system $(S, \mathcal{F})$ is called a *pair-Δ-matroid* if it satisfies (5).

$$\text{For all } X, Y \in \mathcal{F} \text{ and } x, y \in X\Delta Y (x \neq y), \text{ there}$$
$$\text{exist } s, t \in X\Delta Y \text{ such that } X\Delta\{x, y, s, t\} \in \mathcal{F}. \tag{5}$$

It should be noted that s, t in (5) may be identical to x or y. It is easy to see that a Δ-matroid is a pair-Δ-matroid. Thus the class of pair-Δ-matroids includes that of Δ-matroids.

We use a modified rank function for simple expression. The definition is as follows.

Definition 7. Let $(S, \mathcal{F})$ be a discrete system. Define a function r from $S \times S$ to nonnegative integers as (6).

$$r(A, B) = \max_{X \in \mathcal{F}\Delta B} (|A \cap X|) \tag{6}$$

The function r is called the *s-rank function* of $(S, \mathcal{F})$.

Remark. By Definition (7), the s-rank function of a discrete system satisfies (7).

$$0 \leq r(A, B) \leq r(A \cup \{e\}, B) \leq r(A, B) + 1 \leq |A| + 1 \tag{7}$$

And the equation $r(A, B) = |A|$ holds if and only if $A \subseteq F$ for some $F \in \mathcal{F}\Delta B$. Thus the s-rank function is set-theoretical, while the rank function of Definition 2 is polyhedral-theoretical.

These functions are translated mutually by the following equations. (The proof is easy and left for readers.)

$$f(A, B) = r(A \cup B, B) - |B| \qquad (A \cap B = \phi) \tag{8}$$
$$r(A, B) = f(A \backslash B, A \cap B) + |(A \cap B)| \quad \text{(for all } A, B) \tag{9}$$

Inequality (10) is the property of the s-rank function that characterizes the class of Δ-matroids.

$$r(X, Z) + r(Y, Z) \geq r(X \cap Y, Z) + r(X \cup Y, Z) \quad \text{(for all } X, Y, Z) \tag{10}$$

Inequality (10) is just a restatement of the fact that a discrete system is a Δ-matroid if and only if the maximal independent sets, under an arbitrary twisting, compose the basis family of a matroid.

By the s-rank function, pair-Δ-matroids are characterized.

Theorem 8. *Let r be the s-rank function of a pair-Δ-matroid. Then inequality (11) holds. Conversely, if the s-rank function of a discrete system satisfies (11), then the system is a pair-Δ-matroid.*

$$\text{For all } A, B \subseteq S, e_1, e_2, e_3 \in S, \text{ and for all } p, q \in A \ (p \neq q) \text{ that}$$
$$\text{satisfy } r(A, B) = |A| \geq 2, \text{ equations } r(A \cup \{e_i\}, B) = r(A, B)$$
$$(i = 1, 2, 3) \text{ imply } r(A \cup \{e_1, e_2, e_3\}, B\Delta\{p, q\}) \leq r(A, B) + 2. \tag{11}$$

Proof. Let a discrete system $(S, \mathcal{F})$ be a pair-Δ-matroid. Suppose that (11) does not hold for some $A, B, e_1, e_2, e_3, p, q$. We consider a twisted system $(S, \mathcal{F} \Delta B)$, which is also a pair-$\Delta$-matroid. This discrete system has an independent set that includes A, since $r(A, B) = |A|$. It also has an independent set F such that $F \Delta \{p, q\}$ includes $A \cup \{e_1, e_2, e_3\}$, since $r(A \cup \{e_1, e_2, e_3\}, B \Delta \{p, q\}) \geq r(A, B) + 3$. Thus F contains neither p nor q. The condition (5) shows the existence of an independent set that includes A and contains at least one of e_1, e_2, e_3. This implies that $r(A \cup \{e_i\}, B) > r(A, B)$ for $i = 1, 2,$ or 3. This is a contradiction.

Let the s-rank function of a discrete system $(S, \mathcal{F})$ satisfy (11). Suppose that the system is not a pair-Δ-matroid. Let $X, Y \in \mathcal{F}$ with $p, q \in X \Delta Y$ be a counterexample of (5) which minimizes $|X \Delta Y|$. Let $Z = (S \backslash \{p, q\}) \Delta Y$. Then $Y \Delta Z = S \backslash \{p, q\}$ and $X \Delta Z = (S \backslash (X \Delta Y)) \cup \{p, q\}$. Let A denote $X \Delta Z$. Then $r(A, Z) = |A|$. For any $e \in (X \Delta Y) \backslash \{p, q\}$, $r(A \cup \{e\}, Z) = r(A, Z)$. (Suppose not. There exists a subset $P \in \mathcal{F} \Delta Z$ which includes $A \cup \{e\}$. Then $P \Delta Z$ is independent in $(S, \mathcal{F})$. Since $|X \Delta Y|$ is minimized, $P \Delta Z$ and Y with p, q satisfy (5). But it means X, Y with p, q also satisfy (5). A contradiction.) Then $r(A \cup \{e_1, e_2, e_3\}, Z \Delta \{p, q\}) \leq r(A, Z) + 2$ for any three elements e_1, e_2, e_3 in $(X \Delta Y) \backslash \{p, q\}$. This contradicts the existence of $Y \in \mathcal{F}$. $\qquad \Box$

Here we give two examples of discrete systems which are pair-Δ-matroids but not Δ-matroids.

Example 1. Let S be a finite set. If we consider all subsets of S whose cardinality is equal to $4k$ ($k \in \mathbb{Z}^+$) is independent, then a pair-Δ-matroid is obtained.

Example 2. Let S be a set of people. Let $\mathcal{F}$ comprise all subsets of S that contain an even number of women and an even number of students. Then $(S, \mathcal{F})$ is a pair-Δ-matroid.

5 Excluded Minors

We describe the excluded minors [4] of Δ-matroids and pair-Δ-matroids. Before the description of excluded minors, we define the minors of a discrete system. This definition is the same as that of Δ-matroids [3].

Definition 9. Let $M = (S, \mathcal{F})$ be a discrete system, let $S' \subseteq S$, let $X = F \backslash S'$ for some $F \in \mathcal{F}$, and let $\mathcal{F}' = \{X' \subseteq S' | X' \cup X \in \mathcal{F}\}$. The discrete system $M' = (S', \mathcal{F}')$ is called a *minor* of M.

Chandrasekaran and Kabadi proved that minors of a Δ-matroid are Δ-matroids [3]. And it is also proved that minors of a pair-Δ-matroid are pair-Δ-matroids. (The proof is easy.)

We consider the following discrete systems, which do not satisfy (1).

Definition 10. A discrete system $(S, \mathcal{F})$ is a *basic non-Δ-matroid* (BNDM), if it satisfies the following conditions.

1. $|S| \geq 3$.
2. There exist $X, Y \in \mathcal{F}$ and $x \in S$ that satisfy $X \Delta Y = S$ and $x \in X \Delta Z$ for all $Z \in \mathcal{F} \setminus \{X\}$.

Obviously no BNDM is a Δ-matroid. Needless to say, every discrete system that is not a Δ-matroid is not necessarily a BNDM. But the concept of BNDM is important on characterizing whether a discrete system is a Δ-matroid or not. We prove the following proposition.

Theorem 11. *A discrete system is a Δ-matroid if and only if it does not have a BNDM as its minor.*

Proof. If a discrete system have a BNDM as its minor, it is not a Δ-matroid by the fact that the minors of a Δ-matroid are Δ-matroids.

Now we show that a discrete system has a BNDM as its minor if the system is not a Δ-matroid. Let $M = (S, \mathcal{F})$ be a discrete system which is not a Δ-matroid. There exist $X, Y \in \mathcal{F}$ and $x \in X \Delta Y$ such that $X \Delta \{x, y\} \notin \mathcal{F}$, for any $y \in X \Delta Y$. Let X_0, Y_0, x_0 minimize $|X \Delta Y|$ among such X, Y, x.

Trivially, $|X_0 \Delta Y_0| \geq 3$.

Without loss of generality, we assume that $x_0 \in X_0$. No subset $Z \in \mathcal{F}$ except X that satisfies $Z \Delta X \subseteq X \Delta Y$ contains x_0. (The existence of such a subset Z implies either X_0, Z, x_0 minimize $|X_0 \Delta Z|$ or X_0, Y_0, x_0 satisfy (1).)

Let S' be $X_0 \Delta Y_0$ and let $\mathcal{F}'$ be $\{X' \subseteq S' | X' \cup (X_0 \setminus S') \in \mathcal{F}\}$. Let X_0' be $X_0 \cap S'$. Then X_0' is the only subset in $\mathcal{F}'$ that contains x_0. The pair $(S', \mathcal{F}')$, which is a minor of M, is a BNDM. $\square$

The class of pair-Δ-matroids is also characterized by some excluded discrete systems.

Definition 12. A discrete system $(S, \mathcal{F})$ is a *basic non-pair-Δ-matroid* (BNPDM), if it satisfies the following conditions.

1. $|S| \geq 5$.
2. There exist $X, Y \in \mathcal{F}$ and $x, y \in X \Delta Y$ $(x \neq y)$ that satisfy $X \Delta Y = S$ and $\{x, y\} \subseteq X \Delta Z$ for all $Z \in \mathcal{F} \setminus \{X\}$.

Obviously no BNPDM is a pair-Δ-matroid. BNPDMs are the excluded minors of pair-Δ-matroids.

Theorem 13. *A discrete system is a pair-Δ-matroid if and only if it does not have a BNPDM as its minor.* $\square$

The proof of Theorem 13 is similar to that of Theorem 11.

There is an infinite number of BNDMs and BNPDMs. However, they are reduced to a small number of discrete systems, if we employ two operations defined in [2]. One is twisting, described in Section 2 and the other is projection, described below.

Projection (or *restriction*) is an operation that makes $(S', \mathcal{F}')$ from a discrete system $(S, \mathcal{F})$, where $S' \subseteq S$ and $\mathcal{F}' = \{F \cap S' | F \in \mathcal{F}\}$.

With these operations, Δ-matroids are transformed to Δ-matroids [2, 3]. It is easy to see that it is also true of pair-Δ-matroids.

By using twisting and projection, BNDMs are reduced to the discrete systems defined below. (First restrict the support to some subset which comprises three elements including x. Then twist the system.)

Definition 14. A discrete system is a *Minimum Non-Δ-Matroid* (MNDM), if it satisfies the following conditions.

1. The support comprises three elements.
2. The empty set and the support are independent.
3. For some element, the support is the only independent set that contains the element.

Minimum Non Pair-Δ-Matroids are similarly defined.

Definition 15. A discrete system is a *Minimum Non-Pair-Δ-Matroid* (MN-PDM), if it satisfies the following conditions.

1. The support comprises five elements.
2. The empty set and the support are independent.
3. For some two elements, the support is the only independent set that contains both of these elements.

There exists a finite number of MNDMs and MNPDMs. With these discrete systems, Theorem 11 and Theorem 13 are rephrased as Corollary 16.

Corollary 16. *A discrete system is a Δ-matroid (pair-Δ-matroid) if and only if it has no minor that can be reduced to a MNDM (MNPDM) with projection and twisting.*

Proof. The "if" part is obvious. If a system has a minor that can be reduced to a MNDM(MNPDM), then the system can be reduced to a discrete system that has the MNDM(MNPDM) as its minor. This fact proves the "only if" part. □

References

1. A. Bouchet. Greedy algorithm and symmetric matroids. *Mathematical Programming*, 38:147–159, 1987.
2. A. Bouchet and W. H. Cunningham. Delta-matroids, jump systems and bisubmodular polyhedra. *SIAM journal on Discrete Mathematics*, 8:17–32, 1995.
3. R. Chandrasekaran and S. N. Kabadi. Pseudomatroids. *Discrete Mathematics*, 71(3):205–217, 1988.
4. A. Recski. *Matroid Theory and its Applications in Electric Network Theory and in Statics*. Springer-Verlag, 1989.
5. T. Takabatake. Weakly greedy algorithm, weak-Δ-matroid and its subclasses. Master's thesis, University of Tokyo, 1995.

On Integer Multiflows and Metric Packings in Matroids

Karina Marcus & András Sebő

e-mail: [Karina.Marcus,Andras.Sebo]@imag.fr
ARTEMIS IMAG, Université Joseph Fourier, BP 53
38041 Grenoble Cedex 9, France

Abstract. Seymour [10] has characterized graphs and more generally matroids in which the simplest possible necessary condition, the "cut condition", is also sufficient for multiflow feasibility. In this work we exhibit the next level of necessary conditions, three conditions which correspond in a well-defined way to minimally non-ideal binary clutters. We characterize the subclass of matroids where the presented conditions are also sufficient for multiflow feasibility, and prove the existence of integer multiflows for Eulerian weights. The theorem we prove uses results from Seymour[10] and generalizes those results and those in Schwärzler, Sebő [7]. We then study the polar of the considered multiflow problems, and characterize the subclass where the integer metric packing theorem holds for bipartite weights; surprisingly, unlike for most of the known multiflow theorems this subclass is not the same as the class where integer multiflow theorems hold for bipartite weights, but is essentially smaller.

1 Introduction

Let M be a binary matroid defined on the finite set $E(M)$ and p a function assigning integer values to the elements of $E(M)$. We think of the negative values of p as representing *demands* and of the nonnegative values as representing *capacities*. Define $F(p) = \{e \in E(M) : p(e) < 0\}$. A *flow problem* is a pair (M, p). It has a *solution* if there exists a *multiflow*, that is a function $\Phi : C_P(M) \to \mathbb{R}_+$ defined on the set $C_P(M)$ of all circuits C of M with $|C \cap F(p)| = 1$ such that

$$\sum_{C \in C_P, C \ni e} \Phi(C) \begin{cases} \leq \ p(e), & \text{if } e \in E(M) - F(p), \\ = -p(e), & \text{if } e \in F(p). \end{cases}$$

A function $m : E(M) \to \mathbb{R}_+$ is a *metric* if $m(e) \leq m(C - \{e\})$ for all circuits C of M and all elements e of C. (We use the notation $m(X) = \sum_{e \in X} m(e)$ for subsets X of $E(M)$.) Δ is a *family of metrics* if for every binary matroid M, $\Delta(M)$ is a set of metrics defined on $E(M)$. For $A \subseteq \mathbb{R}_+$, we will denote the family of all metrics $m : E(M) \to A$ by $\Delta_A(M)$, or simply by Δ_A. A metric m is *bipartite* if $m(C)$ is even for all circuits C of M. The extreme rays of the cone $(\Delta_A(M))$ are called *primitive*.

Let Δ be a family of metrics, and (M, p) be a flow problem. Consider the condition

$$m.p \geq 0 \text{ for all } m \in \Delta(M). \tag{1}$$

It follows easily from LP duality that (1) is necessary for the existence of multi-flows, even if $\Delta(M)$ is the set of all metrics on M, and, in this case, (1) is also sufficient. The question that arises is then the following one: When is (1) being true for a specific family of metrics sufficient to imply that (1) is true for all metrics?

A binary matroid M for which the condition (1) is sufficient for the existence of a solution of (M, p) for arbitrary functions p, will be called *flowing with respect to* Δ. A flow problem (M, p) is *Eulerian* if $p(D)$ is even for all cocircuits D of M. If (1) is sufficient for the existence of an integer solution for all Eulerian problems (M, p), then M is called *cycling with respect to* Δ.

A well known and easy fact to be used throughout (it is a consequence of Farkas' Lemma):

Fact 1. *[7] Let M be a matroid and $A \subseteq \mathbb{Z}_+$. M is flowing with respect to Δ_A if and only if $\Delta_{\mathbb{Z}_+} \subseteq$ cone $(\Delta_A(M))$.* $\qquad\qquad\square$

The polar problem of the multiflow problem could be seen as the packing of a metric m into a set of primitive metrics $\Delta_A(M)$, that is we want to write m as $\lambda_1 m_1 + \ldots + \lambda_k m_k$, $\lambda_i \in \mathbb{Z}_+, m_i \in \Delta_A(M)$, $1 \leq i \leq k$. From Fact 1 it follows that if M is flowing with respect to Δ_A, then a metric m on $E(M)$ may be always written as a – fractional – sum of metrics in A. So now we are interested in a packing with integer coefficients, but with no further hypotheses this seems to be too restrictive. So we ask an integer packing whenever a given metric m is bipartite (see the analogy with "cycling"); if a binary matroid M has this property for all bipartite metrics, then we say that it is *packing with respect to* Δ_A. If the coefficients in the packing are integer multiples of $1/2$, we say that M is *half packing*.

The problem of packing metrics in graphs has been raised in several papers in the past: For the case of cut-metrics, Karzanov [2] and Schrijver ([5], [6]) have proved the existence of integer "polars" of several well-known multiflow theorems, and Karzanov in [4] proves the existence of an integer packing of $bip(2, 3)$-*metrics* and cuts for graphs with a demand-set adjacent to at most five vertices. Given an undirected graph $G = (V, E)$ and a partition of V in 5 possibly classes A_1, A_2 and B_1, B_2, B_3, such that $A_1 \cup A_2$ and $B_1 \cup B_2 \cup B_3$ are non-empty, define a metric $m : E \to \mathbb{Z}_+$, a bip(2, 3)-metric, as follows:

$$m(x, y) = \begin{cases} 1, & \text{if } x \in A_i, \, y \in B_j, \\ 2, & \text{if } x \in A_i, \, y \in A_j \ (i \neq j) \text{ or } x \in B_i, \, y \in B_j \ (i \neq j), \\ 0, & \text{if } x, y \in A_i \text{ or } x, y \in B_i. \end{cases}$$

We shall denote by $\mathcal{C}(M)$ the set of cycles (that is, disjoint union of circuits) of the matroid M and by $\mathcal{C}^*(M)$ the set of cocycles. We refer to Welsh [11] for the basic concepts and facts of matroid theory.

In section 2 we give an overview of the multiflow problem in binary matroids and its relation to metrics; in section 3 we study the K_5- and F_7-metrics, showing that both are primitive and that the condition (1) restricted to K_5- and F_7-metrics is sufficient for the existence of a multiflow in a certain class of matroids.

In section 4 we show that $M(K_5)$ and F_7 are packing and that under certain hypotheses we can get a half packing matroid out of a special 2-sum of packing matroids - and that is the best that we can get.

2 Multiflows

The incidence vector χ_D of a cocycle D of M is called a *cut-metric*, and $\Delta_{(CC)}(M)$ denotes the set of all *cut-metrics* of the binary matroid M. We say that (M, p) satisfies the so-called *cut-condition* if and only if

$$m.p \geq 0 \text{ for all } m \in \Delta_{(CC)}(M). \tag{CC}$$

Seymour's following result (see [10]) tells us that the metrics in $\Delta_{(CC)}$ are sufficient to describe the flowingness with respect to $\Delta_{\{0,1\}}$ and characterizes the related class of matroids.

Theorem 2. *For a binary matroid M the following are equivalent:*
(i) M is cycling with respect to $\Delta_{(CC)}$;
(ii) M is flowing with respect to $\Delta_{\{0,1\}}$;
(iii) M has no F_7, R_{10} or $M(K_5)$ minor. □

F_7 is the Fano matroid on 7 elements, $M(K_5)$ is the graphic matroid of the complete graph on 5 nodes, and R_{10} is a special matroid on 10 elements used to characterize regular matroids [8], that can be represented by the node-edge incidence matrix of the complete bipartite graph $K_{3,3}$, plus a column of 1.

Schwärzler and Sebő [6] have shown that extending the cut condition to a larger class of metrics, called $CC3$-metrics, a statement similar to Seymour's holds for a larger class of matroids. We will deduce the following sharper form in Sect. 3, where $CC3$ is replaced by the cut-condition or either of two conditions which correspond to the only primitive metrics in $CC3$.

Theorem 3. *For a binary matroid M the following are equivalent:*
(i) M is cycling with respect to $\Delta_{(CC,F_7,K_5)}$;
(ii) M is flowing with respect to $\Delta_{\{0,1,2\}}$;
(iii) M has no $AG(2,3)$, S_8, R_{10}, $M(H_6)$, $M(K_5) \oplus_2 F_7$, $M(K_5) \oplus_2 M(K_5)$, $F_7 \oplus_2 F_7$ minor. □

Here H_6 is the graphic matroid in Figure 1 (a), $AG(2, 3)$ is the representation of a projective plane and S_8 can be represented as the node-edge incidence matrix of the graph in Figure 1 (b), with a column with the circled elements. The definition of 2-sum $M_1 \oplus M_2$ of binary matroids is given in [10].

3 The two conditions

Let $\Delta_{K_5}(M)$ (respectively $\Delta_{F_7}(M)$) be the class of metrics $m \in \Delta_{\{0,1,2\}}$ such that, if we contract the elements e with $m(e) = 0$, we obtain a $M(K_5)$ (respectively F_7), probably with some parallel elements, with the weights on each

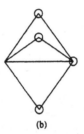

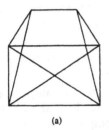

Fig. 1. H_6 and S_8

element of a parallel class defined below. For the K_5, if we denote by $\{1, 2, 3, 4, 5\}$ the set of vertices, and by ij the edge between the vertices i and j, then we have

$$m(ij) = \begin{cases} 2, & \text{if } ij \in \{12, 23, 13, 45\}, \\ 1, & \text{otherwise.} \end{cases}$$

If C is a three-element circuit of $\mathcal{C}(F_7)$, then we define

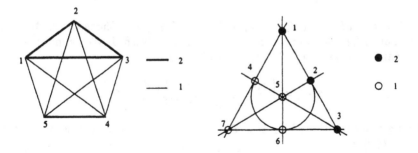

Fig. 2. K_5- and F_7-metrics

$$m(e) = \begin{cases} 2, & \text{if } e \in C, \\ 1, & \text{otherwise.} \end{cases}$$

Lemma 4. *The K_5- and F_7-metrics are primitive.*

Proof. We will show that the F_7-metric is an extreme ray of the cone $\Delta_{\mathbb{Z}_+}$ (for K_5 the proof works in the same way, and is well known, see for example Karzanov [3]). If it is not primitive, then m can be decomposed in a sum of primitive metrics, and the equalities $m(C - e) = m(e)$, $e \in C \in \mathcal{C}(F_7)$, satisfied by the F_7-metric, must be satisfied by any primitive metric in the decomposition. We check that the only solution to the system formed by these equalities is the F_7-metric, and its positive multiples.

To facilitate our task, let x_i denote the value of the function x on the element i, following Figure 2. Then we have the equalities:

$$\left.\begin{array}{l} x_1 = x_4 + x_7 = x_5 + x_6 \\ x_2 = x_5 + x_7 = x_4 + x_6 \end{array}\right\} \Rightarrow x_5 = x_4, \quad x_6 = x_7,$$

and in the same way we obtain that $x_4 = x_7$, $x_5 = x_6$ and so $x_4 = x_5 = x_6 = x_7$ and $x_1 = x_2 = x_3 = 2x_4$, and this corresponds to the F_7-metric, proving that it is the only primitive metric in the decomposition. □

Now we prepare the proof of the implication (iii) $\Rightarrow$ (i) of the Theorem 3. A twofold application of Seymour's 'Splitter Theorem' gives the following [10].

Proposition 5. *Every binary matroid with no $AG(2,3), S_8, R_{10}$ or $M(H_6)$ minor may be obtained by 1- and 2-sums from matroids cycling with respect to $\Delta_{(CC)}$ and copies of F_7 and $M(K_5)$.* □

And we can use it to prove that

Proposition 6. *Any 2-sum $M_1 \bigoplus_2 M_2$ of a matroid M_1 cycling with respect to $\Delta_{(CC,K_5,F_7)}$ and a matroid M_2 cycling with respect to $\Delta_{(CC)}$ is cycling with respect to $\Delta_{(CC,K_5,F_7)}$.*

Proof. Let $E(M_1) \cap E(M_2) = \{f\}$ and $M = M_1 \bigoplus_2 M_2$. Choose $p : E(M) \to \mathbb{Z}$ such that (M, p) is Eulerian and (CC, K_5, F_7) is satisfied. We define functions $p_i : E(M_i) \to \mathbb{Z}$ ($i \in \{1, 2\}$) in the following way:

$$p_i(e) = \begin{cases} p(e), & \text{if } e \in E(M_i) - f, \\ (-1)^{i-1}q, & \text{if } e = f, \end{cases}$$

where $q = \min\{p(D - f) : f \in D \in C^*(M_2)\}$. Let D_0 be a cocycle of M_2 with $p(D_0 - f) = q$.

Claim 1. p_i *($i \in \{1, 2\}$) is an Eulerian function.*

Proof. Let D_i be a cocycle of M_i. If $f \notin D_i$, then $p_i(D_i) = p(D_i) \equiv \text{mod } 2$, because D_i is also a cocycle of M. If $f \in D_i$, then

$$p_i(D_i) = p_i(D_i - f) + p_i(f)$$
$$\equiv p(D_i - f) + p(D_0 - f) \equiv p(D_i \triangle D_0) \text{ mod } 2,$$

because $D_i \triangle D_0$ is a cocycle of M. □

Claim 2. (M_2, p_2) *satisfies (CC).*

Proof. Let $D \in C^*(M_2)$. If $f \notin D$, then again D is a cocycle of M and $p_2(D) = p(D) \geq 0$, because we assumed that (CC, K_5, F_7) and so in particular (CC) is satisfied for (M, p). If $f \in D$, then the definition of q implies the following inequality: $p_2(D) = p_2(D - f) + p_2(f) = p(D - f) - p(D_0 - f) \geq 0$. □

Claim 3. (M_1, p_1) *satisfies* (CC, K_5, F_7).

Proof. We have to show that $pm_1 \geq 0$ for every choice of $m_1 \in \Delta_{(CC,K_5,F_7)}(M_1)$.

If m_1 is a CC-metric, then everything works as in Claim 2. Otherwise we associate to m_1 a metric $m \in \Delta_{(CC,K_5,F_7)}(M)$ defined as

$$m(e) = \begin{cases} m_1(e), & \text{if } e \in E(M_1) - f, \\ m_1(f), & \text{if } e \in D_0, \\ 0, & \text{otherwise.} \end{cases}$$

It is not difficult to see that if m_1 is a K_5- or F_7-metric on M_1, then m is a K_5- or F_7-metric on M. And so we have that

$$\begin{aligned}
p_1 m_1 &= \sum_{e \in E(M_1)} p_1(e) m_1(e) \\
&= \sum_{e \in E(M)} p_1(e) m_1(e) + p_1(f) m(f) \\
&= \sum_{e \in E(M) - D_0} p(e) m(e) + p(D_0 - f) m(f) \\
&= \sum_{e \in E(M)} p(e) m(e) \\
&\geq 0,
\end{aligned}$$

because (M, p) satisfies (CC, K_5, F_7). Thus Claim 3 is proved. $\square$

As M_1 (respectively M_2) was assumed to be cycling with respect to $\Delta_{(CC,K_5,F_7)}$ (respectively $\Delta_{(CC)}$), the above claims guarantee the existence of integer flows ϕ_i in (M_i, p_i) ($i \in \{1, 2\}$). ϕ_i consists of a list of cycles of $C_{p_i}(M_i)$. Suppose without loss of generality that precisely the first k_i cycles of each list contain the element f. It follows from the definition of a flow that $k_i \leq q = k_2$. After deleting the first $k_2 - k_1$ cycles from the second list ϕ_2, the union of the two lists contains exactly k_1 cycles of $C(M_1)$ and k_1 cycles of $C(M_2)$ passing through the element f. Build k_1 pairs (C_1, C_2) ($C_i \in C(M_i)$) of the cycles passing through f and replace each pair by $C_1 \triangle C_2$. It is easy to see that the list of cycles obtained in this way represents an integer flow of (M, p). $\square$

Proof. (of Theorem 3:)

(i) $\Rightarrow$ (ii) trivial.

(ii) $\Rightarrow$ (iii) There are several ways of proving this implication. [7] checks it by showing multiflow problems that have no solution, but whose multiflow functions satisfy (1) for $\Delta_{\{0,1,2\}}$. We show that there are primitive metrics for these matroids that are not in $\Delta_{\{0,1,2\}}$, which is a shorter and easier way of proving the implication.

Let S_8, R_{10} and $AG(2,3)$ be represented by the matrices in Figure 3. We will prove the result for the S_8 case, the other ones follow similarly.

$$\begin{bmatrix} 1 & 0 & 0 & 0 & 1 & 1 & 1 & 1 \\ 0 & 1 & 0 & 0 & 1 & 0 & 0 & 1 \\ 0 & 0 & 1 & 0 & 0 & 1 & 0 & 1 \\ 0 & 0 & 0 & 1 & 0 & 0 & 1 & 1 \end{bmatrix} \quad \begin{bmatrix} 1 & 1 & 1 & 1 & 1 & 0 & 0 & 0 & 0 \\ 1 & 1 & 1 & 0 & 0 & 0 & 1 & 1 & 0 \\ 1 & 0 & 0 & 1 & 1 & 0 & 1 & 1 & 0 & 1 \\ 0 & 1 & 0 & 1 & 0 & 1 & 1 & 0 & 1 & 1 \\ 0 & 0 & 1 & 0 & 1 & 1 & 0 & 1 & 1 & 1 \end{bmatrix} \quad \begin{bmatrix} 0 & 0 & 0 & 1 & 0 & 1 & 1 & 1 \\ 0 & 0 & 1 & 1 & 1 & 0 & 1 & 0 \\ 0 & 1 & 1 & 1 & 0 & 0 & 0 & 1 \\ 1 & 1 & 1 & 1 & 1 & 1 & 1 & 1 \end{bmatrix}$$

Fig. 3. Matrix representation of S_8, R_{10} and $AG(2,3)$.

Now let $m := (2,1,1,1,1,1,1,3)$. For this metric m we have the following equalities arising from the inequality $m(e) \le (C - e)$, where C is a circuit in S_8:

$$m_1 = m_2 + m_5 \qquad m_1 = m_3 + m_6 \qquad m_1 = m_4 + m_7$$
$$m_8 = m_2 + m_3 + m_7 \qquad m_8 = m_2 + m_4 + m_6 \qquad m_8 = m_3 + m_4 + m_5$$
$$m_8 = m_5 + m_6 + m_7.$$

These equations are affinely independent and all solutions for this system are vectors of the form $(2a, a, a, a, a, a, a, 3a)$, $a \ge 0$, which is exactly the extreme ray of the cone of metrics $\Delta_{\mathbb{Z}_+}(S_8)$ defined by m. Therefore m is primitive, but is is not a $(0,1,2)$-vector, so, by Fact 1, S_8 is not flowing with respect to $\Delta_{\{0,1,2\}}$.

In the same way we can show that $m = (3,3,1,1,3,1,1,1,1,1)$ and $m = (1,1,1,4,1,1,1,1)$ define extreme rays of the cone of metrics $\Delta_{\mathbb{Z}_+}(R_{10})$ and $\Delta_{\mathbb{Z}_+}(AG(2,3))$, respectively, proving that they are not flowing with respect to $\Delta_{\{0,1,2\}}$.

A primitive metric for H_6 is represented in Figure 4, and one can check that it is primitive in the same way as in the cases above.

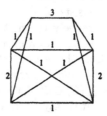

Fig. 4. H_6

Now let $M(K_5) \bigoplus_2 M(K_5)$ be as in Figure 5, the marker is indicated in dashed line, and let $m : E(M(K_5) \bigoplus_2 M(K_5)) \to \mathbb{Z}$ be as follows

$$m(x) := \begin{cases} 4, & \text{if } x \in \{j, l, o, r\}, \\ 2, & \text{if } x \in \{a, c, i, k, m, n, p, q\}, \\ 1, & \text{otherwise.} \end{cases}$$

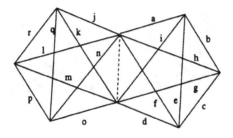

Fig. 5. $M(K_5) \bigoplus_2 M(K_5)$

The following are tight inequalities for m

$$
\begin{array}{llll}
r = p + q & r = m + k & j = n + q & j = k + f + d \\
j = k + g + h & l = p + n & l = m + f + d & l = m + g + h \\
o = k + q & o = p + m & o = n + f + d & o = g + h + n \\
a = f + e & a = h + b & i = b + g & i = d + e \\
c = e + b & c = f + h & c = d + g
\end{array}
$$

and it is not difficult to find that the solutions for the system are

$$
s(x) := \begin{cases}
4\delta, & \text{if } x \in \{j, l, o, r\}, \\
2\delta, & \text{if } x \in \{a, c, i, k, m, n, p, q\}, \\
1\delta, & \text{otherwise,}
\end{cases}
$$

where $\delta \geq 0$.

Let $F_7 \bigoplus_2 M(K_5)$ and $F_7 \bigoplus_2 F_7$ be as in Figure 6, the markers are indicated by dashed lines, and $m_1 : E(F_7 \bigoplus_2 M(K_5)) \to \mathbb{Z}$ and $m_2 : E(F_7 \bigoplus_2 F_7) \to \mathbb{Z}$ be as follows

$$
m_1(x) := \begin{cases}
4, & \text{if } x \in \{g, h, i, m\}, \\
2, & \text{if } x \in \{e, f, j, k, l, n, o\}, \\
1, & \text{otherwise.}
\end{cases}
$$

$$
m_2(x) := \begin{cases}
4, & \text{if } x \in \{a, b, e\}, \\
2, & \text{if } x \in \{c, d, f, l, m\}, \\
1, & \text{otherwise.}
\end{cases}
$$

We can check in the same way as above that both metrics m_1 and m_2 are primitive, and as they are not $(0, 1, 2)$-vectors, and together with the Fact 1 they show that the matroids are not flowing with respect to $\Delta_{\{0,1,2\}}$.

(iii) $\Rightarrow$ (i) Now we know that K_5 and F_7 are flowing with respect to $\Delta_{\{CC, K_5, F_7\}}$ (see [4] and [7]). Using these results with Proposition 6 we get easily the desired implication. $\qquad\square$

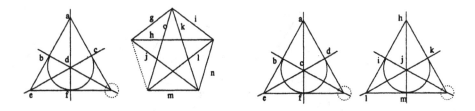

Fig. 6. $F_7 \bigoplus_2 M(K_5)$ and $F_7 \bigoplus_2 F_7$

4 Packing matroids

We prove now two statements showing that K_5 and F_7 are packing with respect to $\Delta_{\{CC,K_5,F_7\}}$. The first is in fact a consequence of a theorem of Karzanov ([4]), but for the sake of completeness and because of the analogy of our proof for K_5 and F_7 we include a simple proof. We say that we *can subtract* a metric m_2 from m_1 if $m_1 - m_2$ is a metric. If both m_1 and m_2 are bipartite, then obviously $m_1 - m_2$ is also bipartite.

Lemma 7. *In the matroid $M(K_5)$ every bipartite metric can be expressed as a positive integer sum of metrics in $\Delta_{(CC,K_5)}(M(K_5))$.*

Proof. Let m be a bipartite metric on $M(K_5)$. We want to write it as an integer sum of cuts and K_5-metrics. By Theorem 3 and Fact 1 we know that m can be expressed as $m = \nu_1 \chi_{C_1} + \ldots + \nu_n \chi_{C_n} + \lambda_1 m_1 + \ldots + \lambda_k m_k$, where C_i is a cut, m_i is a K_5-metric, and $\nu_i, \lambda_i > 0$.

Claim 1. *Let D be a cut on K_5. If $(m - \chi_D)(C - e) - (m - \chi_D)(e) < 0$, for a circuit C and $e \in C$, then there exists a triangle T (a circuit of cardinality 3), and an element $f \in T$, such that $(m - \chi_D)(T - f) - (m - \chi_D)(f) < 0$.*

Proof. Let $l = m - \chi_D$. If $|C| = 4$, then there exists a triangle T such that $|T \cap C| = 2, T \setminus C = \{f\}, e \in T \cap C$. Let $T' = C \Delta T$. Then $0 > l(C - e) - l(e) = l(T - e) - l(e) + l(T' - f) - l(f)$ and so we have that either $l(T - e) - l(e) < 0$ or $l(T' - f) - l(f) < 0$.

If $|C| = 5$, then there exists a triangle T, $|T \cap C| = 2, T \setminus C = \{f\}, e \in T \cap C$ and a circuit $C' = C \Delta T$, $|C'| = 4$, such that $0 > l(C - e) - l(e) = l(T - e) - l(e) + l(C' - f) - l(f)$. Now either we get directly the conclusion, or we use the previous case. □

Claim 2. *If C_i is a cocircuit, then χ_{C_i} can be subtracted from m.*

Proof. If χ_{C_i} cannot be subtracted from m, then by Claim 1, there is a triangle T and $e \in T$, such that

$$(m - \chi_{C_i})(T - e) - (m - \chi_{C_i})(e) < 0. \tag{$*$}$$

As $m(T-e) - m(e) = 0$ implies that $\chi_{C_i}(T-e) - \chi_{C_i}(e) = 0$, we may suppose that $m(T-e) - m(e) \geq 2$. As $\chi_{C_i}(T-e) - \chi_{C_i}(e) \leq 2$, we cannot have (*). $\quad\square$

Claim 3. *If m_i and m_j are different K_5-metrics, then $m_i + m_j$ can be written as a sum of cut-metrics.*

Proof. The Figure 7 shows the sum of two K_5-metrics, which has the following cut decomposition (we use the numeration of Figure 2): $\{15, 25, 35, 14, 24, 34\}$ $+\{15, 25, 45, 13, 23, 34\} + \{12, 13, 15, 24, 34, 45\} + \{12, 14, 15, 23, 34, 35\}$. The other cases are similar. $\quad\square$

 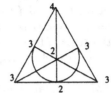

Fig. 7. Sums of two K_5-metrics and two F_7-metrics

The three Claims above imply that we can obtain from a fractional decomposition for m an integer decomposition for m consisting on cut-metrics and K_5-metrics. $\quad\square$

Lemma 8. *In the F_7 matroid every bipartite metric may be expressed as a positive integer sum of metrics in $\Delta_{(CC,F_7)}$.*

Proof. Let m be a bipartite metric on F_7. We proof as above:

Claim 1. *If C_i is a cocircuit, then χ_{C_i} can be subtracted from m.*

Proof. χ_{C_i} may be subtracted from m if and only if

$$m(C-e) - m(e) \geq \chi_{C_i}(C-e) - \chi_{C_i}(e), \quad \text{for all } e \in C \in \mathcal{C}(F_7).$$

$m(C-e) - m(e) = 0$ implies that $\chi_{C_i}(C-e) - \chi_{C_i}(e) = 0$. If $m(C-e) - m(e) \geq 2$, then $|C_i \cap C| \equiv 0 \bmod 2$, because F_7 is a binary matroid. So we have only to consider when $|C_i \cap C| = 4$. In this case $C = C_i$ and $e \in C_i \cap C$, so $\chi_{C_i}(C-e) - \chi_{C_i}(e) \leq 2$, and we have the desired inequality. $\quad\square$

Claim 2. *If m_i and m_j are different F_7-metrics, then $m_i + m_j$ can be written as a sum of cut metrics.*

Proof. In Figure 7 we show the sum of two F_7-metrics, and its decomposition in metric-cuts is: $\{1,2,4,5\} + \{1,3,5,7\} + \{1,2,6,7\} + \{1,3,4,6\} + \{2,3,4,7\}$ (with the same numeration of Figure 2). All the others are symmetric. □

Now putting together Claims 1 and 2 we obtain that m is an integer sum of cut-metrics and perhaps one F_7-metric, with an integer coefficient too – because as we can always subtract the cut-metrics from m, repeating this procedure we will end up with a F_7-metric and necessarily with an integer coefficient. □

We will introduce now a concept that will help us prove some new characterizations of flowing and packing matroids. We say that M has *the sums of circuits property* (see [9]) if the following are equivalent for all $p : E(M) \to \mathbb{Z}_+$:
(i) There is a function $\alpha : \mathcal{C}(M) \to \mathbb{R}_+$ such that $\sum \alpha(C)\chi_C = p$.
(ii) For every cocircuit D and $f \in D$, $p(f) \leq p(D - \{f\})$.
In [10] Seymour characterized matroids that have the sums of circuits property - they are the duals of those flowing with respect to $\Delta_{(CC)}$, and conjectured the following result, proved by Alspach, Goddyn and Zhang ([1]).

Theorem 9. *If M is a binary matroid and has no F_7^*, R_{10}, $M^*(K_5)$ or $M(P_{10})$ minor, and p satisfies (ii) and is Eulerian, then there is an integral α satisfying (i).* □

$M(P_{10})$ is the graphic matroid of the Petersen graph. Dualizing this result, we get a class of packing matroids:

Corollary 10. *If M is a binary matroid and has no F_7, R_{10}, $M(K_5)$ or $M^*(P_{10})$ minor, then M is packing with respect to $\Delta_{(CC)}$.* □

Notice that this is the class of matroids cycling with respect to $\Delta_{(CC)}$, except for those containing $M(P_{10})$ as minor. We know, by Lemmas 7 and 8, that the $M(K_5)$ and F_7 matroids are packing with respect to $\Delta_{(CC,K_5,F_7)}$. We would like to join these classes of matroids and obtain something "bigger". The 2-sum of matroids could help us in this direction, but the packing property is not preserved by the 2-sum. We show in Figure 8 an example of a bipartite metric on a 2-sum of matroids that are each packing, but its decomposition in primitive metrics is half-integer.

Notice that in the Proposition 6 we do not "2-sum" two matroids $M(K_5)$ or F_7, or a $M(K_5)$ with an F_7, so we might suppose that the matroid $M = M_1 \oplus_2 M_2$ is in fact a "big" 2-sum of several matroids packing with respect to $\Delta_{(CC)}$ and one $M(K_5)$ or (exclusive) one F_7. This fact will be used in the following proposition, which is not a recursive result, but shows a way of decomposing a metric in such a "big" 2-sum.

Lemma 11. *A matroid M resulting from 2-sums of one copy of $M(K_5)$ or F_7, and matroids packing with respect to $\Delta_{(CC)}$ is half packing with respect to $\Delta_{(CC,K_5,F_7)}$.*

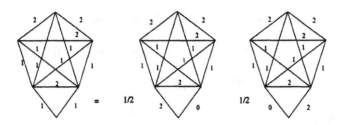

Fig. 8. A half integer metric-packing

Proof. We will find a half integral metric packing in M from integral metric packings of each piece of the 2-sums.

Let $M = R_1 \bigoplus_2 R_2 \bigoplus_2 \ldots \bigoplus_2 R_n$ where R_1 is either a $M(K_5)$ or a F_7, and each R_i, $2 \leq i \leq n$, is a matroid packing with respect to $\Delta_{(CC)}$ that is being "2-summed" with R_1. Let $f_i = E(R_1) \cap E(R_i)$, $2 \leq i \leq n$.

Choose $m : E(M) \to \mathbb{Z}_+$ such that m is a bipartite metric. Let $m'_1 : E(R_1) \to \mathbb{Z}_+$ be such that:

$$m'_1(e) := \begin{cases} m(e), & \text{if } e \in E(R_1) - \{f_j, 2 \leq j \leq n\}, \\ q_i, & \text{if } e = f_j, \end{cases}$$

where $q_i := \min\{m(C - f_i) : C \in \mathcal{C}(R_i)\}$, $2 \leq i \leq n$. Now we define functions $m_i : E(R_i) \to \mathbb{Z}_+$, $1 \leq i \leq n$, in the following way: $m_1(e) := \min\{m'_1(X) : X = \{e\}$ or $X = C - e$ for some $C \in \mathcal{C}(R_1)$ with $e \in C\}$ and

$$m_i(e) := \begin{cases} m(e), & \text{if } e \in E(R_i) - f_i, \\ m_1(f_i), & \text{otherwise.} \end{cases}$$

Let C_i^0 be the circuit of $\mathcal{C}(R_i) \cup \mathcal{C}(R_1 \bigoplus_2 \ldots \bigoplus_2 R_{i-1} \bigoplus_2 R_{i+1} \bigoplus_2 \ldots \bigoplus_2 R_n)$ with $f_i \in C_i^0$ and $m(C_i^0 - f_i) = m_i(f_i)$, $2 \leq i \leq n$.

Claim 1. m_i, $2 \leq i \leq n$, is a bipartite metric.

Proof. Let C_i be a cycle of R_i. If $f_i \notin C_i$, then $m_i(C_i - e) \leq m_i(e)$, $\forall e \in C_i$, and $m_i(C_i) = m(C_i) \equiv 0 \mod 2$, because C_i is also a cycle of M.

If $f_i \in C_i$, then $m_i(C_i - f_i) - m_i(f_i) = m(C_i) - m(C_i^0) \geq 0$, and for $e \neq f_i$, $m_i(C_i - e) - m_i(e) = m(C_i - e) + m(C_i^0) - m(e) \geq 0$, because of the definition of C_i^0; and $m_i(C_i) = m_i(C_i - f_i) + m_i(f_i) = m(C_i - f_i) + m(C_i^0 - f_i) \equiv m(C_i \triangle C_i^0) \equiv 0 \mod 2$, because $C_i \triangle C_i^0$ is a cycle of M. □

Claim 2. m_1 is a bipartite metric.

Proof. By the definition of m_1 as an induced metric, we have only to verify that m_1 is bipartite. Let C be a cycle of R_1 and $L := \{j : \exists f_j \in C\}$, then

$$m_1(C) = m_1(C \backslash \{f_j : j \in L\}) + \sum_{j \in L} m_1(f_j)$$

$$= m(C \backslash \{f_j : j \in L\}) + \sum_{j \in L} m(C_j - f_j)$$

$$\equiv m(C \triangle_{j \in L} C_j^0)$$

$$\equiv 0 \bmod 2,$$

because $C \triangle_{j \in L}(C_j^0)$ is a cycle of M. $\qquad\qquad\qquad\qquad\qquad\qquad$ □

As R_1 (respectively R_i, $2 \le i \le n$) was assumed to be packing with respect to $\triangle_{(CC,K_5,F_7)}$ (respectively to $\triangle_{(CC)}$), the above Claims guarantee the existence of cocircuits $C_1^i, \ldots, C_{r_i}^i \in C^*(R_i)$ such that $\sum_{j=1}^{r_i} \chi_{C_j^i} = m_i$ and cocircuits $C_1^1, \ldots, C_{r_1}^1 \in C^*(R_1)$, and K_5- or F_7-metrics $l_1, \ldots, l_t$ such that $\sum_{j=1}^{r_1} \chi_{C_j^1} + \sum_{j=1}^t l_j = m_1$.

Now we make the packing of cocircuits, K_5- and F_7-metrics for M and f. We can already put in the list the cocircuits C_j^1 ($j = 1, \ldots, r_1$).

We will consider two cases:

Case I : $l_i(f_j) = 1$, $i \in \{1, \ldots, t\}$, $j \in \{2, \ldots, n\}$

To each K_5- or F_7-metric l_i such that $l_i(f_j) = 1$, we associate a cocircuit C_k^j, $k \in \{1, \ldots, r_j\}$, where $f_j \in C_k^j$ and we join l_i to C_k^j to create a K_5- or F_7-metric l_i' in the matroid M. It is not difficult to see how to "glue" l_i and C_k^j: the elements in C_k^j will replace f_j, creating perhaps some parallel elements in the new K_5- or F_7-metric. We will replace l_i by l_i', if there is some j' such that $l_i(f_{j'}) = 2$; otherwise l_i' goes to the list of the packing.

Case II : $l_i(f_j) = 2$, $i \in \{1, \ldots, t\}$, $j \in \{2, \ldots, n\}$

In this case, if we have, for every circuit $K \in C(R_j)$ with $f_j \in K$,

$$m_j(K - f_j - e) - m_j(e) \ge 0, \quad \text{for all } e \in K,$$

then we can simply contract f_j in both matroids R_1 and R_j and find a packing of cocircuits for l_i in $R_j / \{f_j\}$ in R_1 and a new packing of cocircuits in $R_j / \{f_j\}$ for m_j, and add this last packing to the list. If the corresponding K_5- or F_7-metric has already been transformed into cocircuits, we have only to find a packing in $R_j / \{f_j\}$.

For these metrics l_i that have been transformed, and for all f_j such that $l_i(f_j) = 2$, but f_j has not been contracted, there is a natural association between cocircuits in the decomposition of l_i and cocircuits in the decomposition of m_j which contain f_j; the symmetric difference of such pairs of cocircuits is a cocycle in M, and we put them all in the list.

Now we have only metrics l_i not yet decomposed for which there exists a f_j with $l_i(f_j) = 2$ and one can not contract f_j as above.

To each l_i and each R_j for which we have $l_i(f_j) = 2$, as $m_1(f_j) = m_j(f_j)$, we know that we can associate two cocircuits $C_1^j, C_2^j \in C(R_j)$, such that $f_j \in C_1^j \cap C_2^j$. We now build two lists L_k, $k = 1, 2$ of all these C_j^k, for all j with $l_i(f_j) = 2$. We will transform the metric l_i and the lists L_1, L_2 into two single K_5- or F_7-metrics in the following way ($k = 1, 2$):

$$l_i^k(e) := \begin{cases} l_i(e), & \text{if } e \neq f_j, \; \forall j \in \{2, \ldots n\}, \\ 2, & \text{if } e \in C \in L_k, \; e \neq f_j, \\ 0, & \text{otherwise.} \end{cases}$$

It is not difficult to see that l_i^1, l_i^2 are K_5- or F_7-metrics in M, and that

$$\frac{1}{2}l_i^1 + \frac{1}{2}l_i^2 = l_i + \sum_{C \in L_1} \chi_{C-f_j} + \sum_{C \in L_2} \chi_{C-f_j}.$$

And this completes the packing.

At the end we add to the list the metrics corresponding to the remaining cocircuits that do not contain the marker. $\quad\square$

Remark: In the second case of the proof, if $l_i^1 = l_i^2$, then the sum $\frac{1}{2}l_i^1 + \frac{1}{2}l_i^2$ is integer, and so, the result of the decomposition may be integer.

As a conclusion we get that

Theorem 12. *If a binary matroid M has no $M^*(P_{10})$ minor, then M is cycling with respect to $\Delta_{(CC,F_7,K_5)}$ if and only if M is half-packing with respect to this set of metrics.*

Proof. If M is cycling with respect to $\Delta_{(CC,F_7,K_5)}$, then, by Theorem 3 and Proposition 5, M may be obtained by 1- and 2-sums from matroids cycling with respect to $\Delta_{(CC)}$ and one copy of F_7 or (exclusive) one copy of $M(K_5)$. From Corollary 10 and the hypothesis that M has no $M^*(P_{10})$ minor, we conclude that M may be obtained by 1- and 2-sums from matroids packing with respect to $\Delta_{(CC)}$ and one copy of F_7 or (exclusive) one copy of $M(K_5)$.

Now, using Lemma 11 one sees that M is half-packing with respect to $\Delta_{(CC,F_7,K_5)}$.

In the other direction, from the proof of Theorem 3 one can conclude that the matroids $AG(2,3)$, S_8, R_{10}, $M(H_6)$, $M(K_5) \oplus_2 F_7$, $M(K_5) \oplus_2 M(K_5)$ and $F_7 \oplus_2 F_7$ are not half-packing with respect to $\Delta_{(CC,F_7,K_5)}$, since there is a primitive metric for each of them that is not in $\Delta_{\{0,1,2\}}$. $\quad\square$

As $M^*(P_{10})$ is not graphic, for graphic matroids we get the following sharper result:

Theorem 13. *For a graph G the following are equivalent:*
(i) G is cycling with respect to $\Delta_{(CC,K_5)}$;
(ii) G is flowing with respect to $\Delta_{\{0,1,2\}}$;
(iii) G has no H_6, $K_5 \oplus_2 K_5$ as minor;
(iv) G is half-packing with respect to $\Delta_{(CC,K_5)}$. $\quad\square$

We will see that the example given in Fig. 8 is essentially the only one, where we 2-sum a $M(K_5)$ or a F_7 with a matroid consisting on two circuits C_1, C_2 such that $C_1 \cap C_2 = \{f\}$, where f is the marker of the 2-sum, and $|C_1| \geq 3$. We call this configuration a $\bar{K}_5$ and a $\bar{F}_7$, respectively.

For the characterization we will need first the following lemma.

Lemma 14. *The matroid M, resulting from the 2-sum of a matroid M_1 that is packing with respect to $\Delta_{(CC,K_5,F_7)}$ with a matroid M_2, that is a circuit, is metric packing with respect to $\Delta_{(CC,K_5,F_7)}$.*

Proof. If M_1 does not contain $M(K_5)$ or F_7 as a minor, the result is trivial. So let M_1 contain one of $M(K_5)$ and F_7 as minor. Given a bipartite metric m on the matroid M, we will find an integral decomposition for m as a sum of cocircuits, F_7- or K_5-metrics. Let $M_1 \cap M_2 = \{f\}$.

As in the proof of Lemma 11, we define two metrics $m_1 : E(M_1) \rightarrow \mathbb{Z}_+$ and $m_2 : E(M_2) \rightarrow \mathbb{Z}_+$ such that

$$m_i(e) = \begin{cases} m(e), & \text{if } e \in E(M_i) - f, \\ q, & \text{if } e = f, \end{cases}$$

where $q = \min\{m(C - f) : C \in \mathcal{C}(M_1) \cup \mathcal{C}(M_2)\}$.

With a similar reasoning to the proof of Lemma 11 it can be seen that m_i, $i = 1, 2$, is a bipartite metric, and so there are cocircuits $C_1, \ldots, C_r \in \mathcal{C}^*(M_2)$ such that $\sum_{j=1}^r \chi_{C_j} = m_2$, and we assume that the first k_2 cocircuits contain f; and there are cocircuits $D_1, \ldots, D_s \in \mathcal{C}^*(M_1)$, and K_5- or F_7-metrics $l_1, \ldots, l_t$ such that $\sum_{j=1}^r \chi_{D_j} + \sum_{j=1}^t l_j = m_1$. We suppose that the first k_1 cocircuits contain f, and that the first k_3 F_7- or K_5-metrics l_i are such that $l_i(f) = 1$. Notice that $k_2 = k_1 + k_3 + 2(t - k_3)$.

To each l_j, $1 \leq j \leq k_3$, and to each D_i, $1 \leq i \leq k_1$, we associate a cocircuit $C_k \in \mathcal{C}^*(M_2)$, in each of them we replace f by C_k, and the result is clearly a cocircuit, a K_5- or an F_7-metric.

Now we associate to each l_i, $k_3 + 1 \leq i \leq t$, two cocircuits C_j, C_k, and replace all of them by B_1 and B_2 defined as follows. Let $B_1 = C_j \triangle C_k$, B_1 is a cocircuit in M_2, and so in M. Let $B_2 = E(M_2) - (C_j \triangle C_k)$; if $|B_2| \equiv 0 \mod 2$, then since M_2 is a circuit, B_2 is a cocycle in M_2, and we replace in l_i the element f with $B_2 - f$. If $|B_2| \equiv 1 \mod 2$, then we decompose l_i with $l_i(f) = 0$ into cocircuits, and put them in the list, with a coefficient 2. We add $B_2 - f$ to the list, with a coefficient 2, if $B_2 - f \neq \emptyset$, since in this case $B_2 - f$ is a cocycle in M_2 and so in M. Proceeding this way we get an integer packing of cocircuits and F_7- or K_5-metrics for m. □

Combining Corollary 10 and Lemma 14 one gets the following:

A binary matroid M flowing with respect to $\Delta_{\{0,1,2\}}$ is packing with with respect to $\Delta_{(CC,F_7,K_5)}$ if and only if M has no $M^*(P_{10})$, $\bar{K}_5$, $\bar{F}_7$, R_{10} minor. □

Just before submitting this article, the authors have realized that in fact a characterization of packing matroids follows. (A matroid is *packing* if it is integer packing with respect to its primitive metrics.)

For a binary matroid M the following statements are equivalent (provided R_{10} is packing):
(i)M is packing,
(ii) M is packing with respect to $\Delta_{(CC,F_7,K_5,R_{10})}$,
(iii) M has no $M^*(P_{10})$, $\bar{F}_7$, $\bar{K}_5$, $\bar{R}_{10}$ minor.

The matroid $\bar{R}_{10}$ is defined in the same way as $\bar{F}_7$, $\bar{K}_5$. The fact that R_{10} is packing is being checked (if not, that makes only trivial changes to the theorem).

References

1. X. FU AND L. GODDYN, *Matroids with the circuit cover property*, (1995). pre-print.
2. A. V. KARZANOV, *Metrics and undirected cuts*, Mathematical Programming, 32 (1985), pp. 183–198.
3. ———, *Half-integer five-terminus flows*, Discrete Applied Mathematics, 18 (1987), pp. 263–278.
4. ———, *Sums of cuts and bipartite metrics*, European Journal of Combinatorics, 11 (1990), pp. 473–484.
5. A. SCHRIJVER, *Distances and cuts in planar graphs*, Journal of Combinatorial Theory, Series B, 46 (1989), pp. 46–57.
6. ———, *Short proofs on multicommodity flows and cuts*, Journal of Combinatorial theory, Series B, 53 (1991), pp. 32–39.
7. W. SCHWÄRZLER AND A. SEBŐ, *A generalized cut-condition for multiflows in matroids*, Discrete Mathematics, 113 (1993), pp. 207–221.
8. P. SEYMOUR, *Decomposition of regular matroids*, Journal of Combinatorial Theory, 28 (1980).
9. P. D. SEYMOUR, *Graph theory and related topics*, Graph Theory and Related Topics, (1979).
10. ———, *Matroids and multicommodity flows*, European Journal of Combinatorics, 2 (1981), pp. 257–290.
11. D. J. A. WELSH, *Matroid Theory*, Academic Press, London, 1976.

Optimum Alphabetic Binary Trees

T. C. Hu and J. D. Morgenthaler

Department of Computer Science and Engineering,
School of Engineering,
University of California, San Diego CA 92093–0114, USA

Abstract. We describe a modification of the Hu–Tucker algorithm for constructing an optimal alphabetic tree that runs in $O(n)$ time for several classes of inputs. These classes can be described in simple terms and can be detected in linear time. We also give simple conditions and a linear algorithm for determining, in some cases, if two adjacent nodes will be combined in the optimal alphabetic tree.

1 Introduction

Binary trees and binary codes have many applications in various branches of science and engineering, and the subjects have been studied extensively for more than forty years. In 1952, Huffman [7] discovered the algorithm for constructing an optimum variable-length binary code or the binary tree with minimum weighted path length known as the Huffman's tree. In 1959, Gilbert and Moore [3] considered an alphabetic constraint which restricts the ordering of leaves in the binary tree and they found an algorithm for constructing an optimum alphabetic binary tree in $O(n^3)$ time. Their technique is based on dynamic programming and is illustrated in many computer science textbooks. In 1971, their solution technique was refined to $O(n^2)$ time by Knuth [9], and a new algorithm was discovered by Hu and Tucker [6] which needs $O(n \log n)$ time and $O(n)$ space. The implementation of the Hu-Tucker algorithm in $O(n \log n)$ time was pointed out by Knuth [10], and several improvements over the original algorithm was reported in Hu [4]. A similar algorithm was discovered by Garsia and Wachs in 1977 [2]. But the time complexity remained at $O(n \log n)$. The applicability of Huffman's algorithm and the Hu-Tucker algorithm were generalized to other cost functions in [5]. Recently, the equivalence of alphabetic binary search tree and alphabetic binary code is noted by [1, 13]. Actually, Huffman's algorithm can be implemented in linear time if the weights of leaves are already sorted as noted by Larmore [12]. Also, testing the optimality of a given alphabetic binary tree is shown to be linear in several cases by Ramanan [15]. Thus, the search for a new linear algorithm for constructing an optimum alphabetic binary tree seems to be a natural one. Recently Klawe and Mumsy [8] and Przytycka and Larmore [14] have both obtained algorithms that run in less than $O(n \log n)$ time for special cases of weight sequences. Here we basically achieve the same results without introducing as much new terminology. Also, we have found simple conditions for two adjacent nodes to be combined in the optimal alphabetic tree which can be detected in $O(n)$ time.

2 Problem and Known Results

We use the definitions in Knuth [10] and consider an extended binary tree where every internal node has two children and every leaf has no children. The internal node is represented by a circle and the leaf by a square. The root of the extended binary tree (binary tree for brevity from now on) is said to be at the level zero, and its two children at the level one. Given a sequence of n leaves with positive weights w_i, the problem is to construct an alphabetic binary tree with min $\sum w_i l_i$ where l_i is the level of the leaf with weight w_i, and the ordering of leaves in the sequence is to be maintained.

Due to the alphabetic constraint, a square node (a leaf) can only be combined with its left or right neighboring square, or a circular node whose descendants are a subsequence of squares next to the given square. Two circular nodes can be combined only if they are both roots of subtrees with leaves from two adjacent subsequences.

A related problem is to find the min-cost forest built on a sequence of positive-weight leaves where the forest has exactly k circles. (The alphabetic tree is a special case with $k = n - 1$.)

We shall use w_i to denote the node with weight w_i or the node itself. Also, w_{ij} to denote the parent of w_i and w_j with its weight $w_i + w_j$. The parent will occupy the position of the left child. A node w_{bc} is said to be crossed over by another node w_{ad} if one of the children of w_{bc} is between the children of w_{ad}.

Two nodes in a sequence are called a compatible pair if they are adjacent in the sequence or if all nodes between them are circular nodes. Among all compatible pairs in a weight sequence, the one with minimum weight is called the minimum compatible pair (mcp). To break the ties among weights, we shall adopt the convention that the node on the left has smaller weight. A pair of nodes (w_b, w_c) is a **local minimum compatible pair (lmcp)** if

$$w_a > w_c \text{ for all nodes } w_a \text{ compatible with } w_b$$
$$w_b \leq w_d \text{ for all nodes } w_d \text{ compatible with } w_c$$

Hu-Tucker algorithms can be described in three steps:

1. **Combination.** Keep combining lmcp until a tree **T** is obtained.
2. **Level Assignment.** Find the level number of every leaf in tree T, say $l_1, l_2, \ldots, l_n$.
3. **Reconstruction.** Use a stack algorithm to construct an alphabetic binary tree based on $l_1, l_2, \ldots, l_n$.

Note that both steps 2 and 3 take linear time while step 1 takes $O(n \log n)$ time. A weight sequence is called an increasing sequence if

$$w_1 \leq w_2 \leq \ldots \leq w_n$$

A decreasing sequence if

$$w_1 > w_2 > \ldots > w_n$$

A valley sequence if

$$w_1 > w_2 > \ldots > w_i \leq w_{i+1} \leq \ldots \leq w_n$$

A bimonotonal increasing sequence if

$$w_1 + w_2 \leq w_2 + w_3 \leq w_3 + w_4 \leq \ldots \leq w_{n-1} + w_n$$

A bimonotonal decreasing sequence if

$$w_1 + w_2 > w_2 + w_3 > w_3 + w_4 > \ldots > w_{n-1} + w_n$$

A bimonotonal valley if it contains a single lmcp.

Lemma 1. *Let w_a be an arbitrary node in a weight sequence consisting of circles and squares, and w_i be the smallest node compatible to w_a. If any lmcp is combined in the weight sequence and w_d becomes compatible with w_a, then $w_i \leq w_d$. In other words, the smallest node compatible to any node w_a will always remain the same. In particular, an lmcp in a weight sequence will remain as other lmcp are successively combined.*

Proof. See Hu [4].

Lemma 2. *If in a weight sequence $\ldots w_a, w_b, w_c, w_d, \ldots$ where w_{bc} is crossed over by w_{cd} in the construction of T, then $l_{bc} \geq l_{ad}$, i.e., the four nodes are either at the same level or the two nodes w_b and w_c are at lower levels than w_a and w_d.*

Proof. See Hu [4].

3 Summary of Key Observations

We show that for weight sequences in several classes, the optimum alphabetic binary tree can be computed in linear time. We present these classes from the simplest to the most complicated.

A binary tree is called a complete binary tree where all its leaves are at the same level or two adjacent levels. Obviously, if all leaves are of the same weight in a weight sequence and $n = 2^k$, then the optimum alphabetic binary tree is a complete binary tree with all leaves at the same level. If $2^k < n < 2^{k+1}$, then all the leaves will be at two adjacent levels with $2(n - 2^k)$ leaves at the lower level and $(2^{k+1} - n)$ leaves at the higher level.

Our first observation is that for n leaves of almost the same weight, the optimum alphabetic binary tree is again a tree whose leaves are at two adjacent levels. We call such a weight sequence an almost uniform sequence. (Precise definition to be given later in Section 4.) In addition, given a sequence of n nodes of almost uniform weight (circle or square), with $n = 2^k$, the optimum alphabetic tree is again a complete binary tree where these circles and squares occupy a single level.

The second observation is that if the n leaves form an increasing sequence, we can perform the combination phase of the Hu-Tucker algorithm in linear time by queuing the circular nodes as they are created. Compatible circular nodes are placed at the end of a sorted queue as they are formed. Only the bottom two nodes of this queue needs to be examined to determine the next lmcp. This is essentially the same modification reported by Larmore [12] that allows Huffman's algorithm to run in linear time on a sorted sequence.

We can also perform the combination phase for a bimonotonal increasing sequence in linear time, since the bimonotonal condition guarantees that the leftmost square node will always be smaller than the node two to the right.

Finally, a bimonotonal valley sequence consisting of a bimonotonal decreasing sequence followed by a bimonotonal increasing sequence is still linear, since we are guaranteed a single queue of circular nodes by the presence of a single initial lmcp. Figure 1 shows the operation of the modified combination phase on a bimonotonal valley sequence.

At each combination in Fig. 1, the lmcp is found among the six nodes consisting of: the bottom two circular nodes in the queue, the two square nodes to the left of the queue, or the two square nodes to the right of the queue. For example, after the final step shown in Fig. 1, the next lmcp cannot include the circular nodes with weights 11 or 13 due to construction of the queue. Thus, determining which nodes to combine at each step requires only constant time for a bimonotonal valley sequence.

Any weight sequence can be broken into a set of bimonotonal valley sequences in linear time. The only thing that prevents the modified algorithm from running in linear time for an arbitrary weight sequence is the need to "merge" two queues when the mountaintop (defined in Section 4) between their valleys combines. Each merge requires linear time, but as $O(n)$ queues could be merged at one time, the merging may require $O(n \log n)$ time. In the next section, we describe a class of initial weight sequences that do not require $O(n)$ queues to be merged, and thus require only linear time for construction of their optimum alphabetic binary tree.

4 Linear Classes of Weight Sequences

The previous section gave a modification of the Hu-Tucker algorithm that runs in linear time for a bimonotonal valley sequence. By restricting the number of queues of circular nodes that need to be merged at one time, the algorithm also runs in linear time for a larger class of initial sequences.

When two valleys are adjacent, the point at which the bimonotone sequence changes from increasing to decreasing is called the *mountaintop*. This concept is the dual of lmcp, and could be called the *locally maximum compatible pair*. More formally:

Definition 3. A mountaintop is the point between the two adjacent square nodes whose combined weight is locally the greatest. The weight of the mountaintop is the greater weight of these two nodes.

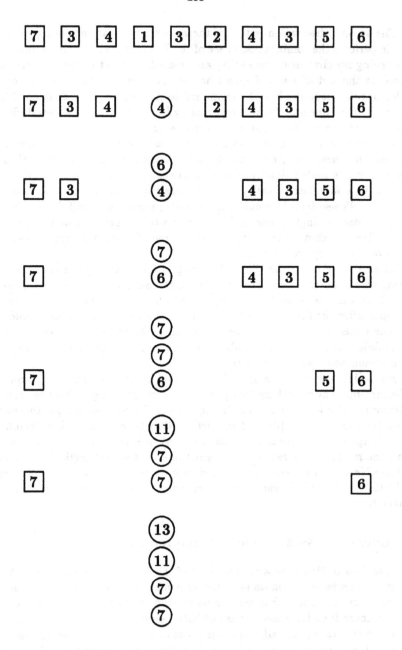

Fig. 1. First seven steps in the combination phase for the initial bimonotonal valley sequence 7, 3, 4, 1, 3, 2, 4, 3, 5, 3, 6

Adjacent mountaintops are the closest mountaintop on either the left or right of a node. If we extend the initial sequence by adding two special nodes of infinite weight, one at each end, then every node, circular or square, has two adjacent mountaintops (except these two new infinite weight nodes). A mountaintop *combines* (forming circular nodes) when both square nodes have combined. It is at the point in the algorithm when a mountaintop combines that queues of circular nodes are merged.

Clearly, if the number of valleys in the weight sequence is bounded by a constant k, the modified Hu-Tucker algorithm runs in $O(n \log k) = O(n)$ time. We may be required to merge all k queues at once if the $k - 1$ mountaintops have the same weight.

An interesting property of a queue of circular nodes is that they are all *almost uniform*. We can use this property to broaden the class of sequences where Hu-Tucker runs in linear time.

Definition 4. A set of nodes is almost uniform if any two nodes have combined weight greater than or equal to any weight in the set. Node set W is almost uniform if and only if

$$\forall w_1, w_2, w_3 \in W, \quad w_1 + w_2 \geq w_3$$

Lemma 5. *A set of compatible circular nodes created during phase I of the Hu-Tucker algorithm is almost uniform.*

Proof. Assume that the two smallest weight circular nodes sum to a total weight less than the weight of the largest circular node. Then the largest node was not created from an lmcp, since the two smallest weight nodes are locally minimum, contrary to the Hu-Tucker algorithm.

The queues of two adjacent valleys combine when the largest square node (w_T) adjacent to their common mountaintop combines. When w_T combines, a new circular node with weight greater than that of the last node in either queue is formed. In fact, the largest circular node on either queue (w_c) must have weight less than or equal to twice the weight of the mountaintop.

$$w_c \leq 2w_T$$

Let the children of w_c be w_x and w_y. Then by the fact that w_T has not yet combined and the definition of mountaintop, $w_x \leq w_T$ and $w_y \leq w_T$. After w_T combines, all circular nodes in the two queues will have weight at least w_T.

It is well known that an optimum alphabetic binary tree can be created in linear time from a valley sequence. Now, let us consider two adjacent valley sequences with three mountaintops, e.g.

$$w_L > w_1 > \cdots > w_i < \cdots < w_T > w_j > \cdots > w_k < \cdots < w_R$$

Then, a queue of circular nodes will be created in linear time in the valley between w_L and w_T and similarly a queue of circular nodes will be created in the valley between w_T and w_R.

If w_L and w_R are both much larger than w_T, then w_T is combined into a circular node first. And we will have two queues of circular nodes between w_L and w_R, with the parent of w_T being the largest circular node. When this happens, the next combination is the smallest node in one queue with the smallest node in the other queue (or possibly with the second smallest node in its own queue). In other words, each combination takes constant time inside the valley bounded by the two mountaintops w_L and w_R. Effectively, we have a single valley, if $w_T \leq \min(w_L/2, w_R/2)$.

If we have k adjacent valleys between w_L and w_R, then we may have k queues of circular nodes. To find the next combination, we can select the two smallest nodes among the front elements in the k queues, say α_1 and β_1 (with $\alpha_1 < \beta_1$) and then compare the next node in the α queue (α_2) with β_1 to determine $\min(\alpha_1 + \alpha_2, \alpha_1 + \beta_1)$.

In other words, as long as k is constant and is not a function of n, we can successively combine nodes in linear time. We say that the k valleys are nested inside the big valley w_L and w_R if the largest mountaintop among the inside mountaintops is less than half $\min(w_L, w_R)$.

Of course, all the comments about valley sequences apply equally to bimonotonal valley sequences.

Definition 6. Two valleys v_1 and v_2 are nested inside a larger valley V with mountaintops of weight w_L and w_R if the mountaintop separating v_1 and v_2 has weight $w \leq \min(w_L, w_R)/2$.

Definition 7. A proper valley has no nested valleys. A k-proper valley has at most k nested valleys, either proper valleys or k-proper valleys.

Defined recursively, a k-proper valley is either:

1. a proper valley
2. a valley with at most k nested adjacent proper valleys
3. a valley with at most k nested adjacent k-proper valleys

We now expand the class of weight sequences for which the modified Hu-Tucker algorithm requires only linear time to the set of all k-proper valleys.

Theorem 8. *The optimum alphabetic binary tree for an initial weight sequence W of n nodes that forms a k-proper valley can be computed in $O(n)$ time for any constant k. Further, we can determine whether a sequence is a k-proper valley is $O(n)$ time.*

Proof. We consider each case separately.

Case 1: W is a proper valley. Since it has no nested valleys, W consists of a single bimonotonal valley, with a single lmcp. At each step in the combination phase, only one lmcp exists. When newly created circular nodes are placed in a queue, only six nodes (four square, two circular) could participate in that lmcp. Thus, each combination requires constant time, and the entire combination phase requires $O(n)$ time. As the remaining phases are also linear, the entire tree is constructed in linear time.

Case 2: W is a valley with at most k nested adjacent proper valleys. Each nested proper valley requires linear time to combine all contained square nodes. When a mountaintop is combined, the adjacent compatible queues are merged. This requires at most k merges, each needing at most $O(n)$ comparisons. Once all queues have been merged, this case resembles Case 1.

Case 3: W is a valley with at most k nested adjacent k-proper valleys. Recursively, each nested k-proper valley can be handled in time linear in its size. Again, combining adjacent valleys by merging queues of circular nodes is linear, and must only be done k times.

4.1 Permanent Circular Nodes

During the construction of an optimum k-circle forest, a given circle in the $(k-1)$ circle forest may be split up in the optimum k-circle forest. For example, consider the initial sequence 4, 2, 3, 4. The optimum 1-circle forest combines the nodes weighing 2 and 3; these are not combined in the 2-circle or 3-circle forests.

Would there exist a circle which will never be split up once it is formed? We shall call such a circle a permanent circular node (a permanent node). When we can detect such a circular node, then the circular node can be treated as a single square node in the weight sequence during later combinations.

Lemma 9 Gilbert & Moore. *Let $w_a, w_1, w_2, \ldots, w_k, w_d$ be a weight sequence where*

$$\sum_{i=1}^{k} w_i \leq \min(w_a, w_d)$$

then the subsequence $w_1, w_2, \ldots, w_k$ will form a binary tree whose root is a permanent node.

Proof. See [3].

Lemma 10. *Let w_a, w_b, w_c, w_d be any consecutive four nodes in a sequence and*

1. $w_a > w_c$
2. $w_b \leq w_d$
3. $(w_b + w_c) < \max(w_a, w_d)$

Then the parent of w_b and w_c is a permanent node.

Proof. Conditions 1 and 2 imply that w_b and w_c are a local minimum compatible pair. Without loss of generality, assume that $w_a > w_d$.

Once w_{bc} is formed as a circular node, both w_a and w_{bc} are compatible with w_d and $w_a > w_{bc}$. Thus, w_d combines with w_{bc} before combining with w_a.

If w_a should be combined with another node to its left, there cannot be a new node with its weight smaller than w_{bc} which would be compatible with w_d, by lemma 1.

If w_d should be combined with another node to its right, that circular node would also be compatible with w_a and w_{bc}.

Thus, the node w_d or its ancestor will never cross over w_{bc} to combine with w_a or its ancestor. Node w_{bc} will combine with a node adjacent to itself, and will not be crossed over. In other words, w_{bc} is a permanent node.

To show the power of identifying permanent nodes, consider the following weight sequence.

$$\infty, \ 2, \ 3, \ 4, \ 2, \ 1, \ 8, \ 10, \ \infty$$

Here $w_a = \infty$, $w_b = 2$, $w_c = 3$, $w_d = 4$, and we shall show it as

$$\infty \ \ \underline{2 \ \ 3} \ \ 4 \ \ 2 \ \ 1 \ \ 8 \ \ 10 \ \ \infty$$

After 2 and 3 are combined into 5, we move forward until the condition of permanent node is again satisfied.

$$\infty \ \ 5 \ \ \underline{4 \ \ 2 \ \ 1 \ \ 8} \ \ 10 \ \ \infty$$

And then move backward (after $2 + 1 = 3$)

$$\infty \ \ \underline{5 \ \ 4 \ \ 3 \ \ 8} \ \ 10 \ \ \infty$$
$$\underline{\infty \ \ 5 \ \ 7 \ \ 8} \ \ 10 \ \ \infty$$
$$\infty \ \ \underline{12 \ \ 8 \ \ 10} \ \ \infty$$
$$\underline{\infty \ \ 12 \ \ 18} \ \ \infty$$
$$\infty \ \ 30 \ \ \infty$$

Another example is

$$\infty \ \ 4 \ \ 5 \ \ \underline{6 \ \ 7 \ \ 6+\epsilon \ \ 5} \ \ 4 \ \ \infty$$
$$\infty \ \ 9 \ \ 6 \ \ 7 \ \ \underline{6+\epsilon \ \ 5 \ \ 4} \ \ \infty$$
$$\infty \ \ 9 \ \ 6 \ \ 7 \ \ 6+\epsilon \ \ 9 \ \ \infty$$

And at this stage, the sequence is almost uniform and no more permanent nodes can be created.

Lemma 11. *We can identify and combine all permanent nodes in $O(n)$ time.*

Proof. In a scan from left to right of the initial weight sequence, at each node B, label its neighbors

$$A \quad B \quad C \quad D$$

and do the following:

If $A > C$, $B \leq D$ and $B + C < \max(A, D)$ replace nodes B and C with new square node $B + C$ and back up to consider node A again, else do nothing. If the node has no left neighbor or less than two right neighbors, do nothing.

At each step, the number of nodes to the right of the node being considered either decreases by one (do nothing case), or remains unchanged. If it remains unchanged, the number of nodes to the left decreases by one. Thus, we can perform at most $2n$ total steps before the scan terminates.

Since permanent nodes can be found and combined in linear time, we can expand our class of weight sequences with optimum alphabetic trees computable in linear time to any sequence which after finding and combining permanent nodes is a k-proper valley.

5 Conclusion

We have shown that a modification of the Hu–Tucker algorithm will run in linear time for many classes on input sequences. By scanning a given input sequence for mountaintops and counting adjacent groups of almost uniform mountaintops, membership in such a class can also be determined in linear time.

We have also given a linear algorithm for finding and combining permanent circular nodes detectable on the initial sequence in a four node moving window.

References

1. Arne Andersson. A note on searching in a binary search tree. *Software–Practice and Experience*, 21(10):1125–1128, 1991.
2. A. M. Garsia and M. L. Wachs. A new algorithm for minimum cost binary trees. *SIAM Journal on Computing*, 6(4):622–642, 1977.
3. E. N. Gilbert and E. F. Moore. Variable length binary encodings. *Bell System Technical Journal*, 38:933–968, 1959.
4. T. C. Hu. *Combinatorial Algorithms*. Addison-Wesley, Reading, MA, 1982.
5. T. C. Hu, D. J. Kleitman, and J. K. Tamaki. Binary trees optimum under various criteria. *SIAM Journal on Applied Mathematics*, 37(2):246–256, 1979.
6. T. C. Hu and A. C. Tucker. Optimal computer search trees and variable-length alphabetic codes. *SIAM Journal on Applied Mathematics*, 21(4):514–532, 1971.
7. D. A. Huffman. A method for the construction of minimum redundancy codes. *Proceedings of the IRE*, 40:1098–1101, 1952.
8. M. M. Klawe and B. Mumey. Upper and lower bounds on constructing alphabetic binary trees. In *Proceedings of Fourth Annual ACM-SIAM Symposium on Discrete Algorithms*, pages 185–193, 1993.
9. D. E. Knuth. Optimum binary search trees. *Acta Informatica*, 1:14–25, 1971.
10. D. E. Knuth. *The Art of Computer Programming, Volume III: Sorting and Searching*. Addison-Wesley, Reading, MA, 1973.
11. L. L. Larmore. A subquadratic algorithm for constructing approximately optimal binary search trees. *Journal of Algorithms*, 8(4):579–591, 1987.
12. L. L. Larmore. Height restricted optimal binary trees. *SIAM Journal on Computing*, 16(6):1115–1123, 1987.
13. N. Nakatsu. An alphabetic code and its application to information retrieval. *Transactions of the Information Processing Society of Japan*, 34(2):312–19, 1993.
14. T.M. Przytycka and L.L. Larmore. The optimal alphabetic tree problem revisited. In *Proceedings of 21st International Colloquium on Automata, Languages, and Programming*, pages 251–262. Springer-Verlag, July 1994.
15. P. Ramanan. Testing the optimality of alphabetic trees. *Theoretical Computer Science*, 93(2):279–301, 1992.

Block Codes
for
Dyadic Phase Shift Keying

Patrick Solé, *
Macquarie University,
School of MPCE,
2109 NSW,
Australia
Jean-Pierre Tillich
Dept. of Mathematics, University of British Columbia,
Vancouver V6T 1Z2,
Canada

Abstract. Codes over finite cyclic groups of order a power of 2 for the hermitian distance are used in PSK modulation. In this paper they are approximated by codes over the dyadic integers. The idea is to interpret the hermitian norm as an eigenvalue of the coset graph attached to the dual code. Since the infinite coset graph of a given dyadic code is a projective limit of the finite coset graphs of the finite projected codes, the spectra of the finite codes converge weakly towards the spectrum of the coset graph of the dyadic code. This entails that the probability of error of a code over a large cyclic ring is well approximated for large alphabet size by a measure attached to its dyadic limit.

1 Introduction

Dyadic cyclic codes were introduced recently [2] as a tool to study cyclic codes over cyclic groups of order a power of 2. They proved useful to derive structure theorems and compute idempotents [2]. It was unclear so far how to relate the metric properties of the codes on these finite alphabets to the metric properties of the dyadic codes. More generally, with every dyadic code one can associate an infinite sequence of codes by reduction mod 2^r. The main theme of this work is to show how to obtain information on the metric properties of the latter codes from information on the former code.

In this work we propose to put a metric on the dyadic codes (not necessarily cyclic) which is the hermitian metric on the vectors of group characters of the dyadic rationals. This metric is consistent with the same type of metric on the codes over the finite cyclic groups . The latter has already been investigated in relation with PSK modulation [26].

---

* On leave of absence from CNRS, I3S, France

It will be useful to interpret these hermitian weights as eigenvalues of coset graphs. In this interpretation the coset graph of a dyadic code plays the same role w.r.t. the coset graphs of its truncated codes that the homogeneous tree plays w.r.t. the regular graphs [20]. In particular the spectrum of the finite graphs converge weakly for large alphabet size towards the spectrum of the infinite graph. This gives a limit law on the probability of error of a code over a cyclic ring of large size used for error detection.

The material is organized as follows. Section 2 shows the link between hermitian weights and eigenvalues. Section 3 develops the spectral theory of the coset graph of the dyadic code, and contains the main results on the probability of error and the minimum distance. Finally in section 4 we give the motivating examples of cyclic codes and the enumerative problems opened up by this work.

2 Hermitian Distance and Coset Graphs

Let M, n denote nonnegative integers with $M > 1$. A *group code* C of length n over Z_M is any subgroup of $(Z_M)^n$. A generating matrix G_C can be attached to such a code, which is a k times n matrix over Z_M, whose rows form a generating set of C. The *dual* code hereby denoted by $C^\perp$ is understood with respect to the standard scalar product:

$$C^\perp = \{y \in (Z_M)^n | \forall x \in C \ x.y = 0\}$$

We map the alphabet Z_M in the complex plane and one-to-one with the $M-$PSK constellation by using the usual additive character ϕ_M defined as

$$\phi_M(x) = \exp(\frac{2\pi i x}{M}).$$

This map is extended componentwise to vectors of size n. We define the hermitian weight of a vector y of Z_M for $M > 1$ by the formula

$$h_M(y) = \phi_M(y) + \phi_M(-y),$$

or, more concretely

$$h_M(y) = 2 \sum_{i=1}^{n} \cos(\frac{2\pi y_i}{M}).$$

Note that the squared hermitian norm of $\phi_M(y)$ seen as a complex $n-$dimensional vector is

$$h_M(0) - h_M(y) = 4 \sum_{i=1}^{n} (\sin(\frac{\pi y_i}{M}))^2.$$

More generally the squared hermitian distance of $\phi_M(y)$ to $\phi_M(z)$ is $h_M(0) - h_M(y - z)$. The following special cases are of interest

- $M = 2$ when $h_2(y) = 2n - 4w_H(y)$, with $w_H(.)$ the Hamming weight.
- $M = 3$ when $h_3(y) = 3n - 2w_H(y)$, with $w_H(.)$ the Hamming weight.

- $M = 4$ when $h_4(y) = 2n - 2w_L(y)$, with $w_L(.)$ the Lee weight.
- $M > 4$ when no such simple interpretation exists.

For a discussion of these quantities in the context of modulation theory see [26].

We now give a combinatorial interpretation of $h_M(y)$. With every code C as above we associate a graph $\Gamma(C)$ defined on the cosets of $C^\perp$, two vertices being connected by an edge iff they differ by a coset of minimum Lee weight one if $M > 2$ or by two edges iff they differ by a coset of Hamming weight one if $M = 2$. This graph can be described by using a generating matrix of the code. Since G_C is a parity-check matrix of $C^\perp$, it is readily seen that this coset graph is the Cayley graph over the abelian group $G_C Z_M^n$ and generators the $2n$ group elements $\pm G_C e_i$, where $(e_i)_{1 \leq i \leq n}$ is the canonical basis of Z_M^n (i.e there is an edge between two group elements x and y if and only if they differ by some $G_C e_i$). This graph is regular of degree $2n$ on $|C|$ vertices.

Theorem 1 *With the above notation the eigenvalues of $\Gamma(C)$ are all the $h_M(y)$ with y running over C.*

Proof: Since $\Gamma(C)$ is an abelian Cayley graph, its eigenvalues λ are given by the formula

$$\lambda = \sum_{a \in S} \psi(a).$$

where ψ ranges over the group of characters of $G_C Z_M^n$, and S is the generating set of the Cayley graph, that is $\{\pm G_C e_i\}$. There are as many characters as group elements of $G_C Z_M^n$, and those are therefore in one-to-one correspondance with the codewords of C. The point is that it is easily checked that all the functions f_y over the group $G_C Z_M^n$ associated to a codeword y which are given by

$$f_y(G_C x) = \phi_M(x.y)$$

are different characters of $G_C Z_M^n$, and therefore the set of eigenvalues of $\Gamma(C)$ is actually the set $\{\lambda_y | y \in C\}$, where λ_y the eigenvalue associated to the codeword $y = (y_1, y_2, \cdots, y_n)$ is defined by

$$
\begin{aligned}
\lambda_y &= \sum_{a \in S} f_y(a) \\
&= \sum_{1 \leq i \leq n} f_y(G_C e_i) + \sum_{1 \leq i \leq n} f_y(-G_C e_i) \\
&= \sum_{1 \leq i \leq n} \phi_M(e_i.y) + \sum_{1 \leq i \leq n} \phi_M(-e_i.y) \\
&= \sum_{1 \leq i \leq n} \phi_M(y_i) + \sum_{1 \leq i \leq n} \phi_M(-y_i) \\
&= h_M(y)
\end{aligned}
$$

□

As in [26] denote by $d^2(C)$ the minimum hermitian distance of C. For future use we remark that if $H_M(C)$ is the maximum of $h_M(y)$ over y in C then for a group code or a geometrically uniform code C we have the relation

$$d^2(C) = 2n - H_M(C).$$

3   Dyadic Codes and Universal Covers

3.1   Dyadic Codes

In this section and the following we assume henceforth that M is a power of 2. Recall that the dyadic integers Z_{2^∞} can be thought of as formal power series with coefficients in $\{0,1\}$ where multiplication and addition are carried out as if these formal power series were base 2 integers (with possibly an infinite number of digits). A group code C of length n over Z_{2^∞} is defined by a $k \times n$ generating matrix G with entries in Z_{2^∞}, and the set of codewords is $\{xG | x \in Z_{2^\infty}^k\}$ and is therefore a Z_{2^∞}-submodule of $Z_{2^\infty}^n$.

We aim here at studying the sequence of codes over $Z_2, Z_4, Z_8, \ldots, Z_{2^r}, \ldots$ what we denote by $C_1, C_2, \cdots, C_r$, obtained by reducing the words of a group code C over Z_{2^∞} mod $2, 2^2, \cdots, 2^r$ respectively. It will be convenient to introduce the ring homomorphisms

$$
\begin{array}{ccc}
P_r : Z_{2^\infty} & \longrightarrow & Z_{2^r} \\
\sum_{i=0}^{\infty} a_i 2^i & \longmapsto & \sum_{i=0}^{r-1} a_i 2^i
\end{array}
$$

By using the same notation for their canonical extension to $(Z_{2^\infty})^n$, we have $C_1 = P_1(C), C_2 = P_2(C), \cdots, C_r = P_r(C)$.

The relationships between the metrics of the codes C_i and the code C will follow from considerations on the coset graphs of those codes. We define the coset graph $\Gamma(C)$ as in the case of a code over a cyclic ring of preceding section, by using either the definition of a dual of a dyadic code given in [2] as

$$C^\perp = \{y \in Z_{2^\infty} | \forall x \in C \; x.y = 0.\}$$

or by defining this graph as the Cayley graph over the group $GZ_{2^\infty}^n$, with generating set $\pm Ge_i$, where $(e_i)_{1 \leq i \leq n}$ is the canonical basis of $Z_{2^\infty}^n$.

For instance, if C is the 2-adic repetition code i.e the rank 1 code with generator matrix the all-one vector, $\Gamma(C)$ consists of infinitely many copies of the two way infinite path (each edge repeated n times).

Actually this dyadic coset graph $\Gamma(C)$ consists of infinitely many copies of a smaller connected Cayley graph over the group GZ^n, with generating set $\pm Ge_i$, where $(e_i)_{1 \leq i \leq n}$ is the canonical basis of Z^n. To avoid cumbersome notations, from now on $\Gamma(C)$ will actually denote this connected component.

3.2 Relationships between the coset graphs

These cosets graphs $\Gamma(C)$, $\Gamma(C_i)$ are linked together by the following properties

1. The set of vertices $V_1, V_2, \cdots$ of the Cayley graphs $\Gamma(C_1), \Gamma(C_2), \cdots$ can be seen as the projections by $P_1, P_2, \cdots$ respectively of the set of vertices of $\Gamma(C)$ or $\Gamma'(C_i)$.

2. Generating matrices of $C_1, C_2, \cdots$ denoted by $G_1, G_2, \cdots$ may be obtained from a generating matrix of C by projecting by $P_1, P_2, \cdots$ the coefficients of this generating matrix on $Z_2, Z_{2^2}, \cdots$ respectively. And therefore an edge between two vertices x and $x \pm Ge_i$ in $\Gamma(C)$ becomes an edge between x and $x \pm G_r e_i$ in $\Gamma(C_r)$, when we apply the projection P_r to those vertices.

Clearly for every coset graph $\Gamma(C_i)$, the number of cycles of length l which starts at a given vertex x does not depend on x. Let $\mu_l(C_i)$ be this number of cycles. We can deduce from the previous remarks that

Lemma 1.
$$\mu_l(C_i) \geq \mu_l(C)$$

Lemma 2. *The graph $\Gamma(C)$ is the limit of $\Gamma(C_r)$ as r tends to infinity.*

Proof: We use here the definition of the limit of a graph which is given in [11], the limit of a sequence of graphs $\mathcal{G}_1, \mathcal{G}_2, \cdots$ which share the same set of vertices V, and whose set of edges is denoted by $E_1, E_2, \cdots$ respectively is the graph $\mathcal{G}$ with set of vertices V, and with set of edges $\cup_{i=1}^{\infty} \cap_{j=i}^{\infty} E_j$. This definition may be extended to graphs which do not share the same set of vertices by defining a sequence of injective mappings $\varphi_1, \varphi_2, \cdots$ from the set of vertices of $\mathcal{G}_1, \mathcal{G}_2, \cdots$ respectively to a common set of vertices V which will be the set of vertices of the limit. Let us denote from now on $G, G_1, G_2, \ldots$, the generating matrices of $C, C_1, C_2, \ldots$ respectively. In our case V is the group GZ^n and we may define a sequence of mappings $\varphi_1, \varphi_2, \cdots$ from the set of vertices of $\Gamma(C_1), \Gamma(C_2), \cdots$ to V, by mapping 0 to 0 and extending this mapping in a neighborhood of the origin such as to preserve the edges. More precisely

$$\varphi_i : G_i Z_{2^i}^n \rightarrow GZ^n$$
$$G_i x \mapsto Gx$$

this for every $x = (x_1, x_2, \cdots, x_n) \in Z_{2^i}^n$ such that there exists no other $y = (y_1, y_2, \cdots, y_n) \in Z_{2^i}^n$ for which $\sum_{1 \leq i \leq n} |y_i| \leq \sum_{1 \leq i \leq n} |x_i|$, $G_i x = G_i y \mod 2^i$ and $Gx \neq Gy$. For all the other $G_i x$ for which this condition is not met, we define φ_i in an arbitrary way, with the constraint that φ_i stays injective. It is readily seen that for a given element x in Z^n, $\varphi_i(G_i x)$ will be Gx for i sufficiently large. Therefore for sufficiently large i, the mappings φ_i map an edge of E_i between $G_i x$ and $G_i x \pm G_i e_j$ to an edge in $\Gamma'(C)$ between Gx and $Gx \pm Ge_j$, and every edge of $\Gamma'(C)$ is such a mapping of edges of $\Gamma(C_i), \Gamma(C_{i+1}), \ldots$. $\square$

We will see later on that this implies that in some sense the spectra of $\Gamma(C_1), \Gamma(C_2), \ldots$ "tend" to the spectrum of $\Gamma(C)$. Actually the spectra of the graphs $\Gamma(C_1), \Gamma(C_2), \ldots$ are nested, this can be proved by using the following lemma :

Lemma 3. *The graphs $\Gamma(C_1), \Gamma(C_2), \cdots$ form a tower of coverings (that is for each $i > 0$, $\Gamma(C_{i+1})$ is a covering of $\Gamma(C_i)$), and all these finite graphs are covered by $\Gamma(C)$.*

Proof: Recall that the graph $\mathcal{G}$ is a covering of the graph $\mathcal{G}'$, if there exists a surjective mapping ϕ from the vertices of $\mathcal{G}$ to the vertices of $\mathcal{G}'$ such that all the neighbors of a given vertex $\phi(x)$ of $\mathcal{G}'$ are precisely the $\phi(y)$'s, where the y's are the neighbors of x in G. The projections P_i show that the graph $\Gamma(C)$ is a covering of the graphs $\Gamma(C_i), \Gamma(C_2), \ldots$, and the mapping

$$\phi_i : G_{i+i}Z_{2^{i+1}}^n \rightarrow G_i Z_{2^i}^n$$
$$G_{i+1}x \;\mapsto\; G_{i+1}x \bmod 2^i$$

show that the graph $\Gamma(C_{i+1})$ is a covering of the graph $\Gamma(C_i)$. $\qquad\square$

4 Relationships between the metric of the dyadic code and its finite projections

We define here the spectrum of the infinite graph $\Gamma(C)$ with vertex set V as the spectrum of the adjacency operator acting on the functions $L^2(V)$ as an infinite matrix. To that spectrum S can be attached Plancherel measure α_C (the so called spectral measure of the adjacency operator [11]) defined in the following way:

$$\sum_{i=0}^{+\infty} \mu_i(\Gamma(C))z^i = \int_{-2n}^{2n} \frac{d\alpha_C(\lambda)}{1 - z\lambda}.$$

In words the measure α_C is the inverse Stieltjes transform of the moment generating function. See [23, equ. 4.19, 4.20] for details. We can define similarly the spectrum of $\Gamma'(C)$, and it is obvious that those graphs have the same Plancherel measure.

In the same way we attach to the spectrum S_r of the finite graph $\Gamma(C_r)$ the discrete masses measure

$$d\alpha_{C_r} = \frac{1}{|C_r|} \sum_{\lambda \in S_r} \mu(\lambda)\delta_\lambda,$$

with $\mu(.)$ denoting multiplicity. It is elementary to check the formula

$$\sum_{i=0}^{+\infty} \mu_i(\Gamma(C_r))z^i = \int_{-2n}^{2n} \frac{d\alpha_{C_r}(\lambda)}{1 - z\lambda},$$

whose RHS is indeed the finite sum

$$\frac{1}{|C_r|} \sum_{\lambda \in S_r} \frac{\mu(\lambda)}{1 - z\lambda}$$

We see that if $x \in S$ then $d\alpha_C(x)/dx$ is intuitively like the multiplicity $\mu(\lambda)$ of $\lambda \in S_r$.

The next result shows that the spectrum of the infinite graph is a good approximation of the spectra of the finite graphs for large alphabet size. Denote by $A_C(\lambda)$ (resp. $A_{C_r}(\lambda)$) the cumulative distribution function $\int_{-2n}^{\lambda} d\alpha_C(\nu)$ (resp. $\int_{-2n}^{\lambda} d\alpha_{C_r}(\nu)$.)

Theorem 2 *For large r the c.d.f. $A_{C_r}(\lambda)$ converge towards $A_C(\lambda)$ at every point of continuity of $A_C(\lambda)$.*

Proof:This is a consequence of Lemma 2 and Theorem 4.12 of [23]. $\qquad \square$
Recall that the probability of error per symbol-up to a factor 2^r- for a code C_r used for error detection is by the union bound

$$P_r(g) = \int_{-2n}^{2n} g(\nu) d\alpha_{C_r}(\nu),$$

where $g(\nu)$ is the probability of an error of Hermite weight $2n - \nu$, a quantity which depends on the channel model used. A most natural model fitting with the Hermite weight is to assume an Additive White Gaussian Noise of variance σ^2, which yields

$$g(\nu) = erf(\frac{\nu}{\sigma}),$$

where erf is the so-called error-function, the tail of the Gaussian:

$$erf(x) := \int_x^{+\infty} \exp(-\frac{t^2}{2}) dt.$$

With these notations we can formulate an information-theoretic corollary of the preceding theorem.

Corollary 1 *For r large the probability of error $P_r(e, g)$ goes to*

$$P(e, g) = \int_{-2n}^{2n-e} g(\nu) d\alpha_C(\nu).$$

Proof:This follows from the preceding Theorem by use of the extended Helly-Bray Lemma [19, p.183]. In words vague convergence entails convergence in the sense of distributions. $\qquad \square$
One should note that in order to use this corollary one does not necessarily have to compute the Plancherel measure explicitly. Actually it is sufficient to calculate the exponential generating function of the number of cycles what we denote by $f_C^{exp}(z)$:

$$f_C^{exp}(z) = \sum_{i=0}^{\infty} \mu_i(\Gamma(C))\frac{z^i}{i!} = \int_{-2n}^{2n} e^{\lambda z} d\alpha_C(\lambda)$$

As a matter of fact if we denote by $\hat{g}(\theta)$ the Fourier transform of $g(x)$, i.e $\int_{-\infty}^{\infty} e^{i\theta x} g(x)$, then we get

Fact 1

$$P(g) = \frac{1}{2\pi} \int_{-\infty}^{\infty} \hat{g}(\theta) f_C^{exp}(-i\theta) d\theta$$

which is Parseval's equality.

For example with the aforementioned choice of the probability g this gives us

$$P(g) = \frac{1}{2\pi} \int_{-\infty}^{\infty} e^{2in\theta} e^{-\frac{\sigma^2 \theta^2}{2}} f_C^{exp}(-i\theta) d\theta$$

We now study the sequence of minimum distances, even though the probability of error is a more fundamental invariant.

Theorem 3 *The spectra S_r are nested and the sequence of minimum distances is nonincreasing.*

Proof: This follows from lemma 3, and theorem 4.5 given in section 4 of [5] which says that if G is a finite covering of G', then the spectrum of G' is contained in the spectrum of G. □

We will show that this sequence converges to zero. We will rely on the following graph-theoretic lemma.

Lemma 1 *Let A denote a family of abelian Cayley graphs of given degree Δ. Their second largest eigenvalue λ_2 goes to Δ when the number of vertices v goes to infinity. More precisely*

$$0 \leq \Delta - \lambda_2 = O\left(\Delta \left(\frac{\log(v)}{v^{\frac{1}{\Delta}}}\right)^2\right)$$

Proof: Let D denote the diameter of a graph in A. There is a lower bound on D say $D \geq B_\Delta(v)$ which is of order $v^{\frac{1}{\Delta}}$ [7]. Besides, there is an upper bound on the diameter D as a function of the second eigenvalue λ_2 (see [4])

$$D \leq \lfloor \frac{\cosh^{-1}(v-1)}{\cosh^{-1}\frac{3\Delta - \lambda_2}{\Delta + \lambda_2}} \rfloor + 1$$

Getting rid of D between these two bounds we obtain the stated bound . By Perron-Frobenius $\lambda_2 \leq \Delta$. □

From this by using Theorem 4.13 of [23] one can infer that the spectral radius of $\Gamma(C)$ is $2n$.

Theorem 4 *For large r the (squared) hermitian distances $d^2(C_r)$ go to zero. More explicitly,*

$$0 \leq d^2(C_r) = O\left(\frac{n(\log |C_r|)^2}{|C_r|^{1/n}}\right).$$

Proof: We apply the preceding Lemma, with A the class of all $\Gamma(C_r)$, with r an arbitrary integer, $\Delta = 2n$, $v = |C_r|$ and $\lambda_2 = 2n - d^2(C_r)$. □

5 The computation of the generating function of the number of cycles in $\Gamma(C)$

5.1 The basic tool : harmonic analysis on the lattice Z^n

In order to estimate the probability of error for a code C_r used in error detection, we know from corollary 1 and fact 1 that we only need to compute the exponential generating function $f_C^{exp}(z)$ of the number of cycles of the infinite coset graph $\Gamma(C)$. We are going to do this in the sequel for several families of infinite graphs. The fundamental tool we are going to use in order to solve this question is Fourier analysis on the lattice Z^n. We bring in lattices by making the following observation, call $g_1, g_2, \ldots, g_n$ the n column vectors which form a generating matrix of our code C. Clearly the set of n-tuples of integers $(a_1, a_2, \ldots, a_n)$ such that $a_1 g_1 + a_2 g_2 + \ldots + a_n g_n = 0$ is a sublattice $\mathcal{L}$ of Z^n. $\mu_k(C)$ is the number of times the expression $\epsilon_1 g_{i_1} + \epsilon_2 g_{i_2} + \ldots + \epsilon_k g_{i_k}$ is equal to 0, where the ϵ_i's are equal to ± 1. We can view this sequence of ϵ_i's as a walk of length l on the lattice Z^n. Each step of the walk consists of going from one point x in the lattice Z^n to a neighbor $x \pm e_1$, (where $(e_i)_{1 \leq i \leq n}$ is the canonical basis of Z^n) on the lattice Z^n, by associating to each g_i the vector e_i. This walk starts at the origin and goes successively from the origin to $\epsilon_1 e_{i_1}$, then from $\epsilon_1 e_{i_1}$ to $\epsilon_1 e_{i_1} + \epsilon_2 e_{i_2}$, and ends at the last step at $\epsilon_1 e_{i_1} + \epsilon_2 e_{i_2} + \ldots + \epsilon_l e_{i_l}$. In other words $\mu_l(C)$ is the number of walks of length l over Z^n whose endpoints are elements of the sublattice $\mathcal{L}$- so if we denote by $a_l(x)$ be the number of walks of length l from the origin to x in the lattice Z^n, we have

Fact 2 $\mu_l(C) = \sum_{x \in \mathcal{L}} a_l(x)$

This remark enables us to bring in harmonic analysis on Z^n in order to give a formula for $\mu_l(C)$. Let us recall some basic facts about harmonic analysis on Z^n and its relationship with the number of walks between two points in this lattice (see [28] for more details). Let $\mathcal{A}_l(\theta)$ be the Fourier series with coefficients $a_l(x)$, that is

$$\sum_{x \in Z^n} a_l(x) e^{i<x.\theta>}$$

For example

$$\mathcal{A}_1(\theta) = \sum_{x \in Z^n} a_1(x) \exp^{i<x.\theta>} = 2 \sum_{i=1}^{n} \cos \theta_i$$

(with $\theta = (\theta_1, \theta_2, \ldots, \theta_n)$). It is straightforward to check that

$$\mathcal{A}_l(\theta) = (\mathcal{A}_1(\theta))^l$$

From there we obtain the following expression for $a_l(x)$

$$a_l(x) = \frac{1}{(2\pi)^n} \int_{[-\pi,\pi]^n} e^{-i<x,\theta>} (\mathcal{A}_1(\theta))^l \, d\theta_1 d\theta_2 \ldots d\theta_n$$

$$= \frac{1}{\pi^n} \int_{[-\pi,\pi]^n} e^{-i<x,\theta>} \left(\sum_{i=1}^{n} \cos\theta_i \right)^l d\theta_1 d\theta_2 \ldots d\theta_n$$

We are interested in calculating in our case the following sum

$$\mu_l(C) = \sum_{x \in \mathcal{L}} a_l(x) = \sum_{x \in \mathcal{L}} \frac{1}{(\pi^n} \int_{[-\pi,\pi]^n} e^{-i<x,\theta>} \left(\sum_{i=1}^{n} \cos\theta_i \right)^l d\theta_1 d\theta_2 \ldots d\theta_n$$

Basically two approaches can be used. The first is the simple remark that when we consider the exponential generating function of $\mu_l(C)$ we may observe that

$$f_C^{exp}(z) = \sum_{l=0}^{\infty} \frac{\mu_l(C)z^l}{l!}$$

$$= \sum_{x \in \mathcal{L}} \frac{1}{(2\pi)^n} \int_{[-\pi,\pi]^n} e^{-i<x,\theta>} e^{2z \sum_{i=1}^{n} \cos\theta_i} d\theta_1 d\theta_2 \ldots d\theta_n$$

$$= \sum_{(x_1,x_2,\ldots,x_n) \in \mathcal{L}} i^{-(x_1+x_2+\ldots+x_n)} J_{x_1}(2iz) J_{x_2}(2iz) \ldots J_{x_n}(2iz)$$

where $J_n(x)$ denotes the Bessel function of the first kind of order n, i.e. $J_n(x) = \frac{i^{-n}}{2\pi} \int_0^{2\pi} e^{ix\cos\theta} e^{in\theta}$ (see [21, 30]).

In other words in our case since we are essentially interested in $f_C^{exp}(-iz)$

Theorem 5

$$f_C^{exp}(-iz) = \sum_{(x_1,x_2,\ldots,x_n) \in \mathcal{L}} i^{-(x_1+x_2+\ldots+x_n)} J_{x_1}(2z) J_{x_2}(2z) \ldots J_{x_n}(2z)$$

We can use this formula for example if we want to approximate the exponential generating function by restricting the summation on the lattice points around the origin. This method is particularly well suited to "sparse" lattices $\mathcal{L}$. When $\mathcal{L}$ is "dense" we can use a duality result which is essentially Poisson formula in order to calculate $\mu_l(C)$ and the associated generating function. First of all let us recall Poisson formula

Lemma 4. *For a function of rapid decay f (for a definition of this notion see for example [16] p.138) and a lattice $\mathcal{L}$ of rank k, for which we assume that the first k coordinates of points of this lattice form a lattice of rank k (a similar formula holds when another set of k coordinates forms a lattice of full rank).*

$$\sum_{x \in \mathcal{L}} f(x) = \frac{1}{(2\pi)^{n-k} V(\mathcal{L})} \int_{R^{n-k}} \sum_{(\theta_1,\theta_2,\ldots,\theta_k,\theta_{k+1},\ldots,\theta_n) \in 2\pi\mathcal{L}'} \hat{f}(\theta) d\theta_{k+1} \ldots d\theta_n$$

where $\hat{f}(\theta) = \hat{f}(\theta_1,\theta_2,\ldots,\theta_n) = \int_{R^n} e^{i<x,\theta>} f(x) dx$ is the Fourier transform of f,
$V(L)$ is the volume of $E/\mathcal{L}$, where E is the vector space spanned by $\mathcal{L}$,
$\mathcal{L}' = \{x \in R^n | < x,y >\in Z \text{ for all } y \in \mathcal{L}\}$.

(for a proof of this lemma see the appendix). Let us pause here a moment in order for a few remarks

1. when the rank of the lattice is n we have Poisson formula

$$\sum_{x\in\mathcal{L}} f(x) = \frac{1}{V(\mathcal{L})} \sum_{\theta\in 2\pi\mathcal{L}'} \hat{f}(\theta)$$

In this case $\mathcal{L}'$ coincides with the usual notion of the dual lattice of $\mathcal{L}$.

2. Actually this formula holds with weaker hypotheses on the function f, for example we can use it with $f(x) = a_l(x)$ and in this case we obtain

Theorem 6

$$\sum_{x\in\mathcal{L}} a_l(x) = \frac{1}{(2\pi)^{n-k}V(\mathcal{L})} \int_{[-\pi,\pi]^{n-k}} g_l(\theta_{k+1},\theta_{k+2},\ldots,\theta_n)\,d\theta_{k+1}\ldots d\theta_n$$

where $g_l(\theta_{k+1},\theta_{k+2},\ldots,\theta_n)$ is the finite sum

$$\sum_{(\theta_1,\theta_2,\ldots,\theta_n)\in 2\pi\mathcal{L}'\cap[-\pi,\pi]^n} (2\cos(\theta_1) + 2\cos(\theta_2) + \ldots + 2\cos(\theta_n))^l$$

In order to have a handy tool, we point out that the following description of $\mathcal{L}'$ may be used (for details about what follows we refer to the appendix). Let $u_1, u_2, \ldots, u_k$ be a basis of the lattice $\mathcal{L}$, and A be the matrix with columns $u_1, u_2, \ldots, u_k, e_{k+1}, \ldots, e_n$. Then A is nonsingular, let us call this inverse B, and B^t its transpose. Then it can be checked that, θ denoting the vector $(\theta_1, \theta_2, \ldots, \theta_n)$, and $\theta' = (\theta_1', \theta_2', \ldots, \theta_n')$ the vector $-B^t\theta$

$$g_l(\theta_{k+1},\theta_{k+2},\ldots,\theta_n) = \sum_{(\theta_1,\theta_2,\ldots,\theta_k)\in(2\pi Z)^k|\theta'\in[-\pi,\pi]^n} (2\cos\theta_1' + 2\cos\theta_2' + \ldots + 2\cos\theta_n')^l$$

An additional attractive feature of theorem 6 is that by using standard estimates of integrals of the form $\int f^n(x)dx$ for large values of n, we easily derive good estimates of $\mu_l(C)$ for l large. We just sketch how we can find an equivalent of $\mu_l(C)$, for a method which would give the whole power series expansion of $\mu_l(C)$ we refer to [8]. We rely on the following lemma whose proof is classical, and which can be found in a context similar to ours in [18] chapter 1, when proving the local central limit theorem.

Lemma 5. Let $f(x)$ be a function of n variables, which attains its maximum at 0 and nowhere else, and assume that around 0, $f(x) = f(x_1, x_2, \ldots, x_n) = 1 - a_1 x_1^2 - a_2 x_2^2 - \ldots a_n x_n^2 + O(|x|^4)$, where the a_i are positive constants. Then

$$\int_A f^l(x)dx = \frac{1}{l^{n/2}} \int_{R^n} e^{-a_1 x_1^2 - a_2 x_2^2 - \ldots - a_n x_n^2}\,dx_1\ldots dx_n\,(1+O(1/l)) = \sqrt{\frac{\pi^n}{l^n a_1 a_2 \ldots a_n}}\,(1+O(1/l)).$$

Where A is a compact set which contains 0 in its interior.

This enables us to find an equivalent of $\mu_l(C)$, and in practice even if in some cases we do not get a "simple" expression for the exponential generating function, we can always calculate the first terms $\mu_l(C)$ by using theorem 5, and estimate the remaining ones with the previous lemma, and obtain from that good estimates of the exponential generating function.

5.2 Basic examples : Z^n and the repetition code

The simplest example occurs when for a generating matrix of our dyadic code C of length there is no relationship of the form $a_1g_1 + a_2g_2 + \ldots + a_ng_n = 0$, where the g_i's are the columns of the generating matrix, and the a_i's are integers which are not all equal to 0. In this case the Cayley graph $\Gamma(C)$ is isomorphic to the lattice Z^n, and from theorem 5 we know that the exponential generating function of the number of cycles of a given length of $\Gamma(C)$ is given by

$$f_C^{exp}(-iz) = (J_0(2z))^n$$

In this case the lattice associated to the code had only one point, let us consider now the converse example where the lattice associated to the code is very "dense", this happens for instance for the repetition code. For such a code of length n, the generator matrix is the row vector $(g_1, g_2, \cdots, g_n)$ of length n where all the g_i's are equal to 1.

Thus in this case the set of n-tuples of integers $(a_1, a_2, \ldots, a_n)$ such that $a_1g_1 + a_2g_2 + \ldots + a_ng_n = 0$ is the sublattice $\mathcal{L}$ of Z^n of n-tuples which satisfy $a_1 + a_2 + \cdots + a_n = 0$. This sublattice is generated by the $n-1$ vectors $e_1 - e_2, e_2 - e_3, \cdots, e_{n-1} - e_n$, and the first $n-1$ coordinates of points of this lattice form a lattice of rank $n-1$. A straightforward calculation shows that $g(\theta_n)$ as defined in theorem 6 is equal to

$$2n \cos(\alpha_n)$$

Therefore

$$\mu_l(C) = \frac{1}{2\pi} \int_0^{2\pi} (2n \cos x)^l \, dx$$
$$f_C^{exp}(-iz) = J_0(2zn)$$

5.3 An application to Irreducible Cyclic Codes

An important class of dyadic codes for which we wish to apply our methods is the class of cyclic codes. We will consider irreducible cyclic codes first. Such dyadic codes may be described by using a primitive n^{th} root of 1 denoted by ζ_n, and the cyclotomic extension $K = Q_2(\zeta_n)$ of Q_2 (the fraction field of Z_{2^∞}). Denote by k the degree of this extension over Q_2, and by Tr the associated trace function down to Q_2. Let $n = 2^l n_0$, where n_0 is odd. Then ([13] Chapter 15, Section 3), the degree of the extension is $n_1 2^{l-1}$, where n_1 is the smallest exponent such that $2^{n_1} \equiv 1 \bmod n_0$. We will denote in the sequel such a code by C_n.

Then $y \in C_n$ can be parametrized in the following way.

$$y_i = Tr(\alpha \zeta_n^i),$$

for α ranging over K. In other words the generator matrix is the k by n matrix with entries over Q_2

$$(1 \quad \zeta_n \quad \cdots \quad \zeta_n^{n-1})$$

(where the ζ_n^i's are represented by column vectors over Q_2 of size the degree of the extension $Q_2(\zeta_n)$). To apply theorem 5 or Fact 2 and 6 we need to find a basis of the sublattice $\mathcal{L}$ of Z^n whose points represent nontrivial relations $\sum_{i=0}^{n-1} a_i \zeta_n^i = 0$ between the generators ζ_n^i's of the code. A relation of this kind is given by a polynomial $P(x)$ over Z such that $P(\zeta_n) = 0$, and therefore by a multiple of the nth cyclotomic polynomial $\Phi_n(x)$ over Q. Recall here that (see chapter VI of [9] for instance)

$$\Phi_n(x) = \prod_{d|n} (x^d - 1)^{\mu(n/d)}$$

where μ is the Möbius function $\mu(1) = 1, \mu(i) = (-1)^r$ if n is a product of r distinct primes, and 0 otherwise. These remarks enable us to find all these relations explictly, and to deduce from that a basis for $\mathcal{L}$, and henceforth to calculate the generating function of the number of cycles.

Primitive Case Here $n = 2^m - 1$, the extension field K is the unique unramified extension of Q_2 of degree m, and the associated cyclic code is of dimension m. The code C_1 is the Simplex code of length n in cyclic form. The code C_2 is related for m odd to the Kerdock code [12].

Quadratic Residue Codes In some small cases the quadratic residue codes are irreducible cyclic. Here n is a prime congruent to $\pm 1 \bmod 8$. The cases $n = 7, 23$ are treated in [2]. For $n = 23$ it is known that C_1 is the dual of the perfect binary Golay code and that C_2 is the dual of the cyclic lifted Golay [1].

Prime length Let us calculate the number of cycles of the graph $\Gamma(C_n)$, and the associated generating function, when we have an irreducible cyclic code C_n of prime length n. In this case the nth cyclotomic polynomial is $\Phi_n(x) = 1 + x + \ldots + x^{n-1}$ and therefore the only nontrivial relations between the generators are of the kind

$$\alpha(1 + \zeta_n + \ldots + \zeta_n^{n-1}) = 0$$

In other words the sublattice $\mathcal{L}$ is generated by the vector $e_1 + e_2 + \ldots + e_n$

By applying Theorem 5 we have the following expression for the exponential generating function and if we apply lemmas 2 and 6 we obtain that

$$f_{C_n}^{exp}(-iz) = \sum_{l=-\infty}^{\infty} i^{-l}(J_l(2z))^n$$

Thereafter we can estimate this sum by using standard results on Bessel functions (see [21, 30]). Classical results on convolution products may be used too. Let $F_z(x)$ be the function $e^{2z\cos x}$, and $G_z(x)$ be the function equal to F_z on the interval $[0, 2\pi]$ and 0 elsewhere. By using standard Fourier series considerations the convolution product of F_z with the $n - 1$th convolution product of G_z satisfy

$$f_{C_n}^{exp}(z) = F_z * \underbrace{G_z * \ldots * G_z}_{n-1 \text{ terms}}(0) = \sum_{i=\infty}^{\infty} [f_i(z)]^n$$

where $f_i(z) = \frac{1}{2\pi} \int_0^{2\pi} e^{2z\cos\theta} e^{in\theta} d\theta$. Several approaches may be used to estimate this convolution product. Either we use well known central-limit estimates of iterated convolutions (see sec. 11.2 and 11.3 in [25] for instance) or we use Fourier analysis, i.e if we denote by $h_z(x)$ the convolution product $G_z * \ldots * G_z$

$$h(x) = \frac{1}{2\pi} \int_{-\infty}^{\infty} e^{-ixu} \hat{h}(u) du$$

and the Fourier transform of $h_z(x)$ can be computed explicitly :

$$\hat{h}_z(u) = (\hat{G}_z(u))^{n-1}$$
$$= \left(\int_0^{2\pi} e^{iux} e^{2z\cos x} dx \right)^{n-1}$$

Hence

$$f_{C_n}^{exp}(z) = \int_0^{2\pi} e^{2z\cos\theta} H(\theta)$$

where $H(\theta) = \frac{1}{2\pi} \int_{-\infty}^{\infty} e^{-i\theta u} \hat{h}(u)$.

An approach which is essentially equivalent is to note that theorem 6 can be used in conjunction with lemma 5 in order to estimate the generating function. First of all a straightforward application of theorem 6 gives us

$$\mu_l(C_n) = \frac{1}{(2\pi)^{n-1}} \int_{[-\pi,\pi]^{n-1}} (2(\cos x_1 + \cos x_2 + \ldots + \cos x_{n-1} + \cos(x_1 + x_2 + \ldots + x_{n-1})))^l dx_1$$

Lemma 5 can be applied here and we derive from it

$$\mu_l(C_n) = \frac{n^{(n-2)/2)}}{(2\pi)^{(n-1)/2}} \frac{(2n)^l}{l^{(n-1)/2}} (1 + O(1/l)).$$

From that, we can compute easily a good approximation of the generating function.

Composite length When the length of the irreducible cyclic code is not a prime, the calculation of the number of cycles and of the associated generating functions becomes more complicated. The nth cyclotomic polynomial $\Phi_n(x)$ is of degree $\varphi(n)$ where φ is Euler totient function. The rank of the lattice is $n - \varphi(n)$, and therefore greater than 1 when n is composite. We can still apply theorem 6, and find explictly a basis for $\mathcal{L}$.

Nevertheless in some special cases graph theoretical tools may be used too. When the length is a power of a prime number p, the coset graph $\Gamma_{\mathcal{C}_{p^m}}$ is the Cartesian product of p^{m-1} coset graphs $\Gamma(\mathcal{C}_p)$ (for a definition of this product see [5] for example). This can be checked easily by understanding the structure of the lattice which is associated to $\mathcal{C}_{p^m}$. This lattice is namely the direct sum of p^{m-1} lattices which are isomorphic to the lattice associated to $\mathcal{C}_p$. This can be seen if we notice that

$$\Phi_{p^m}(x) = \Phi_p(x^{p^{m-1}}).$$

From [23] we deduce that the Plancherel measure associated to the coset graph $\Gamma(\mathcal{C}_{p^m})$ is the the p^{m-1}th convolution product of the Plancherel measure of $\Gamma(\mathcal{C}_p)$. This implies that there is a simple relationship between the exponential generating functions of these graphs

$$f_{\mathcal{C}_{p^m}}^{exp}(z) = \left(f_{\mathcal{C}_p}^{exp}(z)\right)^{p^{m-1}}.$$

Another case where the structure of the coset graph can be decomposed in simpler structures, is the case of even length, say $n = 2^m n_1$, where n_1 is odd. In this case

$$\Phi_{2^m n_1}(x) = \Phi_{n_1}(-x^{2^{m-1}}).$$

From that it can be checked that $\Gamma(\mathcal{C}_{2^m n_1})$ can be obtained from the cartesian product of 2^{m-1} graphs $\mathcal{C}_{n_1}$ by doubling each edge. Therefore

$$f_{\mathcal{C}_{2^m n_1}}^{exp}(z) = \left(f_{\mathcal{C}_{n_1}}^{exp}(2z)\right)^{2^{m-1}}.$$

Now let us us consider an example which shows how the calculation is performed in the general case. For instance let us see how this works with $n = 15$.

$$\Phi_{15}(x) = x^8 - x^7 + x^5 - x^4 + x^3 - x + 1$$

From that it is easy to construct a basis of relations between the generators of the code

$$\zeta_{15}^8 - \zeta_{15}^7 + \zeta_{15}^5 - \zeta_{15}^4 + \zeta_{15}^3 - \zeta_{15} + 1 = 0$$
$$\zeta_{15}^9 - \zeta_{15}^8 + \zeta_{15}^6 - \zeta_{15}^5 + \zeta_{15}^4 - \zeta_{15}^2 + \zeta_{15} = 0$$
$$\zeta_{15}^{10} - \zeta_{15}^9 + \zeta_{15}^7 - \zeta_{15}^6 + \zeta_{15}^5 - \zeta_{15}^3 + \zeta_{15}^2 = 0$$
$$\zeta_{15}^{11} - \zeta_{15}^{10} + \zeta_{15}^8 - \zeta_{15}^7 + \zeta_{15}^6 - \zeta_{15}^4 + \zeta_{15}^3 = 0$$
$$\zeta_{15}^{12} - \zeta_{15}^{11} + \zeta_{15}^9 - \zeta_{15}^8 + \zeta_{15}^7 - \zeta_{15}^5 + \zeta_{15}^4 = 0$$
$$\zeta_{15}^{13} - \zeta_{15}^{12} + \zeta_{15}^{10} - \zeta_{15}^9 + \zeta_{15}^8 - \zeta_{15}^6 + \zeta_{15}^5 = 0$$
$$\zeta_{15}^{14} - \zeta_{15}^{13} + \zeta_{15}^{11} - \zeta_{15}^{10} + \zeta_{15}^9 - \zeta_{15}^7 + \zeta_{15}^6 = 0$$

If we use theorem 6 with this lattice we obtain

$$\mu_l(C) = \frac{1}{(2\pi)^8} \int_{[0,2\pi]^8} (g(x_1,\ldots,x_8))^l dx_1 \ldots dx_8$$

where

$$g(x_1, x_2, \ldots, x_8) = 2 \left(\sum_{i=1}^{7} \cos y_i + \sum_{i=1}^{8} \cos x_i \right)$$

and the y_i's are given by

$$y_1 = x_1 - x_3 + x_4 - x_5 + x_7 - x_8$$
$$y_2 = x_1 - x_2 - x_5 + x_6 - x_8$$
$$y_3 = -x_3 - x_8$$
$$y_4 = -x_2 - x_7$$
$$y_5 = -x_1 - x_6$$
$$y_6 = -x_1 + x_3 - x_4 - x_7 + x_8$$
$$y_7 = -x_1 + x_2 - x_4 + x_5 - x_6 + x_8$$

6 Acknowledgement

We thank Gilles Lachaud and Greg Lawler for helpful discussions and giving us [10, 17, 18]. The first author thanks Branka Vucetic and Richard Blahut for helpful discussions.

APPENDIX

A   Proof of Theorem 5

Let us assume that the restriction to the first kth coordinates of the lattice $\mathcal{L}$ is a lattice of full rank. Let $u_1, u_2, \ldots, u_k$ be a basis of this lattice, and A the matrix with columns $u_1, u_2, \ldots, u_k, e_{k+1}, \ldots, e_n$. A is nonsingular, let us denote its inverse by B, and the transpose of it B^t. First of all let us observe that the set of lattice points of $\mathcal{L}$ is given by the set $\{Ax | x = (x_1, x_2, \ldots, x_k, 0, \ldots, 0), x_1, x_2, \ldots, x_k \in Z\}$. The "dual" lattice $\mathcal{L}'$ can be easily seen to be

$$B^t(Z^k \times R^{n-k}).$$

From this we may notice that

$$\sum_{x \in \mathcal{L}} f(x) = \sum_{x \in \mathcal{L}_k} f(Ax) \tag{1}$$

where $\mathcal{L}_k$ is the lattice $Z^k \times 0^{n-k}$. Let $\Pi_{\mathcal{L}}$ be the sum of Dirac functions $\sum_{x \in \mathcal{L}} \delta_x$. If we denote by $< T, f >= T(f)$ the action of the distribution T on the test function f, then it easy to check that using equality (1) we obtain

$$\sum_{x \in \mathcal{L}} f(x) = < \Pi_{Z^k} \otimes \delta_{\underbrace{(0, \ldots, 0)}_{n-k \text{ terms}}}, f(Ax) >$$

$$= < \mathcal{F}(\Pi_{Z^k} \otimes \delta_{0, \ldots, 0}), \mathcal{F}^{-1}(f(Ay)) >$$

($\mathcal{F}$ denotes here the Fourier transform, $\mathcal{F}^{-1}$ the inverse Fourier transform, and $\otimes$ the direct product of distributions— see [16] for the background of the material which is presented here)

It is well known (see [16]) that

$$\mathcal{F}(\Pi_{Z^k} \otimes \delta_{0, \ldots, 0})(\theta_1, \ldots, \theta_k, \theta_{k+1}, \ldots, \theta_n) = \mathcal{F}(\Pi_{Z^k})(\theta_1, \ldots, \theta_k)\mathcal{F}(\delta_{0, \ldots, 0})(\theta_{k+1}, \ldots, \theta_n)$$

$$= (2\pi)^k \Pi_{(2\pi Z)^k}(\theta_1, \ldots, \theta_k) 1_{R^{n-k}}(\theta_{k+1}, \ldots, \theta_n)$$

($1_{R_{n-k}}$ is the function which associates 1 to every input in R^{n-k}). Moreover

$$\mathcal{F}^{-1}(f(Ax))(u) = \frac{1}{(2\pi)^n} \int_{R^n} e^{-i<u,x>} f(Ax) dx$$

$$= \frac{1}{(2\pi)^n V(\mathcal{L})} \int_{R^n} f(y) e^{-i<u, By>}$$

$$= \frac{1}{V(\mathcal{L})} \hat{f}(-B^t u)$$

Putting these facts together we get

$$\sum_{x \in \mathcal{L}} f(x) = \frac{1}{(2\pi)^{n-k}V(\mathcal{L})} \int_{R^{n-k}} \sum_{(\theta_1,\theta_2,\dots,\theta_k) \in (2\pi Z)^k} \hat{f}(-B^t(\theta_1,\theta_2,\dots,\theta_k,\theta_{k+1},\dots,\theta_n)d\theta_{k+1}d\theta_n$$

$$= \frac{1}{(2\pi)^{n-k}V(\mathcal{L})} \int_{R^{n-k}} \sum_{(\theta_1,\theta_2,\dots,\theta_n) \in 2\pi \mathcal{L}'} \hat{f}(\theta_1,\theta_2,\dots,\theta_k,\theta_{k+1},\dots,\theta_n)d\theta_{k+1}d\theta_n$$

References

1. A. Bonnecaze, R. A. Calderbank, P. Solé, "Unimodular constructions of unimodular lattices" IEEE IT 41 (1995) 366-376.
2. R.A. Calderbank and N.J.A. Sloane, "Modular and p-adic cyclic codes" Des. Codes Crypt.(1995) 21-35.
3. J.W.S. Cassels, A. Froehlich, *Algebraic Number Theory*, Academic Press (1967).
4. F. R. K. Chung, V. Faber, and T. A. Manteuffel, "An upper bound on the diameter of a graph from eigenvalues associated with its laplacian", SIAM J. Disc. Math., Vol. 7, No. 3, (1994), 443-457.
5. D. Cvetkovic, M. Doob, H. Sachs, *Spectra of graphs, Theory and applications*, Pure Applied Math. 9, Academic Press, New York, 1979.
6. C. Delorme, P. Solé , "Diameter, Covering number, Covering radius and Eigenvalues" Europ. J. of Comb., Vol.12,(1991) 95-108.
7. C.K. Wong, D.K. Coppersmith, "A combinatorial problem related to multimodule organization", J. of the ACM 21, 3 (1974) 392-402.
8. C. Domb, Proc. Camb. Phil. Soc. 50, (1954), 589-591.
9. A. Fröhlich and M. J. Taylor, *Algebraic number theory*, Cambridge Studies in advanced mathematics 27, 1991.
10. R. Godement *Adèles et Idèles* Cours de l' IHP 1964-65 Paris VII Prépublication.
11. C.D. Godsil, B. Mohar, "Walk Generating Functions and Spectral Measures of Infinite Graphs" Lin. Alg. and its Appl. 107 (1988) 191-206.
12. R.Hammons, V. Kumar, R.A. Calderbank, N.J.A. Sloane and P. Solé, "The Z_4-linearity of Kerdock, Preparata, Goethals and related codes", IEEE IT -40 (1994) 301-319.
13. H. Hasse, *Number Theory*, Springer Velag, 1980.
14. S. Iyanaga, *The theory of numbers* North Holland (1975).
15. D. S. Jones, *The theory of generalized functions*, Cambridge University Press, 2nd edition 1982.
16. R.P. Kanwal, *Generalized Functions: Theory and Technique*, Vol. 171 in Mathematics in Science and Engineering, Academic Press 1983.
17. G. Lachaud, "Une présentation adélique de la série singulière et du problème de Waring". L'Enseignement Mathématique T. XXVIII-Fasc.1-2 Janvier-Juin (1982) 139-169.
18. G. Lawler, *Intersection ofRandom Walks*, Probability and its applications, Birkhäuser, 1991.
19. M. Loève, *Probability Theory I*, Springer GTM 45 (1977).
20. A. Lubotzsky, R.Phillips, P. Sarnak, "Ramanujan Graphs", Combinatorica 8 (1988) 261-277.
21. N.W. McLachlan, *Bessel Functions for Engineers*, Oxford at the Clarendon Press, 1934.

22. F.J. MacWilliams, N.J.A. Sloane, *The Theory of Error- Correcting Codes* North-Holland (1977).
23. B. Mohar, W. Woess, "A survey on spectra of infinite graphs", Bulletin of the L.M.S. 21 (1989) 209-234.
24. A. Nilli, "On the second eigenvalue of a graph", Discr. Math. 91 (1991) 207-210.
25. A. Papoulis, *The Fourier integral and its applications*, McGraw-Hill Company, Inc, 1962.
26. P. Piret, "Codes on the unit circle" IEEE IT-32 (1986) 760-767.
27. J. Proakis, *Digital Communications*,McGraw-Hill (1989) second edition.
28. F. Spitzer, *Principles of Random Walk*, D. Van Nostrand Company, Inc, 1964.
29. A. Weil,*Basic Number Theory* Springer (1973).
30. G.N. Watson, '*A Treatise on the Theory of Bessel Functions*, 2nd Ed. Cambridge Univ. Press, 1966.

Zigzag Codes and Z-Free Hulls [*]

Bruno Patrou

LaBRI-Unité de recherche associé au CNRS n° 1304,
Université Bordeaux I, 351, cours de la Libération, 33405 Talence Cedex, France.
Email: patrou@labri.u-bordeaux.fr

Abstract. We extend the notion of free hull to the zigzag operation in order to build the z-free hulls of a language. We present some properties of those z-free hulls and, specially, we show that a language can have several z-free hulls whereas the free hull is always unique. We also study the defect theorem which becomes false for the zigzag case. Then we are interested by a problem closed to the defect theorem and we solve it in some particular cases.

1 Introduction

The zigzag operation $^\uparrow$ is an extension of the $*$ operation introduced in [1]. Let X be a language on an alphabet Σ. Every word $w \in X^*$ is obtained by a run on w with left-right steps in X, while a word $w' \in X^\uparrow$ can be obtained by a run on w' with left-right and also right-left steps in X. For instance if $X = \{abc, bc, bca\}$, the word $abca$ belongs to $X^\uparrow$ (but not to X^*); indeed it is obtained by the left-right step abc, the right-left step bc and the left-right step bca. That is the word $abca$ has a *zigzag factorization* (or z-factorization) on X.

As for $*$ operation, one can define the following. A language C is a z-*code* if every word in A^* has at most one z-factorization on C [1]. A language M is a z-*submonoid* of A^* if $M = X^\uparrow$ for some language X, called z-*generator* of M [7]. Every z-submonoid M has a least z-generator called z-root(M) and M is z-*free* if its z-root is a z-code.

Concerning the $*$ operation, the notion of stability defined in [8], gives a direct characterization of free submonoids. From this notion, it was shown in [9] that the intersection of an arbitrary family of free submonoids is still a free submonoid. As a consequence, the intersection of all free submonoids containing a language X is the smallest free submonoid containing X. It is called the *free hull* of X. If X^* is a free submonoid then it coincides with its free hull. Now if X is not a code and if Y is the root of its free hull then we have an interesting relationship between X and Y given by the following result known as the *defect theorem*:

[*] This work was supported by the ASMICS program

Theorem 1 [2]. *Let X be a finite language on an alphabet Σ and Y be the root of the free hull of X. If X is not a code then:*

$$Card(Y) < Card(X).$$

We are interested here in the notions corresponding to the free hull and the defect theorem for the zigzag operation. In [5], the authors proposed a definition of z-*stability* we will recall, and proved the equivalence between z-free z-submonoids and z-stable z-submonoids. We see, from this definition, that the intersection of z-free z-submonoids is not necessarily a z-free z-submonoid. So we define the notion of z-free hull and we see on an example that there exist languages having several z-free hulls.

Then, we study the set of z-free hulls of a given language X, denoted by Z-$Hull(X)$. If X is a finite language we prove that Z-$Hull(X)$ is also finite. We give an example of a rational language having an infinity of z-free hulls some of them being non algebraic languages and, if we denote by $H(X)$ the free hull of X, we prove that Z-$Hull(X) = Z$-$Hull(H(X))$.

At last, we study the defect theorem. In general it is not verified, that is there exist z-free hulls of a language X whose z-roots have greater cardinality than z-$root(X^{\uparrow})$. Then the question becomes: "does always exist at least one z-free hull which verifies the defect theorem ?". We solve the problem for two-elements languages and for a certain class of three-elements languages, but the general case is still open.

Section 2 contains the preliminaries. In the next section we give the definitions of stability and z-stability, and we introduce the problem. In section 4 we present some properties of z-free hulls and, in section 5, we study the defect theorem extended to the zigzag operation.

2 Preliminaries

Let Σ be an alphabet (for all the paper the symbol Σ will denote an alphabet). As usual, Σ^* is the *free monoid* of all finite words over Σ, the *empty word* is denoted by ε and $\Sigma^+ = \Sigma^* \setminus \{\varepsilon\}$. The *concatenation* of two words $u, v \in \Sigma^*$ is denoted by uv.

For any language X, X^* denotes the *submonoid* of Σ^* generated by X and we denote the smallest generator of X^* by $root(X^*) = (X^* \setminus \varepsilon) \setminus (X^* \setminus \varepsilon)^2$. A *factorization* on X of a word w is a tuple $(x_1, ..., x_n)$ of words in X such that $x_1...x_n = w$. A language $X \subseteq \Sigma^*$ is a *code* if every word $w \in \Sigma^*$ has at most one factorization on X. If X is a code then X^* is a *free* submonoid of Σ^*.

For any alphabet Σ, we denote by $\overline{\Sigma}$ a disjoint alphabet in bijection with Σ. For every $a \in (\Sigma \cup \overline{\Sigma})$, we denote by $\overline{a}$ the element associated with a in order that $a = \overline{\overline{a}}$. For every word $x = a_1 \cdots a_n \in (\Sigma \cup \overline{\Sigma})^*$, $\overline{x}$ is the word $\overline{a}_n \cdots \overline{a}_1$. For every subset X of Σ^*, we denote $\overline{X} = \{\overline{x} : x \in X\}$ and we use the free monoid

M_X generated by $X \cup \overline{X}$. Then a word in the free monoid M_X is denoted by a tuple $(x_1, \ldots, x_n)$ where every x_i is in $(X \cup \overline{X})$.

We use a canonical morphism φ_X from M_X to $(\Sigma \cup \overline{\Sigma})^*$ such that: $\forall x_1, x_2 \in M_X$, $\varphi_X(x_1, x_2) = \varphi_X(x_1).\varphi_X(x_2)$. Let $w \in (\Sigma \cup \overline{\Sigma})^*$; we denote by $red_\Sigma(w)$ the canonical representative of the class of w in the free group generated by Σ.

We denote by $\overset{l}{\longmapsto}$ the relation defined for all $u, v \in M_X$ by $u \overset{l}{\longmapsto} v$ if $u = (f, \alpha, g)$, $v = (f, g)$ with $red_\Sigma(\varphi_X(\alpha)) = \varepsilon$. We call l-reduction the reflexo-transitive closure of $\overset{l}{\longmapsto}$. Then with every word $w \in M_X$ is associated a set $red_l(w)$ of l-reduced words. Note that the l-reduction is confluent iff X is a zigzag-code [4].

A *zigzag calculus* (z-calculus), $C_X(w_1, u, w_2)$, of a word $u \in \Sigma^*$ with context $(w_1, w_2) \in \Sigma^* \times \Sigma^*$ is a tuple $(x_1, \ldots, x_n) \in M_X$ such that :

1. $red_\Sigma(x_1 \ldots x_n) = u$
2. $\forall \, 1 \leq i \leq n, \ \varepsilon \leq red_\Sigma(w_1 x_1 \ldots x_i) \leq w_1 u w_2$

When furthermore :

3. $\forall \, 1 < i < n, \ \varepsilon < red_\Sigma(w_1 x_1 \ldots x_i) < w_1 u w_2$

the z-calculus is called *strict*.

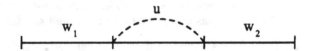

Fig. 1. Representation of a z-calculus of u with context (w_1, w_2)

A *z-decomposition* on X of a word u is a z-calculus of u with context $(\varepsilon, \varepsilon)$. A *z-factorization* is a l-reduced z-decomposition; that is a z-decomposition $(x_1, \ldots, x_n)$ such that :

$$\forall \, 1 \leq i < j \leq n, \ red_\Sigma(x_1 \ldots x_i) \neq red_\Sigma(x_1 \ldots x_j).$$

In the sequel $f_{w,X}$ (or simply f_w) denotes a z-factorization of a word w on X, it is drawn with full line, while a z-calculus is drawn with dashed line (see Figure 1).

Example 1 . Let $X = \{aba, ba, baa\}$. The tuple $(\overline{ba}, baa, aba, \overline{ba})$ is a z-calculus of the word aa with context (ba, ba) on X (see figure 2). The tuple $(ba, baa, ba, \overline{aba}, \overline{ba}, baa, ba, ba)$ is a z-decomposition of the word $babaabab a$ on X (but not a z-factorization because $red_\Sigma(ba, baa, ba, \overline{aba}, \overline{ba}) = red_\Sigma(ba)$). At last, the tuple $(aba, \overline{ba}, baa)$ is a z-factorisation of the word $abaa$.

Fig. 2. Examples of z-calculus, z-decomposition and z-factorisation

Let $u, v \in \Sigma^*$, u is a z-prefix on X of uv if there exists a z-calculus $C_X(\varepsilon, u, v)$. If the z-calculus $C_X(\varepsilon, u, v)$ is strict (on the right context), u is a *strict* z-prefix of uv. We denote by $Z\text{-}pref_X(w)$ (resp. $Z\text{-}pref\text{-}strict_X(w)$) the set of words $u \in \Sigma^*$ such that u is a z-prefix (resp. strict z-prefix) on X of the word w.

Example 1 [continued]. From the z-calculus $(aba, \overline{ba})$ we can deduce that a is a z-prefix on X of any word in $aba\Sigma^*$ and a strict z-prefix of any word in $aba\Sigma^+$.

The set $X^{\uparrow}$ of words having a z-factorization on X is called the *z-submonoid* of Σ^* generated by X [7]. Of course, $X^{\uparrow}$ is a submonoid of Σ^* which contains X^* and the intersection of two z-submonoids is still a z-submonoid. Let L be a z-submonoid of Σ^*, we call z-*root* of L the set of words having exactly one z-factorization on L. A language X is a *z-code* if every word of $X^{\uparrow}$ has exactly one z-factorization on X [1]. A z-submonoid L is *z-free* if z-$root(L)$ is a z-code.

3 Stability and zigzag stability

In this section we recall the definitions of stability and z-stability, and, from them, we explain why the notions of free hull and z-free hull are different.

Definition 2 [2]. Let M be a submonoid on Σ^*. Then M is stable if for all $u, v, w \in \Sigma^*$,

$$u, vw, uv, w \in M \Rightarrow v \in M.$$

In [8] it is proved that a submonoid M is free iff it is stable. From this, it is possible to conclude (see [9]) that, for a given language X, the intersection of all free submonoids containing X is the smallest free submonoid containing X. It is unique and it is called the *free hull* of X (we denote it by $H(X)$).

In [3] the authors give a construction to build the set $H(X)$ from X. The principle is the following: we set $S = X^*$. As long as we have an occurrence of four words $u, vw, uv, w \in S$ we add the word v to S and when we have no more such occurrences, we have $S = H(X)$. This construction also permits to show that $H(X)$ is a rational language if X is rational.

Now, we look at the zigzag case:

Definition 3 [5]. Let M be a z-submonoid on Σ^*. Then M is z-stable if it satisfies the three following conditions for all $u_1, u_2, u_3, u_4, u_5 \in \Sigma^*$:

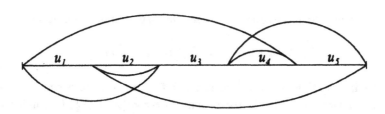

Fig. 3. First case for z-stability.

1. $u_1u_2, u_2, u_2u_3u_4u_5 \in M$ and $u_1u_2u_3u_4, u_4, u_4u_5 \in M \Rightarrow u_2u_3u_4 \in M$. (fig.3)
2. $u_1u_2u_3, u_2u_3, u_2u_3u_4u_5 \in M$ and $u_1u_2u_3u_4, u_3u_4, u_3u_4u_5 \in M \Rightarrow u_2u_3u_4 \in M$. (fig.4)

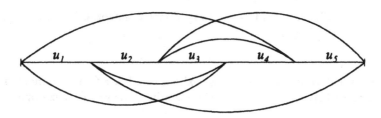

Fig. 4. Second case for z-stability.

3. $u_1u_2u_3, u_3, u_3u_4u_5 \in M$ and $u_1u_2u_3u_4, u_2u_3u_4, u_2u_3u_4u_5 \in M \Rightarrow u_2$ is a z-prefix on M of $(u_2u_3u_4)$. (fig.5)

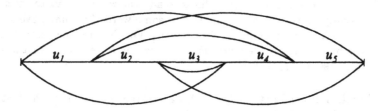

Fig. 5. Third case for z-stability.

In [5] the authors proved that a z-submonoid is z-stable iff it is z-free, but the intersection of z-free z-submonoids is not always a z-free z-submonoid as it is shown by the following example:

Example 2 (due to C.Choffrut). Let $X_1 = \{ab, a, bba\}$ and $X_2 = \{ab, b, baa\}$. It is easy to verify that X_1 and X_2 are z-codes and then $X_1^{\uparrow}$ and $X_2^{\uparrow}$ are z-free z-submonoids.

But, if we study the intersection of those two z-submonoids, we obtain:

$$\{ab, bbaa, bbaab, abbaa\} \subseteq z\text{-}root(X_1^{\uparrow} \cap X_2^{\uparrow}).$$

Then the intersection $X_1^{\uparrow} \cap X_2^{\uparrow}$ is not a z-free z-submonoid since the word $bbaabbaa$ has two different z-factorizations on $z\text{-}root(X_1^{\uparrow} \cap X_2^{\uparrow})$ (see figure 6).

Fig. 6. A word having two different z-factorizations

Then we deduce it does not always exist a smallest z-free z-submonoid containing a given language X. In fact, if we continue the previous example:

Example 2 [continued]. If we set $X = \{bbaa, bbaab, ab, abbaa\}$, $Y_1 = \{ab, a, bbaa\}$ and $Y_2 = \{ab, b, bbaa\}$, the languages $Y_1^{\uparrow}$ and $Y_2^{\uparrow}$ can be proved to be two minimal z-free z-submonoids containing X. We call them z-free-hulls of X.

In fact, if we try to build a z-free-hull in a way similar to [3] for the free hull, there is no problem with the cases 1 and 2 of the definition of z-stability. When these occurrences appears, it suffices to add the word $u_2 u_3 u_4$ to the set S. But, considering the third case, we must add some words (we may have several possibilities) to have u_2 z-prefix in S of $u_2 u_3 u_4$. On the previous example, the word a must be z-prefix of the word ab, so we can add the word a or the word b since $ab \in X$ (so $a = Red_{\Sigma}(ab\bar{b})$).

Another definition of z-stability which was proved equivalent to the first one is given in [6]. This definition is used in the next section.

Definition 4. Let M be a z-submonoid of Σ^*, M satisfies the property of z-stability if (see figure 7):
$\forall u, v, w \in \Sigma^*$:

$$\left. \begin{array}{l} uv \in Z\text{-}pref\text{-}strict_M(uvw), \quad w \in M \\ u \in Z\text{-}pref\text{-}strict_M(uvw), \quad vw \in M \end{array} \right\} \Rightarrow v \in Z\text{-}pref\text{-}strict_M(vw)$$

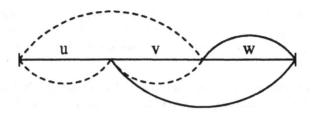

Fig. 7. Another definition of z-stabitity

4 Z-free hulls

We have seen in the previous section that there exist languages having no smallest z-free z-submonoid containing them. Now, we prove that there always exist minimal z-free z-submonoids containing a given language.

Definition 5. Let $X \subseteq \Sigma^*$. We denote by $Z\text{-}Free(X)$ the set of z-free z-submonoids containing X. This set is not empty since $\Sigma^\dagger$ is a z-free z-submonoid.

Proposition 6. Let $X \subseteq \Sigma^*$. Each element $M \in Z\text{-}Free(X)$ contains a minimal element of $Z\text{-}Free(X)$.

Proof. The set $Z\text{-}Free(X)$ is (partially) ordered for the relation of inclusion. To show the result, it is sufficient to prove (from Zorn's lemma) that any chain (i.e. any totally ordered subset) of elements of $Z\text{-}Free(X)$ admits a greatest lower bound in $Z\text{-}Free(X)$. Let C be a decreasing chain of $Z\text{-}Free(X)$ indiced by a totally ordered set K. We set:

$$\widehat{M} = \bigcap_{M \in C} M.$$

The intersection of an arbitrary family of z-submonoids is a z-submonoid, so $\widehat{M}$ is a z-submonoid. Now, let us suppose $\widehat{M}$ is not z-free. Then $\widehat{M}$ is not z-stable. So, from the second definition of z-stability, we know there exist three words $u, v, w \in \Sigma^*$ with $u, uv \in Z\text{-}pref\text{-}strict_{\widehat{M}}(uvw)$, $w, vw \in \widehat{M}$ and such that $v \notin Z\text{-}pref\text{-}strict_{\widehat{M}}(vw)$. Of course, we have $\forall M \in C$, $u, uv \in Z\text{-}pref\text{-}strict_M(uvw)$, $w, vw \in M$ and then $v \in Z\text{-}pref\text{-}strict_M(vw)$. Now, we have a finite number (say n) of strict l-reduced z-calculi on Σ^*, of v with the context (ε, w). To each z-calculus it corresponds a finite set of factors of vw (those used in the z-calculus). So we have n finite sequences of factors: $E_1, ..., E_n$. For all $i \in \{1, .., n\}$, $E_i \not\subseteq \widehat{M}$ (else v would have a strict z-calculus on the context (ε, w)). Then, $\forall i \in \{1, .., n\}$, $\exists k_i \in K$ such that $\forall j \geq k_i$: $E_i \not\subseteq M_j$ (because $E_i \not\subseteq \widehat{M} \Rightarrow \exists M_{k_i}$ such that $E_i \not\subseteq M_{k_i}$ and so $\forall j \geq k_i$, $E_i \not\subseteq M_j$). Let $k = max(k_i)_{1 \leq i \leq n}$; then $\forall i \in \{1, .., n\}$, $E_i \not\subseteq M_k$ and M_k is not z-free, which is a contradiction. $\square$

Definition 7. Let $X \subseteq \Sigma^*$. We denote by $Z\text{-}Hull(X)$ the set of minimal elements of $Z\text{-}Free(X)$, and we say that these elements are the z-free hulls of X.

Remark. For any language $X \subseteq \Sigma^*$, we have $Z\text{-}Free(X) = Z\text{-}Free(z\text{-}root(X))$ and then $Z\text{-}Hull(X) = Z\text{-}Hull(z\text{-}root(X))$. Indeed, this directly proceeds from the result in [7] which says that $X^\dagger$ is the smallest z-submonoid containing X.

We now propose to study the cardinality of $Z\text{-}Hull(X)$. First we prove, in corollary of the following proposition, that $Z\text{-}Hull(X)$ is finite if $z\text{-}root(X^\dagger)$ is finite.

Proposition 8. *Let $X \subseteq \Sigma^*$ with $z\text{-}root(X^\dagger)$ having a finite cardinality. If m is a longest word of $z\text{-}root(X^\dagger)$ and $M \in Z\text{-}Hull(X)$ then $\forall w \in z\text{-}root(M)$, we have: $|w| \leq |m|$.*

Proof. Let us suppose there exists a word $v \in z\text{-}root(M)$ such that $|v| > |m|$. Then $\forall u \in z\text{-}root(X^\dagger)$, there exists a z-factorization $(u_1, ..., u_n)$ of u on $z\text{-}root(M)$ such that $\forall i \in \{1, .., n\}$, $u_i \neq v$. But then $M' = ((z\text{-}root(M)\backslash\{v\})^\dagger$ which is also a z-free z-submonoid contains $z\text{-}root(X^\dagger)$ and then contains X (from the previous remark), which is a contradiction with the minimality of M. $\square$

Corollary 9. *Let $X \subseteq \Sigma^*$. If $z\text{-}root(X^\dagger)$ has a finite cardinality then $Z\text{-}Hull(X)$ also has a finite cardinality, and any element M of $Z\text{-}Hull(X)$ is constructible since $z\text{-}root(M)$ is finite.*

Now we give an example of a language which has an infinity of z-free hulls.

Proposition 10. *There exist rational languages having an infinity of z-free hulls. Some of these z-free hulls can be non algebraic languages.*

Proof. We give an example of such a language.
Let $X = abcd^*ef + bcd^*ef + bcd^*efg + abcd^*e + cd^*e + cd^*efg$. The language X is not a z-code because of, for example, the word $abcdddefg$ which has two different z-factorizations on X (see figure 8).

Fig. 8. A word of $X^\dagger$ having two z-factorizations

To build the z-free hulls of X we introduce the following notation: we denote by I any subset of $\mathbb{N}$ and by $\overline{I}$ the complement of I in $\mathbb{N}$. Then the set of z-free hulls of X contains the following languages: $\{Y_I^\dagger : I \subseteq \mathbb{N}\}$ where:

$$Y_I = bcd^*ef + abcd^*e + cd^*e + cd^*efg \cup \{bcd^ie : i \in I\} \cup \{cd^ief : i \in \overline{I}\}.$$

Indeed, $\forall I \subseteq \mathbb{N}$, Y_I verifies the three following points:

- $X \subseteq Y_I^\uparrow$. Specially, the words $abcd^nef$ and bcd^nefg admit the following z-factorizations on Y_I:
  If $n \in I$:

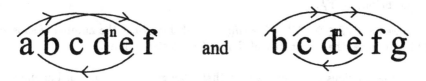

Fig. 9. z-factorizations on Y_I

If $n \in \bar{I}$:

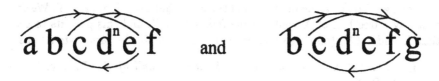

Fig. 10. z-factorizations on Y_I

- Y_I is a z-code. Indeed, the only words which would be able to have two z-factorizations on Y_I have the following form:

Fig. 11.

But, to have this configuration, we must have $n \in I \cap \bar{I}$ and this is, of course, impossible.
- It does not exist a language $Y \subseteq \Sigma^*$ (Y z-code) such that $X \subseteq Y^\uparrow \subset Y_I^\uparrow$. If the z-code Y exists, it contains necessarily the subset $E = bcd^*ef + abcd^*e + cd^*e + cd^*efg$ because $E \subset X \Rightarrow \forall u \in E$, $u \in Y^\uparrow$ and as we have $E \subset Y_I$, if there exists a word $u \in E$ having a non trivial z-factorization (i.e. using more than one step) on Y: $f_{u,Y} = (x_1, .., x_n)$, $n \geq 2$ then $\exists j \in \{1, .., n\}$ such

that $x_j \notin Y_I^\uparrow$ (since Y_I is a z-code).

Now, let us suppose there exists $i \in I$ such that $bcd^i e \notin Y$. In this case, we easily verify that to build a z-factorization of the word $abcd^i ef (\in X)$ on Y we must use, at least, one word u which does not belong to $Y_I^\uparrow$. The same situation occures if there exists $i \in \overline{I}$ such that $cd^i ef \notin Y$. The word $bcd^i efg$ cannot be z-factorized on Y without using, at least, one word which does not belong to $Y_I^\uparrow$.

So we have $\forall I \in \mathbb{N}$, $Y_I^\uparrow \in Z\text{-}Hull(X)$, and then, since the number of subsets of $\mathbb{N}$ is infinite, the cardinality of $Z\text{-}Hull(X)$ is also infinite.

On the previous example we note that, though X is a rational language, it has an infinity of z-free hulls. Some of them are non rational and even non algebraic like the following one :

If I is the set of prime numbers, then $\{bcd^i e : i \in I\}$ is not an algebraic language as Y_I and $Y_I^\uparrow$. $\square$

Now, we look at some properties of the z-free hulls of a language X. We show that any z-free hull of X contains the free hull of X, and this permits to prove that X and $H(X)$ (and also $root(H(X))$) have exactly the same z-free hulls.

Lemma 11. Let $X \subseteq \Sigma^*$ and $H(X)$ be the free hull of X. Then $\forall M \in Z\text{-}Hull(X)$, $H(X) \subseteq M$.

Proof. M is a z-free z-submonoid containing X, so M is z-stable. The property of stability is a particular case of the property of z-stability (it corresponds, in the first definition, to the first point with $u_2 u_4 = \varepsilon$), so M is also a stable and then a free submonoid containing X. Then M necessarily contains $H(X)$ from the definition of the free hull. $\square$

Proposition 12. Let $X \subseteq \Sigma^*$ and $H(X)$ be the free hull of X. Then $Z\text{-}Hull(X) = Z\text{-}Hull(H(X))$.

Proof. Let $M \in Z\text{-}Hull(X)$. From the previous lemma, $H(X) \subseteq M$. Let us suppose there exists a z-free z-submonoid M' such that: $H(X) \subseteq M' \subseteq M$. In this case we have: $X \subseteq M'$ which implies $M' = M$ and then $M \in Z\text{-}Hull(H(X))$. Now, if $M \in Z\text{-}Hull(H(X))$, M is a z-free z-submonoid such that $X \subseteq H(X) \subseteq M$. Let us suppose there exists a z-free z-submonoid M' such that: $X \subseteq M' \subseteq M$. As M belongs to $Z\text{-}Hull(H(X))$, we have for any z-free z-submonoid $M'' \subseteq M'$, $H(X) \not\subseteq M''$. From the previous lemma, $M'' \notin Z\text{-}Hull(X)$. We have a contradiction and so $M \in Z\text{-}Hull(X)$. $\square$

Corollary 13. Let $X \subseteq \Sigma^*$ and Y be the root of $H(X)$. Then $Z\text{-}Hull(X) = Z\text{-}Hull(Y)$.

Proof. $Z\text{-}Hull(Y) = Z\text{-}Hull(H(Y))$. Now $H(Y) = Y^* = H(X)$ so, we have $Z\text{-}Hull(Y) = Z\text{-}Hull(H(X)) = Z\text{-}Hull(X)$. $\square$

5   On a defect conjecture

Concerning the notion of free hull there exists a result, known as the defect theorem, which gives an interesting relationship between a language X and the root of its free hull. We recall it:

Theorem 14 [2]. *Let $X \subseteq \Sigma^*$ and Y be the root of the free hull of X. If X is a finite language but not a code then:*

$$Card(Y) < Card(X).$$

We would like to obtain an equivalent relation for the notion of z-free hull but the idea of the proof for the free hull case (see [2]) does not work in our situation. In fact, the theorem becomes false, in general, for the notion of z-free hull, as it is shown in the following proposition:

Proposition 15. *There exists languages having lower cardinality than the z-roots of some of their z-free hulls.*

Proof. We give an example of such a language. Let us consider the five-elements language $X = \{abcde, bcde, bcdef, abc, def\}$. By a demonstration closed to the proof of proposition 10 we can verify that the six-elements language $Y = \{abc, def, bcde, bcd, cd, c\}$ is the z-root of a z-free hull of X, and its cardinality is greater than the cardinality of X. The idea which permits to build this example is described on the figure 12 (here the dashed lines does not represent z-prefixes but the three added words). It consists to have the word bc z-prefix of the word $bcde$ by adding the three words bcd, cd and c. Then these three words belong to Y whereas only two words ($abcde$ and $bcdef$) of X are suppressed. □

Fig. 12. A key word to construct Y

On the other hand, there is still an open problem: "given a language X, we do not know, in the general case, if there always exists at least one z-free hull of X whose z-root has cardinality strictly lower than cardinality of X". In the previous proposition, the language $\{abc, def, bcde, bc\}^\dagger$ is a z-free hull of X whose z-root has a cardinality lower than the cardinality of X.

We conjecture that the answer of this problem is positive and we solve it for two-elements languages and for some three-elements languages which are degenerated cases. To prove it, we need the following result:

Proposition 16 [4]. *Let $X = \{u, v\}$ be a two-elements language in Σ^*. Then,*

$$X \text{ is a z-code iff } X \text{ is a code.}$$

Proposition 17. *Let $X \subseteq \Sigma^*$. If X is not a z-code and if X verifies one of the two following properties:*

- *X is a two-elements language,*
- *X is a three-elements language and X is not a code,*

then X has a unique z-free hull M, and $Card(z\text{-}root(M)) < Card(X)$

Proof.

- If X is a two-elements language but not a z-code, then X is not a code from the previous proposition. Then, from the defect theorem, we know that the root of the free hull of X contains only one word. A one-word language is trivially a z-code, then $H(X)$ is a z-free z-submonoid. By proposition 6 and lemma 11 we deduce that $H(X)$ is the only z-free hull of X.
- If X is a three-elements language and is not a code then, from the defect theorem, we have: $Card(root(H(X))) \leq 2$. So $root(H(X))$ is a two-elements code (at most) and then it is also a z-code from the previous proposition. By corollary 13 we can conclude that $[root(H(X))]^\dagger$ is the only z-free hull of X and its z-root contains, at most, two words. □

References

1. M. Anselmo, *Automates bilatères et codes zigzag*, Thèse L.I.T.P. 90-27 (1990).
2. J. Berstel and D. Perrin, *Theory of codes*, Academic Press, (1985).
3. J. Berstel, D. Perrin, J. Perrot and A. Restivo, *Sur le théorème du défaut*, J. Algebra 60, 169-180
4. Do Long Van, B. Le Saëc and I. Litovsky, *On coding morphisms for zigzag codes*, R.A.I.R.O. Theoretical Informatics and Applications 26-6 (1992), 565-580.
5. Do Long Van, B. Le Saëc and I. Litovsky, *Stability for the zigzag submonoids*, Theoretical Computer Science 108 (1993), 237-249.
6. B. Le Saëc, I. Litovsky and B. Patrou, *A more efficient notion of the zigzag stability*, To appear in R.A.I.R.O. Theoritical Informatic and Applications.
7. M. Madonia, S. Salemi and T. Sportelli, *On z-submonoids and z-codes*, R.A.I.R.O. Theoretical Informatics and Applications 25-4(1991), 305-322.
8. M.P. Schutzenberger, *Une théorie algébrique du codage*, Séminaire Dubreil-Pisot, 1955-1956, Exposé No 15.
9. B. Tilson, *The intersection of free submonoids is free*, Semigroup forum, 4 (1972), 345-350.

Contiguity Orders

Vincent BOUCHITTÉ[1], Abdelmajid HILALI[2],
Roland JÉGOU[3], Jean-Xavier RAMPON[4]

[1] LIP, CNRS URA 1398, Ecole Normale Supérieure de Lyon, 46 allée d'Italie, 69364 Lyon Cédex 07, France: email: vbouchit@lip.ens-lyon.fr
[2] Groupe L.M.D.I.-I.M.I., Université Claude Bernard-Lyon 1, 43, boulevard du 11 Novembre 1918, 69622 Villeurbanne Cédex, France; email: hilali@lan1.univ-lyon1.fr
[3] Centre SIMADE, Ecole des Mines de Saint-Etienne, 158 Cours Fauriel, 42023 Saint-Etienne Cédex 2, France; email: jegou@emse.fr
[4] IRISA, Campus de Beaulieu, 35042 Rennes Cédex, France; email: rampon@irisa.fr

Abstract. This paper is devoted to the study of contiguity orders i.e. orders having a linear extension L such that all upper (or lower) cover sets are intervals of L. This new class appears to be a strict generalization of both interval orders and N-free orders, and is linearly recognizable. It is proved that computing the number of contiguity extensions is $\#P$-complete, and that the dimension of height one contiguity orders is polynomially tractable. Moreover the membership is a comparability invariant on bi-contiguity orders.

1 Introduction

In [2] we introduced the class of lower-contiguity orders and we showed that such orders can be represented by translating line-segments in the plane. This latter problem has been defined and investigated first by Rival and Urrutia in [17], see also [3] for recent results on this topics. Roughly speaking a lower-contiguity order $P = (X, \leq_P)$ (resp. an upper-contiguity order) admits a linear extension L such that each lower-cover set (resp. upper-cover set) appears as an interval of L. This extension allows the encoding of the Hasse diagram of P in $\mathcal{O}(|X|)$. This notion is closely related to the notion of interval hypergraph. Every interval hypergraph $\mathcal{H} = (X, \mathcal{E})$ can be seen as an height one upper-contiguity order $P = (X \cup \mathcal{E}, \leq_P)$ where $e <_P x$ whenever $x \in e$, $x \in X$ and $e \in \mathcal{E}$, they are also called convex graphs by Brandstädt in [4]. Then upper-contiguity orders give a natural generalization of interval hypergraphs to arbitrary height orders. This generalization is extended by the notion of ordered interval hypergraphs introduced by Quilliot and Chao [16].

In Section 2, we give a characterization of upper-contiguity orders in terms of interval hypergraphs. This characterization leads to a linear time recognition algorithm and shows the complexity of such orders. While computing the dimension is difficult for contiguity orders it becomes polynomially tractable on height one contiguity orders for which the dimension is proved to be less than or equal to 3. In Section 3, we deal with bi-contiguity orders which are both upper–contiguity and lower–contiguity orders. This class contains interval orders, N–free orders and planar lattices. An interesting feature of this class is the membership comparability invariance.

Throughout this paper, all orders are assumed to be finite. In order to present our results, we have to recall some basic facts and to fix some notation related to an order $P = (X, \leq_P)$.

Let $x, y \in X$ with $x \neq y$. We say that x and y are *comparable* in P, when either $x \leq_P y$ or $y \leq_P x$ holds, otherwise they are said to be *incomparable* in P. The undirected graph $G = (X, E)$ where two vertices are joined by an edge iff the corresponding pair of elements of P is comparable, is called the *comparability graph* of P.

We say that x is *covered* by y (or y *covers* x) and we denote it by $x \prec_P y$ if $x <_P y$ and $\forall z \in X$ with $x \leq_P z \leq_P y$ we have $z = x$ or $z = y$. This binary relation defines the *Hasse diagram* of P.

A *chain* in P is a subset A of X such that every pair of distinct elements of A are comparable. The *length* of a chain in P is one less than its cardinality. The *height* of P is the maximum length of a chain in P.

An element $x \in X$ is called a *minimal* element (respectively, *maximal* element) if there is no element $y \in X$ with $y <_P x$ (respectively, $x <_P y$). We denote by $Min(P)$ (resp. $Max(P)$) the set of minimal (resp. maximal) elements in P.

We denote by $P^- = (X, \leq_{P^-})$ the *dual order* of P, that is $\forall x, y \in X, (x \leq_{P^-} y) \iff (y \leq_P x)$.

For $A \subseteq X$ we denote by $P_{|A}$ the *order restriction* of P on A. That is, $P_{|A} = (A, \leq_{P_{|A}})$ and $\forall x, y \in A, (x \leq_{P_{|A}} y) \iff (x \leq_P y)$. Let $A \subseteq X$, we denote by $P - A$ the order $P_{|(X-A)}$.

An order $Q = (X, \leq_Q)$ is said to be an *order extension* of P if $\forall x, y \in X$, $x \leq_P y \implies x \leq_Q y$. A *linear extension* L of P is an order extension of P which is a total order, then L can be written $x_1 < x_2 < \cdots < x_n$ if $|X| = n$. We denote by $\mathcal{L}(P)$ the set of all linear extensions of P. Let $G = (X, U)$ be an acyclic directed graph. A permutation σ of X such that $(x, y) \in U$ implies $\sigma(x) <_\sigma \sigma(y)$, is also called a linear extension of G.

We denote by $U_P(x)$ (resp. $D_P(x)$) the upset (resp. downset) of x in P, $U_P(x) = \{y \in X, x \leq_P y\}$ (resp. $D_P(x) = \{y \in X, y \leq_P x\}$) and by $UC_P(x)$ (resp. $LC_P(x)$) the upper cover (resp. lower cover) set of x in P, $UC_P(x) = \{y \in X, x \prec_P y\}$ (resp. $LC_P(x) = \{y \in X, y \prec_P x\}$).

2 Contiguity orders

Definition 1. An order $P = (X, \leq_P)$ is an upper–contiguity (resp. a lower–contiguity) order, if there exists $L \in \mathcal{L}(P)$ such that for any $x \in X$, $UC_P(x)$ (resp. $LC_P(x)$) is an interval of L, that is if $x \prec_P a$ and $x \prec_P b$ (resp. $a \prec_P x$ and $b \prec_P x$) and if $a \leq_L b$ then $\forall c$ such that $a \leq_L c \leq_L b$ we have $x \prec_P c$ (resp. $c \prec_P x$). Such an extension is called an upper–contiguity (resp. a lower–contiguity) extension of P.

Directly from the above definition we can deduce the following property on upper–contiguity extensions.

Property 2. *Any upper–contiguity order has an upper–contiguity extension starting with all its minimal elements.*

Since the dual of an upper–contiguity order is a lower–contiguity order, all properties and definitions given on upper–contiguity orders are dually available on lower–contiguity orders. The property of upper–contiguity is not stable by duality (see Figure 1). When an order is either an upper–contiguity order or a lower–contiguity order, it will be called a contiguity order.

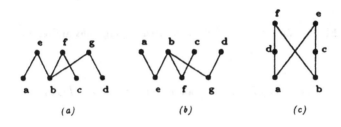

Fig. 1. (a) an upper–contiguity order; (b) a non upper–contiguity order; (c) nor upper–contiguity neither lower–contiguity order.

2.1 Characterization of upper–contiguity orders

In this section, we present a characterization of upper–contiguity orders in terms of interval hypergraphs (see Duchet [8] for a study of such hypergraphs). In [16], Quilliot and Chao gave a characterization of ordered interval hypergraphs leading to an $\mathcal{O}(|X|^3)$ recognition algorithm. Upper–contiguity order appears to be a subclass of ordered interval hypergraphs. Our characterization leads to a linear time recognition algorithm of such orders given by their transitive reduction.

An *hypergraph* $\mathcal{H} = (X, \mathcal{E})$ on a set X of *vertices*, is a family $\mathcal{E}$ of subsets of X, called the *edges*. Every hypergraph $\mathcal{H} = (X, \mathcal{E})$ can be represented by its incidence bipartite graph $B(\mathcal{H}) = (\mathcal{E}, X, E)$ where $ex \in E$ iff $e \in \mathcal{E}$ and $x \in X$ such that $x \in e$. Then the connected components of $B(\mathcal{H})$ are exactly those of $\mathcal{H}$. An hypergraph, $\mathcal{H}$ is an *interval hypergraph* if there exists a permutation $\sigma = (x_1, \cdots, x_n)$ of X such that each edge of $\mathcal{H}$ is an interval of σ. Every total order $" < "$ on X, such that $\mathcal{E}$ is a family of intervals of $(X, <)$ is said to be induced by $\mathcal{H}$. Given an order $P = (X, \leq_P)$, the *upper cover hypergraph* associated to P is $\mathcal{UC}(P) = (X - Min(P), \mathcal{E})$ where $\mathcal{E} = \{UC_P(x), x \in X\}$. The *connection graph* of $\mathcal{UC}(P)$ is the directed graph $\mathcal{C}(P)$ whose vertices are the connected components of $UC(P)$ (i.e. $\{\mathcal{H}_i, \ i \in I$ where $\mathcal{H}_i = (X_i, \mathcal{E}_i)\}$) and such that $(\mathcal{H}_i, \mathcal{H}_j)$ is an arc iff $i \neq j$ and there exists $x \in X_i$, $y \in X_j$ such that $x <_P y$.

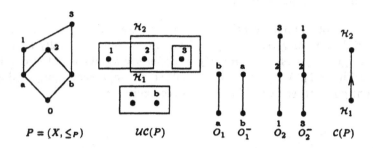

Fig. 2. An order P and its associated hypergraphs and graphs.

Theorem 3. *Let* $P = (X, \leq_P)$ *be an order, the two following conditions are equivalent:*

(i) *P is an upper–contiguity order*
(ii) *P satisfies the properties:*
    (a) *Each connected component $\mathcal{H}_i = (X_i, \mathcal{E}_i)$ of $UC(P)$, is an interval hypergraph, and there exists a total order on X_i induced by $\mathcal{H}_i$ which is a linear extension of $P_{|X_i}$.*
    (b) *$C(P)$ is acyclic.*

Proof. Let $\{\mathcal{H}_i, i \in \{1, \cdots, k\}\}$ with $\mathcal{H}_i = (X_i, \mathcal{E}_i)$ be the set of connected components of $UC(P)$. Assume that $P = (X, \leq_P)$ is an upper–contiguity order and that L is one of its upper–contiguity extension. By definition of L, each $\mathcal{H}_i$ is clearly an interval hypergraph and $L_{|X_i}$ is a linear extension of $P_{|X_i}$. Since the X_i's are pairwise disjoint intervals of L, the sequence $\sigma(1) < \cdots < \sigma(k)$, defined by $\sigma(i) < \sigma(j)$ if $X_{\sigma(i)}$ is on the left of $X_{\sigma(j)}$ in L, satisfies: $(X_{\sigma(i)}, X_{\sigma(j)})$ is in $C(P)$ implies $\sigma(i) < \sigma(j)$. Thus $C(P)$ is acyclic.
Let $P = (X, \leq_P)$ be an order satisfying the two conditions of part (ii). Without loss of generality, let us denote by O_i a total order on X_i induced by $\mathcal{H}_i$ which is a linear extension of $P_{|X_i}$. Since $\forall \ i, j \in \{1, \cdots, k\}$ with $i \neq j$ we have $X_i \cap X_j = \emptyset$, it is then clear that for each linear extension $(\sigma(1), \cdots, \sigma(k))$ of $C(P)$, the total order $O_{\sigma(1)} < \cdots < O_{\sigma(k)}$ is a linear extension of $P - Min(P)$. Thus, for any total order $O(Min(P))$ on $Min(P)$, the total order $O(Min(P)) < O_{\sigma(1)} < \cdots < O_{\sigma(n)}$ is a linear extension of P. By construction such a total order is an upper–contiguity extension of P.

Since the recognition of interval hypergraphs can be achieved in linear time using the PQ–tree data structure introduced by Booth and Lueker in [1], the above characterization alow to state:

Corollary 4. *Upper-contiguity orders are linearly recognizable in the size of their Hasse diagrams.*

Proof. Let $P = (X, \leq_P)$ be an order given by its upper covering relations. The set of elements is stored in an array where each element is pointing on its upper cover list. In order to compute the connected components of $\mathcal{UC}(P)$ we have to build the incidence bipartite graph of $\mathcal{UC}(P)$. This graph $B(\mathcal{UC}) = (U, V, E)$ is defined by $U = \mathcal{E}$, $V = X - Min(P)$ and $\forall (u, v) \in U \times V$, $(u, v) \in E$ iff $v \in u$. The data structure representing $B(\mathcal{UC})$ is in fact the same as the one representing P. Thus one can compute the connected components of $\mathcal{UC}(P)$ in linear time. Let $\mathcal{H}_i$ be a connected component of $\mathcal{UC}(P)$ taken as input for the PQ-tree construction algorithm. If the PQ-tree cannot be produced then P is not an upper-contiguity order. Otherwise the PQ-tree encodes all the compatible permutations. If one of them is a linear extension of $P_{|X_i}$ then it is easy to see that for any other compatible permutation σ, σ or σ^- is also a linear extension of $P_{|X_i}$. These operations can be done in linear time on all the connected components of $\mathcal{UC}(P)$. It remains to compute $C(P)$ and to check that this graph is acyclic, which can be done in linear time.

2.2 Structure of the connection graph

The previous characterization provide an interesting tool in order to capture the actual complexity of upper-contiguity orders. Indeed,

Theorem 5. *Any order can be obtained as the transitive closure of the connection graph of some upper-contiguity order.*

Proof. To an order $P = (X, \leq_P)$ we associate the order $P' = (X', \leq_{P'})$ defined as follows. For $UC_P(x) = \{y_1, \cdots, y_k\}$ we set $A(x) = \{x_1, \cdots, x_k\}$ if $k \geq 1$ and $A(x) = \{x\}$ if $k = 0$. The underlying set of P' is then $X' = X_0 \cup (\bigcup_{x \in X} A(x))$ where $X_0 = \{x_0, x \in Min(P)\}$. Now, for all $x \in Min(P)$ we set $UC_{P'}(x_0) = A(x)$ and for every $x \prec_P y_j$ we set $UC_{P'}(x_j) = A(y_j)$ (see Figure 3). By construction the connected components of $\mathcal{UC}(P')$ are exactly the sets $A(x)$. Note that this implies that P' is an N-free order (see Section 3.1 for a formal definition). We also denote by $A(x)$ the element of $C(P')$ corresponding to the set $A(x)$. If $x \prec_P y$ then there exists $x' \in A(x)$ such that $UC_{P'}(x') = A(y)$ so $(A(x), A(y))$ is an arc of $C(P')$. Conversely suppose that $(A(x), A(y))$ is an arc of $C(P')$ then there exists $x' \in A(x)$ such that $x' \leq_{P'} y'$ for every $y' \in A(y)$. For any $y' \in A(y)$ there exists a chain $(x' = z_1, \cdots, z_k = y')$ where $z_i \prec_{P'} z_{i+1}$, $1 \leq i < k$. If we denote by t_i the element of P such that $z_i \in A(t_i)$ then $(x = t_1, \cdots, t_k = y)$ is a chain of P, so $x \leq_P y$.

Corollary 6. *Computing the number of upper-contiguity extensions of an order is #P-complete.*

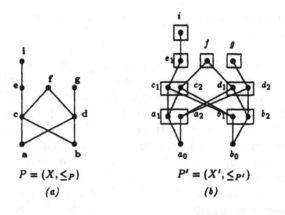

Fig. 3. An order P and its associated order P'.

Proof. Let $P = (X, \leq_P)$ be an order with a minimal element, say x^1, and such that $X = \{x^1, \cdots, x^n\}$. Let P' be the order associated to P as previously defined. By construction the connected components of $\mathcal{UC}(P')$ are exactly the sets $A(x)$. We also denote by $A(x)$ the element of $C(P')$ corresponding to the set $A(x)$. Any upper–contiguity extension of P' can be written $(x_0^1, \pi_{i_1}(x^{i_1}), \pi_{i_2}(x^{i_2}), \cdots, \pi_{i_n}(x^{i_n}))$ where $(A(x^{i_1}), \cdots, A(x^{i_n}))$ is any linear extension of $C(P')$ and $\pi_{i_j}(x^{i_j})$ is any permutation of the set $A(x^{i_j})$. Every linear extension of P induces exactly
$$\prod_{x \in X} |UC_P(x)|!$$
distinct upper–contiguity extensions of P'. Moreover two different linear extensions of P induce two disjoint sets of upper–contiguity extensions of P'. So the number of upper–contiguity extensions is exactly the product $|\mathcal{L}(P)| \prod_{x \in X} |UC_P(x)|!$. The result follows since the factor $\prod_{x \in X} |UC_P(x)|!$ is polynomially tractable in the size of P and since the computation of the number of linear extensions of an order is $\#P$–complete (see Brightwell and Winkler [5]).

2.3 Dimension of height one upper–contiguity orders

A *realizer* of P is a subset of $\mathcal{L}(P)$ such that the intersection of all its elements is P. The minimum cardinality of a realizer is called the *dimension* of P and denoted by $dim(P)$. For an up to date monograph on finite poset dimension theory see Trotter [21].

Height one upper–contiguity orders can be seen as bipartite graphs having the consecutive one's property for their adjacency matrix. That is, an height one order is an upper–contiguity order iff its upper cover hypergraph is an interval hypergraph. The class of height one upper–contiguity orders is a strict generalization of that of height one 2–dimensional orders. Indeed, 2–dimensional orders have been characterized, by Spinrad *et al.* in [20], as being orders whose adjacency matrix has both the consecutive one's property and the enclosure property. That is there exists an ordering τ of $Max(P)$ such that for every $x \in P, UC_P(x)$ is an interval of τ and if $UC_P(x) \subseteq UC_P(y)$ then $UC_P(y) - UC_P(x)$ is also an interval of τ. Height one upper–contiguity orders are also called convex graphs in Brandstädt [4] and the jump number problem has been proved to be polynomially tractable by Dahlhaus [6].

Theorem 7. *The dimension of height one upper–contiguity orders is at most 3.*

Proof. Let $P = (X, \leq_P)$ be an upper–contiguity order of height one, let τ be an upper–contiguity extension of P putting $Min(P)$ below $Max(P)$. We denote $UC_P(x)$ by the interval $[L_x, R_x]$. We associate $R_{x'}$ to R_x, which is the rightmost element R_z according to τ in $Max(P)$, such that there exists z in $Min(P)$ with $[L_x, R_x] \subseteq [L_z, R_z]$. We define, on the same underlying set as P, the order P_1 by setting $x \prec_{P_1} y$ iff $y \in [L_x, R_{x'}]$. Clearly P_1 is an extension of P. P_1 is a two dimensional order since $UC_{P_1}(x)$ is an interval of τ for every x in $Min(P_1)$ and $UC_{P_1}(y) - UC_{P_1}(x)$ is also an interval whenever $UC_{P_1}(x) \subseteq UC_{P_1}(y)$. In order to do this, we prove that $R_{x'} = R_{y'}$. By assumption we have $R_{x'} \leq_\tau R_{y'}$. Moreover, by definition there exists z in $Min(P_1)$ such that $[L_y, R_{y'}] \subseteq [L_z, R_z]$ with $R_{y'} = R_z$ (z can be y), but then $[L_x, R_x]$ is also contained in $[L_z, R_z]$ so $R_{x'} \geq_\tau R_{y'}$. We now define P_2 an other extension of P by adding the comparabilities $z \prec_{P_2} x$ for every $z \in [R_x, R_{x'}]$ and this for every $x \in Min(P)$. That is, all comparabilities added to P in order to obtain P_1 are reversed in P_2. We claim that P_2 is acyclic. Indeed, assume that $(x_1, y_1, x_2, y_2, \cdots, x_k, y_k, x_1)$ is a directed cycle of P_2 with $x_i \in Min(P)$ and $y_i \in Max(P)$ for $1 \leq i \leq k$. Since (y_i, x_{i+1}) has been added in P_2 then we have $y_{i+1} <_\tau y_i$ for $1 \leq i \leq k$ (where $x_{k+1} = x_1$ and $y_{k+1} = y_1$), that is a cycle in τ: a contradiction. We get a realizer of P by taking a realizer of P_1 and any linear extension of P_2, so $dim(P) \leq 3$.

Since orders of dimension two are polynomially recognizable (Spinrad gives in [18] an $\mathcal{O}(n^2)$ recognition algorithm) the dimension of height one upper–contiguity orders is polynomially tractable.

Note that the previous theorem has been obtained independently by Sritharan [19].

3 Bi–contiguity orders

In this paragraph we study a strict subclass of contiguity orders for which the contiguity property holds for both upper covers and lower covers. However, this restriction is not drastic since interval orders and N–free orders still belong to this class. Moreover we obtain the membership comparability invariance.

Definition 8. An order $P = (X, \leq_P)$ is a bi–contiguity order if it is both an upper–contiguity and a lower–contiguity order.

3.1 Some classes of bi–contiguity orders

The first relevant class we speak about is that of N–free orders which yields that the computation of the dimension is an NP–complete problem for bi–contiguity orders. The NP–completeness of the N–free orders dimension has been shown in [13].

$P = (X, \leq_P)$ is an N-free order iff it has no "N" as subdiagram, where "N" is the ordered set on four elements $\{a, b, c, d\}$ defined by $a <_P c, b <_P c$ and $b <_P d$. As it has been established by Habib and Jégou in [10] P is an N–free order iff it can be obtained by a sequence of parallel and quasi–series compositions from the one–element order. Given $P_1 = (X_1, \leq_{P_1})$ and $P_2 = (X_2, \leq_{P_2})$ be orders where $X_1 \cap X_2 = \emptyset$. We define the parallel composition, denoted by $P_1 \oplus P_2$, as being the disjoint sum of P_1 and P_2. If $A \subseteq Max(P_1)$ and $B \subseteq Min(P_2)$, the quasi–serie composition, denoted by $(P_1, A) \otimes (P_2, B)$, generates the order P on $X_1 \cup X_2$ obtained by setting $UC_P(x) = B \ \forall x \in A, UC_P(x) = UC_{P_1}(x) \ \forall x \in X_1 - A$ and $UC_P(x) = UC_{P_2}(x) \ \forall x \in X_2$.

Using Theorem 3, we are now able to show that N–free orders are bi–contiguity orders giving by the same way a new proof of the lower–contiguity of N–free orders stated in [2].

Proposition 9. *N-free orders are bi-contiguity orders.*

Proof. Let $P = (X, \leq_P)$ be an N–free order. By duality it is sufficient to establish the upper–contiguity of P. We proceed by induction using the parallel and quasi-series characterization. The only point to check is the compatibility of the upper–contiguity property with the quasi–serie composition. Assume that $P = (P_1, A) \otimes (P_2, B)$ and let L_1 be an upper–contiguity extension of P_1 and let L_2 be an upper–contiguity extension of P_2 starting with B. Then $L_1 < L_2$ is a suitable upper–contiguity extension for P.

The second class we are concerned with is that of interval orders. This implies that the computation of jump number is NP–complete for bi–contiguity orders. The NP–completeness of the jump number problem for interval orders has been shown in [14].

$P = (X, \leq_P)$ is an *interval order* iff we can associate to X a collection $(I_x)_{x \in X}$ of non void intervals of the real line such that $x \leq_P y \Longleftrightarrow I_x$ *lies strictly on the left of* I_y (i.e. $\forall a \in I_x, \forall b \in I_y$, we have $a <_R b$ where "$<_R$" is the usual order on real numbers). Equivalently P is an interval order iff there exists a linear extension of P such that, according to this numbering, the downsets are linearly ordered with regard to inclusion (see for example the survey of Möhring [15]).

Proposition 10. *A linear extension of an order P, such that according to this numbering, the downsets are linearly ordered with regard to inclusion, is an upper–contiguity extension of P.*

Proof. Let L be a linear extension of $P = (X, \leq_P)$ satisfying: $x \leq_L y \Longrightarrow$ $D_P(x) \subseteq D_P(y))$. Assume that L is not an upper–contiguity extension of P. Then, there exists $x, y \in X$ such that $x \notin UC_P(y)$ and $a <_L x <_L b$ where a is the leftmost (in L) upper cover of y in P while b is its rightmost one. Since $a <_L x$, $D_P(a) \subseteq D_P(x)$ then $y <_P x$ and therefore $\exists z \in UC_P(y)$ such that $z \leq_P x$. So, $a \leq_L z <_L x <_L b$ and we have a contradiction with $b \in UC_P(y)$ because $z \in D_P(x)$ and $D_P(x) \subseteq D_P(b)$.

It is then straightforward to get the bi–contiguity property of interval orders using this relationship between the upset inclusion property and the lower–contiguity property. As a consequence we get a new proof of the lower–contiguity of interval orders stated in [2].

Corollary 11. *Interval orders are bi–contiguity orders.*

Lattices and 2–dimensional orders are not bi–contiguity orders, for the first class take the eight element boolean algebra and for the second one see the poset given in Figure 1.(c). However, 2–dimensional lattices are bi–contiguity orders. Recall that a lattice is 2–dimensional iff it is planar. We will use definitions on planar orders and lattices mentioned by Kelly and Rival in [12].

Let $P = (X, \leq_P)$ be an order, an element y in P is said to be *doubly irreducible* if it has both an unique upper cover and an unique lower cover in P. P is *planar* if it has a planar embedding, where a planar embedding $e(P)$ of P consists of:

(i) An injection $x \longmapsto \overline{x}$ from X to $\mathbb{R}^2$ such that $\overline{x}$ is strictly below $\overline{y}$ whenever $x <_P y$.
(ii) Straight line segments $\overline{xy}$, connecting $\overline{x}$ and $\overline{y}$ whenever $x-<_P y$. These segments do not intersect, except possibly at their endpoints.

For each $x \in X$, a planar embedding $e(P)$ induces a strict linear ordering L_x on the set $UC_P(x)$ defined by $z <_{L_x} t$ iff $\widehat{\overline{x}\overline{z}} > \widehat{\overline{x}\overline{t}}$, for $z, t \in UC_P(x)$. As usual, $\widehat{\overline{x}\overline{z}}$ is the angle taken counterclockwise when the Euclidean plane is centered on $\overline{x}$. In a planar embedding of a lattice, there exits two maximal chains from the infimum to the supremum such that all the elements of the lattice belong to the interior of the Jordan curve defined by these two chains. One of these chains is the right boundary while the other is the left boundary (the words left and right are understood as usual).

We summarize some properties stated in [12] in the following:

(i) *Every finite planar lattice with at least 3 elements has a doubly irreducible element distinct from the top and the bottom on the right boundary of any planar embedding.*
(ii) *Let y be a doubly irreducible element in a planar lattice P, then $P - \{y\}$ is still a planar lattice.*

(iii) *Every non infinite face of $e(P)$ consists in two paths of $e(P)$ whose bounds are the only elements common to both its left and right boundary and whose interior is empty.*

If P is planar lattice with planar embedding $e(P)$ and y is a doubly irreducible element in P on the right boundary such that x–$<_P$ y–$<_P$ z, then P has a planar embedding $e'(P)$ such that $\overline{x}$, $\overline{y}$ and $\overline{z}$ lie on the same line–segment and for any t the strict linear ordering L_t of $UC_P(t)$ remains the same. Note that $e'(P)$ can be obtained by horizontally translating to the right $\overline{x}$ and $\overline{z}$ or $\overline{y}$. Thus $P - \{y\}$ has a planar embedding denoted $e^*(P - \{y\})$ where $e^*(P - \{y\})$ is $e'(P)$ in which $\overline{y}$, $\overline{xy}$ and $\overline{yz}$ have been deleted and $\overline{xz}$ has been added if necessary.

Theorem 12. *Planar lattices are bi–contiguity orders.*

Proof. Let $P = (X, \leq_P)$ be a planar lattice. Using duality, it is sufficient to show that for any planar embedding $e(P)$ there exists an upper–contiguity extension of P such that the ordering induced on each upper cover set is exactly the strict linear ordering induced by $e(P)$. If $|X| \leq 3$ this is clearly true. Assume that $|X| = n$ with $n \geq 4$. There exists a doubly irreducible element y on the right boundary of the planar embedding $e(P)$ of P. We denote by x and z its lower and upper covers. By induction hypothesis, let L be an upper–contiguity extension of $P - \{y\}$ satisfying the lemma with $e(P - \{y\}) = e^*(P - \{y\})$. If $UC_P(x) = \{y\}$, we obtain L' by inserting y immediately after x in L. It is sufficient to prove that L' is an upper–contiguity extension. This property comes from the upper–contiguity of L' on $P - \{y\}$ and from that x is necessarily the rightmost element of $UC_P(t)$ in L whenever $x \in UC_P(t)$. Now if $|UC_P(x)| \geq 2$, consider the non infinite face of $e(P)$ having y in its right boundary and let t be the upper cover of x on its left boundary. L' is obtained by inserting y immediately after t in L. Since L is an upper–contiguity extension of $P - \{y\}$ and t is the rightmost element of $UC_{P-\{y\}}(u)$ in L whenever $t \in UC_{P-\{y\}}(u)$ then L' is an upper–contiguity extension of P.

3.2 Comparability invariance of bi–contiguity orders

A parameter or a function is said to be a *comparability invariant* if it has the same value for all orders with the same comparability graph (see Kelly [11] for a survey on comparability graphs and comparability invariants). An efficient tool for proving comparability invariance, in the finite case, has been given by Dreesen *et al.* in [7]: "A parameter α of a finite order is a comparability invariant if and only if for each pair of finite orders P and Q and for each element x of P, we have $\alpha(P_x^Q) = \alpha(P_x^{Q^-})$". where P_x^Q is the order obtain by substituting x by Q in P.

A subset H of X is an *autonomous* set of P if $\forall\ h_1, h_2 \in H$, $U_P(h_1) - H = U_P(h_2) - H$ and $D_P(h_1) - H = D_P(h_2) - H$. Given an autonomous set H of P, $P = P^H = \sum_{x \in P'} Q_x$ with $P' = (X', \leq_{P'})$ where $X' = \{z\} \cup (X - H)$ for a

$z \notin X$ (note that $P^H = P_z'^H$), $\forall x, y, t \in X' - \{z\}$, $(x \leq_{P'} y) \Longleftrightarrow (y \leq_P x)$, $(z <_{P'} t) \Longleftrightarrow (\forall h \in H, h <_P t)$ and $(t <_{P'} z) \Longleftrightarrow (\forall h \in H, t <_P h)$. The order family is $\{Q_x = (Y_x, \leq_{Q_x}), x \in X'\}$ where $Q_z = (H, \leq_{P_{|H}})$ and otherwise $Y_x = \{x\}$. We denote by P^{H^-} the order obtained by substituting Q_z by its dual Q_z^- in the previous order family.

Lemma 13. *Let $P = (X, \leq_P)$ be an upper–contiguity order and H be an autonomous set of P, then there exists an upper–contiguity extension L of P such that:*

(i) $Min(P_{|H})$ is an interval of L.
(ii) $H - Min(P_{|H})$ is an interval of L.

That is, L can be written as $L^s < L_{|Min(P_{|H})} < L^m < L_{|(H-Min(P_{|H}))} < L^e$.

Proof. (i) Let L be an upper–contiguity extension of P starting with all minimal elements of P. If $Min(P_{|H}) \subseteq Min(P)$ there is nothing to prove. Otherwise we write $L = (x_1, \cdots, x_n, a_1, \cdots, a_k, y_1, \cdots, y_m)$ with $Min(P_{|H}) \subseteq \{a_1, \cdots, a_k\}$ and $a_1, a_k \in Min(P_{|H})$. If $Min(P_{|H}) = \{a_1, \cdots, a_k\}$ we are done. If not, let $a_i \in \{a_2, \cdots, a_{k-1}\} - Min(P_{|H})$, i being minimal with this property. We denote by l the smallest subscript greater than i such that a_{l+1} is in $Min(P_{|H})$. Assume that $L' = (x_1, \cdots, x_n, a_i, \cdots, a_l, a_1, \cdots, a_{i-1}, a_{l+1}, \cdots, a_k, y_1, \cdots, y_m)$ is not an upper–contiguity extension of P. There are only two cases. First, the sequence $a_i, \cdots, a_l$ is carried into an interval representing an upper cover set $UC_P(y)$. Clearly $a_1 \in UC_P(y)$ and since H is an autonomous set, $a_k \in UC_P(y)$. Thus, $a_i, \cdots, a_l \in UC_P(y)$ because L is an upper–contiguity extension of P, which is a contradiction. Secondly, the sequence $a_i, \cdots, a_l$ is carried out of an interval representing an upper cover set $UC_P(z)$. Clearly $a_{l+1} \in UC_P(z)$ and since H is an autonomous set $a_1, \cdots, a_{i-1} \in UC_P(z)$, which is a contradiction.

(ii) Let L be an upper–contiguity extension of P satisfying (i), then we can write $L = L^s < L_{|Min(P_{|H})} < L^m < L_1 < L_1' < L_2 \cdots < L_{k-1}' < L_k < L''$ with

$$\bigcup_{i \in \{1, \cdots, k\}} L_i = H - Min(P_{|H}), \quad \left(\bigcup_{i \in \{1, \cdots, k-1\}} L_i' \right) \cap H = \emptyset \text{ and } L'' \cap H = \emptyset.$$

Since H is an autonomous set of P, then on one hand elements of any L_i' are incomparable with elements of H and on the other hand there is no element x in X having upper covers in both $\bigcup_{i \in \{1, \cdots, k\}} L_i$ and $L'' \cup \left(\bigcup_{i \in \{1, \cdots, k-1\}} L_i' \right)$.

So $L' = L^s < L_{|Min(P_{|H})} < L^m < L_1 \cdots < L_k < L_1' \cdots < L_{k-1}' < L''$ is an upper–contiguity extension of P.

Theorem 14. *Belonging to the bi–contiguity order class is a comparability invariant.*

Proof. Given a bi–contiguity order P, using the result given by Dreesen *et al.*, it is sufficient to prove that P^{H^-} is a bi–contiguity order for any autonomous set H of P. Let $L = L^s < L_{|Min(P_{|H})} < L^m < L_{|H-Min(P_{|H})} < L^e$ be an upper–contiguity extension of P satisfying Lemma 13. Since P is a bi–contiguity order, it is clear that $P_{|H}$ is also a bi–contiguity order. Thus, there exists Γ a lower–contiguity extension of H ending with all the maximal elements of H. We claim that $L' = L^s < \Gamma^-_{|Max(P_{|H})} < L^m < \Gamma^-_{|H-Max(P_{|H})} < L^e$ is an upper–contiguity extension of P^{H^-}. The claim is achieved since Γ^- is an upper–contiguity extension of H^- and

(i) $\forall\, x \in X - H$, for x such that $UC_P(x) \cap Min(P_{|H}) = \emptyset$, we have $UC_{P_{H^-}}(x) = UC_P(x)$; otherwise, we have $UC_{P_{H^-}}(x) = (UC_P(x) - Min(P_{|H})) \cup Max(P_{|H})$.

(ii) $\forall\, x \in H$, for $x \notin Min(P_{|H})$, we have $UC_{P_{H^-}}(x) = LC_P(x)$; otherwise, we have $UC_{P_{H^-}}(x) = UC_P(y)$ for any $y \in Max(P_{|H})$.

A dual construction ensures the existence of a lower–contiguity extension for the order P^{H^-}.

References

1. K.S. Booth, G.S. Lueker, Testing for the Consecutive Ones Property, Interval Graphs and Graph Planarity Using PQ–Tree Algorithms. Journal of Computer and System Science 13, 335–379 (1976).

2. V. Bouchitté, R. Jégou, J.X. Rampon, Ordres représentables par des translations de segments dans le plan. C.R. Acad. Sci. Paris, t. 315, Série I, p. 1427–1430, 1992.

3. V. Bouchitté, R. Jégou, J.X. Rampon, Line–Directionality of Orders. Order 10: 17–30, 1993.

4. A. Brandstädt, The Jump Number Problem for Biconvex Graphs and Rectangle Covers of Rectangular Regions. Lecture Notes in Computer Science N° 380, pp. 68–77, Springer–Verlag 1990.

5. G. Brightwell, P. Winkler Counting Linear Extensions. Order 8: 225–242, 1991.

6. E. Dahlhaus, The Computation of the Jump Number of Convex Graphs. Lecture Notes in Computer Science N° 831, pp. 176–185, Springer-Verlag 1994.

7. B. Dreesen, W. Poguntke, P. Winkler Comparability Invariance of the Fixed Point Property. Order 2: 269–274, 1985.

8. P. Duchet, Représentations, noyaux en théorie des graphes et hypergraphes. Doctorat d'État ès Science Paris VI, 1979.

9. M.R. Garey, D.S. Johnson, Computers and Intractability: A Guide to the Theory of NP-Completeness. W.H. Freeman and Company, 1979.

10. M. Habib, R. Jégou, N–free Posets as Generalizations of Series–Parallel Posets. Discrete Applied Mathematics 12 (1985) 279–291.

11. D. Kelly, Comparability Graphs. I. Rival (ed.), Graphs and Order, 3–40. 1985 by D.Reidel Publishing Company.

12. D. Kelly, I. Rival, Planar Lattices. Canadian Journal of Mathematics, Vol. XXVII N° 3, (1975), 636–665.

13. H.A. Kierstead, S.G. Penrice, Computing the Dimension of N–Free Ordered Sets is NP–complete. Order 6: 133–136, 1989.

14. J. Mitas, Tackling the Jump Number of Interval Orders. Order 8: 115–132, 1991.

15. R.H. Möhring, Computationally Tractable Classes of Ordered Sets. I. Rival (ed.), Algorithms and Order, 105–193. 1989 by Kluwer Academic Publishers.

16. A. Quilliot, S.X. Chao, Algorithmic Characterizations of Interval Ordered Hypergraphs and Applications. Discrete Applied Mathematics 51 (1994) 159–170.

17. I. Rival et J. Urrutia, Representing Orders on the Plane by Translating Convex Figures. Order 4: 319–339, 1988.

18. J. Spinrad, On Comparability and Permutation Graphs. SIAM Journal of Computing Vol. 14, N° 3, August 1985, 658–670.
19. J. Spinrad, *Private communication*.
20. J. Spinrad, A. Brandstädt, L. Stewart, Bipartite Permutation Graphs. Discrete Applied Mathematics 18 (1987) 279–292.
21. W.T. Trotter, Combinatorics and Partially Ordered Sets, Dimension Theory. The Johns Hopkins University Press. 1992.

Worst–Case Analysis for On–Line Data Compression *

József Békési[1], Gábor Galambos[1], Ulrich Pferschy[2], Gerhard J. Woeginger[3]

[1] JGYTF, Department of Computer Science, P.O. Box 396,
H-6720 Szeged, Hungary
[2] Universtät Graz, Institut für Statistik und Operations Research,
Herdergasse 11,
A-8010 Graz, Austria
[3] TU Graz, Institut für Mathematik B, Steyrergasse 30,
A-8010 Graz, Austria

Abstract. On–line text–compression algorithms are considered, where compression is done by substituting substrings of the text according to some fixed dictionary (code book). Due to the long running time of optimal compression algorithms, several on–line heuristics have been introduced in the literature. In this paper we continue the investigations of Katajainen and Raita [4]. We complete the worst–case analysis of the longest matching algorithm and of the differential greedy algorithm for several types of special dictionaries and derive matching lower and upper bounds for all variants of the problem in this context.

Keywords: On–line algorithm, data compression, worst–case analysis.

1 Introduction

Certain progresses made in computer technology during the last few years — both in the field of hardware and in the field of software — strongly require large amounts of data to be moved between various components or to be stored in bounded capacity devices with high speed and reliability. All these operations need *data transfers*, either between two computers or between two parts of the same computer.

Obviously, the longer the length of the transmitted bit–string, the longer will be the transmission time and the larger the probability of making a mistake during the transfer. Essentially, there are two possibilities to increase the performance of the transfer: Either to use better (and more expensive) hardware, or to *compress* the data before the transfer.

One possibility to perform a compression on a given source–string is to substitute pieces of the string with the help of a *dictionary*. A dictionary consists

* This research was partially supported by the Spezialforschungsbereich F003 "Optimierung und Kontrolle", Projektbereich Diskrete Optimierung, by the Christian Doppler Laboratorium für Diskrete Optimierung and by a grant from the Hungarian Academy of Sciences (OTKA, No. T 016349).

of pairs of strings over a finite alphabet (*source-word, code-word*), which are used to replace substrings in the source-string. We will consider only methods which use a *static dictionary*, i.e. a fixed dictionary that cannot be changed or extended during the encoding–decoding procedure. Our aim is to translate (encode) the source-string with the help of the dictionary strings into a code-text with minimal length; in other words, we want to find a space-optimal encoding procedure.

The above setup is equivalent to the problem of finding a *shortest path* in a related directed edge–weighted graph (cf. Schuegraf and Heaps [5]): For a source-string $S = s_1 s_2 \ldots s_n$, we define a graph $N = (V, A)$ on the vertex-set $V = \{v_0, v_1, \ldots, v_n\}$. There is an edge $(v_i, v_{i+d}) \in A$ iff in the dictionary there exists a pair (source-word, code-word) such that the source-word consists of d characters that exactly match the original source-string in the positions $i + 1, \ldots, i + d$. The weight of this edge is the number of bits in the corresponding code-word. It is easily seen that a shortest path from v_0 to v_n in the graph N corresponds to an optimal compression of the source-string S.

In case that the graph has many *cut vertices* (i.e. vertices which divide the original problem into independent subproblems) and in case that all these sub-problems are reasonably small, we can indeed solve the problem efficiently and compute the *optimal* encoding. However, in general this will hardly be the case and a shortest path algorithm cannot be applied, as it takes too much time and space to compute an optimal solution for very long strings. Therefore, *heuristics* have been developed to derive near optimal solutions.

The earlier developed heuristics (for example the *longest fragment first heuristic (LFF)* cf. Schuegraf and Heaps [5] and a combined approximation algorithm cf. Katajainen and Raita [3]) have not been deeply analyzed and only experimental results on their performance have been reported.

A very fast type of heuristic is given by an *on-line* approach: An *on-line data compression algorithm* starts at the source vertex v_0, examines all outgoing edges and chooses one of them according to some given rule. Then the algorithm continues this procedure from the vertex reached via the chosen edge. There is no possibility to undo a decision made at an earlier time, and no backtracking is allowed.

Of course, an on-line heuristic will generate only a suboptimal compression. In order to know how much worse its performance can possibly be compared to the optimal compression a *worst-case analysis* has to be performed.

The *worst-case behaviour* of a heuristic is generally measured by its *asymptotic worst-case ratio* which is defined as follows: Let $D = \{(w_i, c_i) : i = 1, \ldots, k\}$ be a static dictionary and consider an arbitrary data compression algorithm A. Let $A(D, S)$ resp. $OPT(D, S)$ denote the compressed string produced by algorithm A resp. the optimal encoding for a given source-string S. The length of these codings will be denoted by $\|A(D, S)\|$ resp. $\|OPT(D, S)\|$. Then the *asymptotic worst–case ratio* of algorithm A is defined as

$$R_A(D) = \lim_{n \to \infty} \sup \left\{ \frac{\|A(D,S)\|}{\|OPT(D,S)\|} : S \in S(n) \right\},$$

where $S(n)$ is the set of all text–strings with exactly n characters.

The first worst–case analysis for an on–line data compression method was performed by Katajainen and Raita [4]. They analyzed two simple on–line heuristics, the *longest matching* and the *differential greedy* algorithm which will be defined exactly in Section 2.

In this paper, we will continue the investigations of Katajainen and Raita. We answer several problems that remained open and provide matching upper and lower bounds for special dictionary types. Furthermore, the *prefix* property will be introduced and analyzed.

The paper is organized as follows: Section 2 gives definitions and a general result. The prefix property is dealt with in Section 3. In Section 4 we investigate the differential greedy algorithm. Among other results, we show some improvements on the results in [4] for those cases where the given worst–case bounds were not tight. Some concluding remarks are presented in Section 5.

2 Preliminaries

Four parameters are used in the literature to investigate and state asymptotic worst–case ratios:

$$Bt(S) = \text{length of a symbol of the source–string } S \text{ in bits}$$
$$\text{(identical for all symbols of } S)$$
$$lmax(D) = \max\{|w_i| \; i = 1, \ldots, k\}$$
$$cmin(D) = \min\{\|c_i\| \; i = 1, \ldots, k\}$$
$$cmax(D) = \max\{\|c_i\| \; i = 1, \ldots, k\},$$

where $|w_i|$ denotes the length of a string w_i in characters and $\|c_i\|$ the length of a code word c_i in bits. If the meaning is clear from the context, we will simply denote the bit length of each input character by Bt and omit the reference to the dictionary by using $lmax$ instead of $lmax(D)$.

Not surprisingly, the worst–case behaviour of a heuristic strongly depends on the features of the available dictionary. The following types of dictionaries will be examined in this paper:

A dictionary is called *general* if it contains all symbols of the input alphabet as source–words (this ensures that every heuristic will in any case reach the sink of the underlying graph and thus will terminate the encoding with a feasible solution). In this paper we will only deal with general dictionaries.

A general dictionary is called

1. **code uniform** dictionary, if all code–words are of equal length (i.e. $\|c_i\| = \|c_j\|$, $1 \leq i, j \leq k$),

2. **nonlengthening** dictionary, if the length of any code–word never exceeds the length of the corresponding source–word (i.e. $\|c_i\| \leq |w_i|Bt$, $1 \leq i \leq k$),

3. **suffix** dictionary, if with every source–word w also all of its proper suffixes are source–words (i.e. if $w = \omega_1\omega_2\cdots\omega_q$ is a source–word $\Rightarrow \omega_h\omega_{h+1}\cdots\omega_q$ is a source–word for all $2 \leq h \leq q$),

4. **prefix** dictionary, if with every source–word w also all of its proper prefixes are source–words (i.e. if $w = \omega_1\omega_2\cdots\omega_q$ is a source–word $\Rightarrow \omega_1\omega_2\cdots\omega_h$ is a source–word for all $1 \leq h \leq q - 1$).

Two on–line heuristics were analyzed in [4] with respect to their worst–case behaviour:

- The **longest matching** heuristic LM chooses at each vertex of the underlying graph the longest outgoing arc, i.e. the arc corresponding to the encoding of the longest substring starting at the current position.

- The *greedy* heuristic, which we will call **differential greedy** DG (introduced by Gonzalez–Smith and Storer [2]) chooses at each position the arc implied by the dictionary entry (w_i, c_i) yielding the maximal "local compression", i.e. the arc maximizing $|w_i|Bt - \|c_i\|$. Ties are broken arbitrarily.

Katajainen and Raita [4] analyzed the worst–case behaviour of LM for dictionaries that are code uniform / nonlengthening / suffix and they derived tight bounds for all eight combinations of these properties. For the DG heuristic bounds for general and nonlengthening dictionaries were derived, the suffix property remained open for the most part.

We will first show a general upper bound which is valid for any compression algorithm.

Theorem 1 *Let D be a **general** dictionary. Then for any encoding algorithm A*

$$R_A(D) \leq \frac{(lmax - 1)cmax}{cmin}.$$

Proof. We consider source–strings where the A–path and an OPT–path are vertex disjoint with common endvertices v and $\bar{v}$. Obviously, any given string can be partitioned into a number of sub-strings with this property and the worst–case ratio for every sub-string carries over to their combination. The number of characters between these endvertices is denoted by j and the number of vertices on the OPT–path between them by r. Because the two paths are vertex disjoint and each arc "consumes" at least one character the number of arcs of the A–path between v and $\bar{v}$ is at most $j - r$. However, the arcs on the optimal path have to cover the distance between v and $\bar{v}$. Hence, we have

$$r \geq \left\lceil \frac{j}{lmax} \right\rceil.$$

This yields

$$R_A(D) \leq \frac{(j-r)cmax}{(r+1)cmin}$$

$$\leq \left(\frac{j}{r} - 1\right)\frac{cmax}{cmin}$$

$$\leq \frac{(lmax - 1)cmax}{cmin}.$$

□

Examples where this general bound is attained by the Longest Matching and the Differential Greedy algorithm are given by Katajainen and Raita in [4], where an identical upper bound was shown for the Differential Greedy heuristic in a significantly longer proof.

Setting $cmax = cmin$ and using arguments as above we get the following

Corollary 2 *Let D be a* **code–uniform** *dictionary. Then for any encoding algorithm A*

$$R_A(D) \leq lmax - 1.$$

□

3 Prefix Dictionaries

Katajainen and Raita conjectured in [4] that "the coding result for prefix dictionaries can be weaker than the bounds derived for suffix dictionaries".

In this section, we will derive tight worst–case bounds for all four combinations of the property *prefix* with the properties code uniform / nonlengthening. We will show that both the longest matching algorithm and the differential greedy heuristic can behave as bad as possible for prefix dictionaries: All bounds for *prefix* dictionaries with additional properties $\mathcal{P}$ are the same as the corresponding bounds for *general* dictionaries with properties $\mathcal{P}$; in other words, adding the prefix property does not improve any worst–case bound neither for LM nor for DG.

Theorem 3 *Let D be a* **prefix** *dictionary. Then*

$$R_{LM}(D) \leq \frac{(lmax - 1)cmax}{cmin}, \quad R_{DG}(D) \leq \frac{(lmax - 1)cmax}{cmin}$$

and this bound can be attained.

Proof. The upper bound follows immediately from Theorem 1. A matching lower bound is derived by choosing the 3–symbol alphabet $\{u, v, w\}$ and the following prefix dictionary:

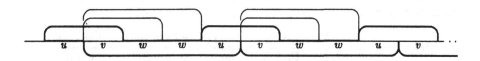

Fig. 1. Illustration for the dictionary and string S_i defined in the proof of Theorem 3 (with $lmax = 4$). The optimal code path runs beyond the horizontal line, the heuristic path above.

source-word	u	v	w	uv	vw^j $j=1,...,lmax-2$	$vw^{lmax-2}u$
code-word	a	b	c	d	e_j	f
weight	$cmax$	$cmax$	$cmax$	$cmax$	$cmax$	$cmin$

For $i > 0$, we consider the strings $S_i = u(vw^{lmax-2}u)^i$ with length $n = i\, lmax+1$. Visualizing the corresponding network (see Figure 1 for an illustration), it can be checked that $OPT(D, S_i) = af^i$ and $LM(D, S_i) = DG(D, S_i) = (dc^{lmax-2})^i a$. Hence

$$R_{LM}(D) \geq \lim_{n\to\infty} \frac{\|LM(D,S_i)\|}{\|OPT(D,S_i)\|} = \lim_{i\to\infty} \frac{i(lmax-1)cmax + cmax}{i\, cmin + cmax}$$
$$= \frac{(lmax-1)cmax}{cmin},$$

and the same inequality holds for DG. $\qquad\square$

Theorem 4 *Let D be a **prefix** and **nonlengthening** dictionary. Then*

$$R_{LM}(D) \leq \frac{lmax\, Bt}{cmin}, \quad R_{DG}(D) \leq \frac{lmax\, Bt}{cmin}$$

and this bound can be attained.

Proof. To derive the upper bound, we follow the proof of Theorem 1 and modify the weights of the edges in an appropriate way (we take into account that a code–word corresponding to a unit–length source–word cannot be longer than Bt).

To get a matching lower bound example, we use the dictionary given in the proof of Theorem 3 and modify the weights as follows:

source-word	u	v	w	uv	vw^j $j=1,...,lmax-2$	$vw^{lmax-2}u$
code-word	a	b	c	d	e_j	f
weight	$cmin$	Bt	Bt	$2Bt$	$(j+1)Bt$	$cmin$

Thereby we get

$$R_{LM}(D) \geq \lim_{n \to \infty} \frac{\|LM(D, S_i)\|}{\|OPT(D, S_i)\|} = \lim_{i \to \infty} \frac{i\, lmax\, Bt + cmin}{(i+1)cmin} = \frac{lmax\, Bt}{cmin},$$

and the same holds for DG. □

Theorem 5 *Let D be a* **prefix** *and* **code uniform** *dictionary. Then*

$$R_{LM}(D) \leq lmax - 1\,, \quad R_{DG}(D) \leq lmax - 1$$

and this bound can be attained.

Proof. The upper bound follows immediately from Corollary 2. Setting $cmin = cmax$ in the proof of Theorem 3 the lower bounds are easily deducted. Details are left to the reader. □

It can be checked easily that the above bound also holds for **prefix,** '**nonlengthening** and **code uniform** dictionaries.

4 Improvements and New Bounds for the Differential Greedy Heuristic

Obviously, for code uniform dictionaries DG and LM yield identical coding results. Hence, this property will not be considered in the following. The case of prefix dictionaries was treated in the previous section.

For nonlengthening, suffix dictionaries, Katajainen and Raita proved the result stated in the following proposition. Among other open problems they asked for the exact value of the worst–case ratio.

Proposition 6 (Katajainen and Raita [4]) *Let D be a* **nonlengthening,** **suffix** *dictionary. Then*

$$R_{DG}(D) \leq \frac{\min\{lmax\, Bt, 2cmax - Bt\}}{cmin}.$$

□

In the next theorem we prove that this upper bound is indeed best possible.

Theorem 7 *For infinitely many quadruples of positive integers Bt, $lmax$, $cmin$ and $cmax$ with $cmin \leq Bt$, $cmin \leq cmax$ and $cmax \leq lmax\, Bt$, there exists a* **nonlengthening,** **suffix** *dictionary D such that*

$$R_{DG}(D) \geq \frac{\min\{lmax\, Bt, 2cmax - Bt\}}{cmin}.$$

Proof. **Case I:** If $lmax\,Bt \geq 2cmax - Bt + 1$, let us suppose that $cmax = \alpha Bt$. (The nonlengthening property requires $cmax \leq \alpha Bt$.) This implies $2\alpha - 1 < lmax$. We consider the following dictionary:

source word	u^j $j=1,..,\alpha-1$	u^α	uw^j $j=1,..,\alpha-1$	w^j $j=1,..,\alpha-2$	$w^{\alpha-1}$	$w^{2(\alpha-1)}u$	$w^j u$ $j=1,..,2\alpha-3$ $j\neq\alpha-1$	$w^{\alpha-1}u$
code	a_j	b	c_j	d_j	e	f	g_j	h
weight	Bt	$2Bt$	$(j+1)Bt$	jBt	$cmax - Bt$	$cmin$	$cmin$	$cmax$

For $S_i = u^\alpha(w^{2(\alpha-1)}u)^i$, we get $OPT(D, S_i) = bf^i$ and $DG(D, S_i) = a_{\alpha-1}(c_{\alpha-1}e)^i a_1$.

There are some points where the choice of DG is not unique. As DG chooses $u^{\alpha-1}$, there also is another possible candidate u^α. The two differences computed by DG are $(\alpha - 1)Bt - Bt$ resp. $\alpha Bt - 2Bt$. W.l.o.g. we suppose that DG chooses $u^{\alpha-1}$. In a similar way the algorithms chooses $uw^{\alpha-1}$ instead of u and $w^{\alpha-1}$ instead of $w^{\alpha-1}u$. With this the worst–case ratio is given by

$$R_{DG}(D) \geq \lim_{n\to\infty} \frac{\|DG(D, S_i)\|}{\|OPT(D, S_i)\|}$$
$$= \lim_{i\to\infty} \frac{i(2cmax - Bt) + 2\,Bt}{i\,cmin + 2Bt}$$
$$= \frac{2cmax - Bt}{cmin}.$$

Case II: If $lmax\,Bt \leq 2cmax - Bt$, then we consider the following dictionary:

source-word	u	w^j $j=1,...,lmax-1$	uw	$w^j u$ $j=1,...,lmax-2$	$w^{lmax-1}u$
code-word	a	b_j	c	d_j	e
weight	Bt	jBt	$2Bt$	$(j+1)Bt$	$cmin$

For $i \geq 1$, we define strings $S_i = u(w^{lmax-1}u)^i$ of length $n = i\,lmax + 1$. With this we get $DG(D, S_i) = (cb_{lmax-2})^i a$ and $OPT(D, S_i) = ae^i$ (breaking ties again w.l.o.g.). Calculating the worst–case ratio we have

$$R_{DG}(D) \geq \lim_{n\to\infty} \frac{\|DG(D, S_i)\|}{\|OPT(D, S_i)\|} = \lim_{i\to\infty} \frac{i\,lmax\,Bt + Bt}{Bt + i\,cmin} = \frac{lmax\,Bt}{cmin}.$$

$\square$

Concerning the analysis of the DG heuristic for *suffix* dictionaries, the authors in [4] mention that "the behaviour of the heuristic is inherently more difficult to analyze". The only upper bounds that were known for this type of dictionary are the *same as for general dictionaries*; they are summarized in the following proposition. Deriving tighter bounds was posed as another open problem.

Proposition 8 (Katajainen and Raita [4]) *Let D be a suffix dictionary. Then*

$$
R_{DG}(D) \leq
\begin{cases}
\dfrac{cmin + (lmax - 1)cmax}{cmin + (lmax - 1)Bt} & \begin{array}{l} \text{if } (lmax - 1)^2\, cmax\, Bt < cmin^2 \text{ and} \\[4pt] \lfloor (cmax - cmin)/Bt \rfloor \geq lmax - 1 \end{array} \\[18pt]
\dfrac{(lmax - 1)cmax}{cmin} & \text{otherwise.}
\end{cases}
$$

□

In the next theorem we give tight bounds for the asymptotic worst–case behaviour of the DG heuristic for suffix dictionaries. (Note that any general dictionary with $lmax = 2$ is a suffix dictionary.)

Theorem 9 *Let D be a suffix dictionary with $lmax \geq 3$. Then*

$$
R_{DG}(D) \leq
\begin{cases}
\dfrac{2cmax - Bt}{cmin} & \text{if } cmax \leq 3/2\, Bt \\[18pt]
\dfrac{(2cmax + Bt)^2}{8Bt\, cmin} & \text{if } 3/2\, Bt < cmax \leq (lmax - 3/2)Bt \\[18pt]
\dfrac{(lmax - 1)(2cmax - (lmax - 2)Bt)}{2cmin} & \text{if } (lmax - 3/2)Bt < cmax.
\end{cases}
$$

All bounds can be attained.

Proof. Let $N = (V, A)$ be the network for a string S and the dictionary D as defined in the introduction. Let v_i and v_j be two consecutive *cutvertices* of N. This implies that both of them lie on every optimal path and on every DG–path.

First, we will prove an upper bound on $R_{DG}(D)$. W.l.o.g. we may assume that the optimal path has only a single edge from v_i to v_j with (maximum) length $lmax$ and (minimum) weight $cmin$. We introduce the following notations. We assume that the DG–path consists of the sequence $(v_i, v_{i_1}, v_{i_2}, \ldots, v_{i_{k+1}}, v_j)$. The length resp. weight of an edge $(v_{i_p}, v_{i_{p+1}})$, $1 \leq p \leq k$, is denoted by t_p resp. c_p.

Since the given dictionary has the suffix property, at vertex v_{i_p} the algorithm DG has to choose between the two edges $(v_{i_p}, v_{i_{p+1}})$ and (v_{i_p}, v_j). The latter has weight at most $cmax$. Since the DG heuristic chooses at each stage vertex $v_{i_{p+1}}$ instead of v_j,

$$
t_p Bt - c_p \geq \left(\sum_{l=p}^{k} t_l + 1 \right) Bt - cmax
$$

must hold. Summing up over all p we get

$$\sum_{p=1}^{k} c_p \le k\, cmax + Bt \sum_{p=1}^{k} t_p - Bt \sum_{p=1}^{k} \sum_{l=p}^{k} t_l - k\, Bt. \tag{1}$$

First, we will estimate the third term in the right hand side

$$\sum_{p=1}^{k} \sum_{l=p}^{k} t_l = \sum_{p=1}^{k} pt_p. \tag{2}$$

We denote

$$T = \sum_{l=1}^{k} t_l + 1 \quad \le lmax - 1.$$

It is easy to verify that the minimum of (2) is attained iff $t_1 = T - k$ and $t_p = 1$, $p = 2, \ldots, k$. Hence, we get

$$\min_{t_l} \sum_{p=1}^{k} \sum_{l=p}^{k} t_l = (T - k) + \sum_{i=2}^{k} i = T + \frac{(k+1)(k-2)}{2}$$

Substituting this result into (1) yields

$$\sum_{p=1}^{k} c_p \le \max_{1 \le k \le lmax-2} \left\{ k\, cmax - Bt(k+1) - Bt\frac{(k+1)(k-2)}{2} \right\}$$

$$= \frac{1}{2} \max_{1 \le k \le lmax-2} \{2k\, cmax - (k^2 + k)Bt\}. \tag{3}$$

This expression is a parabolic function in k and becomes maximum for $k = \frac{2cmax - Bt}{2Bt}$. Depending on the feasible range for k, we distinguish three cases:

1. If $cmax \le \frac{3}{2}Bt$, the maximum is taken at $k = 1$.
2. If $\frac{3}{2}Bt < cmax \le (lmax - \frac{3}{2})Bt$, the maximum is taken at $k = \frac{2cmax - Bt}{2Bt}$.
3. If $(lmax - \frac{3}{2})Bt < cmax$, the maximum is taken at $k = lmax - 2$.

Substituting the corresponding values of k into the right hand side of (3) and assigning the weight $cmax$ to the final edge skipping vertex v_j, we arrive at the desired results for the upper bounds.

In the second part of the proof we show that all the above upper bounds indeed are tight.

Case I: For $cmax \le \frac{3}{2}Bt$, we consider the following dictionary.

source–word	u	w	uw	$w^j u$ $j=1,\ldots,lmax-2$	$w^{lmax-1} u$	w^j $j=1,\ldots,lmax-1$
code–word	a	b	c	d_j	d_{lmax-1}	e_j
weight	$cmax - Bt$	$cmax$	$cmax$	$cmax$	$cmin$	$cmax - Bt$

We compress the string $S_i = u(w^{lmax-1}u)^i$ of length $n = i\,lmax + 1$. Since $OPT(D, S_i) = ad^i_{lmax-1}$ and $DG(D, S_i) = (ce_{lmax-2})^i a$,

$$
\begin{aligned}
R_{DG}(D) &\geq \lim_{n \to \infty} \frac{\|DG(D, S_i)\|}{\|OPT(D, S_i)\|} \\
&= \lim_{i \to \infty} \frac{i(2cmax - Bt) + cmax - Bt}{(cmax - Bt) + i\,cmin} = \frac{2\,cmax - Bt}{cmin}
\end{aligned}
$$

Case II: If $\frac{3}{2}Bt < cmax \leq (lmax - \frac{3}{2})Bt$, we suppose that Bt is even, $cmin \leq \frac{Bt}{2}$ and that $cmax = (2\alpha + 1)\frac{Bt}{2}$ for some integer α, $1 \leq \alpha \leq lmax - 4$. We construct a dictionary on $lmax$ letters. The letters of the alphabet are denoted by $u, v, w_1, \ldots, w_{lmax-2}$. Let $k = \frac{2cmax - Bt}{2Bt}$

source-word	u	uv	v^j $j=1,\ldots,lmax-2-k$	$v^{lmax-1-k}$
code-word	a	b	c	d_0
weight	$cmax$	$cmax$	$cmax$	$\frac{1}{2}Bt$

source-word	w_j $j=1,\ldots,k-1$	$v^j w_1 \ldots w_{k-1}u$ $j=1,\ldots,lmax-1-k$	$v^{lmax-k} w_1 \ldots w_{k-1}u$	$w_j \ldots w_{k-1}u$ $j=1,\ldots,k-1$
code-word	d_j	e_j	e_{lmax-k}	f_j
weight	$(2j+1)\frac{Bt}{2}$	$cmax$	$cmin$	$cmax$

and the source-string $S_i = u(v^{lmax-k}w_1 \ldots w_{k-1}u)^i$. It is easy to check that $OPT(D, S_i) = ae^i_{lmax-k}$. Following the DG-path one can see that at the beginning the DG algorithm chooses the edge uv. At the endvertex of this edge there are several edges: One skips the string v and the others skip the strings v^j, $j = 1, \ldots, lmax - 1 - k$. So DG chooses the edge $v^{lmax-1-k}$. On the remaining part of the string (until it meets again a letter u) the DG algorithm has to decide between the edges $w_j \ldots w_{k-1}u$, $j = 1, \ldots, k-1$ and the single character edge w_j. In each vertex it resolves the arising tie by choosing w_j. In this way we get $DG(D, S_i) = (bd_0 d_1 \ldots d_{k-1})^i a$ which yields

$$
\begin{aligned}
R_{DG}(D) &\geq \lim_{n \to \infty} \frac{\|DG(D, S_i)\|}{\|OPT(D, S_i)\|} \\
&= \lim_{i \to \infty} \frac{i\left(cmax + \frac{Bt}{2}\sum_{j=0}^{k-1}(2j+1)\right) + cmax}{cmax + i\,cmin} \\
&= \lim_{i \to \infty} \frac{i\left(cmax + \frac{Bt}{2}\frac{(2cmax-Bt)^2}{4Bt^2}\right) + cmax}{cmax + i\,cmin} \\
&= \frac{(2cmax + Bt)^2}{8Bt\,cmin}
\end{aligned}
$$

Case III: Finally, for $(lmax - \frac{3}{2})Bt < cmax$ we again consider a suffix dictionary on $lmax$ letters called $u, w_1, \ldots, w_{lmax-1}$.

source–word	u	w_j $j=1,\ldots,lmax-1$	uw_{lmax-1}	$w_j \ldots w_1 u$ $j=lmax-2,\ldots,1$	$w_{lmax-1} \ldots w_1 u$
code–word	a	b_j	c	d_j	e
weight	$cmax$	$cmax - j\,Bt$	$cmax$	$cmax$	$cmin$

For the source–string $S_i = u(w_{lmax-1} \ldots w_1 u)^i$ of length $n = i\,lmax + 1$, we get $OPT(D, S_i) = ae^i$ and $DG(D, S_i) = (cb_{lmax-2}b_{lmax-3} \ldots b_1)^i a$. Hence

$$
\begin{aligned}
R_{DG}(D) &\geq \lim_{n \to \infty} \frac{\|DG(D, S_i)\|}{\|OPT(D, S_i)\|} \\
&= \lim_{i \to \infty} \frac{i\left(cmax + \sum_{j=1}^{lmax-2}(cmax - j\,Bt)\right) + cmax}{cmax + i\,cmin} \\
&= \lim_{i \to \infty} \frac{i(lmax - 1)(2cmax - (lmax - 2)Bt) + cmax}{2cmax + 2i\,cmin} \\
&= \frac{(lmax - 1)(2cmax - (lmax - 2)Bt)}{2cmin}
\end{aligned}
$$

which completes the proof. □

5 Conclusions

We analyzed two *on–line* data compression algorithms with static dictionaries from a worst–case point of view.

Four properties of dictionaries were investigated: code uniform, nonlengthening, suffix and prefix. The case of a dictionary that is prefix and suffix at the same time does neither seem practical nor interesting to us because in the generation of a dictionary usually only one of these two properties is sought. Thus there remain twelve "reasonable" combinations of these properties.

The analysis of the worst–case behaviour of the longest matching algorithm on eight of these dictionary types which are not prefix was performed by Katajainen and Raita [4]. The remaining four prefix types were examined in this paper. Now for all twelve types tight bounds on the worst–case ratio of LM are known.

The differential greedy algorithm behaves identically to LM in case that the dictionary is code uniform. Thus, there remain six dictionary types for which the behaviour of DG has to be analyzed. The two types that are prefix were analyzed in the current paper. The two types that are neither prefix nor suffix were analyzed in [4]. These four types turned out to be easy to analyze. For the case of a suffix nonlengthening dictionary, [4] gave an upper bound and we provided the matching lower bound. For the case of a suffix dictionary, rather involved (but matching) lower and upper bounds were derived in Section 4 of the current paper.

Essentially, we see four lines for future research.

1. Identify other reasonable properties for dictionaries and investigate the consequences of these properties on the worst–case behaviour of *LM* and *DG*.
2. Construct better approximation algorithms that exploit the special properties of some dictionary types (e.g. it would be interesting to have a heuristic able to deal more efficiently with suffix dictionaries).
3. Develop better heuristics by relaxing the on–line condition. E.g. the next $3lmax$ characters could be used to decide on the next replacement.
4. Extend the analysis to dynamic (adaptive) dictionaries. E.g. there is nothing known on the worst–case performance of the Ziv–Lempel algorithm [6].

Recently, a new algorithm, the *fractional greedy heuristic*, was introduced and analyzed by the authors in [1].

Acknowledgement: Gábor Galambos gratefully acknowledges the hospitality of the TU Graz during his visiting position at the Institute of Mathematics.

References

1. J. Békési, G. Galambos, U. Pferschy, G.J. Woeginger, The fractional greedy algorithm for data compression, Report 292, Institute of Mathematics, Graz, 1994, to appear in *Computing*.
2. M.E. Gonzalez-Smith and J.A. Storer, Parallel algorithms for data compression, *Journal of the ACM* **32**, 1985, 344–373.
3. J. Katajainen and T. Raita, An approximation algorithm for space-optimal encoding of a text, *The Computer Journal* **32**, 1989, 228–237.
4. J. Katajainen and T. Raita, An analysis of the longest matching and the greedy heuristic in text encoding, *Journal of the ACM* **39**, 1992, 281–294.
5. E.J. Schuegraf and H.S. Heaps, A comparison of algorithms for data base compression by use of fragments as language elements, *Inf. Stor. Ret.* **10**, 1974, 309–319.
6. J. Ziv and A. Lempel, A universal algorithm for sequential data compression, *IEEE Trans. Inf. Theory* **23**, 1977, 337–343.

Gossiping in Cayley Graphs by Packets[*]

Jean-Claude Bermond, Takako Kodate and Stéphane Pérennes

Laboratoire I3S, Université de Nice - Sophia Antipolis
Bât. Essi, 930 Route des Colles, B.P.145
06903 Sophia Antipolis Cedex, France
emails: {bermond, kodate, sp}@unice.fr

Abstract. *Gossiping* (also called *total exchange* or *all-to-all communication*) is the process of information diffusion in which each node of a network holds a packet that must be communicated to all other nodes in the network. We consider here gossiping in the *store-and-forward, full-duplex* and Δ-*port* (or *shouting*) model. In such a model, the protocol consists of a sequence of rounds and during each round, each node can send (and receive) messages from all its neighbors. The great majority of the previous works on gossiping problems allows the messages to be freely concatenated and so messages of arbitrary length can be transmitted during a round. Here we restrict the problem to the case where at each round communicating nodes can exchange exactly one packet. We give a lower bound of $\frac{N-1}{\delta}$, where δ is the minimum degree, and show that it is attained in Cayley symmetric digraphs with some additional properties. That implies the existence of an optimal gossiping protocol for classical networks like hypercubes, k-dimensional tori, and star-graphs. Furthermore we introduce the notion of complete rotation in Cayley symmetric digraphs which appears to be an interesting property worth of study in itself.

1 Introduction

Gossiping (also called total exchange or all–to–all communication) in distributed systems is the process of distribution of information known to each processor to every other processor of the system. We consider here the store and forward model where the protocol consists in rounds ; during each round the message is transmitted from one node only to adjacent nodes.

The gossiping problem appears as subroutine in a large class of parallel or distributed computation problems, such as linear system solving, Discrete Fourier Transform, and sorting, where both input and output data are required to be distributed across the network [3]. Gossiping has been widely studied under various communication models (see the surveys [8, 9, 11] and the book [4]).

However most of the previous work on gossiping has considered the case in which the packets known to a processor at any given time during the execution

---

[*] That work was supported by the French GDR/PRC PRS and by the Galileo cooperation project with the University of Salerno

of the gossiping protocol can be freely concatenated and the resulting (longer) message can be transmitted in a constant amount of time, that is, it has been assumed that the time required to transmit a message is independent from its length. While this assumption is reasonable for short messages, it is clearly unrealistic when the size of the messages becomes large. Indeed most of the gossiping protocols proposed in the literature require the transmission, in the last rounds of the execution of the protocol, of messages of size $\Theta(N)$, where N is the number of nodes in the network.

Here we consider the gossiping problem under the restriction that communicating nodes can exchange exactly one packet at each round and where a node can communicate with all its neighbors. The case where a node can only communicate with one of its neighbors has been considered in [1, 2].

1.1 Communication Models

We model a communication network by a symmetric digraph $G = (V, E)$ where the node set V represents the set of processors and the arc set E represents the communication links between processors. Here we suppose that if a node x can directly communicate with a node y, then the converse is true. So we have a symmetric digraph where if $(x, y) \in E$ then $(y, x) \in E$. Initially, each node holds a packet that must be transmitted to all the other nodes of the network by a sequence of calls between adjacent processors. The restriction considered in the introduction implies that during each call, communicating nodes can exchange only one packet. We therefore see the gossiping protocol as a sequence of *rounds*. During each round, we suppose that each processor can communicate *with all its neighbors*, more exactly it can send a packet to all its neighbors and receive a packet from all its neighbors. Furthermore we allow each node to send *different packets* to *different neighbors* at each round.

We will denote by $g_{F_*}(1, G)$ the minimum gossip time, that is the minimum number of rounds to complete the gossiping process in the network G subject to the above condition. This model is often referred as F_* or *full-duplex Δ-port* or *shouting model*. Other models also popular restrict a node to communicate only with one of its neighbors during each round (F_1 or *telephone model*) or do not allow both emission and reception (*half-duplex model* denoted H_* or H_1 according one node can send to all its neighbors or only one.) The problem of estimating $g_{F_1}(1, G)$ and more generally $g_{F_1}(p, G)$ where one allows to exchange up to a fixed number p of packets at each round has been considered in [2]. The similar problem for $g_{H_1}(p, G)$ has been considered in [1]. Analogous problems on bus networks have been considered in [10]. In [12], a similar problem is considered for the toroidal mesh as they limit the size of the buffers. However they use a linear time model (of the form $\beta + L\tau$) and allow pipelining.

1.2 Graph Definitions

We use the definitions of the book [4].

In what follows:

- N will denote the number of vertices of G.
- D (or $D(G)$) will denote the *diameter* of a graph G, that is the maximum of all over the minimum distances between every pair of vertices.
- δ (or $\delta(G)$) will denote *minimum degree* of a graph G, that is the minimum over the degrees of all vertices of G.

Let us now present some of the classical networks for which we want to determine $g_{F_*}(1, G)$. We give their definitions in terms of graphs, but we recall that in our model, we will always work with the symmetric digraphs associated obtained by replacing each edge $[x, y]$ by the two opposite arcs (x, y) and (y, x).

Definition 1. The *cartesian sum* (also called *cartesian product* or *box product*) denoted by $G \square G'$ of two graphs $G = (V, E)$ and $G' = (V', E')$, is the graph whose vertices are the pairs (x, x') where x is a vertex of G and x' is a vertex of G'. Two vertices (x, x') and (y, y') of $G \square G'$ are adjacent if and only if $x = y$ and $[x', y']$ is an edge of G', or $x' = y'$ and $[x, y]$ is an edge of G.

Definition 2. *The k-dimensional toroidal mesh* (or *the torus*) is the cartesian sum of k cycles of orders $p_1, p_2, \ldots, p_k$ and is denoted by $TM(p_1, p_2, \ldots, p_k) = C_{p_1} \square C_{p_2} \square \cdots \square C_{p_k}$.

If all $p_i \geq 3$, it is a regular graph of degree $2k$. Its order is $p_1 \times p_2 \times \cdots \times p_k$ and its diameter is $\sum_{i=1}^{k} \lfloor \frac{p_i}{2} \rfloor$. When $p_1 = p_2 = \cdots = p_k$, we will use the abbreviated notation $TM(p)^k$ and suppose in what follows that $p \geq 3$.

Definition 3. When $p = 2$, $TM(2)^k = \underbrace{K_2 \square K_2 \square \cdots \square K_2}_{k \text{ times}}$ is known as *the hypercube of dimension k*, denoted by $H(k)$. $H(k)$ can be therefore defined as the graph whose vertices are words of length k over the two-letter alphabet $\{0, 1\}$ and whose edges connect two words which differ in exactly one coordinate. A vertex $x_0 x_1 \cdots x_i \cdots x_{k-1}$ is thus joined to the vertices $x_0 x_1 \cdots \overline{x_i} \cdots x_{k-1}$ with $i = 0, 1, \ldots, k-1$. An edge between two vertices which differ in the ith coordinate will be called an edge of dimension i, or of type e_i.

In fact, these graphs are a particular case of a more general class of graphs.

Definition 4. Let $\mathcal{G}$ be a group and $\mathcal{S} = (s_0, s_1, \ldots, s_{d-1})$ be a set of generators of $\mathcal{G}$ not containing the identity. *The associated Cayley digraph* is the graph whose vertices are the elements of $\mathcal{G}$ and whose arcs are the couples $(x, s_i x)$ for $x \in \mathcal{G}$ and $s_i \in \mathcal{S}$.

As we restrict our attention to *symmetric digraphs*, we will always suppose that $\mathcal{S} = \mathcal{S}^{-1}$, that is if $s \in \mathcal{S}$, then $s^{-1} \in \mathcal{S}$. Some authors call them "Cayley graphs" as they identify the arcs $(x, s_i x)$ and $(s_i x, x)$ with the edge $[x, s_i x]$.

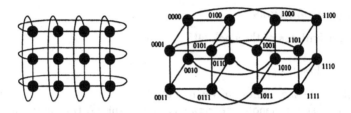

Fig. 1. Toroidal mesh $TM(3,4)$ and hypercube $H(4)$

Example 1. The symmetric k-dimensional torus $TM(p)^k$ is a Cayley digraph. The group $\mathcal{G}$ is $\mathbf{Z}_p^k$ with elements $(x_0x_1\cdots x_{k-1})$ and the $2k$ generators are the "canonical basis" $\pm e_i$ where $e_i = (0\cdots1\cdots0) = (x_0x_1\cdots x_{k-1})$ with $x_j = 0$ for $j \neq i$, $x_i = 1$. In the case $k = 2$, we have the 4 generators $s_0 = e_0 = (1,0)$, $s_1 = e_1 = (0,1)$, $s_2 = -e_0 = (-1,0)$, and $s_3 = -e_1 = (0,-1)$.

Example 2. In the case of $p = 2$, that is for the hypercube $\mathcal{G} = \mathbf{Z}_2^k$ with only k generators as $e_i = -e_i$.

Example 3. *The symmetric "star-graph"*, denoted by $S(k)$, is the Cayley digraph whose vertices are the permutations of a k-element set, and where the generators are the $k-1$ transposition exchanging respectively 1 and i, for $2 \leq i \leq k$. We will denote a permutation π by the word $(\pi(1)\pi(2)\ldots\pi(k))$. For example in figure 2, the permutation identity $e = (1234)$ is joined by an arc to the permutations (2134), (3214) and (4231). $S(k)$ is regular of degree $k - 1$, its order is $k!$ and its diameter is $\lfloor\frac{3(k-1)}{2}\rfloor$.

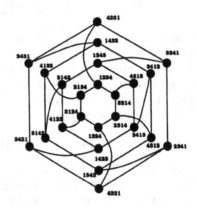

Fig. 2. Star-graph $S(4)$

1.3 Lower Bounds

If we have no restriction on the size of the messages, that is if we can exchange an arbitrary number of packets during each call, then it is well known that in the model F_*, one can gossip in $D(G)$ rounds. It is sufficient to use the greedy protocol where at each round each node send all the messages it has just received during the precedent round to all its neighbors.

Theorem 5. *For a graph* G,

$$g_{F_*}(\infty, G) = D(G) \tag{1}$$

So we have a first immediate lower bound.

Proposition 6.

$$g_{F_*}(1, G) \geq D(G) \tag{2}$$

The following lower bound is often more appropriate.

Proposition 7. *For a graph* G *having* N *vertices and minimum degree* δ,

$$g_{F_*}(1, G) \geq \lceil \frac{N-1}{\delta} \rceil \tag{3}$$

Proof. Let u be a vertex of minimum degree δ. During each round, u can receive only one packet from each of its neighbors and so at most δ packets. At the end of the protocol, u should have received $N - 1$ packets, so the protocol needs at least $\lceil \frac{N-1}{\delta} \rceil$ rounds.

The same proof gives:

Proposition 8. *For a digraph* G *with* N *vertices and minimum in-degree* d,

$$g_{F_*}(1, G) \geq \lceil \frac{N-1}{d} \rceil \tag{4}$$

Remark. We can sometimes obtain a better lower bound in digraphs by considering an edge cut, separating a set S from its complement $V \setminus S$. Let $m^+(S, V \setminus S)$ be the size of this cut; we need at least $\frac{|S|}{m^+(S,V\setminus S)}$ rounds. The bound of (3) corresponds to the particular case where $V \setminus S = \{x\}$ and $m^+(S, V \setminus S) = d$.

1.4 Results

In this paper, we shall show that for k-dimensional tori (and in particular 2-dimensional tori), hypercubes and star-graphs, the lower bound (3) is attained.

That will follow from a more general property valid for Cayley symmetric digraphs with special automorphisms.

2 Gossiping in Cayley symmetric digraphs

In what follows G will always denote a Cayley symmetric digraph of in and out-degree d associated to a group $\mathcal{G}$ with identity element denoted e and a set of d generators $\mathcal{S} = (s_0, s_1, \ldots, s_{d-1})$. The arc $(x, s_i x)$ will be said of dimension i.

If A and B are two subsets of G, the set $\{ab \mid a \in A, b \in B\}$ will be denoted by AB; if $x \in G$ the set $\{ax, a \in A\}$ will be denoted by Ax.

2.1 Balanced sequence of sets

Definition 9. A sequence of sets $\{S_t\}_{t=0,\ldots,T}$ containing elements of G of length T is said to be *balanced* if for each $t = 0, \ldots, T-1$ there exist d vertices $x_0^t, x_1^t, \ldots, x_{d-1}^t$ (not necessarily distinct) in S_t such that $S_{t+1} \subset S_t \cup (\bigcup_{i=0,\ldots,d-1}(s_i x_i^t))$.

The balanced property means that from S_t we can reach the vertices of $S_{t+1} \setminus S_t$ by using at most one arc in each dimension. In term of communications, suppose that S_0 is reduced to one vertex x (the initiator of a broadcast), then S_t represents the set of vertices which know the information of x at time t. The information is then forwarded to vertices of $S_{t+1} \setminus S_t$ by using at most once each generator. We can then associate to the sequence of sets $\{S_t\}_{t=0,\ldots,T}$ a *broadcast tree* by considering at round t only the arcs from S_{t-1} to S_t. If we label these arcs with t, the balanced property means that for any t, there exists at most one arc of dimension i.

Figure 3 and figure 4 give such sequences of sets for different tori, the sequence of sets are in fact implicitly defined by their associated broadcast tree. The label t of a vertex x indicates that x belongs to $S_t \setminus S_{t-1}$. So S_t consists of all the vertices with label less than or equal to t.

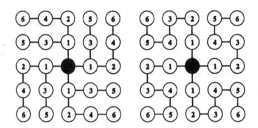

Fig. 3. Balanced set and broadcast tree in toroidal mesh $TM(5)^2$

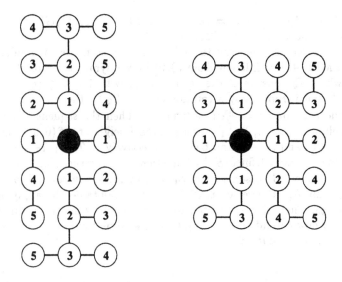

Fig. 4. Balanced set and broadcast tree in $TM(3,7)$ and $TM(4,5)$

Lemma 10. *Let G be a Cayley symmetric digraph with a balanced sequence of sets $\{S_t\}_{t=0,\ldots,T}$ with $S_0 = \{e\}$, then we can broadcast concurrently the packet of x to $S_T x$ in T rounds for each vertex $x \in V$.*

Proof. The result is proved by induction on t. The proposition is true for $t = 0$. For an arbitrary vertex x, we suppose that vertices in $S_t x$ have received the packet of x at time t. Then the nodes $x_0^t x, x_1^t x, \ldots, x_{d-1}^t x$ which are in $S_t x$ send the packet of x to the vertices $s_i x_i^t x$, hence vertices of $S_{t+1} x = S_t x \cup (\bigcup_{i=0,\ldots,d-1}(s_i x_i^t x))$ will have received the packet of x at time $t + 1$.

Now we have to check that there is no communication conflict. As a vertex y sends the information of x along the arc s_i at time t if and only if $y = x_i^t x$, y uses the arc s_i at time t only to send the information originated from $(x_i^t)^{-1} y$. Hence the different calls can be performed in parallel.

From this lemma, we easily deduce the following proposition.

Proposition 11. *Let G be a Cayley digraph and $\{S_t\}_{t=0,\ldots,T}$ be a balanced sequence of sets with $S_0 = \{e\}$, and $S_T = V$ then the gossip time of G is at most T.*

Example 4. The examples of figure 3 and figure 4 show respectively that we can gossip in 6 rounds in $TM(5)^2$ and in 5 rounds in $TM(3,7)$ and $TM(4,5)$. These protocols are optimal as by (3): $g_{F_*}(1, TM(p_1, p_2)) \geq \lceil \frac{p_1 p_2 - 1}{4} \rceil$.

For the torus $TM(2p+1)^2$, we can easily exhibit a balanced sequence of sets S_t such that $|S_t| = 4t + 1$ and so $g_{F_*}(1, G) \leq p^2 + p$ as for $T = p^2 + p$,

$|S_T| = 4p^2 + 4p + 1 = (2p+1)^2$. As by (3): $g_{F_*}(1, G) \geq \lceil \frac{N-1}{4} \rceil = p^2 + p$, we conclude that : $g_{F_*}(1, TM(2p+1)^2) = p^2 + p$.

A very simple way to find the sets S_t consists in dividing the vertices of $TM(2p+1)^2$ different from $(0,0)$ into 4 symmetric subsets T_0, T_1, T_2 and T_3, where T_0 consists of the vertices (i, j) with $1 \leq i \leq p$, $0 \leq j \leq p$ and T_1 is obtained from T_0 by a rotation ω of $\frac{\pi}{2}$, T_2 by the rotation ω^2 of π, and T_3 by the rotation ω^3 of $\frac{3\pi}{2}$ (see figure 5). Then the sequence S_t can be defined as follows: $S_0 = (0,0)$, S_1 consists of the 4 neighbors of $(0,0)$. They are $(1,0)$ $(0,1)$ $(-1,0)$ and $(0,-1)$ and belong respectively to T_0, T_1, T_2 and T_3. Then suppose that we have defined S_t by induction. To construct S_{t+1}, choose in T_0 a vertex y_0 connected to some x_0^t of $S_t \cap T_0$ by an arc in some dimension i_0. Then we put in S_{t+1} the vertex y_i of T_i, $0 \leq i \leq 3$, obtained by applying the rotation ω^i to y_0. Then y_i is connected to $x_i^t = \omega^i(x_0^t)$ by an arc in dimension $i_0 + i$. We will now extend that idea to Cayley symmetric digraph by using the notion of "*complete rotation*".

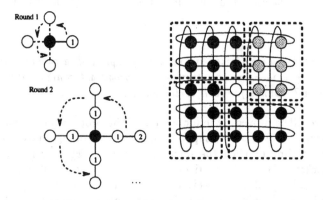

Fig. 5. Algorithm for the toroidal square mesh $TM(5)^2$ and its four parts.

2.2 Complete rotation

Definition 12. An automorphism ω of a Cayley symmetric digraph $G = (\mathcal{G}, \mathcal{S})$ is *a complete rotation* if $\omega(e) = e$ and $\forall x \in G$ and $\forall i = 0 \cdots d-1$, $\omega(s_i x) = s_{i+1}\omega(x)$ (where the indices are taken modulo d).

Remark. A complete rotation is a particular case of what is called in the literature an *Adam automorphism* (automorphism such that $\omega(xy) = \omega(x)\omega(y)$ for any couple of elements x and y). It has the additional property of acting cyclically on the set of generators.

Definition 13. A vertex $x \neq e$ will be called a *fixed point* for a complete rotation ω if there exists some i, $1 \leq i \leq d-1$ such that $\omega^i(x) = x$. The set of fixed points of ω will be denoted by F_ω.

Remark. The orbit of $x \neq e$ under the action of ω is of length d except when x is a fixed point of ω in which case the orbit is degenerated.

Example 5. In the torus $TM(p)^2$, consider the automorphism ω which associates to $x_0 x_1$ the vertex $(-x_1)x_0$ (which corresponds to a rotation of $\frac{\pi}{2}$ in the plane). Then ω is a complete rotation and we have $\omega(s_0) = \omega(1,0) = (0,1) = s_1$, $\omega(s_1) = (-1,0) = s_2$ and $\omega(s_2) = (0,-1) = s_3$.

If p is odd, then $F_\omega = \emptyset$.

If p is even, then $F_\omega = \{(0,\frac{p}{2}), (\frac{p}{2},0), (\frac{p}{2},\frac{p}{2})\}$ as $\omega(\frac{p}{2},\frac{p}{2}) = (-\frac{p}{2},\frac{p}{2}) = (\frac{p}{2},\frac{p}{2})$, $\omega^2(0,\frac{p}{2}) = (0,-\frac{p}{2}) = (0,\frac{p}{2})$ and $\omega^2(\frac{p}{2},0) = (-\frac{p}{2},0) = (\frac{p}{2},0)$.

More generally, in the torus $TM(p)^k$, the automorphism ω which associates to $x_0 x_1 \cdots x_{k-1}$ the vertex $(-x_{k-1})x_0 \cdots x_{k-2}$ is a complete rotation. Indeed $\omega^i(e_0) = e_i$ for $0 \leq i \leq k-1$. Then $\omega^k(e_0) = \omega(e_{k-1}) = -e_0$ and $\omega^{k+i}(e_0) = -e_i$. One can check that ω satisfies the definition if we rank the generators of $TM(p)^k$ in the order $s_i = e_i$, $s_{i+k} = -e_i$ for $0 \leq i \leq k-1$. We will characterize the set of fixed points later on, but we can note that if p is odd, then $F_\omega = \emptyset$.

Example 6. For the hypercube $H(k)$, we consider similarly the automorphism ω which associates to $x_0 x_1 \cdots x_{k-1}$ the vertex $x_{k-1} x_0 \cdots x_{k-2}$ (recall that $-x_i = x_i$). ω is clearly a linear one to one mapping of Z_2^k onto itself. Moreover if $s_i = e_i$, we have $\omega(s_i) = s_{i+1}$.

F_ω consists of the vertices such that the associated word is periodic, that is x can be written as $uu \cdots u$ where u is repeated r times with r dividing k and u being a word of length $q = \frac{k}{r}$.

So if d is a prime number, then F_ω is reduced to the vertex $111 \cdots 1$.

Lemma 14. *Suppose that there exists in G a complete rotation ω and let $\Gamma^*_{G \setminus F_\omega}(e)$ be the connected component of $G \setminus F_\omega$ containing e, then G admits a balanced sequence of sets $\{S_t\}_{t=0,\ldots,T}$, with $S_0 = \{e\}$ and $S_T = \Gamma^*_{G \setminus F_\omega}(e)$ with $T = \frac{|\Gamma^*_{G \setminus F_\omega}(e) - 1|}{d}$.*

Proof. We construct inductively a sequence of sets using the following algorithm, which generalizes the construction given for the torus in example 4.

- $S_0 = \{e\}$.
- If S_t does not contain all the vertices of $\Gamma^*_{G \setminus F_\omega}(e)$, choose a vertex x in $\Gamma^*_{G \setminus F_\omega}(e)$ adjacent to S_t, and let $S_{t+1} = S_t \cup \{\omega^j(x); \ j = 0, \ldots, d-1\}$.
- If $S_t = \Gamma^*_{G \setminus F_\omega}(e)$, then stop.

Clearly, the algorithm stops when $S_T = \Gamma^*_{G \backslash F_\omega}(e)$. The construction implies that $\forall t : \omega(S_t) = S_t$, as $S_0 = \omega(S_0)$ and $\omega(S_{t+1}) = \omega(S_t \cup \{\omega^j(x); \ j = 0, \ldots, d-1\}) = S_t \cup \{\omega^j(x); \ j = 0, \ldots, d-1\} = S_{t+1}$. S_t being invariant under rotation none of the vertices $\omega^j(x)$ can belong to S_t when $x \notin S_t$. Hence $|S_{t+1}| = |S_t| + |\{\omega^j(x); \ j = 0, \ldots, d-1\}|$. As we always choose $x \notin F_\omega$, $|\{\omega^j(x); \ j = 0, \ldots, d-1\}| = d$, and $|S_{t+1}| = |S_t| + d$. The cardinality of S_T is now clearly $1 + dT$, and we get $T = \frac{|\Gamma^*_{G \backslash F_\omega}(e)| - 1}{d}$.

Now let us prove that the sequence is balanced. Suppose that the added vertex x is joined to some vertex x_i^t of S_t along dimension i, that is $x = s_i x_i^t$. Then $\omega^j(x) = s_{i+j} \omega^j(x_i^t)$. As S_t is invariant under ω, the element $\omega^j(x_i^t)$ is in S_t. Let us call it x_{i+j}^t. Then $\bigcup_{j=0,\ldots,d-1} \omega^j(x) = \bigcup_{j=0,\ldots,d-1}(s_{i+j} x_{i+j}^t)$. Therefore the sequence is balanced.

Corollary 15. *If a Cayley symmetric digraph G admits a complete rotation ω such that $F_\omega = \emptyset$, then $g_{F_*}(1, G) = \frac{N-1}{d}$.*

Proof. As $F_\omega = \emptyset$, $\Gamma^*_{G \backslash F_\omega}(e) = V$ and so by lemma 14, $S_T = V$ with $T = \frac{N-1}{d}$. The result follows from proposition 11.

Proposition 16. *For the k-dimensional torus $TM(2p+1)^k$, we have*

$$g_{F_*}(1, TM(2p+1)^k) = \frac{(2p+1)^k - 1}{2k} \tag{5}$$

Proof. That follows from (4) and corollary 15.

Example 7. Lemma 14 enables us also to conclude in many other cases. For example, consider the hypercube $H(k)$ with k prime. F_ω is reduced to the vertex $111 \cdots 1$.
By lemma 14, we can construct a balanced sequences of sets $\{S_t\}_{t=0,\ldots,T}$ with $S_T \Rightarrow V - \{111 \cdots 1\}$ and $T = \frac{2^k - 2}{k}$. We can easily extend this sequence by adding $\{111 \cdots 1\}$ in S_{T+1} to obtain $S_{T+1} = V$. So we have a gossiping protocol in $\frac{2^k - 2}{k} + 1$ rounds. This result is optimal as, k being a prime, $2^k - 2 \equiv 0 \mod k$ so $\lceil \frac{N-1}{k} \rceil = \frac{2^k - 2}{k} + 1$ and therefore $g_{F_*}(1, H(k)) = \lceil \frac{N-1}{k} \rceil$ for k prime.

Example 8. Similarly let us consider the 2-dimensional torus $TM(2p)^2$. By lemma 14, we have a balanced sequence of sets $\{S_t\}_{t=0,\ldots,T}$, with $S_T = V \setminus F_\omega$ with $T = \frac{4p^2 - 4}{4} = p^2 - 1$ as $|V| = 4p^2$ and $|F_\omega| = 3$ (see example 5). But again one can easily extend this sequence by adding the 3 vertices of F_ω, which have all their 4 neighbors in $S_T = V \setminus F_\omega$. We can for example join one of them say (p, p) via generator $s_0 = (1, 0)$; another say $(p, 0)$ via generator $s_1 = (0, 1)$ and the last one $(0, p)$ via generator $s_2 = (-1, 0)$. So we obtain $S_{T+1} = V$ and have an optimal gossip protocol in $p^2 = \lceil \frac{N-1}{4} \rceil$ rounds (see figure 6 where the elements of F_ω are in light grey and the corrections as above).

In fact, these results are particular cases of a more general proposition.

Let us recall that a subset A of vertices is said to be *independent* (or *stable*) if there is no arc between any couple of vertices of A. A is said to be *separating* (or a *vertex cut set*) if the deletion of vertices of A disconnects G.

Lemma 17. *If a Cayley symmetric digraph G admits a complete rotation such that F_ω is an independent and not separating set of G, then $g_{F_*}(1, G) = \lceil \frac{N-1}{d} \rceil$.*

Proof. As F_ω is not a separating set, $\Gamma^*_{G\backslash F_\omega}(e) = V \backslash F_\omega$. So we can construct from lemma 14 a sequence with $S_0 = \{e\}$, $S_{t_0} = V \backslash F_\omega$, and $|S_{t_0}| = 1 + t_0 d$. After this step any vertex in $V \backslash S^{\bullet}_{t_0+t} \subset F_\omega$ will be adjacent to $V \backslash F_\omega \subset S_{t_0+t-1}$ along any direction as F_ω is an independent set. So we can continue the construction by including d vertices in S_{t_0+t}, as long as there exist d vertices in $V \backslash S_{t_0+t-1}$. If there are less than d vertices in $V \backslash S_{t_0+t-1}$, we add in S_{t_0+t} all the remaining vertices. Let T be the value such that $S_T = V$. Therefore for $t < T$, $S_t = 1 + td$ and so $T = \lceil \frac{N-1}{d} \rceil$.

We have reduced the problem to check if F_ω is an independent, not separating set. Clearly in the digraph of examples 7 and 8, the set F_ω was independent and not separating set.

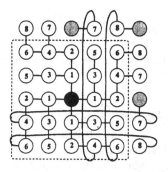

Fig. 6. Balanced sets in $TM(6)^2$

Lemma 18. *Let G be a Cayley symmetric digraph with complete rotation ω; If any pair of vertices (y, y') such that $d(y, y') = 2$, has at most two common neighbors, then F_ω is an independent set.*

Proof. a) If x and y are two adjacent vertices and if ω^t fixes both x and y, then t is a multiple of d. Indeed let $y = s_i x$, then $\omega^t(y) = s_{i+t} \omega^t(x) = s_{i+t} x = y = s_i x$. So $s_{i+t} = s_i$ which implies that t is a multiple of d.

b) Now let x and y be two adjacent vertices in F_ω. As x and $y \in F_\omega$, there exist p and q such that $\omega^p(x) = x$, $\omega^q(y) = y$, with p and q proper dividers of d. As x and y are adjacent, $p \neq q$ by a).

c) Let $y' = \omega^p(y)$ and $x' = \omega^q(x)$. As $p \neq q$ and p and q are proper dividers of d by a), $x' \neq x$ and $y' \neq y$.

Furthermore y' is a neighbor of x as $y' = \omega^p(y) = \omega^p(s_i x) = s_{i+p}\omega^p(x) = s_{i+p}x$ and so $d(y, y') = 2$. Similarly x' is a neighbor of y and y'.

d) By the hypothesis of the lemma, x and x' are the only neighbors of y and y'. So repeating the argument of c) with x' and y', we obtain that $\omega^q(x')$ is a neighbor of y and y' and so $\omega^q(x') = x$ and similarly $\omega^p(y') = y$.

e) Therefore $\omega^{2q}(x) = x$ and $\omega^{2p}(y) = y$. By b), $\omega^{2p}(x) = x$ and $\omega^{2q}(y) = y$. So by a), $2p$ and $2q$ are multiple of d. As p and q are proper dividers of d, the only possibility is that d is even and $p = q = \frac{d}{2}$ contradicting $p \neq q$.

Corollary 19. *For any complete rotation ω of $TM(p)^k$, $H(k)$ and $S(k)$, F_ω is an independent set.*

Proof. For $TM(p)^k$ and $H(k)$, that follows from the fact that there is a unique C_4 containing any pair of vertices at distance 2.

For $S(k)$, that follows from the fact that there is no C_4 as the girth of $S(k)$ is 6.

Proposition 20. *For the hypercube $H(k)$,*

$$g_{F_*}(1, H(k)) = \lceil \frac{2^k - 1}{k} \rceil \tag{6}$$

Proof. By lemma 17 and lemma 18, it remains to prove that F_ω is not a separating set of $H(k)$.

a) First let us prove that if $x \notin F_\omega$, x has at most two neighbors in F_ω. For that, let f be the application which associates to a vertex x of $H(k)$ the complex $f(x) = \sum_{i=0}^{k-1} x_i \theta^i$ where θ is a primitive root of 1 of order k. If $x \in F_\omega$, then $f(x) = 0$. Indeed x is of the form $uu \cdots u$ where u is repeated r times with r dividing k and u being a word of length $q = \frac{k}{r}$. Then $f(x) = \sum_{i=0}^{r-1} \theta^{iq} \varphi(u) = 0$ where $\varphi(u) = \sum_{i=0}^{q-1} u_i \theta^i$. Note also that f is nearly linear as $f(s_i x) = \varepsilon \theta^i + f(x)$ where $\varepsilon \in \{-1, 1\}$ depending on s_i.

Now let $x \notin F_\omega$ and $s_i x$ and $s_j x$ be two neighbors of x in F_ω. Then $f(s_i x) = f(s_j x) = 0$ which implies $\varepsilon \theta^i + f(x) = \varepsilon \theta^j + f(x)$, that is $\theta^i = \pm \theta^j$. So either $i = j$ or $\theta^{i-j} = -1$ which implies k even and $i = j + \frac{k}{2}$. In summary either k is odd and x has at most one neighbor in F_ω, or k is even and x has at most two neighbors in F_ω.

b) We prove now that $V \setminus F_\omega$ is connected. For this we show inductively on j that from e one can reach all the vertices in $V \setminus F_\omega$ at distance at most j. We start from vertex e, then as the vertices at distance 1 are not in F_ω we can reach these vertices. Vertices at distance 2 which are not in F_ω can be reached as they have one neighbor at distance 1, vertices at distance $j > 2$

have at least three neighbors at distance $j - 1$; so by a) one of them is not in F_ω and can be reached from e.

Lemma 21. *For the k-dimensional torus $TM(p)^k$,*

$$g_{F_*}(1, TM(p)^k) = \lceil \frac{p^k - 1}{2k} \rceil \tag{7}$$

Proof. By lemma 17 and lemma 18, it remains to prove that F_ω is not separating. When k is even, the proof looks like that of the hypercube. When k is odd, the proof relies on the analysis of F_ω which contains *"pseudo periodic words"*. Indeed, if $\omega^q(x) = x$, then we have $x_i = x_{i+q} = \cdots = x_{i+rq}$ for r such that $i+rq < k$ plus $x_i = -x_{i+(r_0+1)q}$ where $i + r_0 q < k \le i + (r_0 + 1)q$. So some fixed points are of the form $u_1 u_2 u_1 u_2 \cdots u_1 u_2 u_1$ with $(-u_2)(-u_1) = u_1 u_2$. For example, for $k = 5$, $x = a(-a)a(-a)a$ is a fixed point as $\omega^2(x) = x$ (here $u_1 = a$ and $u_2 = -a$). The full proof is too tedious and specific to be included here and can be found in [14].

Proposition 22. *For the star-graph $S(k)$,*

$$g_{F_*}(1, S(k)) = \lceil \frac{k! - 1}{k - 1} \rceil \tag{8}$$

Proof. a) First let us prove that the star-graph admits a complete rotation.
Recall that we denote a permutation π by the word $\pi(1)\pi(2) \cdots \pi(k)$. Let σ be the permutation $(1, 3, 4, \ldots, k, 2)$; σ fixes 1 and acts cyclically on the elements $2, \ldots, k$. Let us define ω as the automorphism which associates to a permutation π the permutation $\sigma^{-1}\pi\sigma$. We have $\omega(e) = e$ and $\omega(\pi\pi') = \omega(\pi)\omega(\pi')$.
Now let the generator s_i, $0 \le i \le k - 2$, denote the transposition exchanging 1 and $i + 2$. Then $\omega(s_i) = \sigma^{-1}s_i\sigma$.
$s_i\sigma = (i + 3, 3, 4, \ldots, i + 2, 1, i + 4, \ldots, k, 2)$ for $i \le k - 3$
$s_i\sigma = (2, 3, 4, \ldots, k, 1)$ for $i = k - 2$
Then $\sigma^{-1}s_i\sigma = s_{i+1}$.
b) By corollary 19, F_ω is an independent set. Now let us note that if a permutation $\pi \in F_\omega$, then necessarily $\pi(1) = 1$. But one can check that the set U of permutations such that $\pi(1) = 1$ is itself not a separating set. Indeed let x be any vertex such that $\pi(1) \ne 1$ and let j be such that $\pi(j) = 1$. We first do the s_{j-2} in order to put 1 in a correct position and then apply other generators different from s_{j-2} so we will never use a vertex in U.

3   Conclusions

In this paper, we have exhibited a way how to achieve optimal gossiping with packet size limited to 1 in Cayley symmetric digraphs having a complete rotation. It will be nice to characterize this class of digraphs; it does not include all the Cayley symmetric digraphs as they should be *arc-transitive*, but it contains many interesting networks like hypercubes, star-graphs, k-dimensional tori. The technics developed in this paper can be easily extend to the case where we allow p packets to be send (and receive) during each round and it can be shown that in the networks considered in the article, one can optimally gossip in $\lceil \frac{N-1}{pd} \rceil$ rounds.

We have also been able to construct balanced sequences of sets for any 2-dimensional torus $TM(p, q)$, so we can optimally gossip in them. We conjecture that optimal gossip exists in any k-dimensional torus $TM(p_1, p_2, \ldots, p_k)$ although the construction of balanced sequence of sets can be tedious. We have also solved the case of 2-dimensional meshes, which are not Cayley graphs.

Finally let us note that many interesting questions remain open. For example, it would be interesting to determine the computational complexity of computing $g_{F_*}(1, G)$ or $g_{F_*}(p, G)$ for a fixed p; it is very likely that it is NP-hard. It will be also interesting to study the same problem with the model H_* (half-duplex) although in that case, the case $p = \infty$ is already difficult [13].
Another direction will be to use the technics of this paper for other communication problems.

Acknowledgment : We thank M.-C. Heydemann and N. Marlin for their useful remarks concerning the article.

Note added : A notion similar to the complete rotation has also been introduced by P. Fragopoulou, S.G. Akl and H. Meijer [5, 6, 7].

References

1. A. Bagchi, E.F. Schmeichel and S.L. Hakimi.: Parallel Information Dissemination by Packets. SIAM Journal of Computing, **23** (1994) 355–372
2. J.-C. Bermond, L.Gargano, A. Rescigno and U. Vaccaro.: Fast Gossiping by Short Messages. Proc. 22nd ICALP 95, Szeged, Hungary, Lecture Notes in Computer Science, Springer–Verlag, **944**(1995) 135–146.
   An extended version should appear in: SIAM Journal of Computing.
3. D. P. Bertsekas and J. N. Tsitsiklis.: Parallel and Distributed Computation: Numerical Methods. Prentice–Hall, Englewood Cliffs, NJ, (1989)
4. J. de RUMEUR.: Communications dans les réseaux de processeurs. Masson Paris, (1994)
5. P. Fragopoulou.: Communication and Fault Tolerance Algorithms on a Class of Interconnection Networks. Ph.D. Thesis, Department of Computing and Information Science, Queen's University, Kingston, ON, Canada, (1995)
6. P. Fragopoulou and S.G. Akl.: A Framework for Optimal Communication on a Subclass of the Cayley Class Based Networks. Proc. 14th Annual IEEE International Phoenix Conference on Computers and Communications (IPCCC-95), Phoenix, AR, U.S.A. (1995) 241–248

7. P. Fragopoulou, S.G. Akl and H. Meijer.: Optimal Communication Primitives on the Generalized Hypercube Network. Journal of Parallel and Distributed Computing, **32(2)** (1996) 173–187

8. P. Fraigniaud and E. Lazard.: Methods and Problems of Communication in Usual Networks. Discrete Applied Mathematics **53** (1994) 79–134

9. S.M. Hedetniemi, S.T. Hedetniemi and A.L. Liestman.: A Survey of Gossiping and Broadcasting in Communication Networks. NETWORKS, **18** (1988) 319–349

10. A. Hily and D. Sotteau.: Communications in Bus Networks. Parallel and Distributed Computing, Lectures Notes in Computer Science, Springer–Verlag, **805** (1994) 197–206

11. J. Hromkovič, R. Klasing, B. Monien and R. Peine.: Dissemination of Information in Interconnection Networks (Broadcasting and Gossiping). To appear in: F. Hsu, D.-Z. Du (Eds.) Combinatorial Network Theory, Science Press & AMS

12. J.-C. König, P.S. Rao, and D. Trystram.: Analysis of Gossiping Algorithms in Torus with Restricted Bufferization Capabilities. Technical Report IMAG Grenoble, (1994)

13. M. Mahéo and J -F. Saclé.: Note on the Problem of Gossiping in Multidimensional Grids. Discrete Applied Mathematics **53** (1994) 287–290

14. N. Marlin.: Rapport de DEA (1996)

On Embedding 2-Dimensional Toroidal Grids into de Bruijn Graphs with Clocked Congestion One

Thomas Andreae[1], Michael Nölle[2], Christof Rempel[1], Gerald Schreiber[2]

[1] Mathematisches Seminar
Universität Hamburg
Bundesstraße 55
D-20146 Hamburg
Germany
[2] Technische Universität Hamburg-Harburg
Technische Informatik I
Harburger Schloßstraße 20
D-21071 Hamburg
Germany

Abstract. For integers m, d, D with $m \geq 3, d \geq 2$, and $D \geq 2$, let $T(m)$ be a 2–dimensional quadratic toroidal grid with side length m and let $B(d, D)$ be the base d, dimension D de Bruijn graph; assume that $|T(m)| = |B(d, D)|$. The starting point for our investigations is the observation that, for m, D even, embeddings $f : T(m) \rightarrow B(d, D)$ with load 1, expansion 1, and dilation $D/2$ can easily be found (and have previously been described in the literature). In the present paper, we pose the question whether or not there exist embeddings $f : T(m) \rightarrow B(d, D)$ with these properties *and with clocked congestion 1*. We prove results implying a positive answer to this question when d is greater than two. For $d = 2$, we do not have a complete answer, but present partial results.

Key words: de Bruijn graphs, toroidal grids, graph embeddings, clocked congestion, dilation, interconnection networks, parallel computers

1 Introduction

The problem of embedding one interconnection network into another one is of significant importance in the context of parallel computers. Among the network topologies which have been proposed and investigated in the literature, toroidal grids (briefly: tori) and de Bruijn graphs are among the most popular ones; see, e.g., [1, 3, 4, 6, 7, 8, 9]. In the present paper, we deal with the problem of constructing "good" embeddings of a 2-dimensional quadratic torus into a de Bruijn graph $B(d, D)$ with the same number of vertices, where the quality of an embedding is measured in terms of its load, expansion, dilation, and clocked congestion. Before we state our results, we introduce some basic definitions and notations.

For graph-theoretic terminology used but not explained here, we refer to [5]. Graphs, as well as digraphs, may have loops and multiple edges. For a graph (or a digraph) G, $V(G)$ denotes its vertex set; $E(G)$ denotes the edge set of a graph G (or the arc set of G if G is a digraph). For a graph, if there is just one edge joining the vertices v and w, we denote this edge by vw. In a similar way, we use the symbols (v, w) (or, alternatively, $v \rightarrow w$) for digraphs. As usual, a *walk* in a graph G is a sequence $(v_0, e_1, v_1, \ldots, e_k, v_k)$ whose terms are alternately vertices and edges of G, such that, for $1 \leq i \leq k$, the ends of e_i are v_{i-1} and v_i. A walk $(v_0, e_1, v_1, \ldots, e_k, v_k)$ is a *path* if $v_i \neq v_j$ for all $i \neq j$. Similarly, the notions of a *directed walk* and a *directed path* are defined. For a graph G, we denote by G^* the digraph that results from G in the following way. For each pair of distinct vertices v, w of G and each edge e joining v with w, choose a pair $\vec{e}_1, \vec{e}_2$ of arcs such that v is the tail and w is the head of $\vec{e}_1$ and, for the arc $\vec{e}_2$, w is the tail and v is the head. Then, assuming that all so chosen arcs are pairwise distinct, we define G^* as the digraph that results from G by replacing each edge e joining distinct vertices by its corresponding arcs $\vec{e}_1, \vec{e}_2$. The construction of G^* reflects the fact that the communication interfaces of most parallel computers of the MIMD (Multiple Instruction Multiple Data) type are bidirectional. The symbol $\mathbb{Z}_r$ denotes the integers modulo r and operations of elements of $\mathbb{Z}_r$ are to be taken modulo r; we put $\mathbb{Z}_r^n := \{(x_{n-1}, \ldots, x_0) : x_i \in \mathbb{Z}_r (i = 0, \ldots, n-1)\}$.

For integers $d, D \geq 2$, the *directed de Bruijn graph* $\vec{B}(d, D)$ has as its vertex set the set $\mathbb{Z}_d^D$ and there exists an arc from $(a_{D-1}, \ldots, a_0)$ to $(b_{D-1}, \ldots, b_0)$ if and only if $b_i = a_{i-1}(i = 1, \ldots, D-1)$. The corresponding *undirected de Bruijn graph* $B(d, D)$ results from $\vec{B}(d, D)$ by ignoring the orientations of the arcs. We stress the fact that $B(d, D)$ contains loops and pairs of parallel edges. If $G = B(d, D)$, then $B^*(d, D)$ denotes the above defined digraph G^*. In the obvious way, we consider $\vec{B}(d, D)$ as a subgraph of $B^*(d, D)$. Let $\vec{e}$ be an arc of $B^*(d, D)$ joining distinct vertices. Then we say that $\vec{e}$ is in *shuffle direction* if $\vec{e} \in \vec{B}(d, D)$ and, otherwise, $\vec{e}$ is in *unshuffle direction*. Further, by definition, loops of $B^*(d, D)$ are both in shuffle and in unshuffle direction.

For $m \geq 3$, the (*2-dimensional quadratic*) *torus* $T(m)$ is the Cartesian product of two cycles of length m, i.e., $V(T(m)) = \mathbb{Z}_m \times \mathbb{Z}_m$ and two vertices (a, b), (c, d) are joined by an edge iff either $a = c$ and b, d differ by one or $b = d$ and a, c differ by one (mod m).

Let G, H be graphs and assume that G is loopless. Let, further, $f : V(G) \longrightarrow V(H)$ be an injective mapping together with a mapping P_f (called *routing*) which assigns to every arc $\vec{e} \in E(G^*)$ with tail v and head w a directed walk $P_f(\vec{e}) = (f(v) = v_0, \vec{e}_1, v_1, \ldots, \vec{e}_k, v_k = f(w))$ of H^*. Then the pair (f, P_f) (and sometimes just the mapping f) is called an *embedding* of G into H, and $\vec{e}_i$ is called the *i-th arc* of $P_f(\vec{e})$ for $P_f(\vec{e}) = (f(v) = v_0, e_1, v_1, \ldots, e_k, v_k = f(w))$. Further, G is the *guest graph* and H is the *host graph* of the embedding (f, P_f). For a given embedding (f, P_f) of G into H and for $\vec{h} \in E(H^*)$, let $\mu_i'(\vec{h})$ denote the number of arcs $\vec{e} \in E(G^*)$ for which $\vec{h}$ is the i-th arc of $P_f(\vec{e})$; put $\mu_i(\vec{h}) = \mu_i'(\vec{h})$ if $\vec{h}$ joins distinct vertices, and $\mu_i(\vec{h}) = \lceil \mu_i'(\vec{h})/2 \rceil$ otherwise. Then we define the *clocked congestion* of (f, P_f) by

$$\text{clocked cong } (f, P_f) := \max\{\mu_i(\vec{h})\},$$

where the maximum is taken over all $\vec{h} \in E(H^*)$ and all $i \in \{1, \ldots, r\}$ with r denoting the maximum length of the walks $P_f(\vec{e})$. A similar notion, called *dynamic congestion*, is considered in [2].

In the following, the walks $P_f(\vec{e})$ are called the *communication paths* of (f, P_f). Informally speaking, the clocked congestion of an embedding is the maximum number of communication paths that, at the same time and in the same direction, pass through any edge h of the host graph H, where we imagine that loops of H can be traversed in two different directions. (In the above definition, the latter is reflected by putting $\mu_i(\vec{h}) = \lceil \mu_i'(\vec{h})/2 \rceil$ for loops $\vec{h}$.) In comparison with the commonly used measure of congestion, the definition of the clocked congestion provides a more realistic model of the communication capabilities of MIMD parallel computers since, in contrast to the usual notion of congestion, different communication paths can traverse the same arc at different points in time without conflict. This reflects the impact of message passing schemes.

The following definitions are standard. Let (f, P_f) be an embedding of G into H. Then the *dilation* of (f, P_f) (which is denoted by dil(f, P_f)) is the maximum length of the communication paths of (f, P_f). The *load* of a mapping $f : V(G) \longrightarrow V(H)$ is the maximum number of vertices of G which are mapped to the same vertex of H, and thus the fact that we consider only embeddings (f, P_f) with injective f can be expressed by saying that we restrict ourselves to embeddings with load 1 (in symbols, load $(f, P_f) = 1$). Finally, the *expansion* of (f, P_f) is defined by putting expansion $(f, P_f) := \frac{|V(H)|}{|V(G)|}$.

For integers m, d, D with $m \geq 3, d \geq 2$, and $D \geq 2$, let $T(m)$ and $B(d, D)$ be a torus and a de Bruijn graph, respectively, with the same number of vertices, i.e., $m^2 = d^D$. Assume that m and D are even and put $n = \frac{D}{2}$. Then $m = d^n$. Let $\varphi : \mathbb{Z}_{d^n} \longrightarrow \mathbb{Z}_d^n$ be a bijective mapping and let the mappings $\varphi_i : \mathbb{Z}_{d^n} \longrightarrow \mathbb{Z}_d (i = 0, \ldots, n-1)$ be defined such that $\varphi(x) = (\varphi_{n-1}(x), \ldots, \varphi_0(x))$. We then obtain a bijective mapping $f_\varphi : V(T_m) \longrightarrow V(B(d, D))$ by putting (for $a, b \in \mathbb{Z}_m$)

$$f_\varphi((a, b)) = \begin{cases} (\varphi_{n-1}(a), \ldots, \varphi_0(a), \varphi_{n-1}(b), \ldots, \varphi_0(b)) \text{ if } a \equiv b \,(\text{mod } 2) \\ (\varphi_{n-1}(b), \ldots, \varphi_0(b), \varphi_{n-1}(a), \ldots, \varphi_0(a)) \text{ otherwise } . \end{cases} \quad (1)$$

Then, based on the assumption that m is even, one can define (in an obvious way) a corresponding routing P_{f_φ} such that dil$(f_\varphi, P_{f_\varphi}) = \frac{D}{2}$, load $(f_\varphi, P_{f_\varphi}) = 1$, and expansion $(f_\varphi, P_{f_\varphi}) = 1$. (For the details concerning P_{f_φ}, we refer to Section 2.)

For one particular choice of φ, namely, for the case in which φ is the mapping κ which corresponds to the base d representation of the members of $\mathbb{Z}_{d^n}$ (see below), the mapping f_φ was previously considered in [1, 6, 7] and has turned out to be one of the basic tools for constructing embeddings of other network topologies into de Bruijn graphs. One of our results states that clocked cong $(f_\varphi, P_{f_\varphi}) = 2$ for this particular choice of φ, and thus it is natural to ask whether or not there exist mappings φ such that

$$\text{clocked cong } (f_\varphi, P_{f_\varphi}) = 1.$$

It is useful to introduce some notational conventions. For a mapping $\varphi : \mathbb{Z}_{d^n} \longrightarrow \mathbb{Z}_d^n$, we throughout use the symbols $\varphi_i (i = 0, \ldots, n-1)$ as in the paragraph before (1). We write $\varphi = (\varphi_{n-1}, \ldots, \varphi_0)$. Let $\kappa = (\kappa_{n-1}, \ldots, \kappa_0)$ be the mapping $\kappa : \mathbb{Z}_{d^n} \longrightarrow \mathbb{Z}_d^n$ which corresponds to the base d representation of the members of $\mathbb{Z}_{d^n}$, i.e., for $a \in \mathbb{Z}_{d^n}$ the numbers $\kappa_i(a)$ are the uniquely determined elements of $\mathbb{Z}_d$ for which $a = \sum_{i=0}^{n-1} \kappa_i(a) d^i$. We use the following convention: whenever one of the symbols a, b, a', b', x, y denotes a member of $\mathbb{Z}_{d^n}$, then we write $\alpha_i, \beta_i, \alpha_i', \beta_i', \xi_i,$ and η_i for $\kappa_i(a), \kappa_i(b), \kappa_i(a'), \kappa_i(b'), \kappa_i(x),$ and $\kappa_i(y)$, respectively.

We now describe our main results. Among the mappings $\varphi : \mathbb{Z}_{d^n} \longrightarrow \mathbb{Z}_d^n$, we consider mappings of a certain type which are defined as follows. For all integers i, j with $0 \le j \le i \le n-1$, let $\lambda_{i,j} \in \mathbb{Z}_d$ and let $\varphi : \mathbb{Z}_{d^n} \longrightarrow \mathbb{Z}_d^n, \varphi = (\varphi_{n-1}, \ldots, \varphi_0)$, be defined by

$$\varphi_i(a) = \sum_{j=0}^{i} \lambda_{i,j} \alpha_j \quad (a \in \mathbb{Z}_{d^n}, \ i = 0, \ldots, n-1). \tag{2}$$

Then φ is called a *mapping of triangular type* and the $\lambda_{i,j} (0 \le j \le i \le n-1)$ are its *coefficients*. We note that the operations in (2) are to be taken modulo d (because the operands are in $\mathbb{Z}_d$ and because we want the $\varphi_i(a)$ to be in $\mathbb{Z}_d$). In a similar manner, whenever the meaning is clear from the context, several other of the identities and operations in this paper are understood to be mod d or mod d^n without explicit mention.

For $d > 2$, our main result characterizes, in terms of the coefficients $\lambda_{i,j}$, the bijective mappings φ of triangular type for which clocked cong $(f_\varphi, P_{f_\varphi})$ is equal to 1. As a consequence, for $d > 2$, one obtains that there exist bijective mappings $\varphi : \mathbb{Z}_{d^n} \longrightarrow \mathbb{Z}_d^n$ of triangular type such that clocked cong $(f_\varphi, P_{f_\varphi}) = 1$.

This solves the above mentioned existence problem for the case in which d is greater than two. For $d = 2$, we do not have a complete solution of this problem, but have obtained the following partial results:

(i) there exist no bijective mappings $\varphi : \mathbb{Z}_{2^n} \longrightarrow \mathbb{Z}_2^n$ of triangular type for which clocked cong $(f_\varphi, P_{f_\varphi})$ is equal to 1,

(ii) for $n = 2$, bijective mappings $\varphi : \mathbb{Z}_{2^n} \longrightarrow \mathbb{Z}_2^n$ with clocked cong $(f_\varphi, P_{f_\varphi}) = 1$ exist, and

(iii) no such mapping exists for $n = 3$.

2 The clocked congestion one condition

Let m, d, D, n, and $\varphi = (\varphi_{n-1}, \ldots, \varphi_0)$ be as in the paragraph preceding (1) and let $f_\varphi : V(T(m)) \longrightarrow V(B(d, D))$ be the bijective mapping defined by (1). We want to define a corresponding routing P_{f_φ}. For this purpose, let $e \in E(T(m))$. Then there are $a, a', x \in \mathbb{Z}_{d^n}$ such that a and a' differ by one (mod d^n) and such that either $e = (a, x)(a', x)$ or $e = (x, a)(x, a')$. Note that, since m is even, we have $a \not\equiv a' \pmod{2}$. If $e = (a, x)(a', x)$, then (by symmetry) we may assume

$a \equiv x \pmod 2$, and we put $v = (a, x), w = (a', x)$; if $e = (x, a)(x, a')$, then we assume $a' \equiv x \pmod 2$ and put $v = (x, a), w = (x, a')$. In either case, one concludes from $a \not\equiv a' \pmod 2$ that

$$f_\varphi(v) = (\varphi_{n-1}(a), \ldots, \varphi_0(a), \varphi_{n-1}(x), \ldots, \varphi_0(x)),$$
$$f_\varphi(w) = (\varphi_{n-1}(x), \ldots, \varphi_0(x), \varphi_{n-1}(a'), \ldots, \varphi_0(a')).$$

We now define P_{f_φ}. (In the following, for notational simplicity, we write P instead of P_{f_φ}.) We specify directed walks ("communication paths") $P(v \to w)$ and $P(w \to v)$ in $B^*(d, D)$ as follows. Let $P(v \to w) = (v_0, \vec{e}_1, v_1, \ldots, \vec{e}_n, v_n)$ with

$$v_i = \begin{cases} f_\varphi(v) \text{ for } i = 0 \\ f_\varphi(w) \text{ for } i = n \\ (\varphi_{n-1-i}(a), \ldots, \varphi_0(a), \varphi_{n-1}(x), \ldots, \varphi_0(x), \varphi_{n-1}(a'), \ldots, \varphi_{n-i}(a')) \\ \text{for } 1 \leq i \leq n - 1 \end{cases}$$

and such that all arcs $\vec{e}_i$ are in shuffle direction. (We express this by saying that $P(v \to w)$ *is in shuffle direction.*) Let further $P(w \to v) = (v_n, \vec{f}_n, v_{n-1}, \ldots, \vec{f}_1, v_0)$, where $P(w \to v)$ traverses the same vertices v_i as $P(v \to w)$, but in opposite order, and where $\vec{f}_i$ is the arc of $B^*(d, D)$ in unshuffle direction corresponding to $\vec{e}_i$ ($i = 1, \ldots, n$). This defines $P = P_{f_\varphi}$. Finally, let $\mathcal{P}$ denote the set of all so defined communication paths $P(v \to w)$ which are in shuffle direction. We claim that the statement

$$\text{clocked cong}(f_\varphi, P) = 1 \tag{3}$$

is equivalent to the following condition (4), which we call the *clocked congestion one condition* (*ccl-condition* for short).

For all $a, a', b, b' \in \mathbb{Z}_{d^n}$ such that $(a, a') \neq (b, b')$, $a' = a - 1$ or $a + 1$, $b' = b - 1$ or $b + 1$ and for all $i \in \{0, \ldots, n - 1\}$, the $(n + 1)$-tuples $(\varphi_i(a), \ldots, \varphi_0(a), \varphi_{n-1}(a'), \ldots, \varphi_i(a'))$ and $(\varphi_i(b), \ldots, \varphi_0(b),$ $\varphi_{n-1}(b'), \ldots, \varphi_i(b'))$ are distinct. $\tag{4}$

We first show that (4) implies (3). Hence, assuming that (4) holds, let $e = vw, \tilde{e} = \tilde{v}\tilde{w}$ be distinct edges of $T(m)$, where the notations are chosen such that $P(v \to w)$ and $P(\tilde{v} \to \tilde{w})$ are in $\mathcal{P}$. For the edge e, we use the symbols a, a', x as in the first paragraph of this section. Analogously, for the edge $\tilde{e}$, the symbols b, b', y are used.

Then $a \equiv x \pmod 2$ if $e = (a, x)(a', x)$, and $a' \equiv x \pmod 2$ if $e = (x, a)(x, a')$. Similarly, $b \equiv y \pmod 2$ if $\tilde{e} = (b, y)(b', y)$, and $b' \equiv y \pmod 2$ if $\tilde{e} = (y, b)(y, b')$. From this, together with $e \neq \tilde{e}$ and $a \not\equiv a' \pmod 2$, one easily finds that $x = y$ implies $(a, a') \neq (b, b')$. Further, one easily obtains that, in order to prove that (3) holds, it is sufficient to show that

$$(\varphi_i(a), \ldots, \varphi_0(a), \varphi_{n-1}(x), \ldots, \varphi_0(x), \varphi_{n-1}(a'), \ldots, \varphi_i(a')) \neq$$
$$(\varphi_i(b), \ldots, \varphi_0(b), \varphi_{n-1}(y), \ldots, \varphi_0(y), \varphi_{n-1}(b'), \ldots, \varphi_i(b')) \tag{5}$$

for all $i \in \{0, \ldots, n - 1\}$.

For $x \neq y$, this clearly holds, and thus we can assume $x = y$. Hence $(a, a') \neq (b, b')$, and we can use (4) to obtain (5).

In order to show that (4) follows from (3), suppose that (3) holds, but (4) does not hold. Then there are $a, a', b, b' \in \mathbf{Z}_{d^n}, i \in \{0, 1, \ldots, n-1\}$ with $(a, a') \neq (b, b'), a' = a + 1$ or $a - 1, b' = b + 1$ or $b - 1$ such that

$$
\begin{aligned}
&\big(\varphi_i(a), \ldots, \varphi_0(a), \varphi_{n-1}(a'), \ldots, \varphi_i(a')\big) = \\
&\big(\varphi_i(b), \ldots, \varphi_0(b), \varphi_{n-1}(b'), \ldots, \varphi_i(b')\big).
\end{aligned} \tag{6}
$$

Pick $z \in \mathbf{Z}_{d^n}$ with $z \equiv a \pmod 2$ and let $P_1 := P\big((a, z) \to (a', z)\big)$. Let $P_2 := P\big((b, z) \to (b', z)\big)$ if $b \equiv z \pmod 2$ and $P_2 := P\big((z, b) \to (z, b')\big)$ otherwise. Then $P_1, P_2 \in \mathcal{P}$. Let further

$$
\begin{aligned}
w_1 &:= \big(\varphi_i(a), \ldots, \varphi_0(a), \varphi_{n-1}(z), \ldots, \varphi_0(z), \varphi_{n-1}(a'), \ldots, \varphi_{i+1}(a')\big) \quad \text{and} \\
w_2 &:= \big(\varphi_{i-1}(a), \ldots, \varphi_0(a), \varphi_{n-1}(z), \ldots, \varphi_0(z), \varphi_{n-1}(a'), \ldots, \varphi_i(a')\big).
\end{aligned}
$$

Then, by definition of P_1, P_2 and because of (6), the $(n-i)$-th arc of both P_1 and P_2 is the uniquely determined arc of $B^*(d, D)$ which has w_1 as its tail and w_2 as its head and which is in shuffle direction. Hence, because of (3), the arc $(a, z) \to (a', z)$ must be identical to one of the arcs $(b, z) \to (b', z)$, or $(z, b) \to (z, b')$ which in either case contradicts the fact that $(a, a') \neq (b, b')$. This shows that (4) follows from (3).

3 Main results

Throughout this section we assume that d, n are integers with $d \geq 2$ and $n \geq 1$. We start with a simple lemma.

Lemma. Let $\varphi : \mathbf{Z}_{d^n} \longrightarrow \mathbf{Z}_d^n$ be of triangular type with coefficients $\lambda_{i,j} (0 \leq j \leq i \leq n - 1)$. For some $i \in \{0, \ldots, n-1\}$ and some $a, b \in \mathbf{Z}_{d^n}$, assume that $gcd(\lambda_{i,i}, d) = 1$, $\varphi_i(a) = \varphi_i(b)$, and $\alpha_j = \beta_j$ for all j with $0 \leq j < i$. Then $\alpha_i = \beta_i$.

Proof. From $\varphi_i(a) = \varphi_i(b)$ and $\alpha_j = \beta_j$ $(0 \leq j < i)$, together with (2), one obtains $\lambda_{i,i}\alpha_i = \lambda_{i,i}\beta_i$, and thus $\alpha_i = \beta_i$, because the computations are mod d and we are assuming $gcd(\lambda_{i,i}, d) = 1$. $\square$

Based on this lemma, we show the following.

Proposition 1. Let $\varphi : \mathbf{Z}_{d^n} \longrightarrow \mathbf{Z}_d^n$ be of triangular type with coefficients $\lambda_{i,j} (0 \leq j \leq i \leq n - 1)$. Then φ is bijective if and only if $gcd(\lambda_{i,i}, d) = 1$ for all $i \in \{0, \ldots, n-1\}$.

Proof. Assume first that $gcd(\lambda_{i,i}, d) = 1 (i = 0, \ldots, n - 1)$ and let $a, b \in \mathbb{Z}_{d^n}$ with $\varphi(a) = \varphi(b)$. Then, by recursive application of the lemma, $\alpha_i = \beta_i$ ($i = 0, \ldots, n - 1$). Hence $a = b$, and thus the "if direction" is settled.

For the proof of the converse, assume $gcd(\lambda_{i,i}, d) = t > 1$ for some i. Let $A = \{a \in \mathbb{Z}_{d^n} : \alpha_j = 0 \text{ for } j = 0, \ldots, i - 1, \alpha_i = 0 \text{ or } \frac{d}{t}\}$. Then it follows from (2), together with $\lambda_{i,i} \frac{d}{t} \equiv 0 \pmod{d}$, that $\varphi_j(a) = 0$ for all $a \in A$ and all $j \in \{0, \ldots, i\}$. Hence $|\varphi(A)| \leq d^{n-i-1}$. On the other hand, we have $|A| = 2d^{n-i-1}$, and thus φ is not bijective. $\quad\square$

Proposition 2. *Let $\varphi : \mathbb{Z}_{d^n} \longrightarrow \mathbb{Z}_d^n$ be of triangular type with coefficients $\lambda_{i,j}$ $(0 \leq j \leq i \leq n - 1)$ and assume that φ satisfies the cc1-condition. Let further r, s be integers with $0 \leq s \leq r \leq n - 1$ and assume that $d > 2$ if $s = 0$. Then*

$$\lambda_{r,0} + \ldots + \lambda_{r,s} \neq -\lambda_{r,0}. \tag{7}$$

Proof. Suppose that (7) does not hold. Then

$$\lambda_{r,s} = \lambda_{r,s-1}(d-1) + \ldots + \lambda_{r,1}(d-1) + \lambda_{r,0}(d-2) \text{ if } s > 0 \tag{8}$$

and

$$2\lambda_{r,0} = 0 \text{ if } s = 0. \tag{9}$$

Let $A = \{a \in \mathbb{Z}_{d^n} : \alpha_j = d - 1 (0 \leq j \leq s - 1), \alpha_j = 0 (s \leq j \leq r)\}$ if $s > 0$ and $A = \{a \in \mathbb{Z}_{d^n} : \alpha_0 = 1, \alpha_j = 0 (1 \leq j \leq r)\}$, otherwise. Put $A' = \{a' \in \mathbb{Z}_{d^n} : a' = a - 1 \text{ or } a + 1 \text{ for some } a \in A\}$ and let $\mu = \lambda_{r,s}$ if $s > 0$ and $\mu = 0$, otherwise. Then it follows from (2), (8), and (9), together with the assumption that $d > 2$ if $s = 0$, that $\varphi_r(a') = \mu$ for all $a' \in A'$. Moreover, it follows from the definition of A, together with (2), that $\varphi_j(a) = \varphi_j(b)$ for all $a, b \in A$ and all j with $0 \leq j \leq r$. Hence

$$|\{(\varphi_r(a), \ldots, \varphi_0(a), \varphi_{n-1}(a'), \ldots, \varphi_r(a')) : a \in A, a' = a - 1 \text{ or } a + 1\}| \leq d^{n-1-r}.$$

On the other hand, there are $2d^{n-1-r}$ distinct pairs (a, a') with $a \in A$, $a' = a - 1$ or $a + 1$, and thus we have obtained a contradiction to the assumption that φ satisfies the cc1-condition. $\quad\square$

The next theorem, its corollary and the subsequent Theorem 6 are the main results of this paper.

Theorem 3. *For $d > 2$, let $\varphi : \mathbb{Z}_{d^n} \longrightarrow \mathbb{Z}_d^n$ be of triangular type with coefficients $\lambda_{i,j}$ $(0 \leq j \leq i \leq n - 1)$. Then φ is a bijective mapping satisfying the cc1-condition if and only if the following statements (i) and (ii) hold:*

(i) $gcd(\lambda_{i,i}, d) = 1$ for all $i \in \{0, \ldots, n - 1\}$
(ii) $\lambda_{i,0} + \ldots + \lambda_{i,j} \neq -\lambda_{i,0}$ for all i, j with $0 \leq j \leq i \leq n - 1$.

Proof. The propositions 1, 2 already cover the "only if" direction. For the converse direction, assume that (i) and (ii) hold. Then, by Proposition 1, φ is bijective, and thus it remains to show that φ satisfies the $cc1$-condition. To this end, suppose that there exist $a, a', b, b' \in \mathbf{Z}_{d^n}$ with $(a, a') \neq (b, b'), a' = a - 1$ or $a + 1$, $b' = b - 1$ or $b + 1$ such that, for some $i \in \{0, \dots, n - 1\}$,

$$\varphi_j(a) = \varphi_j(b)\ (0 \leq j \leq i) \text{ and } \varphi_j(a') = \varphi_j(b')\ (i \leq j \leq n - 1). \tag{10}$$

By recursive application of the lemma, it follows from (10) that

$$\alpha_j = \beta_j\ (0 \leq j \leq i). \tag{11}$$

We separately discuss two cases.

Case 1. $a' = a - 1, b' = b - 1$ or $a' = a + 1, b' = b + 1$.

Then it follows from (11), together with the hypothesis of case 1, that $\alpha'_j = \beta'_j\ (j = 0, \dots, i)$. Consequently, by (10) and recursive application of the lemma, we have $\alpha'_j = \beta'_j\ (j = i + 1, \dots, n - 1)$. Hence $a' = b'$, and thus $a = b$, in contradiction to the assumption $(a, a') \neq (b, b')$. This settles case 1 and thus, by symmetry, it remains to consider the following case.

Case 2. $a' = a + 1, b' = b - 1$.

Then, since $\alpha_0 = \beta_0$ by (11), there exists some $k \in \{0, \dots, i\}$ such that $\alpha'_j = \alpha_j + 1\ (0 \leq j \leq k), \alpha'_j = \alpha_j\ (k < j \leq i), \beta'_0 = \beta_0 - 1, \beta'_j = \beta_j\ (1 \leq j \leq i)$, or such that $\beta'_j = \beta_j - 1\ (0 \leq j \leq k), \beta'_j = \beta_j\ (k < j \leq i), \alpha'_0 = \alpha_0 + 1, \alpha'_j = \alpha_j\ (1 \leq j \leq i)$. In either case, (11) implies $\alpha'_0 - \beta'_0 = 2, \alpha'_j - \beta'_j = 1\ (1 \leq j \leq k), \alpha'_j - \beta'_j = 0(k < j \leq i)$. Further, by (10),

$$\lambda_{i,0}\alpha'_0 + \dots + \lambda_{i,i}\alpha'_i = \varphi_i(a') = \varphi_i(b') = \lambda_{i,0}\beta'_0 + \dots + \lambda_{i,i}\beta'_i$$

and thus

$$\lambda_{i,1}(\alpha'_1 - \beta'_1) + \dots + \lambda_{i,i}(\alpha'_i - \beta'_i) = \lambda_{i,0}(\beta'_0 - \alpha'_0)$$

where the left hand side is understood to be zero if $i = 0$. It follows that $\lambda_{i,1} + \dots + \lambda_{i,k} = -2\lambda_{i,0}$ if $k \geq 1$ and $0 = -2\lambda_{i,0}$, otherwise, which contradicts the assumption that (ii) holds. □

As a consequence of Theorem 3, one obtains the following corollary.

Corollary 4. *For all d, n with $d > 2$ and $n \geq 1$, there exist bijective mappings $\varphi : \mathbf{Z}_{d^n} \longrightarrow \mathbf{Z}_d^n$ of triangular type satisfying the $cc1$-condition.*

Proof. Let $\varphi : \mathbf{Z}_{d^n} \longrightarrow \mathbf{Z}_d^n$ be of triangular type with coefficients $\lambda_{i,0} = 1(i = 0, \dots, n - 1), \lambda_{i,i} = -1\ (i = 1, \dots, n - 1)$, and $\lambda_{i,j} = 0$, otherwise. Then φ satisfies the conditions (i) and (ii) of Theorem 3. (For other examples, see below.) Hence the assertion follows from Theorem 3. □

Given a bijective mapping $\varphi : \mathbf{Z}_{d^n} \longrightarrow \mathbf{Z}_d^n$ satisfying the $cc1$-condition, the next proposition shows how to obtain from φ other mappings φ' having the same nice properties.

Proposition 5. *If* $\varphi : \mathbb{Z}_{d^n} \longrightarrow \mathbb{Z}_d^n, \varphi = (\varphi_{n-1}, \ldots, \varphi_0)$, *is bijective and satisfies the ccl-condition, then the same holds for* $\varphi' = (\varphi'_{n-1}, \ldots, \varphi'_0)$ *with* $\varphi'_i = \sigma_i \circ \varphi_i$ *where the* σ_i *are permutations of* $\mathbb{Z}_d$ $(i = 0, 1, \ldots, n-1)$.

Proof. Clearly φ' is bijective. Assume that φ' does not satisfy the ccl-condition. Then there exists an $i, 0 \leq i < n$, together with $a, b \in \mathbb{Z}_{d^n}$ such that $(a, a') \neq (b, b')$, $a' = a + 1$ or $a - 1, b' = b + 1$ or $b - 1$ and such that

$$\left(\varphi'_i(a), \ldots, \varphi'_0(a), \varphi'_{n-1}(a'), \ldots, \varphi'_i(a')\right) =$$
$$\left(\varphi'_i(b), \ldots, \varphi'_0(b), \varphi'_{n-1}(b'), \ldots, \varphi'_i(b')\right).$$

Hence

$$\sigma_j \circ \varphi_j(a) = \sigma_j \circ \varphi_j(b) \quad \text{for all} \quad j = 0, 1, \ldots, i \quad \text{and}$$
$$\sigma_j \circ \varphi_j(a') = \sigma_j \circ \varphi_j(b') \quad \text{for all} \quad j = i, \ldots, n - 1.$$

Multiplying these equations with σ_j^{-1} from the left side yields $\varphi_j(a) = \varphi_j(b)$ for all $j = 0, 1, \ldots, i$ and $\varphi_j(a') = \varphi_j(b')$ for all $j = i, \ldots, n - 1$. This means that φ violates the ccl-condition, in contradiction to the assumptions on φ. $\square$

By making use of Theorem 3, it is easy to construct triangular type mappings φ which are bijective and satisfy the ccl-condition (for given values of d and n such that $d \geq 3$ and $n \geq 1$). Having already fixed the values of $\lambda_{0,0}, \lambda_{1,0}, \lambda_{1,1}, \lambda_{2,0}, \ldots, \lambda_{k,0}, \ldots, \lambda_{k,l}$, the choice of the next candidate $\lambda_{k',l'}$ with $(k', l') = (k, l + 1)$ if $l < k$, and $(k', l') = (k + 1, 0)$ otherwise, is (according to Theorem 3) only limited by

     (i) $\gcd(\lambda_{k',l'}, d) = 1$ if $k' = l'$

and (ii) $\lambda_{k',0} + \ldots + \lambda_{k',l'} \neq -\lambda_{k',0}$.

For example the following list of coefficients for $n = 3, d = 3$ satisfies this restriction (at each step the forbidden values are given in parentheses):

$$\lambda_{0,0} = 2(0), \ \lambda_{1,0} = 1(0), \ \lambda_{1,1} = 2(0,1), \ \lambda_{2,0} = 2(0), \ \lambda_{2,1} = 0(2), \ \lambda_{2,2} = 1(0,2).$$

Theorem 6. *For* $n \geq 2$, *there exist no bijective mappings* $\varphi : \mathbb{Z}_{2^n} \longrightarrow \mathbb{Z}_2^n$ *of triangular type satisfying the ccl-condition.*

Proof. Let $n \geq 2$ and suppose that $\varphi : \mathbb{Z}_{2^n} \longrightarrow \mathbb{Z}_2^n$ is a bijective mapping of triangular type such that φ satisfies the ccl-condition. Let $\lambda_{i,j}(0 \leq j \leq i \leq n-1)$ denote the coefficients of φ. We claim that

$$\lambda_{i,j} = 0 \text{ for all } i, j \text{ with } 1 \leq j < i \leq n - 1. \tag{12}$$

For the proof of (12), suppose that $\lambda_{i,j} = 1$ for some indices i, j with $1 \leq j < i \leq n-1$ and assume that j is chosen such that $\lambda_{i,k} = 0$ for all k with $1 \leq k < j$. Let h be the least index greater than j such that $\lambda_{i,h} = 1$. (An index h with these properties exists because $\lambda_{i,i} = 1$ by Proposition 1.) But then

$$\lambda_{i,0} + \ldots + \lambda_{i,h} = \lambda_{i,0} = -\lambda_{i,0},$$

which (because $h > 0$) contradicts Proposition 2. This proves (12).

As a consequence of (2), (12), and the fact that $\lambda_{i,i} = 1$ by Proposition 1, one obtains that

$$\varphi_0(x) = \xi_0, \ \varphi_i(x) = \lambda_{i,0}\xi_0 + \xi_i \ (1 \leq i \leq n-1) \tag{13}$$

for all $x \in \mathbf{Z}_{2^n}$. Now, by means of (13), it is easy to find $a, a', b, b' \in \mathbf{Z}_{2^n}$ with $(a, a') \neq (b, b'), a' = a - 1$ or $a + 1, b' = b - 1$ or $b + 1$ such that, for some $i \in \{0, \ldots, n-1\}$, the $cc1$-condition is violated. We just specify the choices of a, a', b, b' and the value of i, leaving all further details to the reader:

$$a = 1, \ b = 1, \ a' = 0, \ b' = 2, \ i = 2 \ \text{ if } n \geq 3$$
$$a = 1, \ b = 3, \ a' = 0, \ b' = 0, \ i = 0 \ \text{ if } n = 2, \ \lambda_{1,0} = 0$$
$$a = 0, \ b = 2, \ a' = 1, \ b' = 1, \ i = 0 \ \text{ otherwise .}$$

$\square$

Theorem 6 settles the case of mappings of triangular type. The general problem whether or not there exist bijective mappings $\varphi : \mathbf{Z}_{2^n} \longrightarrow \mathbf{Z}_2^n$ satisfying the $cc1$-condition remains open. For $n = 1$ and $n = 2$ mappings φ with these properties exist: for $n = 1$, we just choose the identity mapping, while for $n = 2$, the mapping φ defined by $\varphi(a) = (\alpha_1, \alpha_1 + \alpha_0)$ has the desired properties, as can easily be checked. For $n = 3$, however, one can show that no such mappings φ exist: the proof can be carried out by exhaustive case distinctions, where by symmetry it suffices to consider a few cases. We leave the details to the reader.

4 An embedding with clocked congestion two

Among the mappings $f_\varphi : V(T(m)) \to V(B(d, D))$ defined as in (1), probably the first example which comes to mind is the mapping which corresponds to the base d representation of the members of $\mathbf{Z}_{d^n}$, i.e. the mapping f_κ where $\kappa : \mathbf{Z}_{d^n} \to \mathbf{Z}_d^n$ is defined as in the introduction. We have already mentioned that the mapping f_κ was previously considered in [1, 6, 7]. Observe that Theorem 3 immediately implies clocked cong $(f_\kappa, P_{f_\kappa}) \geq 2$. In this section, we prove the following result.

Theorem 7. Let $\kappa : \mathbf{Z}_{d^n} \to \mathbf{Z}_d^n$ be defined as in the introduction. Then clocked cong $(f_\kappa, P_{f_\kappa}) = 2$.

We first prove the subsequent Proposition 8 and then obtain Theorem 7 as a consequence of this proposition.

Proposition 8. Let $a, a', b, b' \in \mathbf{Z}_{d^n}$ be such that $(a, a') \neq (b, b'), a' = a - 1$ or $a + 1, b' = b - 1$ or $b + 1$ and assume that, for some $i \in \{0, \ldots, n-1\}$, $(\alpha_i, \ldots, \alpha_0, \alpha'_{n-1}, \ldots, \alpha'_i) = (\beta_i, \ldots, \beta_0, \beta'_{n-1}, \ldots, \beta'_i)$. Then either $a = b$ or $a' = b'$, where the latter can occur only if $d = 2$.

Proof. By assumption, we have

$$\alpha_j = \beta_j \ (j = 0, \dots, i), \tag{14}$$

$$\alpha'_j = \beta'_j \ (j = i, \dots, n-1). \tag{15}$$

We show that $a \neq b$ implies $a' = b'$ and $d = 2$. Hence let $a \neq b$. Then, because of (14), there exists an index $k, i < k \leq n-1$, such that $\alpha_k \neq \beta_k$. By (15), we have $\alpha'_k = \beta'_k$, and thus $\alpha_k \neq \alpha'_k$ or $\beta_k \neq \beta'_k$. Clearly, it suffices to settle the case in which the former holds, which we assume now. From this, together with $a' = a - 1$ or $a + 1$, we conclude $\alpha_j \neq \alpha'_j (j = 0, \dots, k)$. Hence $\alpha_i \neq \alpha'_i$, which together with (14) and (15) implies $\beta_i \neq \beta'_i$. Hence, because $b' = b - 1$ or $b + 1$, we have $\beta_j \neq \beta'_j (j = 0, \dots, i)$. Thus we have

$$\alpha_j \neq \alpha'_j, \ \beta_j \neq \beta'_j \ (j = 0, \dots, i). \tag{16}$$

If $a' = a - 1, b' = b - 1$ or $a' = a + 1, b' = b + 1$, then (14) implies $\alpha'_j = \beta'_j \ (j = 0, \dots, i)$, from which, together with (15), one concludes $a' = b'$. Consequently, we have $a = b$, in contradiction to the hypothesis $a \neq b$. Hence

$$a' = a - 1, b' = b + 1 \text{ or } a' = a + 1, b' = b - 1. \tag{17}$$

From (14), (16), and (17) one concludes $i = 0$. Hence $a' = b'$ by (15). Moreover, one concludes from $i = 0$, together with (14) and (15), that $\alpha_0 = \beta_0, \alpha'_0 = \beta'_0$. But this is compatible with (17) only if $d = 2$, and thus we have finished the proof of Proposition 8. $\qquad\square$

Proof of Theorem 7. It remains to show clocked cong $(f_\kappa, P_{f_\kappa}) \leq 2$. For this purpose, let $e = vw, \tilde{e} = \tilde{v}\tilde{w}$ be edges of $T(m)$, where the notations are chosen such that $P(v \to w), P(\tilde{v} \to \tilde{w}) \in \mathcal{P}$. (For simplicity, we write P instead of P_{f_κ}; for the definition of $\mathcal{P}$, see Section 2.) We use the symbols a, a', x as in the first paragraph of Section 2. Analogously, for the edge $\tilde{e}$, the symbols b, b', y are used. Assume that the communication paths $P(v \to w), P(\tilde{v} \to \tilde{w})$ have their $(n - i)$-th arc in common. Then

$$(\alpha_i, \dots, \alpha_0, \xi_{n-1}, \dots, \xi_0, \alpha'_{n-1}, \dots, \alpha'_i) = (\beta_i, \dots, \beta_0, \eta_{n-1}, \dots, \eta_0, \beta'_{n-1}, \dots, \beta'_i).$$

Hence $x = y$. Further, Proposition 8 implies that $a = b$ or $a' = b'$. In each case, we have $a \equiv b \pmod 2$ and thus $a \equiv x \pmod 2$ if and only if $b \equiv y \pmod 2$. Hence we either have $e = (a, x)(a', x), \tilde{e} = (b, y)(b', y)$ or $e = (x, a)(x, a')$, $\tilde{e} = (y, b)(y, b')$.

Now suppose that $\widetilde{\tilde{e}}$ is an edge of $T(m)$ such that the corresponding communication path $P(\tilde{v} \to \tilde{w}) \in \mathcal{P}$ has the $(n - i)$-th arc in common with both $P(\tilde{v} \to \tilde{w})$ and $P(v \to w)$. For $\widetilde{\tilde{e}}$, we use the symbols c, c' analogously to the use of a, a' and b, b' for e and $\tilde{e}$, respectively. We may assume that $e = (a, x)(a', x)$ because the case $e = (x, a)(x, a')$ can be treated similarly. Then $\tilde{e} = (b, x)(b', x)$, $\widetilde{\tilde{e}} = (c, x)(c', x)$ with

$$a = b \quad \text{or} \quad a' = b',$$
$$b = c \quad \text{or} \quad b' = c',$$
$$a = c \quad \text{or} \quad a' = c'.$$

Hence $a = b = c$ or $a' = b' = c'$, which implies (because $a' = a - 1$ or $a + 1$, $b' = b - 1$ or $b + 1$, $c' = c - 1$ or $c + 1$) that two of the edges $e, \tilde{e}, \tilde{\tilde{e}}$ must be identical. This shows that clocked $\mathrm{cong}(f_\kappa, P_{f_\kappa}) \leq 2$. □

References

1. T. ANDREAE, M. NÖLLE, and G. SCHREIBER. *Cartesian products of graphs as spanning subgraphs of de Bruijn graphs*, in *Proceedings of the 20th Workshop on Graph-Theoretic Concepts in Computer Science (WG 94)*, Herrsching, June 1994. Lecture Notes in Computer Science **903**, pp. 140–151, Springer (1995).

2. F. ANNEXSTEIN, M. BAUMSLAG, and A. L. ROSENBERG. *Group Action Graphs And Parallel Architectures*, SIAM Journal on Computing **19**, pp. 544–569 (1990).

3. D. BARTH, *Embedding meshes of d-ary trees into de Bruijn graphs*, Parallel Processing Letters **3**, pp. 115–127 (1993).

4. J.C. BERMOND and C. PEYRAT, *De Bruijn and Kautz networks: a competitor for the hypercube?*, in *Hypercube and Distributed Computers*, F. Andre and J. Verjus, Eds., Amsterdam, 1989, pp. 279–294, North Holland, Elsevier Science Publisher.

5. J.A. BONDY and U.S.R. MURTY, *Graph theory with applications*, McMillan, London, 1976.

6. M.C. HEYDEMANN, J. OPATRNY, and D. SOTTEAU, *Embeddings of hypercubes and grids into de Bruijn graphs*, Technical Report 723, LRI, University of Paris-Sud, Orsay, France, December 1991.

7. M.C. HEYDEMANN, J. OPATRNY, and D. SOTTEAU, *Embeddings of hypercubes and grids into de Bruijn graphs*, Journal of parallel and distributed computing **23**, pp. 104–111 (1994).

8. F.T. LEIGHTON. *Introduction to Parallel Algorithms and Architectures*, Morgan Kaufman Publishers, San Mateo, California, 1992.

9. B. MONIEN and H. SUDBOROUGH, *Embedding one interconnection network in another*, Computing Suppl., Springer Verlag, vol. 7, pp. 257–282, 1990.

N-Cube String Matching Algorithm with Long Texts

Fouzia Moussouni [*] - Christian Lavault [*]

LIPN, *CNRS URA 1507*

Université Paris-XIII,

Av. J.-B. Clément 93430 Villetaneuse. France

email: mouf@lipn.univ-paris13.fr

Abstract. In this paper, we present a distributed algorithm which runs on the N-cube and solves the string matching problem. A basic prefix-suffix matching technique is used as a building block in the construction of the algorithm. As opposed to the parallel algorithm so far designed on shared-memory PRAM models, our algorithm runs distributively on fixed N-cube and especially applies to long texts. Given a text $t[1..n]$ and a pattern $x[1..m]$ cut into N non-overlapping pieces each of size $s \leq m$, and taking the preprocessing into account, one can find all occurrences of x in t in time $O(s + \log N)$ with N processors on a MIMD hypercube.

1 Introduction

Let $t[1..n]$ (the text) and $x[1..m]$ (the pattern) be two strings over an alphabet Σ. In the string matching problem, we want to know whether x is a substring of t, and if so whereabouts in t it occurs, i.e where are the starting positions of all the occurrences of x in t.

The first sequential pattern matching algorithm to achieve worst case time optimality is the well-known algorithm designed by Knuth, Morris and Pratt in [13]. Its runs in $O(n)$ time and uses $O(m)$ space. Major extensions and improvements in terms of time-space optimality may be found in [2,3,7].

Since the 80's, the string matching problem was also extensively studied on the PRAM computation model. Z. Galil presented in [5] the first optimal parallel string matching algorithm with $O(\log m)$ time complexity for the CRCW/PRAM. More recently, and by introducing the powerful idea of sampling, the text searching was improved by Vishkin to $O(\log^* m)$ time in [12] and then to constant time by Galil in [4] but with slow preprocessing. Recently, Gąsieniec et al. speeded up the preprocessing to $O(\log \log m)$ time. They designed a new recursive parallel 'divide and conquer' algorithm based on the pseudo-period technique. (see [3])

However, the string matching problem has been somewhat neglected on more realistic models with fixed interconnection networks and non-shared memory (e.g the hypercube). In such interconnection networks, routing must be taken into account.

On the mesh connected array of processors, a parallel pattern matching algorithm was also considered in [1]. It is based on the pseudo-period technique and runs in time $O(\sqrt{n})$ on a $\sqrt{n} \times \sqrt{n}$ mesh computation model.

All present algorithms on such models have a fine-granularity (one symbol of the text allocated to each processor) without achieving optimality. One can use the same techniques (effective simulations, pseudo-period ...) with fewer number of processors. Each processor deals with s successive symbols of the text by simulating s virtual processors in its local memory. But it turns out to be unfeasible when having quite a long text per processor.

More exactly, when having N fixed processors on a parallel computer and a reasonably big ratio $s = \lceil \frac{n}{N} \rceil$, it is more efficient to use sequential computations on such long texts in parallel and perform a very limited communication packets. Without loss of generality we assume that N divides n; if not, then t can be padded on the the right with some symbol not occurring in the alphabet of the pattern x and the text t. So, each processor deals with a text segment T_i of size $s \geq m$ such that $\sum_{i=1}^{N} |T_i| = n$. Some of the N-cube's links connect the end of each segment T_i to the beginning of the next segment T_{i+1}. By contrast with the classical string matching problem, there exist pattern occurrences that span two consecutive segments of the text.

We use a prefix-suffix matching technique as a basic building block in the construction of the algorithm to improve results of [8] for finding all overlapping occurrences imposed by the text segmentation. We show that the overall string matching problem can be solved in time $O(s)$, constant space and linear total work on an MIMD hypercube with no more than N processors.

2 Basic definitions

We recall here, some basic definitions on words used in our algorithm.
The text t is a sequence of n symbols. The notation $|t| = n$ means that n is the length of t.

A pattern x is a string of length $|x| = m$, where $m \leq n$. The pattern x is said to have an occurrence in t at position i, where $i \leq n - m$ if $x_1..x_m = t_i..t_{i+m-1}$. The notation u^k means k concatenated copies of the word u.

The period
Periodicity is a fundamental property on strings which is close connected to string matching problem. A subword u of x is a period of x if $x = u^k.v$, for $k > 0$ and where v is a proper prefix, (possibly empty) of x.

Any word has at least one period (including itself). The shortest period of the word x is the *period* of x. If the period is shorter than a half of x, then x is said to be periodic. Thus, a periodic word is of the form $u^k.v$ where $k \geq 2$ and v is a possibly empty prefix of x.

Lemma 1.
Let p be the length of the period of a word x, and let t be an arbitrary word of length $|t| \geq |x|$. Then the following two facts holds :

1. *if c is any period of x such that $|c| \leq |x| - p$, then c is a multiple of x.*

2. *if x occurs in t at positions i and j such that $i < j$, then $j - i \geq p$. If moreover $j - i \leq |x| - p$ then $j - i$ satisfies the inequality $j - i \geq p$ and x occurs at positions $i + kp$ for any integer k such that $i + k.p \leq j$.*

   Proofs of this lemma can be found in [9]

3 The parallel computation model

Many computation model structures could be thought of to deal with string processing. We have chosen the hypercube for its versatility among other reasons. Moreover, the commercially available parallel machines that exhibit a good performance/cost ratio are often based on the hypercube structure.

Thus, the d-dimensionnal hypercube H_d has $N = 2^d$ processors and $d.2^{d-1}$ links. To each processor corresponds an d-bit labels string. Two adjacent processors have their labels different in precisely one bit. This leads to a maximal $d = log(N)$ distance between two processors. This is a small distance comparatively to the network density.

The hypercube is a symmetric network due to the same interconection model for each processor. The capability (e.g the memory capacity) of processors are also similar. Processors communicate with their neighbours by exchanging paquets.

The hypercube has also a recursive structure which is particularly useful whenever a computational task can be recursively decomposed into parallel sub-tasks to be executed on some subcubes.

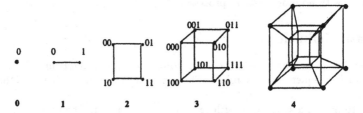

Fig.1. Hypercubes of dimension 0, 1, 2, 3, 4.

Another important property is that the hypercube can simulate many other known networks as $O(N)$-node arrays, rings, meshes, binary trees. . .etc. This way of looking at the hypercube makes it possible to implement simple parallel algorithms based on lines, rings and meshes, without omitting its attractive properties.

To have an excellent task mapping, most of the hypercube string matching algorithms use general simulations on lines, rings (e.g, pipelining or simple LPM algorithm of [1]), and also on meshes with a snake-like ordering (the pseudo-period algorithm).

4 N-cube String Matching with Long Texts

The hypercube can optimally map a two way one-dimensional array of processors. This technique uses a Gray code to map the processing elements (PE), $0...N-1$ such that PE_i is directly connected to PE_{i-1} and PE_{i+1}.

The text t is first broken into non-overlapping s_blocks and then distributed over the PE's such that each PE_i contains subtext $T_i = t_{s.i+1}...t_{s.(i+1)}$. Then we say that position j in the text is at position j' in T_i whenever $j = i.s + j'$ where $s = \lceil \frac{n}{N} \rceil$.

The whole pattern is also broadcasted to all other PE's. By using a spanning tree embedding technique (see [7]) for broadcasting O(m) items to all PE's of an N-cube, the overall time needed for this initialization phase is bounded by $O(m + \log N)$.

This allocation scheme places adjacent segments of the text in adjacent processors. Thus, what we want to do is to find all the occurrences of the pattern across multiple segments by classical computations in each processor. The main difficulty is that the pattern may occur in some subtext T_i as a substring overlapping two adjacent segments, viz. either T_i and T_{i+1} or T_i and T_{i-1}, imposed by the text segmentation.

O.H. Ibarra *et al.* proposed in [8] a technique to alleviate this by reading into PE_i the prefix of size $m-1$ of the subtext stored in PE_{i+1} and then searching for the pattern in this enlarged segment. Using the one dimensional array connection, the corresponding transfer is performed in $O(m)$ time in parallel and with $O(N)$ communication messages of size m.

To avoid such transfer which involves many communication packets, our algorithm performs the matching of some pattern prefixes and suffixes within adjacent segments. Note that a substring of x is overlapping two adjacent segments T_i and T_{i+1}, if there exists a prefix u of x that occurs at the end of T_i and a suffix v of x that occurs at the beginning of T_{i+1} so that $x = u.v$.

The way the algorithm operates depends on whether the pattern has some periods or none (except itself). The possible periodicity of the pattern is the key part of the algorithm.

4.1 The strictly non-periodic case

We define a word x as a strictly non-periodic pattern if x satisfies the following two properties.

(i) x has no periods, but itself and,
(ii) Any prefix or suffix of x is strictly non-periodic.

Pattern : $x = abdrghabe$, is strictly non-periodic.

Pattern : $x = aaabaaabac$, is not strictly non-periodic because there appears one periodicity in the prefix of size 9 (*period* = 4).

According to the period definition, there exists at most one occurrence of a strictly non-periodic pattern in each *m_block* of the text. This fact holds for occurrences that span two adjacent segments.

The algorithm first searches for all occurrences within the subtexts T_i's, by running either the KMP algorithm or the best existing algorithm with linear time and constant space complexities ([2, 5, 6]). It returns the prefix value u of x located at the end of T_i and the suffix value v of x located at the beginning of T_i.

To find occurrences that occurs across two segments T_i and T_{i+1}, one constant size communication message, which carries the prefix length $|u|$ (if any), is sent to PE_{i+1} by each processor PE_i.

Now consider the suffix v located at the beginning of T_{i+1}. When processor PE_{i+1} receives the message carrying the value $|u|$, it checks the equality $|u| + |v| = m$. If the equality holds, one (and only one) occurrence of x exists within the text t and it is located at the starting position in u within the subtext T_i. This completes the proof of the following lemma.

Lemma 2.
Given a text t of size n and a non-periodic pattern x of size m, the above algorithm finds all occurrences of x in text t in time $O(s)$ with constant space and $O(N)$ communication messages, with constant size, on the $O(N\text{-cube})$.

4.2 The periodic case

In this section we sketch the case where the pattern is either partially periodic with the presence of periodicity in some prefixes (or suffixes), or fully periodic with at least one period.

Example

Pattern : $x = aaabaaabcde$ is partially periodic. It has some prefix period pp (e.g $pp = 4$ for the prefixe of size 8).
Pattern : $x = aaabaaa$ is fully periodic. It has $4, 5, 6$ as periods size (or periods).
Pattern : $x = (ababc)^3abab$, is periodic with periods $5, 10, 15, 17, 19$.

Let p be the shortest period of x. Observe that since $p < m$, there exists at most one occurrence of x in each p_block of subtext T_i. Because of the periodicity of x, more than one occurrence may span consecutive segments.

As for the strictly non-periodic case, we search for occurrences that are within the subtexts by running the fastest sequential string matching algorithm on each processor. Now, to find occurrences in the overlapping endpoints of a segment T_i, we need to compute some prefix-suffix information.

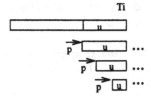

Fig.2. Prefixes of a periodic pattern

We can find naïvely each prefix u and suffix v in linear time and for each pair (u, v) checks if it matches an occurrence of x. A single communication message can be used. But it turns out to be time expensive when having long patterns with short periods. More exactly, we need $O(\frac{m^2}{p})$ time and $O(\frac{m}{p})$ communication packets of size $O(1)$ for each linked pair of segments in the N-cube.

To obtain more efficient results both in terms of time and communication, we combine the period with the longest prefix and the longest suffix of x. These three latter points are essential as shown by the following lemma.

Lemma 3.

Let u be the longest prefix of x that matches the appropriate suffix of T_i starting at position r and v be the longest suffix of x that matches an appropriate prefix of the subtext of processor PE_{i+1}; Let p be the shortest period of x. If $|u|+|v| \equiv m \pmod{p}$ (1) holds, then at least one occurrence of x overlaps T_i and T_{i+1} at each starting positions $j = r + kp$ such that $r \leq j \leq s(i+1)$ and $j + m - 1 \leq s(i+1) + |v|$.

Proof

Since $u = x[1..r]$ is the longest prefix of x (if there is any) located at the end of a processor text T_i of size s, then, there exists an indice a such that $x[1...r] = t_a...t_{s.(i+1)}$.

Recall that if p is the period of x, it is also the period of all prefixes of x, and $x[i] = x[i-p]$ for all $1 \leq i \leq m - p$.

It follows that $u_1 = x[1...r-p]$ is also a suffix of both u and T_i. In general, the prefix $u_k = x[1..r-kp]$ is a suffix of both u and T_i for all k in $[0...\frac{r}{p}]$.

By symmetry, if $v = x[q...m]$ is the longest suffix of x located at the beginning of text T_{i+1}, and since p is also the period length of v, the suffix $v_j = x[q+jp,..,m]$ is the prefix of both v and T_{i+1}. In case of some overlaps, there exists at least one pair (u_k, v_j) such that $|u_k| + |v_j| = m$ \hfill (1).

Now, Knowing that,

- $|u_k| = r - k - p$ and $|v_j| = m - jp - (q-1)$,

- $r = |u|$ and $q = m - |v| + 1$

(1) leads to $|u| + |v| \equiv m \pmod{p}$.

From lemma 3, it follows that the knowledge of the shortest period (if any) of the pattern makes it possible to determine the position of all occurrences that overlap each double segment by only matching the longest prefix and the longest suffix, and then by checking the above condition (1). Moreover, x start at each position $j = s(i+1) - |u| + kp + 1$ for $k \geq 0$ and ends at position $j + m - 1 \leq s(i+1) + |v|$.

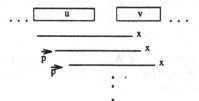

Fig.3. overlapping with regard to the shortest period

Note that in this case, only one constant size message carrying the prefix size $|u|$ is sent to processor PE_{i+1}. An additional linear time (in the pattern size) is used for searching the words u and v which is more insignificant with respect to communication time. Finally, we can combine these facts with the techniques used in the previous non-periodic case algorithm using $O(N)$ communication messages of size $O(1)$.

Remarks

The fact that condition $|u| + |v| \equiv m \pmod{p}$ does not hold is necessary to ensure that no occurrences of x overlap a pair of segments. Moreover, the condition is not sufficient, since for example the pattern may either be periodic with regard to longer periods or non periodic but with periodic prefixes (or suffixes).

Consider the word $x = aaabaaa$, $|x| = 7$ and assume that we have computed all the periods of x. Note that x has 4,5 and 6 as periods. Moreover, assume that the following configuration of the inside texts has been encountered in two adjacent processors of the N-cube.

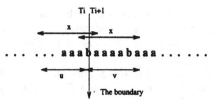

Fig.4. Overlapping occurrences with regard to the period px

From lemma 3, the shift $d = |u| + |v| - m = 5$ is not a multiple of the shortest period $p = 4$. However, there exists two overlapping occurrences of x located at positions r and $r + 5$. This is due to the existence of a larger period $p_x = 5$ that covers the shift d, and again the equality holds with the new period. The place of the occurrences are now depending on the long period p_x. More exactly one occurrence of x is located at position $s(i + 1) - |u| + k.p_x$.

In general, when having more than one period of the pattern, we look for a larger one that may overlay the shift d. Formally, we test if there exists $p_x > p$ such that p_x is a period and $(p_x - d) \bmod p = 0$.

Another similar case to consider, is when the pattern is partially periodic, with at least one periodic prefix (resp. suffix). In such case, we must consider the period of the matched prefix (resp. suffix). Therefore, there is necessary at least one occurrence of x that spans the current subtexts as shown in the following lemma.

Lemma 4.
Let u_{max} be the largest prefix of x that matches a suffix of T_i on processor PE_i at position r, v_{max} be the largest suffix of x that matches a prefix of subtext T_{i+1} on processor PE_{i+1}. The following two facts hold :

1. *if x is periodic with period p and the condition (1) (of Lemma 3 does not hold, and if there exists a greater period p_x ($p_x > p$) such that $p_x \equiv d \pmod p$, with $d = |u_{max}| + |v_{max}| - m$, then,*

   - if $|u_{max}| \geq p_x$ then x occurs in the text t at starting positions $s_j = r + j.p_x$ ($j = 0, 1$) such that $s_j + m - 1 \leq s.(i + 1) + |v_{max}|$.

   - if $|u_{max}| < p_x$ then x occurs at positions $s_j = r + c.p + d \bmod p + j.p$, for $j \geq 0$, $c = \frac{|u|}{p}$, and $s_j + m - 1 \leq s.(i + 1) + |v_{max}|$.

2. *if x is partially periodic since u_{max} (resp. v_{max}) is periodic with period pp and $|u_{max}| + |v_{max}| \equiv m \pmod{pp(resp.sp)}$, holds then x occurs at the position $r + d$ (resp. r) within text t.*

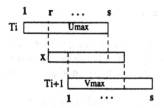

Fig.5. overlapping occurrences of x

Proof (fact 2.)

(1) Recall that according to the shortest period p, the word $x = u^k.v$, where $|u| = p$ and v is a prefix, possibly empty, of u.

(2) Since there exists a greater period p_x for x, such that $p_x \equiv d \pmod{p}$, then p_x is not a multiple of p. It follows according to lemma1 that:

$$p_x > m - p, \text{ and } x = (u_x)^1.v_x,$$
$$\text{where } |u_x| = p_x \text{ and } v_x \text{ is a proper prefix of } u_x.$$

Based on this two observations , and by performing successive decompositions of the pattern, this latter is inevitably of the following form :

$$x = (\sigma^a.\alpha)^k.\sigma^b.v_\sigma \text{ with } b \leq a$$
$$\alpha, \sigma \text{ are subwords of } x,$$
$$\text{and } v_\sigma \text{ is a proper prefix of } \sigma.$$

See Appendix for details.

The period p_x is of the form $k.p + r.|\sigma|$ such that $r.|\sigma| < p$ or equivalently $r \leq b$. So,

- $x = (\sigma^a.\alpha)^k.\sigma^r.\sigma^{b-r}.v_\sigma$

- $u \leftrightharpoons \sigma^a.\alpha$, $|u| = p$

- $u_x = (\sigma^a.\alpha)^k.\sigma^r$

- and $v_x = \sigma^{b-r}.v_\sigma$

Now consider the words u_{max} and v_{max} as the longest prefix value located at the end of text T_i, and the longest suffix value located at the beginning of text T_{i+1}. The concatenation $u_{max}.v_{max}$ is as follows :

$$\mathbf{u\,u \ldots u\,v1 \mid v2\,u\,u \ldots u\,v}$$

Fig.6. the prefix-suffix concatenation

v_1 is a prefixe of u and then $v_1 = \sigma^{o_1}.\alpha_1$, $o_1 \leq a$ and α_1 is a prefix of α.

v_2 is a suffix of u then, $v_2 = \sigma^{o_2}.\alpha$ with $o_2 < a$.

More exactly the value $d = |u_{max}.v_{max}| - m$ exhibit the length of the sequence $S = u^{k_1}.v_1.v_2.u^{k_2}$. Thus we have $d = (k_1 + k_2).p + (o_1 + o_2).|\sigma| + |\alpha| + |\alpha_1|$. If there exists a period p_x such that $p_x = k.p + d \bmod p$, then the value $|\alpha_1| = 0$.

We have also $p_x = k.p + r.\sigma$ which is equal to $(o_1 + o_2).\sigma$ when $d < p$ and $(o_1 + o_2 - a).\sigma$ otherwise. As a consequence of this:

$$u_{max}.v_{max} = u^{k_1}.\sigma^r.u^{k_2}.v$$

Knowing that $p_x > m - p$ is the period, then $x = (u_x)^1.v_x = u^k.\sigma^r.v_x$. To match the word x into the sequence $u_{max}.v_{max}'$, two cases follows:

- **Case 1:** When $|u_{max}| \geq p_x$, u_{max} is enough long to begin an occurrence of x. Then x occurs at positions $s_j = r + j.p_x$, for $(j = 0, 1)$, and $s_j + m - 1 \leq s.(i+1) + |v_{max}|$

- **Case 2:** $|u_{max}| < p_x$. It is obvious to see that the shift d causes problem because an incomplete occurrence of p_x begins at r. Here, we must skip $k_1.p + |\sigma|.r$ positions to reach the possibly first overlapping occurrence of x, and then x occurs on and after the position $r + k_1.p + r.|\sigma|$; more exactly on each position $s_j = r + \lfloor \frac{|u_{max}|}{p} \rfloor.p + d \bmod p + j.p$, where $s_j + m - 1 \leq s.(i+1) + |v_{max}|$.

The following general algorithm implements exactly the ideas presented above. Note that the text algorithm is the same for each processor. The overall parallel algorithm is then symmetric.

Text algorithm
Input:
Sequence $T_i = t[si + 1, .., s(i+1)]$
Pattern x
set_p = set of all periods of x

/* pp is a table of all prefixe periods (if any).
pp[i] contains the period of the prefix of size i.
Array $pp[1...m]$ of integer;

/* sp is a table of all suffixe periods (if any).
sp[i] contains the period of the suffix of size i */
Array $sp[1...m]$ of integer;

Output:
/* $SWITCH[i] = 1$ when x occurs at text position i */ Array $SWITCH[si + 1...s(i + 1)]$
of boolean.

Stage 0
/* In this stage we find pattern occurrences within segment T_i by using a classical matching algorithm */

Run a classical matching algorithm (the best one) to search for the pattern onto text $t_{s.i+1}...t_{s.(i+1)-m+1}$

For each starting position j of an occurrence do $SWITCH[j] = 1$.

Stage 1
/* In this stage we deal with pattern occurrences that overlap processors (PE_i, PE_{i+1}) and (PE_i, PE_{i-1}) */

1. Match the longest prefix u_{max} of x at the end of T_i by running a serial algorithm for pattern matching onto sequence $t_{s.(i+1)-m+2}...t_{s.(i+1)}$.
   - Set $prefix\_length$ as the length of the matched prefix (if any).
2. Match the longest suffix v_{max} of x at the beginning of T_i by running the classical KMP onto sequence $t_{s.i+1}...t_{s.i+m-1}$, but in the reverse direction.
   /* It means that we search for v_{max} by analyzing the first $(m-1)\_block$ of T_i with KMP, from right to left */
   - Set $suffix\_length$ as the length of the matched suffix.
3. if $suffix\_length \neq 0$ then
   Processor PE_i sends to processor PE_{i-1} a message carrying the $suffix\_length$ value and the attached message $SUFFIX$.

Stage 2
Upon receipt of the packet SUFFIX$\{suffix\_length\}$ from PE_{i+1}

- /* a suffix occurs in T_{i+1} */
  if $prefix\_length = 0$ then exit;
- /* a prefix occurs at position r */
  $r = s.(i+1) - prefix\_length + 1$

- if x is periodic with period p then
  $k = 0$;
  if $(prefix\_length + suffix\_length - m)$ mod $p = 0$ then
  /* set the $SWITCH$ array for the overlapping occurrences as in the first case of lemma 3 */
  do
  $SWITCH[r + k.p] = True$;
  $k = k + 1$
  until $r + k.p + m - 1 > s.(i+1) + suffix\_length$

  /* Otherwise, search for a greater period in set_p such that $prefix\_length + suffix\_length - m \equiv p_x$ (mod p) */
  if there exists $p_x \geq p$ such that $(prefix\_length + suffix\_length - m - p_x)$ mod $p = 0$ then,

      if $prefix\_length \geq p_x$
      then,
      k=0
      do
      $SWITCH[r + k.p_x] = True$;
      $k = k + 1$;
      until $k > 1$ and $r + k.p_x + m - 1 > s(i+1) + suffix\_length$

      else /* $prefix\_length < p_x$ */
      $c = \lfloor prefix\_length/p \rfloor$

$k = 0;$
do
$SWITCH[r + c.p + d \bmod p + k.p] = True;$
$k = k + 1;$
until $r + c.p + d \bmod p + k.p + m - 1 > s.(i + 1) + suffix\_length$
$-$ if x is partially periodic and $prefix\_length + suffix\_length \neq m$ then
$SWITCH[r] = 1$ if $sp[|v_{max}|] \neq 0$
$SWITCH[r + d] = 1$ if $pp[|u_{max}|] \neq 0$

return(SWITCH)

With respect to time complexity, an upper bound on local computing is given by $O(s)$, since we run a serial string matching algorithm to match both the word x, the prefix u_{max} and the suffix v_{max}. The space complexity (related to the string matching algorithm) is either linear in the pattern size (e.g using KMP) or constant.

Note that for the hypercube, the communication time significantly dominates the computation time. In our algorithm each processor possibly sends (in case of a matched suffix) one constant size message to its neighbour to check all the occurrences of the pattern in the whole text. Since the communication delay between two adjacents processor is assumed to be bounded by a constant, the overall communication time is constant. This achieves the proof of the following proposition :

Proposition
Taking the preprocessing into account, the overall time complexity of the above N-cube string matching algorithm is $O(\frac{n}{N} + \log(N))$

5 Concluding Remarks

We proposed a new flexible distributed algorithm which solves the string matching problem on a fixed N-cube with long texts stored on processors. Compared to classical parallel string matching algorithms, the text has a broken structure imposed by the distributed computation model with non-shared memory. We must consider here that pattern occurrences can overlap adjacent segments.

For finding all pattern occurrences, we presented a more involved algorithm with (pattern periods assumptions) that takes linear time. Note that, the algorithm will be more complete if a kind of trade-off between the size of the hypercube and the time complexity is proved to show the first optimal algorithm with the smallest possible size of the cube.

We consider two interesting open questions. The first one is to give a more general algorithm to search for long texts as well as small texts allocated to processors. D.K. Kim and K. Park in [11] address a similar problem in the case of an hypertext which is also a nonlinear structure of a text.

The second interesting point concerns the extension of the relevant prefix-suffix technique (illustrated by lemma 3) mainly used to deal with pattern overlapping occurrences, to sequential and incremental string matching problems (see [10]). All pattern occurrences may be derived by matching maximal prefixes and suffixes on independant blocks of the text. One might expect linear algorithm (in the number of comparison).

References

1. B.S. Chlebus and L. Gąsieniec. Optimal pattern matching on meshes. In *11th Ann. Symp. on Theoretical Aspects of Computer Science*, pages 213–224, 1994.
2. M. Crochemore and D. Perrin. Two way string-matching. *Association for Computing Machinery*, 38(3):651–675, 1991.
3. A. Czumaj, Z. Galil, L. Gąsieniec, K. Park, and W. Plandowski. Work-time optimal parallel algorithms for strings problems. Manuscript, 1994.
4. Z. Galil. A constant-time optimal parallel string matching algorithm. In *24th ACM Symp. on Theory of Computing*, pages 69–76, 1992.
5. Z. Galil and J. Seiferas. Time space optimal string matching. *J. Comput. Syst. Sci.*, 26:280–294, 1983.
6. L. Gąsieniec, W. Plandowski, and W. Rytter. Constant-space string matching with smaller number of comparisons: sequential sampling. In *6th Annual Symposium on Combinatorial Pattern Matching*, volume 937 of *Lecture Note in Computer Science*. Springer-Verlag, 1995.
7. C.T. Ho and S.L. Johnsson. Distributed routing algorithms for broadcasting and personalized communication in hypercubes. In *International Conference on Parallel Processing*, pages 640–648, 1986.
8. O.H. Ibarra, T.C. Pong, and S.M. Sohn. String processing on the hypercube. In *IEEE Transactions on Acoustics, Speech and Signal Processing*, volume 38, pages 160–164, january 1990.
9. J. Jájá. *Introduction to Parallel Algorithms*. Addison-Wesley, Reading, Massachusetts, 1992.
10. Z.M. Kedem, G.M. Landau, and K.V. Palem. Optimal parallel suffix-prefix matching algorithm and applications. In *ACM Symposium on Parallel Algorithms and Architectures*, pages 388–398, 1989.
11. D.K. Kim and K. Park. String matching in hypertext. In *6th Annuel Symp. on Combinatorial Pattern Matching*, Lecture Notes in Computer Science, pages 318–329, 1995.
12. U. Vishkin. Deterministic sampling - a new technique for fast pattern matching. *SIAM J.Comput*, 20:22–40, 1991.

Appendix

Pattern decomposition in case of multiple periods

The pattern x is periodic with period p which is the shortest one. It follows :

$$x = u^k.v, k > 0 \text{ and } v \text{ is a prefix of } v$$

(1)

The value $p_x = k.p + r_x, r_x < p$ is also a period, then $p_x > m - p$. According to this long period, x is of the form :

$$x = u_x^1.\beta,$$
where β is also a prefix of u.

(2)

By combining (1) and (2) it follows that $x = u^k.\alpha.\beta$, where α is also a prefix of u. Now supposing that $|\alpha| < |\beta|$, then β is inevitably of the form : $\beta = \alpha.V_\alpha$ and Consequently :

1. $x = u^k.\alpha.\beta$

2. $x = u^k.\alpha.\alpha.V_\alpha$

3. $x = (\alpha.V_\alpha.U_\alpha)^k.\alpha.\alpha.V_\alpha$ since β is a prefix of u.

Now the word $\alpha.\alpha.V_\alpha$ is also a prefix of $u = \alpha.V_\alpha.U_\alpha$ therefore,
4. $x = (\alpha.\alpha.V_\alpha.U_\alpha^{-1})^k.\alpha.\alpha.V_\alpha$.

Again the suffix $\alpha.V_\alpha$ of x (which is equal to β) is a prefix of u. Then :
5. $x = (\alpha.\alpha.V_\alpha.U_\alpha^{-1})^k.\alpha.\alpha.\alpha.V_\alpha^{-1}$, where V_α^{-1} is a prefix of V_α.

The word v which is now equal to $\alpha.\alpha.\alpha.V_\alpha^{-1}$, is also a prefix of u. Then:
6. $x = (\alpha.\alpha.\alpha.V_\alpha^{-1}.U_\alpha^{-2})^k.\alpha.\alpha.\alpha.V_\alpha^{-1}$

$\vdots$

Continue successive decompositions of either word v or word β which are both prefixes of the period u, to achieve the following expected form of x.

$$x = (\alpha^{z_1}.\gamma)^k.\alpha^{z_2}.V_\alpha$$

Where, γ is the nonperiodic part of the period u.

Combinatorics for Multiprocessor Scheduling Optimization and Other Contexts in Computer Architecture

Håkan Lennerstad and Lars Lundberg
University of Karlskrona/Ronneby, Sweden
email: Hakan.Lennerstad@itm.hk-r.se and Lars.Lundberg@ide.hk-r.se

1. Introduction and Definitions

Parallelism is one of the most promising ways to improve computer system performance, and multiprocessor systems are now commercially available. In order to match parallelism in hardware, new parallel programming languages and programming environments have emerged. Many of these languages use the notion of program processes to express parallel execution, i.e. we will end up with a parallel program consisting of a number of processes. The way processes are scheduled to processors affects system performance. In many systems, the most important performance criterion is the completion time of the parallel program.

A *parallel program* consists of a set of sequential processes. A process in a parallel program can be either Blocked, Ready or Running [15]. Processes in the Ready and Running states are referred to as *active* processes. The execution of a process is controlled by two synchronization primitives: *Wait(Event)* and *Activate(Event)*, where *Event* couples a certain Activate to a certain Wait. When a process executes an Activate on an Event, we say that the Event has occurred. If a process executes a Wait on an Event which has not yet occurred, that process becomes blocked until another process executes an Activate on the same Event. However, a process executing a Wait on an Event which already has occurred does not become blocked. Each process can be represented as a list of sequential segments, which are separated by a Wait or an Activate (see figure 2). We assume that, for each process, the length and order of the sequential segments are independent of the way processes are scheduled. All processes are created at the start of the execution. Some processes may, however, be initially blocked by a Wait.

In *multiprocessors with clusters*, processors are connected in a two level hierarchy. At the lowest level processors are connected in equally sized clusters. These clusters are connected via a communication network (see figure 1). Processes may not be relocated from one cluster to another. They may, however, be relocated between the processors in a cluster. We assume that only one parallel program may execute on the system at the same time. Moreover, we disregard overhead for process synchronization, switching and relocation. Under these conditions, the minimal completion time for a program P, using a system with k clusters each containing u processors, is denoted $T(P,k,u)$.

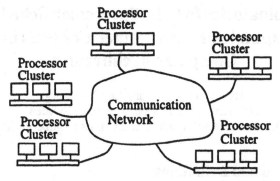

Figure 1: A multiprocessor with clusters.

The left part of figure 2 shows a parallel program consisting of three processes (P1, P2 and P3). Work(t) denotes sequential processing for t time units. Process P3 cannot start its execution before process P2 has executed for one time unit. This dependency is represented with a Wait on Event 1 in P3 and an Activate on the same Event in P2. The right part of the figure shows a graphical representation of P and two schedules resulting in minimal completion time for a system with one cluster containing two processors (T(P,1,2)) and for a system with two clusters containing one processor each (T(P,2,1)), i.e. there are two processors in both systems but in the rightmost case processes may not be relocated.

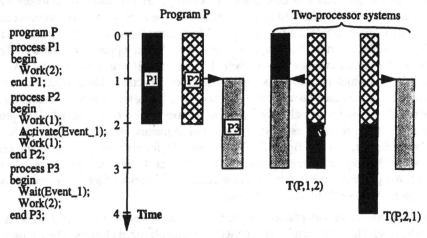

Figure 2: The minimal completion times for two different systems for a parallel program consisting of three processes.

The completion time is affected by the way processes are scheduled. Finding a schedule which results in minimal completion time, i.e. an optimal schedule, is NP-hard [3]. One is therefore interested in obtaining performance bounds for multiprocessor scheduling. In most cases, it is easier to find a schedule which is close to optimal for a system with one cluster than for a system with multiple clusters.

2. Previous Results

The problem of scheduling a set of independent processes on a smaller set of clusters with unit size has been studied extensively. Because this problem is NP-hard, much effort has been directed towards finding near-optimal approximations. The structure of the problem is very similar to the bin packing problem. Consequently, similar approximation techniques have been used.

The well-known LPT (Largest Processing Time first) algorithm always has a completion time within $4/3 = 1.333..$ of the minimal possible time. Coffman et al showed that a modification of this algorithm, called MULTIFIT, guarantees a completion time within 1.22 of the optimal result [1]. This result was further refined by Friesen to 1.20 [2] and then to 1.18 by Langstone [6], using a slightly modified version of the MULTIFIT algorithm.

Hochbaum and Shmoys later developed an approximation algorithm which guarantees a completion time within $(1 + \varepsilon)$ of the optimal result [5]. The computational complexity of this algorithm is $O\left((n/\varepsilon)^{1/\varepsilon^2} \right)$ for n processes. Consequently, this algorithm becomes impractical for small ε.

For systems with one cluster, a very powerful bound exists [4]. This bound, which was developed by Graham, is applicable to all parallel programs, regardless of their synchronization and communication structure. Graham showed that, if overhead for process synchronization is neglected, the execution time of any schedule where no processor is idle when there are processes in the ready state, has a completion time within a factor of 2 of the optimal case.

Non of these performance bounds consider the case of a parallel program with an arbitrary synchronization structure executing on a multiprocessor with clusters. We have calculated a function $g(n,k,u,q) = \max T(P,k,u) / T(P,1,q)$ where the maximum is taken over all programs P in the set P_n (P_n contains all programs with n processes) [7][9][13]. This function gives us an upper bound on the gain of using one large cluster instead of a number of small clusters, using optimal schedules. We have also calculated the function $G(k,u,q) = \max T(P,k,u) / T(P,1,q)$ for all programs P [8].

It turns out that the bound can be improved, i.e. made tighter, if we know more about the parallel program (in the case of $g(n,k,u,q)$ we only knew the number of processes n in the parallel program). If the parallel profile vector $\bar{v}$ is known, the bound can be improved. The parallel profile vector has length n and v_i $(1 \leq i \leq n)$ is the fraction of the completion time where there are exactly i active processes using a system with n processors ($\sum v_i = 1$). If the parallel profile vector is known then we are able to calculate a function $s(n,\bar{v},k,u) = \max T(P,k,u) / T(P,1,n)$ where the maximum is taken over all programs P in P_n with a parallel profile vector $\bar{v}$ [11].

Suppose that the program P in P_n has been executed on a multiprocessor with one cluster containing q processors resulting in a completion time of $T(P,1,1)/\sigma$. In that case we are able to calculate a function $s(n,k,u,q,\sigma) = \max T(P,k,u)/T(P,1,q)$ where the maximum is taken over all programs P in P_n with a completion time of $T(P,1,1)/\sigma$, using a system with one cluster containing q processors. Similarly, we are able to calculate the function $S(k,u,q,\sigma) = \max T(P,k,u)/T(P,1,q)$ where the maximum is taken over all programs P with a completion time of $T(P,1,1)/\sigma$, using a system with one cluster containing q processors [12].

If $n < q$ then $g(n,k,u,q) = g(n,k,u,n)$.
If $n \leq k*u$ then $g(n,k,u,q) = 1$, otherwise $g(n,k,u,q)$ is specified below, where $w = \lfloor n/k \rfloor$ and $n_k = n - kw$:

If $n_k = 0$, then $\quad g(n,k,u,q) = \dfrac{1}{\binom{n}{q}} \displaystyle\sum_{l=1}^{min(w,q)} max\left(1, \dfrac{l}{u}\right) \pi(k,w,q,l)$, otherwise

$$g(n,k,u,q) = \dfrac{1}{\binom{n}{q}} \sum_{l_1 = max\left(0, \left\lceil \frac{q - w(k - n_k)}{n_k} \right\rceil\right)}^{min(w+1,q)} \sum_{l_2 = max\left(0, \left\lceil \frac{q - l_1 n_k}{k - n_k} \right\rceil\right)}^{min(w, q - l_1)}$$

$$\sum_{i = max(l_1, q - l_2(k - n_k))}^{min(l_1 n_k, q - l_2)} max\left(1, \dfrac{l_1}{u}, \dfrac{l_2}{u}\right) \pi(n_k, w+1, i, l_1) \, \pi(k - n_k, w, q - i, l_2)$$

where $\pi(k, w, q, l)$ denotes the number of permutations of q ones in kw slots, which are divided into k sets with w slots in each, such that the set with maximum number of ones has exactly l ones. $\pi(k, w, q, l) = 0$ if $min(q,w) < l$ and $q > kl$, otherwise $\pi(k, w, q, l)$ is given by:

$$\pi(k, w, q, l) = \sum_{I} \binom{w}{i_1} \cdot \ldots \cdot \binom{w}{i_k} \dfrac{k!}{\prod\limits_{j=1}^{b(I)} a(I,j)!}$$

The sum is taken over all sequences of nonnegative integers $I = \{i_1,...,i_k\}$ which are decreasing, i.e. $i_j \geq i_{j+1}$ for all $j = 1,...,k-1$, and for which $i_1 = l$ and $\sum_{j=1}^{k} i_j = q$. The functions $a(I,j)$ and $b(I)$ are defined in the following way:

$a(I,j) = $ the number of occurrences of the j:th distinct integer in I.
$b(I) = $ the number of distinct integers in I.

Equation 1: The formula for g(n,k,u,q).

The obtained functions all have a similar structure. The function $g(n,k,u,q)$ is shown as an example in equation 1. In figure 3, the function $g(n,k) = g(n,k,k,1)$ is plotted for the interval $1 \leq k, n \leq 50$. This plot took approximately one hour to calculate. A similar plot in the interval $1 \leq k, n \leq 100$ takes approximately 12 hours to calculate, i.e. the complexity of the function $g(n,k,u,q)$ grows rapidly.

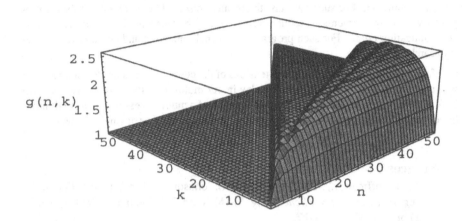

Figure 3: The function $g(n,k) = g(n,k,k,1)$.

The method for obtaining the results above consists of two steps. First, unnecessary programs are eliminated through a sequence of program transformations. Second, within the remaining set of programs, sometimes regarded as matrices, those where all possible combinations of synchronizations occur equally frequently are proven to be extremal. At this stage we obtain a formulation which is simple enough to allow explicit formulas to be derived. More information about this can be found in references 7 to 14.

It turns out that the same method can be used for obtaining worst-case bounds on other NP-hard problems within computer architecture. Hitherto, we have been able to obtain bounds on the maximum gain of increasing the number of memory modules in crossbar networks [14]. We have also obtained worst-case formulas for comparing the optimal performance for cache memories with different associativity [10].

3. Conclusions and Future Work

The obtained results make it possible to evaluate the performance implications of fundamental design decisions in computer architecture, e.g. the maximum performance gain of permitting run-time relocation of processes and the maximum performance gain of using fully associative cache memories. It is also possible to compare the performance of heuristic scheduling and allocation techniques, using the obtained results.

The scheduling results are based on the assumption of zero delay for synchronization and communication. In some systems, intercluster synchronization and communication delays are significant. For such systems, the results presented here are valid only for programs with long segments of sequential processing between the synchronization and communication points. For such programs, the synchronization and communication delays can be neglected.

Our future plans in this area include extensions of the current techniques. Particularly, we will try to include the communication delay in our multiprocessor model. Also, we will look at the problem of comparing the performance of a multiprocessor with a small number of large clusters versus a multiprocessor with a large number of small clusters.

References

[1]   E. G. Coffman Jr., M. R. Garey and D. S. Johnson, *An Application of Bin Packing to Multiprocessor Scheduling*, SIAM Journal of Computing, Vol. 7, No. 1, February 1978, pp. 1-17.

[2]   D. K. Friesen, *Tighter bounds for the multifit processor scheduling algorithm*, SIAM Journal of Computing, 13 (1984), pp. 170-181.

[3]   M. Garey and D. Johnson, *Computers and Intractability*, W.H. Freeman and Company, 1979.

[4]   R. L. Graham, *Bounds on Multiprocessing Timing Anomalies*, SIAM Journal of Applied Mathematics, Vol. 17, No. 2, March 1969, pp. 416-429.

[5]   D. S. Hochbaum and D. B. Shmoys, *Using Dual Approximation Algorithms for Scheduling Problems: Theoretical and Practical Results*, Journal of the ACM, Vol. 34, No. 1, January 1987, pp. 144-162.

[6]   M. A. Langstone, *Processor scheduling with improved heuristic algorithms*, Ph.D. thesis Texas University, Collage Station Texas, 1981.

[7]   H. Lennerstad and L. Lundberg, *An Optimal Execution Time Estimate for Static versus Dynamic Allocation in Multiprocessor Systems*, SIAM Journal of Computing, August 1995.

[8]   H. Lennerstad and L. Lundberg, *Optimal Performance Functions Comparing Process Allocation Strategies in Multiprocessor Systems*, Research Report 3/93. University of Karlskrona/Ronneby, Sweden, 1993.

[9]   H. Lennerstad and L. Lundberg, *Optimal scheduling results for parallel computing*, SIAM News, Vol. 27, No. 7, 1994 (survey article).

[10]   H. Lennerstad and L. Lundberg, *Optimal Worst Case Formulas Comparing Cache Memory Associativity*, Research Report 5/95, University of Karlskrona/Ronneby, Sweden, 1995.

[11]   L. Lundberg and H. Lennerstad, *An Optimal Upper Bound on the Minimal Completion Time in Distributed Supercomputing*, in Proceedings of the 8th ACM Conference on Supercomputing, Manchester, England, July 1994.

[12]   L. Lundberg and H. Lennerstad, *An Optimal Lower Bound on the Maximum Speedup in Multiprocessors with Clusters,* in Proceedings of the First International Conference on Algorithms and Architectures for parallel Processing, Brisbane, Australia, April 1995.

[13]   L. Lundberg and H. Lennerstad, *An optimal bound on the gain of using one large processor cluster instead of a number of small clusters*, in Proceedings of the 8th International Conference on Parallel and Distributed Computing Systems, Orlando, Florida, September 1995.

[14]   L. Lundberg and H. Lennerstad, *Bounding the Maximum Gain of Changing the Number of Modules in Multiprocessor Computers*, Technical Report May 1994 Department of Comp. Engineering, Lund University, Sweden.

[15]   A. Silberschatz, J. Peterson and P. Galvin, *Operating System Concepts (third edition)*, Addison-Wesley Publishing Company, 1991.

Some Applications of Combinatorial Optimization in Parallel Computing

Reinhardt Euler and Laurent Lemarchand

Laboratoire d'Informatique de Brest,
Université de Bretagne Occidentale,
6 av. Le Gorgeu, Brest cedex 29287, France
e-mail: {euler,lemarch}@univ-brest.fr

Abstract. We review recent work in the field of systolic array design, the parallelization of nested loops and of logic synthesis for Field Programmable Gate Arrays (FPGA) to show how combinatorial optimization problems (linear programs, covering, matching, network-flow or stable set problems) arise as natural subproblems. We also point out how their efficient solution together with the application of classical combinatorial methods (double description type methods, computing Smith and Hermite normal forms) has contributed to the development of automatic design tools.

1 Introduction

The automatic generation of systolic architectures, the automatic parallelisation of sequential programs and the automatic synthesis of Field programmable gate arrays (FPGA) represent major research directions in parallel computing. Mathematical models such as systems of recurrence equations provide a theoretical framework rich enough to cover the basic aspects of systolic array design and also the parallelisation of loop nests. Both these subjects will be discussed in section 2 and 3. Our discussion of FPGA synthesis is mainly motivated by the work around the ArMen-machine as it has been conducted in our research laboratory during the last few years. ArMen is a fully-operable parallel computer whose basic constituents are FPGAs of Xilinx type combined with transputers. For a detailed description we refer the reader to [6]. We will not address the combinatorial theory of Placement and Routing for FPGAs, but rather concentrate on logic synthesis aspects.

2 Systems of Recurrence Equations

A *system of (affine) recurrence equations (SARE)* is a finite number of equations defined over a set of *variable functions* $V_1, ..., V_m$ and having the form

$$(2.1) \quad V_i(z) = f_i[V_{i_1}(I_{i_1}(z)), ..., V_{i_{m_i}}(I_{i_{m_i}}(z))], \ i = 1, ..., m,$$

where z is an element of $D = \{x \in \mathbb{Z}^n : Ax \leq b\}$, the *domain* of (2.1), and where $I_1, ..., I_m$ are affine functions over $\mathbb{Z}^n$. For a simple example just consider the product C of two $N \times N$ matrices A, B defined by $c_{i,j} = \sum_{k=1}^{N} a_{ik} b_{kj}$, which gives rise to the following system of recurrence equations:

$$C_{ijk} = 0 \qquad\qquad\qquad k = 0$$
$$C_{ijk} = C_{ij(k-1)} + a_{ik}b_{kj} \qquad 1 \leq k \leq N$$
$$c_{ij} = C_{ijN}$$

The f_i's are supposed to be strict functions not depending on z. In case that $I_{i_j}(z) = z - d_{i_j,i}$ for $d_{i_j,i} \in \mathbb{Z}^n$ and $i \in \{1, ..., m\}$, (2.1) is called *uniform* and the $d_{i_j,i}$'s are called *dependence vectors*. These vectors provide an alternative way to study uniform recurrence equations via the *dependence graph*, a finite, directed graph $D = (V, A)$ whose vertices correspond to the variable functions $V_1, ..., V_m$ and in which vertex v_i is connected to vertex v_j by an arc (valued by d_{i_j}) iff V_j *depends* on V_i, i.e. $V_j(z) = f_j[..., V_i(z - d_{i_j})...]$.

A *schedule* (or *time-function*) is a function $t : \{1, ..., m\} \times D \to I\!N$ such that if $V_l(q)$ is an argument in the evaluation of $V_k(p)$ then $t(k, p) > t(l, q)$. We may thus interpret $t(k, p)$ as the time-step at which $V_k(p)$ is computed. We say that the variable function V_j is *explicitly defined* (or *calculable*) if a schedule exists for its calculation.

Uniform recurrence equations have been studied by Karp, Miller and Winograd [22] in a seminal paper that appeared in 1967, several years before H.T.Kung [23] introduced the concept of a systolic array. Karp et al. addressed the problems of calculability of the V_i's, the existence of time-functions for their computation and, in particular, the problem of characterizing the degree to which such computations can proceed 'in parallel'. Their results are at the heart of any methodology to (automatically) generate systolic arrays, as we will see for the 'projection method' in the next section.

If (2.1) consists of a single uniform equation $V(z) = f[V(z-d_1), ..., V(z-d_m)]$ and $D = I\!N^n$, Karp et al. were able to give necessary and sufficient conditions for the variable function V to be calculable:

Theorem (2.2) The following statements are equivalent:
1) $V(z)$ is calculable.
2) There is no $u \in I\!R_+^m \setminus \{0\}$ such that $\sum_{j=1}^{m} -u_j d_j \geq 0$.
3) The system : $d_j x \geq 1, j = 1, ..., m, x_i \geq 0, i = 1, ..., n$ has a solution.
4) For every $z \in \mathbb{Z}^n$ the following 2 linear programs have a common optimal value $m(z)$:

$$
\begin{array}{ll}
\text{I)} \quad
\begin{array}{l}
z - \sum_{j=1}^{m} u_j d_j \geq 0 \\
u_j \geq 0, j = 1, ..., m \\
\text{maximize } \sum_{j=1}^{m} u_j
\end{array}
&
\text{II)} \quad
\begin{array}{l}
d_j x \geq 1, j = 1, ..., m \\
x_i \geq 0, i = 1, ..., n \\
\text{minimize } zx
\end{array}
\end{array}
$$

It turns out that this approach leads to time-functions that are only piecewise linear over D. The problem of finding linear (or affine) such functions has been addressed by Darte and Robert [12, 13] (see also Feautrier [19]). If $t(x) = \lambda x + \alpha$ and if for $p, q \in D$: $q = p + d_i$ then necessarily $t(q) > t(p)$, i.e. $t(p + d_i) \geq t(p) + 1$ or $t(d_i) \geq 1$, and this has to hold for all dependencies. With dependencies d_i regrouped in a matrix, say Q, the problem of minimizing total execution time then becomes

$$(2.3) \quad \min{}_{(\lambda Q \geq 1)} \max{}_{(Ap \leq b, Aq \leq b)} \lambda p - \lambda q \, ,$$

If we allow λ to be a rational vector, by linear programming duality,

maximize $\lambda p - \lambda q$
subject to $Ap \leq b, Aq \leq b$

is equivalent to

minimize $(\tau_1 + \tau_2)b$
subject to $\tau_1 A = \lambda,$
$\qquad\qquad \tau_2 A = -\lambda$
$\qquad\qquad \tau_1, \tau_2 \geq 0$

and, consequently, (2.3) is equivalent to

$$(2.4) \quad \begin{array}{l} \text{minimize} \quad (\tau_1 + \tau_2)b \\ \text{subject to} \quad \tau_1 AQ \geq 1 \\ \qquad\qquad (\tau_1 + \tau_2)A = 0 \\ \qquad\qquad \tau_1, \tau_2 \geq 0 \, , \end{array}$$

a simple linear program.

The results of Karp et al. [22] have strongly influenced subsequent work in the field such as the design of systolic arrays, nested loop parallelization but also the study of more general systems of recurrence equations. For a recent example, see Andonov-Rajopadhye [1] who study a systolic architecture for the knapsack problem.

3 Polyhedra, Smith and Hermite normal forms and the projection method

Relaxing the integrality constraint on the elements of D yields a *polyhedron* $P = \{x : Ax \leq b\}$, i.e. the intersection of finitely many affine half-spaces. The decomposition theorem for polyhedra (see for instance [37]) states that P is at the same time generated by a finite number of points $x_1, ..., x_s$ and directions $y_1, ..., y_t$, i.e.

$$P = conv\{x_1, ..., x_s\} + cone\{y_1, ..., y_t\}.$$

An algorithmically interesting problem is to determine one representation of P from the other. An overview of existing methods is given by Matheiss and Rubin [26], see also [16]. We just mention two methods that have found particular attention : the double description method [30], [7], and the recent pivoting algorithm [2]. The double description method has been efficiently implemented by H. Le Verge [39] and is at the heart of the polyhedral library as presented in Wilde [42].

We have adapted the code to generate a minimal and irredundant description of the asymmetric traveling salesman polytope on 6 cities, i.e. the convex hull of the incidence vectors of 120 tours, in terms of 319015 linear inequalities (cf. Euler, Le Verge [17], for the symmetric case see also Th. Christof et al. [8]). Observe that combinatorial optimization problems are often defined over a finite (though possibly very large) set of feasible solutions. In order to apply linear programming techniques for their solution one needs a description of the solution set by linear inequalities. For NP-hard optimization problems, complete such descriptions can be obtained only for small instances; however, these results may contribute to better describe the general polyhedron, to detect algorithmically efficient cutting planes for the design of branch and cut methods, etc.

The idea to relate polyhedral theory to systems of recurrence equations and, in particular, to the synthesis of systolic arrays is due to Moldovan [28] and Quinton [33, 32], and has resulted in the so called *projection method*, whose main phases consist in determining

- appropriate time-functions to schedule the calculation of the variable functions;
- appropriate allocation-functions for the mapping of feasible points to processors;
- a way to partition or regroup the processors so that the resulting cells are active at 100 %.

For the case (2.1) of affine recurrence equations the knowledge of the vertices and rays of $P = \{x : Ax \leq b\}$ can be used to characterize time-functions as follows (cf. Mauras et al. [27]) :

Theorem (3.1) A function $t(k, z) = \lambda_k z + \alpha_k$ with $\lambda_k \in \mathbb{Z}^n$, $\alpha_k \in \mathbb{Z}$ is a time-function for $V_k, k = 1, ..., m$, if and only if for any arc of the dependence graph, valued by $(V(i), V(j), D, I)$, the following holds:

i) for any vertex σ of P:
   $\lambda_i \sigma - \lambda_j (I(\sigma)) + \alpha_i - \alpha_j \geq 1$;
ii) for any ray ρ of P
   $\lambda_i \rho - \lambda_j A\rho \geq 0$;

iii) for any vertex σ of P and $k \in \{1, ..., m\}$
   $\lambda_k \sigma + \alpha_k \geq 0$;
iv) for any ray ρ of P and $k \in \{1, ..., m\}$
   $\lambda_k \rho \geq 0$.

Methods of integer linear programming now apply to find a feasible solution.

An appropriate 'allocation-function' is obtained by a projection along a vector s, having integral components prime to each other and satisfying the condition that $\lambda s \neq 0$. This guarantees that no two different points are scheduled at the same time-step on the same processor. The vector s is generally chosen such that the input data are projected onto the boundaries of the array and/or that the number of cells is minimized. By definition, s can be completed to a unimodular matrix S via its associated *Hermite normal form* (see [37] for a general discussion). The domain D is thus projected along s on the hyperplane induced by the $(n-1)$ remaining column vectors of S, and the allocation function a is given by the lower $(n-1) \times n$ submatrix M of S^{-1}. In particular, $S^{-1}S = Id$ implies $Ms = 0$; moreover, for $x \neq y, a(x) = a(y)$ iff $x - y = \mu s$, and since $\lambda s \neq 0$ we have $\lambda(x - y) = \mu\lambda s \neq 0$, and thus both points are scheduled at different time-steps, as required. Now complete λ as a row vector to a unimodular matrix T^{-1} in the same way. The matrix $T = (t', T')$ becomes the time-basis: from $T^{-1}T = Id$ we get $\lambda T' = 0$, i.e. the columns of T' represent a basis of all calculations executed at time 0, whereas $\lambda t' = 1$, i.e. t' is the translation vector from the points calculated at time t to those calculated at time $t + 1$. Finally, consider the $(n-1) \times (n-1)$ right submatrix C of $S^{-1}T$: its *Smith normal form* can be shown to be $\text{diag}(1, ..., 1, c)$, c being equal to λs, the *activity rate* (indicating that a processor is active every c time-steps). This latter submatrix C is at the basis of a subsequent partitioning technique to regroup processors to avoid idleness. Further details can be found in [12], (see also [29]).

Example (taken from Darte [12]): the calculations for the product C of two $N \times N$ matrices A, B are

$$
\begin{aligned}
C_{ijk} &= C_{ij(k-1)} + A_{i(j-1)k}B_{(i-1)jk} & 1 \leq i, j, k \leq N \\
A_{ijk} &= A_{i(j-1)k} & 1 \leq i, j, k \leq N \\
B_{ijk} &= B_{(i-1)jk} & 1 \leq i, j, k \leq N .
\end{aligned}
$$

With time vector $\lambda = \begin{pmatrix} 1 \\ 1 \\ 1 \end{pmatrix}$ and projection vector $s = \begin{pmatrix} 1 \\ 1 \\ 1 \end{pmatrix}$ the resulting systolic array is given in figure 1.

To finish this section let us mention ALPHA, a language to automatically generate and transform systolic architectures, which is based on the methodologies presented in this section (see also Le Verge [40]).

4 Scanning polyhedra and the parallelization of loop nests

Scientific programs spend a large part of their running time in executing DO loops. The growing importance and availability of parallel computers has raised the need for a study of loop nests with respect to inherent parallelism and, in particular, for the development of tools for their automatic parallelisation.

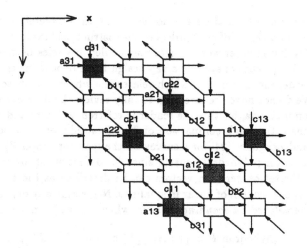

Fig. 1. Systolic array obtained by projection along s (the grey cells are those active at time-step 3).

Definition (4.1) The following program is called a *perfect loop nest*:

for $x_1 = l_1$ to u_1
    for $x_2 = l_2(x_1)$ to $u_2(x_1)$
      .
      .
        for $x_n = l_n(x_1, ..., x_{n-1})$ to $u_n(x_1, ..., x_{n-1})$

          Instruction 1

(4.2)   .

          Instruction k

      endfor

    .
    endfor
endfor

For $i \in \{1, ..., n\}$ the lower and upper loop bounds, $l_i(x_1, ..., x_{i-1})$ and $u_i(x_1, ..., x_{i-1})$, are defined to be the maximum and minimum, respectively, of a finite number of affine functions on $x_1, ..., x_{i-1}$. The set of feasible *iteration vectors* x can thus be described by $D = \{x \in \mathbb{Z}^n : Ax \leq b\}$, the *domain* of the loop nest.

Clearly, D and the instructions are not sufficient to completely define the semantics of the loop nest: different orders of the instructions' evaluation may lead to different results; in other words, there may be dependencies between different iterations. As in the case of systolic array design these dependencies can be represented by a dependence graph, which relates the whole subject to recurrence equations. We do not have the intention to fully review the different nested loop parallelization methods; for this the reader is referred to Darte and Vivien [14], which contains a thorough comparison of 3 well established methods (see also Risset [34], and Lengauer [24] for a survey on the 'polytope model'). Instead we will concentrate on polyhedral aspects since by definition of lower and upper loop bounds the set of iteration vectors can be described as the set of feasible, integral points of a system of linear inequalities. Neglecting any dependencies we are lead to the *polyhedron scanning problem*, which can be formulated as follows:

(4.3)   Given a polyhedron $P = \{x : Ax \leq b\}$ contained in $I\!\!R^n$ produce the set of loop bound expressions $l_1, u_1, ..., l_n, u_n$ such that the loop nest (4.2) will visit (in lexicographic increasing order) all the integer points in P exactly once.

In other words, for i=1,...,n, the bounds l_i and u_i have to be expressed by affine functions in terms of $x_1, ..., x_{i-1}$, only, and this is nothing but the establishment of a (minimal) linear system sufficient to describe the projection of P into $I\!\!R^i$. Figure 2 exhibits an example.

`{i,j,k,N,M|i>=0;-i+M>=0}:`
`  {i,j,k,N,M|j>=0;-j+N>=0}:`
`    {i,j,k,N,M|k>=0;i+j-k>=0}:S`
 	
a. The loop nest in {i, j, k} scan order.
	`{i,j,k,N,M|j>=0;-j+N>=0}:`
	`  {i,j,k,N,M|k>=0;j-k+M>=0}:`
	`    {i,j,k,N,M|i>=0;-i+M>=0;i+j-k>=0}:S`
	
	b. The loop nest in {j, k, i} scan order.

`{i,j,k,N,M|k>=0;-k+N+M>=0}:`
`  {i,j,k,N,M|j>=0;-j+N>=0;j-k+M>=0}:`
`    {i,j,k,N,M|i>=0;-i+M>=0;i+j-k>=0}:S`
 	
c. The loop nest in {k, j, i} scan order.
	`{i,j,k,N,M|i>=0;-i+M>=0}:`
	`  {i,j,k,N,M|k>=0;i-k+N>=0}:`
	`    {i,j,k,N,M|j>=0;-j+N>=0;i+j-k>=0}:S`
	
	d. The loop nest in {i, k, j} scan order.

Fig. 2.

Four different loop nests for the domain $D = \{i, j, k : i \geq 0, \ -i + M \geq 0,$
$j \geq 0, \ -j + N \geq 0,$
(Le Verge et al. [41]) $k \geq 0, \ i + j - k \geq 0\}$

As to a solution of (4.3), the *parametric integer programming algorithm (PIP)* developed by Feautrier [18] proceeds by calculating the lexicographic minimum and maximum of any such projection by an adaptation of the simplex algorithm.

An alternative way is to use Fourier-Motzkin elimination, which by definition consists of a series of projections of P along unit directions. The method is, however, very time consuming and, usually, produces a large set of redundant inequalities. Le Verge et al. [41] have indicated how to make use of the polyhedral library (cf. Wilde [42]) in order to determine the linear system describing one such projection : since Chernikova's algorithm combined with an additional redundancy check is at the heart of this library, the computational effort seems to be high, too.

A more promising direction could be the generation of the projection cone's extreme rays; again, the linear system obtained is not necessary minimal. Balas [3] has indicated a way to obtain such a (minimal) description. It would be interesting to check the computational efficiency of this approach.

5 The problem of technology mapping for FPGAs

Field Programmable Gate Arrays (FPGA) are user-programmable circuits used today for rapid and inexpensive prototyping of specific applications. The high complexity of their architecture makes manual designs time-consuming and error prone. Design automation tools are thus needed to reduce the development time of custom applications.

An LUT-based FPGA is a 2D-Array of programmable Lookup-Tables (LUT). An LUT is a k-input–single-output RAM which can implement any boolean function of up to k variables (e.g $k = 5$ in Xilinx series [44]). Programmable interconnection resources between LUTs complete the FPGA architecture.

FPGAs are used in our parallel architecture ArMen [6]. The machine is composed of a network of Transputers (MIMD layer) and of interconnected FPGAs (logic layer), each of which being also connected to a Transputer and its associated memory. The result is an Armen node, as illustrated in figure 3.

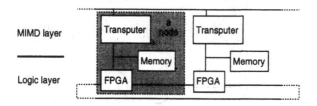

Fig. 3. The ArMen architecture

Applications synthesized on the FPGA part are mainly

- *global operators specified as cellular automata* [4]. These operators and associated data are distributed among the ArMen nodes. Examples include low level image processing and simulation. The cellular automata model leads to regular but eventually large circuits. During logic synthesis, attention must be paid to the area of the resulting circuit;

— *synchronous control circuits* [15]. Examples are global minimum calculation over all processors, deadlock detection and trace collection. Synthesized circuits are smaller than global operators but are unstructured. Since clock constraints imply delay limitations, synthesis must mainly focus on the performance of the circuit.

We review in the following sections combinatorial techniques used in the technology mapping phase for LUT-based FPGAs. Technology mapping is the second phase, after logic optimization, of the logic synthesis process and aims at transforming the circuit such that it can be implemented on one or more FPGAs. After the introduction of some basic definitions, we will describe this phase within the logic synthesis process.

Circuit representation : The input to a logic synthesis problem is a set of m boolean functions

$$f = \{ \ f_i \ : \ B^n \to B \mid B = \{0, 1\}, i \in [1..m] \ \}$$

Example : $f_1(a, b, c) = ab + c\bar{b}$, where a, b, c are boolean variables. Product, sum and complement represent, respectively, logic AND, OR and NOT. Variables in direct or complemented form are called *literals*. *Cubes* are products of literals. The support $sup(f)$ of a function f is the set of distinct variables the function explicitly depends on. In the given example, there are 4 literals $(a, b, c, \bar{b})$, the function is the sum of 2 cubes $(ab, c\bar{b})$ and $sup(f) = \{a, b, c\}$.

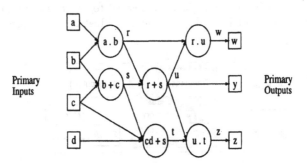

Fig. 4. A boolean network

During the logic synthesis process, the circuit is manipulated as a *boolean network* D (cf figure 4). $D = (V, A)$ is a Directed Acyclic Graph (DAG) built from f as follows : The roots and leaves of D are respectively the *primary inputs* (PIs) and *primary outputs* (POs) of the circuit (squared nodes of figure 4). An internal node $N_i \in V$ is associated to each f_i, and there is an arc (i, j) from N_i to N_j if the output of f_i is an input to f_j. N_i is named by the output of f_i (circled nodes in figure 4). For any node N_i, we denote by $fanin(N_i)$ the *fanin* of N_i, i.e the set of nodes N_i depends on. If f_i is the function associated with N_i, $sup(f_i)$ corresponds to functions associated with nodes of $fanin(N_i)$. The cardinality $|fanin(N_i)|$ is the size of the fanin of N_i. The *fanout* of N_i is the set of nodes which depend on N_i's output. N_i is *fanout free* if $|fanout(N_i)| \leq 1$, and *multiple-fanout* otherwise. A node N_i is *k-feasible* if $|sup(N_i)| \leq k$, and by extension, a network is *k-feasible* if all of its nodes are k-feasible.

Logic synthesis phases : Logic synthesis is embedded in the circuit design process as shown in figure 5. It converts a boolean network D representing a combinational circuit into a feasible network D'. Since a k-LUT can compute any function up to k variables, there is a one-to-one correspondence between the k-feasible nodes of D' and the functions implemented in the LUTs of the FPGA mapping.

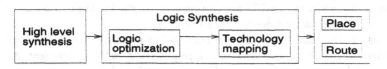

Fig. 5. The circuit design process

Successive phases of the synthesis are :

- logic optimization : boolean properties are used to reduce the global number of literals and the number of nodes [5, 35]. This phase is technology independent.
- decomposition of unfeasible nodes (if any) : various algorithms (Roth-Karp, cube-packing, and-or, disjoint-support ...) are applied to obtain a feasible network [5].
- technology mapping : the goal is to reduce the number of nodes (area optimization) and/or the critical paths (delay optimization), while preserving the feasibility of the network. Nodes are eliminated by reinjection (collapsing) into their fanout. Example : reinject $f = ab + bc$ into $g = f + ac$. The collapsed node is $g' = ab + bc + ac$. Since $|sup(f)| = |sup(g)| = |sup(g')| \doteq 3$, 3-feasibility has been preserved.

According to [35], collapsing operations for technology mapping are grouped into 3 classes. The most important are the covering techniques, that we adress in sections 6 and 7 in the context of combinatorial algorithms for optimizing area and delay. Section 8 presents algorithms for two other reduction techniques, local elimination and merging.

6 Covering for area optimization

Covering considers the global structure of the network to find nodes to be collapsed. Clusters of nodes that can be realized as a single LUT are found, and the network is covered by a subset of these clusters. Nodes of chosen clusters are eliminated by reinjection into their fanout until only one node remains (the root of the cluster). We review in this section combinatorial algorithms used for reducing the number of nodes (LUTs) during the covering phase.

6.1 Exhaustive search

Vismap [43] executes an *exhaustive search* on the solution space to find the area-optimal mapping of a boolean network. Mapping is based on the notion of *visibility*. An arc of the network is visible if it appears as a wire in the final circuit. Conversely, invisible arcs connect nodes that are implemented into a single LUT.

Arcs are ordered and are marked undecided. At each step of the algorithm, states visible and unvisible are successively assigned to the current undecided arc, and next arc is considered in the two cases.

This process can be seen as the traversal of a binary search tree. Left and right sons of a search vertex correspond, respectively, to the assignment of visible and invisible values to the next undecided arc. When branching on a right son, end-nodes of the invisible arc are merged. If the result is an unfeasible node, the search tree can be pruned at this point.

Exhaustive search guarantees an optimal solution but is time-expensive because the solution space is large : if m is the number of undecided arcs, there are 2^m possible assignments for these arcs. Thus, Vismap tries to reduce the search space. 1) Based on the number of PIs that nodes depend on, some arcs can be decided invisible in a pre-processing step. 2) The network can be partitioned into subgraphs before mapping and each subgraph is processed independently. Optimality is then lost.

6.2 Using bin-packing and dynamic programming

Chortle-crf [20] is a technology mapper for optimizing area. The network is first partitioned into a forest of trees by clipping multiple fanout nodes. Complex functions are then decomposed into AND/OR nodes. It uses dynamic programming from leaves to root for finding the optimal mapping for each AND/OR tree. At each step of the mapping, a *Bin Packing Problem (BPP)* is solved to compute the optimal mapping of a node N. The mapping is a combination of N fanin optimal mapping and of the function associated with N.

At the beginning, each leaf is packed into a single LUT. Mapping a node N_i results in a tree of LUTs whose root is denoted $L_i = map(N_i)$. N is processed when $L = \{L_i \mid L_i = map(N_i \in fanin(N))\}$ has been generated. The problem consists now in finding an optimal mapping for the cluster $\{L, N\}$. The algorithm works in two phases.

L is mapped during the first phase. Minimizing the number of LUTs for implementing L is interpreted as a BPP : Items to be packed are the L_i's, associated weights are $|sup(L_i)|$, and the capacity is fixed to k. In figure 6 (a), N is the OR node, L is the set of AND gates, weights are 3,2,2,2,2 from left to right, and $k = 5$. The *First Fit Decreasing algorithm* (FFD) [25] is an efficient heuristic for solving BPP. It orders weights in decreasing value, and for each item, stores it in the first bin it fits, or, if any, creates a new bin to accommodate the current item.

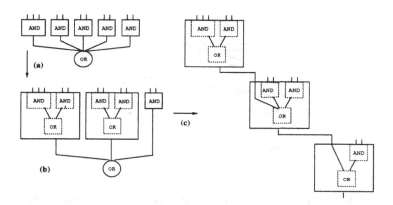

Fig. 6. Mapping of an OR node

The second phase generates an implementation for the remaining logic of N. As shown in figure 6 (b), we have still to implement the OR of the bins packed in the first phase. The bin with the fewest free inputs is closed (here at the left), and a new item of size one is created, corresponding to its output. This new item is packed using the FFD. The process is repeated until there is only one open bin. The last output supplies the original output of N. This process is illustrated in figure 6 (c).

The algorithm finds the optimal number of LUTs needed to map a node, and also minimizes, as a secondary goal, the number of inputs of the root LUT of the mapping. This allows an optimal mapping for the next parent node in the tree.

The complexity of the algorithm is $O(n.logn)$, where n is the number of nodes to pack. The algorithm is proven to be optimal with respect to area for the mapping of a tree if $k \leq 6$ [20]. But since the network is partitioned before mapping, the solution is not globally optimal.

When mapping a node, fanin LUTs must have disjoint support. The cost of packing two gates that share an input is different whether they are placed in the same LUT or not. A solution is to calculate the cost for all sharing possibilities and to choose the best. Because of low cost of the algorithm FFD, this solution is acceptable if the number of shared inputs is limited.

6.3 A Binate Covering Problem formulation

Mis-pga [31] aims at minimizing the area of the mapping. The main elimination phase is realized globally by covering techniques. The clusters used to cover the feasible network are called *supernodes*. A supernode S_i associated with a node N_i contains N_i and some nodes in the transitive fanin of N_i. The constraint is that the supernode must have no more than k inputs, i.e. can be implemented as a single LUT. A supernode S_i cannot contain any primary input, and every node of S_i has a path to N_i which lies entirely in S_i. Figure 7 shows such a cluster.

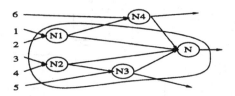

Fig. 7. A supernode rooted at N

The covering algorithm works in two phases : First, all possible supernodes are generated, and second, a subset of them is validated to cover the network. Without any other constraint, the solution would be given by a set covering algorithm. But, if a supernode is chosen, all the signals it uses as inputs must be present as output of at least one other supernode in the cover. This constraint transforms the problem into a *Binate Covering Problem (BCP)*, which can be formulated as a 0-1 Programming Problem as follows :

Given a boolean network $D = (V, A)$ and the set S of supernodes on D,

$$
\begin{array}{lll}
\text{minimize} & \Sigma \, x_i & (1) \\
\text{subject to} & \Sigma_{i \,:\, N_j \in S_i} \, x_i \geq 1 \quad \forall j & (2) \\
& \Sigma_{k \,:\, output(S_k) = y} \, x_k \geq x_i \quad \forall i, \, \forall y \in sup(S_i) & (3) \\
& x_i \in \{0, 1\} \quad \forall S_i \in S & (4)
\end{array}
$$

The objective function (1) aims at minimizing the number of supernodes, i.e the number of LUTs needed for the mapping. The covering constraint (2) is clear. The I/O Availability constraint (3) is illustrated on figure 8 : suppose S_3 is chosen for the cover. y_2 is an input of S_3. Supernodes S_1 and S_4 supply y_2 as output ($k = 1, 4$). Applied to the input y_2 for S_3, the I/O constraint is $x_1 + x_4 \geq x_3$. Since S_3 is chosen, $x_3 = 1$, and the relation is true iff $x_1 = 1$ or $x_4 = 1$, that is S_1 or S_3 is chosen.

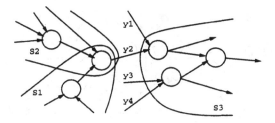

Fig. 8. I/O dependency between supernodes on output y_2

7 Covering for delay optimization

Covering for minimizing delay assumes in general a constant propagation time between LUTs. Delay optimization is thus realized by shortening paths in the network. Recent work focuses on minimizing delays without increasing area too much. Again, heuristics are used [38], or exact techniques for area optimization can be adapted [21]. In the following, we discuss the use of network flow techniques for the mapping of general networks.

7.1 Using network flow techniques

Flowmap [10] addresses the problem of finding a delay-optimized mapping for a feasible network. It is based on *maxflow/mincut* computations.

The algorithm works in two phases on an AND/OR decomposed network : first, a label is assigned to each node of the network. This label corresponds to the optimal depth of an LUT implementing the node, i.e. the length of the critical path from the PIs to this node in the mapping. Labels are associated with cuts in the network. Second, the mapping is done simply according to labels and cuts. We do not detail this greedy phase here.

k-feasible cuts. Let $(X, \bar{X})$ be a cut on a flow network G. The *node cut-size* $n(X, \bar{X})$ is the number of nodes in X adjacent to nodes in $\bar{X}$. $(X, \bar{X})$ is k-feasible iff $n(X, \bar{X}) \leq k$. If $l(N_i)$ is a label associated with each node, the *height* $h(X, \bar{X})$ of a cut is the maximum label in X. The notion of *k-feasible cut* $(X, \bar{X})$ will correspond to an implementation of the nodes in $\bar{X}$ as a single LUT, at a depth (height) $h(X, \bar{X})$ or $h(X, \bar{X}) + 1$.

The main phase consists in computing labels and associated minimal k-feasible cuts. The labeling algorithm proceeds from primary inputs to primary outputs in topological order. For each PI, $l(N) = 0$.

Let us denote G_t the set including a node N_t and all nodes in the transitive fanin of N_t. For flow computations on G_t, a source N_s is connected to PIs, and N_t is the sink. The optimal label (depth) of a node N_t corresponds to a minimal height k-feasible cut $(X, \bar{X})$ of G_t ($l(N_t) = h(X, \bar{X}) + 1$). Figure 9 shows a 3-feasible cut on G_t : N_s is added as a source connected to PIs (squared nodes), labels (depths) are indicated near nodes. The nodes in X adjacent to nodes in

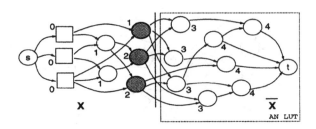

Fig. 9. A 3-feasible cut of height 2

$\bar{X}$ are grayed. $n(X, \bar{X}) = 3$ and $h(X, \bar{X}) = 2$. The nodes of $\bar{X}$ can thus be implemented in a 3-input LUT at a depth of 2 or 3.

There is no algorithm to compute minimal height k-feasible cuts directly. But, it is proved that, if p is the maximal label in the fanin of N_t, then $l(N_t)$ is p or p+1. The problem is thus reformulated as finding a k-feasible cut of fixed height p-1 on G_t.

To find such a cut, the network is transformed two times. First, G_t' is derived from G_t such that a k-feasible cut of height p-1 in G_t corresponds to a k-feasible cut in G_t'. Second, the constraint on nodes is transformed into a constraint on edges (recall that k-feasibility is related to the number of adjacent nodes across a cut). This leads to the derivation of G_t'' from G_t'. A maxflow/mincut algorithm is performed on G_t''. For instance, k+1 steps of an augmenting path algorithm are sufficient to prove that the maximal flow is no greater than k (because of edge values in G_t''). If the flow is less than k, the cut is determined on G_t'', translated back into G_t and $l(N_t) = p$. Otherwise, $l(N_t) = p + 1$, and the associated k-feasible cut of G_t is $(G_t \setminus \{N_t\}, \{N_t\})$.

The algorithm works in $O(k.|V|.|E|)$ time and optimality is guaranteed for every node (not only those along critical paths) of a general network. This overcomes the restriction for the boolean network to be a tree imposed by Chortle-d [21]. Futhermore, the area can be reduced by a bitwise choice of the cuts : minimal height k-feasible cuts are not unique, and intuitively, choosing a cut that maximizes the number of nodes packed in a single LUT will decrease the number of LUTs. That corresponds to maximizing the number of nodes in $\bar{X}_t$ when choosing the cut for G_t.

8   Local Packing Techniques

After the main global phase of covering, local optimization techniques can be used to further reduce the number of nodes. According to the main optimization criterion chosen, some nodes can be considered in priority (for instance those along the critical paths if emphasis is on delays).

8.1 A Stable Set Problem for local elimination

Local elimination consists in collapsing adjacent pairs of nodes while preserving feasibility. In some mappers, candidate nodes are collapsed as soon as they are found [38, 43]. Mis-pga [31] uses a more sophisticated algorithm, with a more global view of the network structure.

It proceeds as follows : according to their number of inputs, a list of candidate couples $< N_i, N_j >$ for collapsing is selected. The collapsed nodes may be infeasible. The cost of the implementation of pairs is computed using the bin-packing algorithm of Chortle-crf (cf 6.2) for two cases : (a) with and (b) without collapsing of nodes. If (a) results in a gain in terms of LUTs (with respect to solution (b)), the couple is stored in a list L with its associated gain g_i.

L is exploited to maximize the gain. Optimality can be reached by solving a *Stable Set Problem (SSP)* on the following graph $G = (L, E)$. Each couple $L_i \in L$ is associated with a node in G, weighted by g_i. E contains edge e_{ij} if L_i and L_j have common members. The formulation of SSP as a 0-1 Programming problem is straightforward.

8.2 A Matching Problem for merging

Merging is the last local elimination of nodes used in technology mapping. It is based on a particular feature of some FPGA architectures like the Xilinx 3000 series. An LUT of these FPGA (called CLB) can implement any functions f and g such that $|sup(f)| \leq 4$ and $|sup(g)| \leq 4$, $|sup(f) \bigcup sup(g)| \leq 5$ and $|sup(f) \bigcap sup(g)| \leq 3$. Thus, a way for minimizing the number of LUTs of the mapping is to find couples of nodes respecting the above merging constraints.

Mis-pga [31] solves the problem of maximizing merged nodes as follows : a merging graph $G = (V, E)$ is constructed from the boolean network $D = (V, A)$. There is an undirected edge $(i, j) \in E$ if nodes i and j are mergeable. Optimizing for area corresponds to solve a *Maximal Cardinality Matching Problem (MCMP)* on G. According to [35], even if $O(n^{2.5})$ algorithms exist for MCMP, they work poorly on the instances generated by merging graphs.

9 Conclusion

In the previous sections we have presented combinatorial techniques that can be used successfully in the different phases of technology mapping for LUT-based FPGAs. The two main optimization criteria considered in technology mapping are area and delay. Optimality with respect to these two criteria is guaranteed, respectively, by Vismap [43] and by Flowmap [10] (if delays between LUTs are considered constant). However, Vismap is time expensive (exponential time) and Flowmap area results are far from being optimal. Recent work [38, 9] tries to reduce these drawbacks by using basic ideas from Flowmap and Vismap combined with heuristics.

Two other problems appear when considering technology mapping as a phase of circuit design (see figure 5):

- Optimality is guaranteed for the input network of technology mapping. Better final results can be obtained if technological constraints are taken into account during the previous phase of logic optimization [11].
- Routability of the mapping is a very important factor for the subsequent phase of placement and routing. Some mappers like Rmap [36] are considering this optimization criterion, however to the detriment of area and delay.

We have tried to show how well known combinatorial optimization problems arise during the attempt to better solve difficult problems in technology mapping for LUT-based FPGAs. We feel that it is worthwhile to have a closer look at the particular structure of these problems in order to obtain further improvements. The recent progress in the field of approximation algorithms could lead to better heuristics. Also, an algorithm like Vismap would certainly benefit from new results in the field of Branch-and-Bound, either by finding new pruning rules or by parallelizing the algorithm. We are currently conducting some work into these directions.

References

1. R. Andonov and S. Rajopadhye. Knapsack on VLSI : from algorithm to optimal circuit. Technical Report 95-80-5, Oregon State University, 1995.
2. D. Avis and K. Fukuda. A pivoting algorithm for convex hulls and vertex enumeration of arrangements and polyhedra. *Discrete Comp. Geom.*, (8):295–313, 1992.
3. E. Balas. Projection with a minimal system of inequalities. Technical Report MSRR-585, GSIA, Carnegie Mellon University, 1992.
4. K. Bouazza, J. Champeau, P. Ng, B. Pottier, and S. Rubini. Implementing cellular automata on the ArMen machine. In P. Quinton and Y. Robert, editors, *Proceedings of the Workshop on Algorithms and Parallel VLSI Architectures II*, pages 317–322, Bonas, France, June 1991. Elsevier.
5. R.K. Brayton, R. Rudell, A. Sangiovanni-Vincentelli, and A.R. Wang. MIS : a multiple-level optimization system. *IEEE transactions on CAD*, 6(6):1062–1081, November 1987.
6. J. Champeau, L. Le Pape, B. Pottier, S. Rubini, E. Gautrin, and L. Perraudeau. Flexible parallel FPGA-based architectures with ArMen. In 37th *Annual Hawaii International Conference on System Sciences*, Hawaii, January 1994.
7. N.V. Chernikova. Algorithm for finding a general formula for the non-negative solutions of a system of linear inequalities. *USSR Computational Mathematics and Mathematical Physics*, 2(5):228–233, 1965.
8. Th. Christof, M. Jünger, and G. Reinelt. A complete description of the traveling salesman polytope on 8 nodes. Technical Report 249, Institut für Mathematik, Universität Augsburg, 1990.
9. J. Cong and Y. Ding. Beyond the combinatorial limit in depth minimization for LUT-based FPGA designs. In *Int. Conf on CAD: digest of Technical Papers*, pages 110–114, 1993. Integration de la synth. logique et du partitionnement. Re-synth logique pour ameliorer le partitionnement. base sur [10].

10. J. Cong and Y. Ding. Flowmap: An optimal technology mapping algorithm for delay optimization in lookup-table based FPGA designs. *IEEE transactions on CAD*, 13(1):1–11, January 1994. Partitionnement pour délais. unit-delay model. Utilisation des techniques de flots pour generer des supernoeuds. optimal par rapport au modele. temps polynomial.

11. J. Cong and Y. Ding. On area/depth trade-off in LUT-based FPGA technology mapping. *IEEE transactions on VLSI*, 2(2):137–148, June 1994. Compromis surface-delais. relaxations a partir d'une sol. opt. (unit-delay) pour ameliorer la surface.

12. A. Darte. *Techniques de parallélisation automatique de nids de boucles*. PhD thesis, ENS Lyon, 1993.

13. A. Darte and Y. Robert. Séquencement des nids de boucles. In M.Cosnard et al., editor, *Algorithmique parallèle*, pages 343–368. Masson, 1992.

14. A. Darte and F. Vivien. A comparison of nested loops parallelization algorithms. Research Report 95-11, LIP, ENS Lyon, 1995.

15. P. Dhaussy, J.-M. Filloque, B. Pottier, and S. Rubini. Global Control Synthesis for an MIMD/FPGA Machine. In D. Buell and K. Pocek, editors, *IEEE Workshop on FPGAs for custom computing machines*, pages 51–58, Napa, California, April 1994. IEEE Computer Society Press.

16. M.E. Dyer. The complexity of vertex enumeration methods. *Math. Oper. Res.*, (8):381–402, 1983.

17. R. Euler and H. Le Verge. Complete linear descriptions of small asymmetric traveling salesman polytopes. *Discrete Applied Mathematics*, (62):193–208, 1995.

18. P. Feautrier. Parametric integer programming. *RAIRO Recherche opérationnelle*, 22:243–268, 1988.

19. P. Feautrier. Some efficient solutions to the affine scheduling problem. Technical Report MIP-9303, MASI-IBP, Université de Versailles-St Quentin, 1993.

20. R. Francis, J. Rose, and Z. Vranesic. Chortle-crf: Fast technology mapping for lookup table-based FPGAs. In *28th ACM/IEEE design automation conference*, pages 227–233, 1991. Partionnement pour surface. prog. dynamique des PI vers les PO d'un arbre. BPP a chaque etape pour creer les blocs. optimal en surface pour les arbres si K ≤ 6.

21. R. Francis, J. Rose, and Z. Vranesic. Technology mapping for lookup table-based FPGAs for performance. In *Int. Conf on CAD*, pages 568–571, 1991. Partitionnement pour délais. unit-delay model. BPP a chaque niveau d'un arbre puis application de l'alg Chortle-crf [20].

22. R.M. Karp, R.E. Miller, and S. Winograd. The organization of computations from uniform recurrence equations. *Journal of the ACM*, 14(3):563–590, 1967.

23. H.T. Kung. Why systolic architectures. *Computer*, 1(15):37–46, 1982.

24. Ch. Lengauer. Loop parallelization in the polytope model. Research Report MIP-9303, University of Passau, 1993.

25. S. Martello and P. Toth. *Knapsack Problems*. Wiley, 1990.

26. T.H. Matheiss and D.S. Rubin. A survey and comparison of methods for finding all vertices of convex polyhedral sets. *Math. of OR*, 2(5):167–185, 1980.

27. C. Mauras, P. Quinton, S. Rajopadhye, and Y. Saouter. Scheduling affine parametrized recurrences by means of variable dependent timing functions. Technical Report 520, IRISA, Campus de Beaulieu, 35042 Rennes cedex, 1990.

28. D.I. Moldovan. On the analysis and synthesis of VLSI systolic arrays. *IEEE Trans. on Computers*, (31):1121–1126, 1982.

29. C. Mongenet, Ph. Clauss, and G.R. Perrin. Implantations optimales d'équations récurrentes affines. In M.Cosnard et al., editor, *Algorithmique parallèle*, pages 332–341. Masson, 1992.

30. T.S. Motzkin, H. Raiffa, G.L. Thompson, and R.M. Thrall. The double description method. In H.W.Kuhn and A.W.Tucker, editors, *Contributions to the theory of games, vol II*, Annals of Mathematical studies (28). Princeton, New Jersey, 1953.

31. R. Murgai, Y. Nishizaki, N. Shenoy, R.K. Brayton, and A. Sangiovanni-Vincentelli. Logic synthesis for programmable gate arrays. In *27th ACM/IEEE design automation conference*, pages 620–625, 1990. optimisation de la surface. La base du partitionnement. max-flow pour la définition des supernoeuds. BCP pour partitionnement, MCMP pour la fusion de sommets.

32. P. Quinton. Automatic synthesis of systolic arrays from uniform recurrence equations. In *11th Annual int. symp. on Computer architecture*, Ann Arbor, Michigan, 1984.

33. P. Quinton and Y. Robert. *Algorithmes et architectures systoliques*. Masson, 1989.

34. T. Risset. *Parallélisation automatique: du modèle systolique à la compilation des nids de boucles*. PhD thesis, ENS Lyon, 1994.

35. A. Sangiovanni-Vincentelli, A. El Gamal, and J. Rose. Synthesis methods for field programmable gate arrays. In *Proceedings of the IEEE*, volume 81, pages 1057–1083, July 1993. Une revue complete des outils de synth. logique et partitionnement FPGA. Pas de circuits sequentiels.

36. M. Schlag, J. Kong, and P.K. Chan. Routability-driven technology mapping for lookup table-based FPGAs. February 1992.

37. A. Schrijver. *Theory of linear and integer programming*. Wiley, 1986.

38. H. Shin and C. Kim. Performance-oriented technology mapping for LUT-based FPGA's. *IEEE transactions on VLSI*, 3(2):323–327, June 1995.

39. H. Le Verge. A note on chernikova's algorithm. Technical Report RR 1662, INRIA, Rocquencourt, France, April 1992.

40. H. Le Verge. *Un environnement de transformations de programmes pour la synthèse d'architectures régulières*. PhD thesis, Université de Rennes, 1992.

41. H. Le Verge, V. van Dongen, and D.K. Wilde. La synthèse de nids de boucles avec la bibliothèque polyédrique. In L. Bougé et al., editor, *Proceedings of RENPAR6*, ENS Lyon, 1994.

42. D.K. Wilde. A library for doing polyhedral operations. Research Report 785, IRISA, University of Rennes, 1993.

43. N.-S. Woo. A heuristic method for FPGA technology mapping based on the edge visibility. In *28th ACM/IEEE design automation conference*, pages 248–251, 1991.

44. Xilinx. *The Programmable Gate Array Data Book*. Xilinx, San Jose (USA), 1994.

On the Parallel Complexity of the Alternating Hamiltonian Cycle Problem

(Extended Abstract)

E. Bampis[1]   Y. Manoussakis[2]   I. Milis[2] *

[1] LaMI, Université d'Evry Val d'Essonne, 91025 Evry Cedex, France
[2] L.R.I., Bât. 490, Université de Paris-Sud, 91405 Orsay Cedex, France
bampis@lami.univ-evry.fr, yannis@lri.lri.fr,
milis@softlab.ece.ntua.gr

Abstract

Given a graph with colored edges, a Hamiltonian cycle is called alternating if its successive edges differ in color. The problem of finding such a cycle, even for 2-edge-colored graphs, is trivially NP-complete, while it is known to be polynomial for 2-edge-colored complete graphs. In this paper we study the parallel complexity of finding such a cycle, if any, in 2-edge-colored complete graphs. We give a new characterization for such a graph admitting an alternating Hamiltonian cycle which allows us to derive a parallel algorithm for the problem. We prove that the alternating Hamiltonian cycle problem is in RNC and it belongs to NC if the perfect matching problem does. In addition, our parallel algorithm improves the complexity of the best known sequential algorithm by a factor of $O(\sqrt{n})$.

1 Introduction and Terminology

Last years problems arising in molecular biology are often formulated using colored graphs, i.e. graphs with colored edges and/or vertices. Given such a graph, original problems correspond to extracting subgraphs such as Hamiltonian and Eulerian paths or cycles colored in a specified pattern [11, 12, 13, 20]. The most natural pattern in such a context is that of alternating coloring, i.e. adjacent edges/vertices having different colors. This type of problems is also encountered in VLSI for compacting a programmable logical array [15]. Colored paths and cycles have applications in various other fields, as in cryptography where a color represents a specific type of transmission or in social sciences where a color represents a relation between two individuals and the notion of alternating colored paths and cycles is related to the balance of a graph [9]. Note also that the concept of alternating colored subgraphs is often used implicitly in graph theory; consider for example a given instance of Edmond's maximum matching algorithm [14]: If we consider the edges of the current matching colored blue and any other edge colored red, then the desired augmenting path (in Edmond's terms) is just an alternating colored one.

---

* Fellow of the European Commission (Program H.C.&M.). Present address: National Technical University of Athens (NTUA), Dpt. of Electr. & Comp. Eng., Division of Comp. Sc., Heroon Polytechniou 9, 15773 Zografou - Athens, Greece.

On the other hand there is a great theoretical interest on problems in colored graphs and a large body of work has been published [3, 5, 6, 8, 21]. Motivated by both the applications and the theoretical interest, we consider here complete graphs with edges colored by two colors (2-edge-colored complete graphs) and we search for a Hamiltonian cycle whose successive edges alternate between the two colors. Such a cycle is known as *alternating Hamiltonian cycle* (AHC). Our study is restricted to complete graphs, since the problem is trivially NP-complete in the case of general 2-edge-colored graphs. The same problem can be also stated in the following way: Given a graph G and its complement $\overline{G}$ find a Hamiltonian cycle whose edges alternate between those of G and those of $\overline{G}$.

In particular, in this paper we deal with the parallel complexity of the AHC problem. We give first a new characterization for a 2-edge-colored complete graph to admit an AHC. Instead of the up to now known conditions based on vertices' degrees, our characterization relies on terms of connectivity of a specified digraph implied by an alternating factor of the initial colored graph. This new characterization allows us to design a parallel algorithm which takes as input an alternating factor, and either finds an AHC or decides that such an AHC does not exist. The algorithm is on the CRCW-PRAM model where concurrent reads and concurrent writes to the same memory location are allowed. Its parallel complexity is $O(\log^4 n)$ time using $O(n^2)$ processors where n is the order of the graph. An alternating factor can be found using a maximum matching algorithm, putting the whole AHC problem in RNC. In addition, as a byproduct of our parallel algorithm we obtain an $O(n^{2.5})$ sequential algorithm for the problem. This result improves the complexity of the best known $O(n^3)$ sequential algorithm [5].

Throughout the paper, K denotes a 2-edge-colored complete graph with an even number of vertices. We assume that the edges of K are colored **red** and **blue**. By $r(v)$ and $b(v)$ we denote the red and the blue degree, respectively, of a vertex v. The edge between the vertices u and v is denoted by uv, and its color by $\chi(uv)$. If A_1 and A_2 are subsets of V, then the set of edges between the vertices of A_1 and A_2 is denoted by A_1A_2, while the edges among the vertices of A_1 are denoted by A_1A_1.

The notation $\chi(A_1A_2)$ is used if and only if all the edges A_1A_2 are monochromatic and represents their common color.

It turns out to be convenient for our presentation to divide the vertices of an alternating cycle, C, of length $2p$ into two **classes** $X = \{x_1, x_2, ..., x_p\}$ and $Y = \{y_1, y_2, ..., y_p\}$ such that $C = x_1, y_1, x_2, y_2, ..., x_p, y_p$ and $\chi(x_iy_i) \neq \chi(y_ix_{i+1})$, for each $i = 1, 2, ..., p$ (the indices are considered modulo p). It is clear that classes X and Y can be interchanged without loss of generality.

Once both classes X and Y are defined, we say that cycle C_1 **dominates** cycle C_2 ($C_1 \Longrightarrow C_2$) if $\chi(X_1 C_2) \neq \chi(Y_1 C_2)$ (recall that the notation $\chi(A_1 A_2)$ is used if and only if all the edges $A_1 A_2$ are monochromatic and by convention assume that in notations $\chi(X_i C_j)$, C_j represents the vertex set of C_j and X_i the class X of C_i).

Obviously, if $C_1 \Longrightarrow C_2$, then $\chi(X_1 C_2)$ implies $\chi(Y_1 C_2)$, and vice versa. By $C_1 \not\Longleftrightarrow C_2$ we denote the fact that neither $C_1 \Longrightarrow C_2$ nor $C_2 \Longrightarrow C_1$.

An alternating factor $F = \{C_1, C_2, ..., C_m\}$ of K is a set of pairwise vertex disjoint alternating cycles covering all the vertices of the graph. It is clear that the existence of an alternating factor is a necessary condition for a graph K to admit an alternating Hamiltonian cycle. An alternating factor is said minimum if there is no other one with smaller cardinality. Clearly F becomes an AHC whenever $m = 1$.

For the sake of completeness, we state next theorem which gives an optimal parallel algorithm for finding a Hamiltonian cycle in a strongly connected [3] semicomplete graph [1] and which will be used in Section 4. We recall that a semicomplete digraph is a digraph with no pair of non-adjacent vertices (i.e. for every pair v, u of vertices either arc (v, u) or arc (u, v) or both are present).

Theorem 1.1 ([1]) *For any semicomplete digraph T, the strong components of T and a Hamiltonian cycle in each strong component can be found by an $O(\log n)$ time, $O(n^2/\log n)$ processors algorithm within the CRCW-PRAM model.*

2 Structural results

M. Bánkfalvi and Z. Bánkfalvi [3] gave the first characterization for the minimality of an alternating factor.

Theorem 2.1 ([3]) *Let K be a 2-edge-colored complete graph with vertex set $V = \{v_1, v_2, ..., v_{n=2p}\}$. Assume that $r(v_1) \leq r(v_2) \leq ... \leq r(v_n)$. Then, the cardinality of the minimum alternating factor F of K is equal to m if and only if there are precisely $m - 1$ distinct numbers k_i, $2 \leq k_i \leq p - 2$ such that for each i, $i = 1, ..., m - 1$,*

$$r(v_1) + r(v_2) + ... + r(v_{k_i}) + b(v_{2p}) + b(v_{2p-1}) + ... + b(v_{2p-k_i+1}) = k_i^2.$$

Next corollary proved in [5], gives a structural translation of Theorem 2.1.

Corollary 2.1 ([5]) *Let $F = \{C_1, C_2, ..., C_m\}$, $m \geq 2$, be an alternating factor of K. If $C_i \Longrightarrow C_j$, $\chi(X_i X_i) = \chi(X_i C_j)$ and $\chi(Y_i Y_i) = \chi(Y_i C_j)$, $1 \leq i < j \leq m$, then F is a minimum alternating factor of K.*

---

[3] A digraph is called *strongly connected* or simply *strong* if for every pair of vertices u and v there is a directed path from u to v and one directed path from v ro u.

In what follows, a structure of an alternating factor satisfying Corollary 2.1 will be referred as a **Bánkfalvi structure**. If F has a Bánkfalvi structure, then K does not contain an AHC.

Häggvist and Manoussakis [16] have proven –in a different phrasing– another condition replacing the degree condition of Theorem 2.1 by a more intuitive one.

Theorem 2.2 ([16]) *Let K be a 2-edge colored complete graph. K contains an AHC if and only if:*
(i) it admits an alternating factor, and
(ii) every pair of vertices is connected with an alternating not necessarily simple path.

Although Theorems 2.1 and 2.2 answer to the decision problem, they do not lead to an algorithm for searching an AHC in a graph K. In the following section, we give a new sufficient condition which relies on terms of connectivity of a digraph implied by an alternating factor of K. Based on this sufficient condition, a new parallel algorithm is presented in Section 4. The main idea is the following: Given an alternating factor $F = \{C_1, C_2, ..., C_m\}$, reduce it to one of minimum cardinality by contracting as far as possible its alternating cycles. Following this direction a series of lemmas for the contraction of alternating cycles in specified cases are stated below. Due to the lack of space we omit their proofs. However, these proofs are available in the full version of the paper. They are based on a transformation from a given alternating factor F of a 2-edge-colored complete graph K to a bipartite tournament $B = (\mathcal{X} \cup \mathcal{Y}, E(B))$ such that alternating cycles of K correspond to directed cycles in B and vice versa. The bipartite tournament B corresponding to F is obtained as following:
For every alternating cycle $C_i \in F$, either $X_i \subset \mathcal{X}$ and $Y_i \subset \mathcal{Y}$, or $X_i \subset \mathcal{Y}$ and $Y_i \subset \mathcal{X}$ depending on the specific domination relations among the cycles in F.
For every pair $x \in \mathcal{X}$, $y \in \mathcal{Y}$ of vertices of B if $\chi(xy)$=red in K, then $(x, y) \in E(B)$; otherwise, if $\chi(xy)$=blue in K, then $(y, x) \in E(B)$.

In such a transformation, it is clear that an alternating cycle C_i of K corresponds to a directed cycle in B and vice versa. Consequently, a cycle obtained by contracting directed cycles in B corresponds to an alternating one in K.

Lemma 2.1 *Let C_1, C_2 be two vertex disjoint alternating cycles. Then, either $C_1 \Longrightarrow C_2$ or $C_2 \Longrightarrow C_1$ or C_1 and C_2 can be contracted to a single alternating cycle in $O(\log n)$ time using $O(n^2)$ processors.*

Proof. Omitted.

Lemma 2.2 *Let $C_1, C_2, ..., C_m$ be m pairwise vertex disjoint alternating cycles. These cycles can be contracted to a single alternating cycle C, in $O(\log n)$ time using $O(n/\log n)$ processors if for each $i = 1, 2, ..., m$ (all indices are considered modulo $(m + 1)$):*
(i) $C_i \Longrightarrow C_{i+1}$.

(ii) Either $C_i \Longrightarrow C_{i+1}$, or else if for some i, $C_i \not\Longleftrightarrow C_{i+1}$, then $C_{i+1} \Longrightarrow C_{i+2}$.
(iii) Either $C_i \Longrightarrow C_{i+1}$ and $\chi(X_i C_{i+1})$=red (resp. blue), or $C_{i+1} \Longrightarrow C_i$ and $\chi(X_{i+1} C_i)$=blue (resp. red).
(iv) $C_i \Longrightarrow C_j$, $1 \leq i < j \leq m-1$, and there is an edge $e \in X_1 X_1$ (resp. $e \in Y_1 Y_1$) such that $\chi(e) \neq \chi(X_i C_j)$ (resp. $\chi(e) \neq \chi(Y_i C_j)$).

Proof. Omitted.

Lemma 2.3 Let $C_1, C_2, ..., C_m$ be m pairwise vertex disjoint alternating cycles forming a Bánkfalvi structure, and C_0 be another alternating cycle such that $C_1 \not\Longleftrightarrow C_0$ and $C_m \not\Longleftrightarrow C_0$. Then $C_1, C_2, ..., C_m$ and C_0 can be contracted to a single alternating cycle in $O(\log n)$ time using $O(n^2)$ processors.

Proof. Omitted

3 Main result

Before stating our main theorem, let us introduce two definitions.

Definition 3.1 Let $F = \{C_1, C_2, ..., C_m\}$ be an alternating factor of K. The **underlying digraph** of F is defined as the semicomplete digraph D with vertex set $V(D) = \{c_1, c_2, ..., c_m\}$ corresponding to the cycle set of F (each cycle C_i is contracted to a single vertex c_i) and arc set $E(D)$ defined as follows:
- If $C_i \Longrightarrow C_j$, then the arc $(c_i, c_j) \in E(D)$.
- Otherwise, if $C_i \not\Longleftrightarrow C_j$, then both $(c_i, c_j) \in E(D)$ and $(c_j, c_i) \in E(D)$. (In this case we say that c_i and c_j are connected by a **symmetric arc**).

In what follows, $C_i \in F$ is called the *underlying cycle* of vertex $c_i \in V(D)$. We show, in Theorem 3.1, that if D is strongly connected then K admits an AHC. If the underlying digraph D is not strongly connected, let $D_1, D_2, ..., D_k$ be its strongly connected components, ordered such that for every pair of vertices $v \in D_i$, $u \in D_j$, $1 \leq i < j \leq k$, no arc (u, v) exists. We focus our interest on its first strongly connected component D_1. Let us denote by C_{D_1} the alternating cycle resulting from the contraction of the cycles involved in D_1, and by $X_{C_{D_1}}$ and $Y_{C_{D_1}}$ the obtained partition classes. By C_i^r let us denote the underlying cycle of some vertex $c_i \in D_r$, $1 \leq r \leq k$.

In the next definition, let us assume, without loss of generality, that the classes X and Y of the alternating cycles are defined such that $\chi(X_i^{r-1} C_s^r)$=red for every cycle C_s^r in D_r, $2 \leq r \leq k$.

Definition 3.2 The first component D_1 of D is a **nice component** if one of the following holds:
(i) there is a cycle $C_{i'}^r$, $3 \leq r \leq k$, such that for some cycle C_i^1, $\chi(X_i^1 C_{i'}^r)$ =blue.
(ii) $X_{C_{D_1}} \neq \cup_{C_i \in D_1} X_i$ and $Y_{C_{D_1}} \neq \cup_{C_i \in D_1} Y_i$.
(iii) there is a blue (resp. red) $X_{C_{D_1}} X_{C_{D_1}}$ (resp. $Y_{C_{D_1}} Y_{C_{D_1}}$) edge.

Theorem 3.1 *A 2-edge-colored complete graph K admits an AHC if and only if it contains an alternating factor and either (a) D is strongly connected or (b) D_1 is nice.*

Proof.

We prove first the <u>only if direction</u>.

Assume by contradiction that K admits an AHC but none of the conditions of Theorem 3.1 holds. By (a), D is not strongly connected. By (b), the first component of D is not nice i.e. $\chi(X_{C_{D_1}} X_{C_{D_1}}) = \chi(X_i^1 C_{i'}^r)$=red and $\chi(Y_{C_{D_1}} Y_{C_{D_1}}) = \chi(Y_i^1 C_{i'}^r)$=blue, for every cycle $C_{i'}^r$ in D_r, $2 \leq r \leq k$. In this case, by Corollary 2.1, C_{D_1} is a cycle of a Bánkfalvi structure and its vertices can not be contained in any alternating Hamiltonian cycle of K a contradiction.

We complete the proof by the <u>if direction</u>.

(a) By induction on the number m of cycles of the factor F. For $m = 2$, since D is strong, $C_1 \nLeftrightarrow C_2$ and therefore C_1 and C_2 can be contracted into a single one by Lemma 2.1. Assume therefore that $m \geq 3$. If D contains a strong proper subdigraph D' with $m' < m$ vertices, it can be contracted into a single cycle C', by induction. The new digraph induced by the cycles of $D - D'$ plus C' remains strong and contains $m - m' + 1 < m$ cycles. Using induction once more the proof is completed. If there is no such a subdigraph, then D is an almost transitive tournament except the arc (c_m, c_1). In this case the Hamiltonian cycle $c_1, c_2, ..., c_m, c_1$ in D implies, by Lemma 2.2(i), an alternating Hamiltonian cycle in K.

(b) Three cases can be distinguished according to Definition 3.2. The idea is to contract, if possible, some cycles in order to obtain a new colored graph admitting a strong underlying digraph D. Then, we can finish using Case (a).

(i) Let $C_{i'}^r$, $3 \leq r \leq k$, be a cycle such that for some cycle C_i^1, $\chi(X_i^1 C_{i'}^r)$=blue. If $r = k$, then we consider a cycle $C_{i''}^2$ in D_2. By Lemma 2.2(iii), we contract the cycles C_i^1, $C_{i''}^2$ and $C_{i'}^k$ into a single one. Since one cycle from the first and one cycle of the last strong component of D participate in the resulting cycle, the new underlying digraph becomes strong. If $r \neq k$ then we consider the cycles C_i^1, $C_{i'}^r$ and an arbitrary cycle $C_{i''}^k$. $\chi(X_i^1 C_{i''}^k)$ is red and these three cycles can be contracted by Lemma 2.2(iii). In the same way as before the obtained cycle makes the new D strong.

(ii) By (a) the cycles included in the first component D_1 of D can be contracted into a single cycle C_{D_1}. If the contraction leads to the mixing of classes of the included F's cycles, then the new cycle does not dominate any cycle in each strong component D_r, $3 \leq r \leq k$. This yields a new strong D.

(iii) If (ii) is not the case, then if there is a $X_{C_{D_1}} X_{C_{D_1}}$ blue edge or a $Y_{C_{D_1}} Y_{C_{D_1}}$ red edge, then we consider the cycle C_{D_1} and one cycle from each strong component of D. This collection of cycles satisfies the hypothesis of Lemma 2.2(iv) and can be contracted to a single cycle. Since a cycle from each strong component of D participates in the new cycle, the new D is strongly connected. □

4   A Parallel Algorithm

The existence of an alternating factor is a necessary condition for a 2-edge-colored complete graph to admit an AHC by Theorem 3.1. In order to find such an alternating factor, we have to apply two times a maximum matching algorithm: Find a red maximum matching M_r in the graph induced by the red edges, and a blue one M_b in the graph induced by the blue edges. If either M_r or M_b is not perfect, it is clear that K has no alternating factor. Otherwise, an alternating factor can be constructed by considering the union of M_r and M_b.

Next algorithm uses a procedure called FIND-AHC(D strong) which finds an AHC in the case where the underlying digraph is strong (such an AHC exists by Case (i) of Theorem 3.1). The details of this procedure are given below.

ALGORITHM HAMILTONIAN CYCLE(K);

1. Find an alternating factor $F = \{C_1, C_2, ..., C_m\}$ of K;
2. If F does not exist then STOP; {K has no AHC }
3. If $m = 1$ then STOP {F is an AHC }
4. Construct D;
5. If D is strongly connected then FIND-AHC(D strong); STOP;
6. Find the strongly connected components $D_1, D_2, ..., D_k$ of D and fix the classes X and Y of the alternating cycles such that the $\chi(X_i^{r-1} C_s^r)$=red for every cycle C_s^r in D_r, $2 \leq r \leq k$;
7. If Definition 3.2(i) holds, then using Theorem 3.1(b-i) construct new D and FIND-AHC(D strong); STOP;
8. FIND-AHC(D_1 strong);
9. If Definition 3.2(ii) holds, then using Theorem 3.1(b-ii) construct new D and FIND-AHC(D strong); STOP;
10. If Definition 3.2(iii) holds, then using Theorem 3.1(b-iii) construct new D and FIND-AHC(D strong); STOP;
11. STOP; {K has no AHC}

The complexity of the above algorithm is determined by the complexity of finding an alternating factor in a 2-edge-colored complete graph. By [19], a perfect matching of a graph containing $n(= 2p)$ vertices and m edges can be found by a randomized algorithm in $O(\log^2 n)$ time using $O(n^{3.5} m)$ processors. Since the number of red or blue edges of a 2-edge-colored complete graph is $O(n^2)$, an alternating factor can be found by a randomized algorithm in $O(\log^2 n)$ time using $O(n^{5.5})$ processors.

If an alternating factor is given, then the complexity of the algorithm is determined by the complexity of the procedure FIND-AHC(D strong).

4.1   Procedure FIND-AHC(D strong)

Recall that the underlying digraph D is a semicomplete digraph and, by Theorem 1.1, we can find in parallel a Hamiltonian cycle in D, since it is strong. Unfortunately, this cycle does not help us to construct in parallel an AHC in K because of the symmetric arcs that may arise in this cycle. We can isolate these arcs as follows: Consider the undirected graph induced by the vertices of D which are extremities of at least one symmetric arc in D. If M is a maximal matching in this graph, then for every $c_i c_j \in M$, we can contract the cycles C_i and C_j using Lemma 2.1(ii). Furthermore, the directed graph induced by $V(D) - V(M)$ is clearly a tournament.

In next lemma, we consider an alternating factor having a tournament as underlying digraph, and we prove that its reduction to a minimum one can be efficiently parallelized.

Lemma 4.1 Let $C_1, C_2, ..., C_m$ be a collection of alternating cycles whose underlying digraph is a tournament T. This collection can be reduced either to a single alternating cycle or to a Bánkfalvi structure, in $O(\log n)$ time using $O(n^2/\log n)$ processors.

Proof. By Theorem 1.1, we can find a Hamiltonian cycle in each strong connected component of T in $O(\log n)$ time using $O(n^2/\log n)$ processors. If T is strong, then a single Hamiltonian cycle is obtained and the alternating cycles $C_1, C_2, ..., C_m$ can be contracted to a single alternating cycle using Lemma 2.2(i).

If T is not strong, we first find a Hamiltonian path $H = \{c_1, c_2, ..., c_m\}$ in T. This can be done in $O(\log n)$ time using $O(n^2/\log n)$ processors, by [4]. This path implies a sequence of the alternating cycles $C_1, C_2, ..., C_m$, where C_i dominates C_{i+1} for $1 \leq i \leq m - 1$. Assume w.l.o.g. that $\chi(X_i C_{i+1})$=red and $\chi(Y_i C_{i+1})$=blue, $1 \leq i \leq m$.
We define now a new tournament T' with the same vertex set as T and arc set as follows: If $C_i \Longrightarrow C_j$ and $\chi(X_i C_j)$=red (resp. blue), then there is an arc (c_i, c_j) in T' (resp. (c_j, c_i)). Clearly, the Hamiltonian path H of T remains unchanged in the new tournament T'.

By Theorem 1.1, we find a Hamiltonian cycle in each strong component of T' in $O(\log n)$ time using $O(n^2/\log n)$ processors. Then, using Lemma 2.2(iii), we contract in parallel the alternating cycles involved in each strong component of T' into a single one. If T' is strong, then a single alternating cycle is obtained and the proof is completed. If T' is not strong, then we obtain a new collection of alternating cycles $C'_1, C'_2, ..., C'_{m'}$, each of which corresponds to a strong component of T'. The contraction of cycles by Lemma 2.2(iii) gives a new cycle with classes that are unions of the corresponding classes of the cycles involved, i.e. there is no mixing of classes. Taking into account the construction of T', we conclude that $\chi(X'_i C'_j)$=red and $\chi(Y'_i C'_j)$=blue, $1 \leq i < j \leq m'$. Finally, we find the minimum p, $1 \leq p \leq m'$, for which there is at least one $X'_p X'_p$ blue edge, or a $Y'_p Y'_p$ red edge and using Lemma 2.2(iv), we contract the cycles $C'_{p+1}, C'_{p+2}, ..., C'_{m'}$, into a single one. By Corollary 2.1, no further reduction is possible. Therefore either a single alternating cycle (if $p = 1$) or a Bánkfalvi structure (otherwise) is obtained. □

Given Lemma 4.1 we can reduce an alternating factor with strong underlying digraph into a Bánkfalvi structure plus some alternating cycles. Since the underlying digraph of a Bánkfalvi structure is a transitive tournament we are in the situation of the next lemma [2].

Lemma 4.2 ([2]) *Let S be a strong semicomplete digraph which contains as an induced subgraph a transitive tournament T. Let $H = t_1, t_2, ... t_k$ be the Hamiltonian path of T and $Q = \{q_1, q_2, ... q_m\}$ the set of vertices in $V(S) - V(T)$. Then there is in S a family of independent cycles $t_i, t_{i+1}, ..., t_j, q_1, q_2, ..., q_l, t_i$, where $t_i, t_{i+1}, ..., t_j$ is a subpath of H and $P = t_j, q_1, q_2, ..., q_l, t_i$ is a minimal path from t_j to t_i covering at least $\lceil k/2 \rceil$ vertices of T. Such a family of cycles can be found in $O(\log n)$ time using $O(n^2)$ processors.*

Using the above lemmas, we can now summarize the procedure FIND-AHC(D strong) as following:

PROCEDURE FIND-AHC(D strong);

1. Construct an undirected graph U, where $V(U) = V(D)$ and $E(U) = \{c_i c_j | c_i c_j$ is a symmetric arc in $D\}$.
2. Find a maximal matching M in U;
3. In parallel, for every $c_i c_j \in M$, contract the corresponding cycles C_i and C_j using Lemma 2.1(ii);
4. Construct new D; If $|V(D)| = 1$ then STOP;
5. Reduce the tournament $V(D) - V(M)$ either to a single cycle or to a Bánkfalvi structure with m alternating cycles using Lemma 4.1;
6. Construct new D; If $|V(D)| = 1$ then STOP;
7. Find a family of independent cycles in D by Lemma 4.2;
8. In parallel, for each cycle in the family, contract the involved alternating cycles as following:
   8.1 If P does not contain consecutive symmetric arcs then apply Lemma 2.2(ii) and go to Step (9);
   8.2 Consider each symmetric arc of P as undirected and find all maximal directed subpaths of P;

8.3 In parallel, for each such maximal path $q_{i_1}, q_{i_2}, ... q_{i_r}$ in P, $r > 3$, contract the involved alternating cycles using Lemma 2.2(ii) and the fact that the arc (q_{i_r}, q_{i_1}) is present (this is due to minimality of P);

8.4 Apply Lemma 2.2(ii) to every symmetric arc whose neighbor in P is not symmetric; (after this step all arcs in P become symmetric)

8.5 Reduce P to a path of length 2 (i.e. with 2 symmetric arcs) by repeating applications of Lemma 2.2(ii) ;

8.6 Apply Lemma 2.3;

9. Construct new D; FIND-AHC (D strong);

Given the complexity of the lemmas used, the complexity of the above procedure is determined by the complexity of the required maximal matching algorithm. The fastest known parallel algorithm for the maximal matching problem has been proposed in [17]. On a CRCW-PRAM this algorithm works in $O(\log^3 n)$ time using $O(n^2)$ processors because the Euler tour can be found on this model in $O(\log n)$ time. Since the recursive depth of the procedure is $O(\log n)$ the whole complexity becomes $O(\log^4 n)$ time using $O(n^2)$ processors. We notice that for a randomized version of procedure FIND-AHC, the maximal matching algorithm used in Step 2 can be replaced by a randomized maximum matching algorithm and thus, in this case, the whole complexity becomes $O(\log^3 n)$ time using $O(n^{5.5})$ processors.

Therefore, we can state the next theorem.

Theorem 4.1 *An AHC in a 2-edge-colored complete graph can be found*
• *by a deterministic $O(\log^4 n)$ time, $O(n^2)$ processors algorithm, if an alternating factor is given, and*
• *by a randomized $O(\log^3 n)$ time, $O(n^{5.5})$ processors algorithm, if an alternating factor has to be found.*

Our parallel algorithm implies also a sequential algorithm for the AHC problem. The complexity of this sequential algorithm is dominated by the complexity of the best known sequential maximum matching algorithm [18] [22].

Theorem 4.2 *There is an $O(n^{2.5})$ sequential algorithm for finding an AHC in a 2-edge-colored complete graph.*

References

1. E. Bampis, M. El Haddad, Y. Manoussakis and M. Santha, *A parallel reduction of Hamiltonian cycle to Hamiltonian path in tournaments*, Journal of Algorithms **19** 3 (1995) 432-440.

2. J. Bang-Jensen, M. El Haddad, Y. Manoussakis and T. M. Przytycka, *Parallel algorithms for the Hamiltonian cycle and Hamiltonian path problems in semicomplete bipartite digraphs*, Preprints No. 15/1993, Institut for Mathematik og Datalogi, Odense Universitet, April 1993, to appear in Algorithmica (1996).

3. M. Bánkfalvi and Z. Bánkfalvi, *Alternating Hamiltonian circuit in two-colored complete graphs*, Theory of Graphs (Proc. Colloq. Tihany 1968), Academic Press, New York, 11-18.

4. A. Bar-Noy and J. Naor, *Sorting, minimal feedback sets and Hamiltonian paths in tournaments*, SIAM J. Discr. Math. **3** (1990) 7-20.

5. A. Benkouar, Y. Manoussakis, V. Th. Paschos and R. Saad, *On the complexity of some Hamiltonian and Eulerian problems in edge-colored complete graphs*, to appear in RAIRO-Operations Research (1996).

6. B. Bollobás and P. Erdös, *Alternating Hamiltonian cycles*, Israel Journal of Mathematics **23** (1976) 126-131.

7. R. Brent, *The parallel evaluation of general arithmetic expressions*, J. ACM **21** (1974) 201-206.

8. C.C. Chen and D. E. Daykin, *Graphs with Hamiltonian cycles having adjacent lines different colors*, J. Combin. Th. (B) **21** (1976) 135-139.

9. D. Cartwright, F. Harary, *Structural balance: A generalisation of Heider's theory*, Psychological Review **73** (1956) 277-293.

10. R. Cole and U. Vishkin, *Approximate and exact parallel scheduling with applications to list tree and graph problems*, In Proc. 27th Annual Symp. FOCS (1986) 478-491.

11. D. Dorniger, *On permutations of chromosomes*, In Contributions of General Algebra 5 (Verlag Hölder-Pichler-Tempsky, Wien, and Teubner-Verlag, Stuttgart), (1987) 95-103.

12. D. Dorniger, *Hamiltonian circuits determining the order of chromosomes*, Disc. Appl. Math. **50** (1994) 159-168.

13. D. Dorniger, W. Timischl, *Geometrical constraints on Bennett's predictions of chromosome order*, Heredity **58** (1987) 321-325.

14. J. Edmonds, *Paths, trees and flowers*, Canad. J. Math. **17** (1965) 449-467.

15. T.C. Hu and Y.S. Kuo, *Graph folding and programmable logical arrays*, Networks **17** (1987) 19-37.

16. R. Häggvist, Y. Manoussakis, *Cycles and paths in bipartite tournaments with spanning configurations*, Combinatorica **9** (1) (1989) 33-38.

17. A. Israeli and Y. Shiloach, *An improved parallel algorithm for maximal matching*, Inf. Proc. Let. **22** (1986) 57-60.

18. S. Micali, V.V. Vazirani, *An $O(|V|^{\frac{1}{2}}|E|)$ algorithm for finding maximum matchings in general graphs*, In Proc. 21th FOCS (1980) 17-23.

19. K. Mulmuley, U. V. Vazirani and V. B. Vazirani, *Matching is as easy as matrix inversion*, Combinatorica **7** (1987) 105-113.

20. P. A. Pevzner, *DNA Physical mapping and alternating Eulerian cycles in colored graphs*, Algorithmica, **13** (1995) 77-105.

21. R. Saad, *Finding a longest alternating Hamiltonian cycle in an edge colored complete graph is not hard*, Rapport de Recherche No. 691, Laboratoire de Recherche en Informatique, Université de Paris-Sud XI, Sep. 1991, to appear in Combinatorics, Probability and Computing.

22. V.V. Vazirani, *A theory of alternating paths and blossoms for proving correctness of the $O(\sqrt{V}E)$ general graph maximum matching algorithm*, Combinatorica **14** (1994) 71-109.

Threshold Graphs and Synchronization Protocols

Rossella Petreschi and Andrea Sterbini

Department of Computer Science, University "La Sapienza",
Via Salaria 113 - 00198 Rome, Italy,
e-mail {petreschi,andrea}@dsi.uniroma1.it

Abstract. This paper is a survey on the synchronization of a system of cooperating processes, when the mutual exclusion graph model and the semaphores are used. Threshold graphs and PV_{chunk} semaphores are explained. Matroidal and matrogenic graphs are presented and their synchronization with a constant number of semaphores for each process are pointed out. Threshold dimension of a graph is explained and a sketched proof of its NP-completeness for $k \geq 3$ and of the polynomiality for $k = 2$ is provided.

The interest in characterizing new classes of graphs not 2-threshold, but synchronizable with a constant number of semaphores, is shown.

Keywords: Synchronization primitives; PV_{chunk}; threshold and matrogenic graphs; 2-threshold dimension.

1   Introduction

In this paper we deal with the problem of ensuring mutual exclusion and synchronization in a system of cooperating processes. Usually, processes may communicate by sending and receiving messages or by manipulating one or more *shared* variables. The standard way to prevent conflicts is to have a designated section of code in each process identified as a *critical section* (here the resources in competition with other processes are used) and to have the processes executing a joint algorithm that controls access to the critical sections.

The algorithm is typically represented within each process by 2 protocols: the *entry* and the *exit* one. The entry (exit) protocol is executed by a process before (after) it is admitted to its critical section. If no process can be constantly waiting to enter its critical section we call the mutual exclusion *lockout-free*. Mutual exclusion will be called *deadlock-free* when at least one process is in the critical section at any time. When two not mutually excluding processes are in their critical sections at the same time we say that *concurrency* is satisfied.

This kind of problems is conveniently described in graph-theoretic terms by representing a system of asynchronous parallel processes as a *mutual exclusion graph* in which a vertex is a process and an edge is a pair of mutually excluding processes. This graph is undirected because the blocking relation is symmetric.

Notice that there exist classes of synchronization problems that are not immediately representable as simple graphs; for example different processes demanding different sizes of a resource presented in a given maximum quantity or asking for a printer unit in a given set without caring which unit will be assigned to them.

The problem of the synchronization has been solved by using several tools that can be differently classified if considered from various viewpoints [5, 34]:

a) only software solutions vs. hardware-assisted solutions:
- the former are based only on algorithms: for example see [37, 7, 11, 1, 39, 22, 6];
- the latter are based on special hardware as the "test-and-set", "exchange", "lock", "increment" and "replace-add" instructions;

b) centralized synchronization protocols (1) vs. distributed protocols with state variables (2) vs. distributed protocols based on message communication (3):
- examples of (1) are semaphores, monitors, critical regions;
- examples of (2) can be seen in [20, 38, 31, 40, 23];
- examples of (3) are token-based synchronization[12], distributed queues [21, 13, 18, 14], time-stamping (some of the previous and [27]);

c) solutions hand-crafted by the programmer (1) vs. semantic-based solutions (2):
- example of (1) are the semaphores;
- example of (2) are monitors, critical regions, serializers, concurrent languages.

In this paper we point the attention on the synchronization when the semaphore-based approach is used (i.e. b1, c1).

Since 1968, Dijkstra's work on control problems in concurrent processing have introduced synchronization primitives dubbed PV. A generalization of this system of primitives, the PV_{chunk}, has been introduced by Vantilborgh and van Lamsweerde [36] and formalized by Henderson and Zalkstein [17]. In this case a label c_i is assigned to each process p_i and a shared variable, S, is initialized to a value t. For each process p_i, the entry and exit protocols consist in decrementing and incrementing the shared variable S by c_i, respectively. The variable S is not allowed to assume values less than 0. The computation is an atomic action, so only one process at a time will actually change the value of S, and a process p_i is busy waiting until the operations on S are executed. Several processes can be in their critical sections at the same time if and only if their corresponding labels sum up to no more than t. Dijkstra's operations coincide with PV_{chunk} when the labels c_i and the value t are all equal to 1.

It is well known that a single PV semaphore can synchronize a system of processes represented by an exclusion graph that is *complete*. The systems of processes synchronizable by the PV_{chunk} primitives are represented by a class of graphs called *threshold* graphs, (see Sec. 2.2). When the exclusion graph is a general graph, more complex protocols have been investigated. In particular, since it is known that a general graph can be covered with k threshold graphs (trivially one graph for each edge or one graph for each vertex), we might proceed by using PV_{chunk} operations on k shared variables. Any process would check all the variables and could enter its critical section after executing the entry protocols on each one.

However, in order to avoid that the number of PV_{chunk} operations to depend on the dimension of the mutual exclusion graph, it is worth to investigate the minimum number k of threshold graphs needed to cover the edges of G, the *threshold dimension k*. It is possible to see that in general this protocol is not efficient since the problem of checking k-thresholdness of a graph is polynomial just for $k \leq 2$ while is NP-complete for any constant $k \geq 3$. Therefore it is important to find classes of mutual exclusion graphs that are synchronizable by a composition of a fixed number of PV_{chunk} for each process, even if they are not 2-threshold.

The motivation for this research is not only theoretical, since PV_{chunk} and multiple PV_{chunk} (or something very similar) have been implemented in several operating systems such as UNIX.

2 Preliminary Definitions

2.1 Semaphores

The semaphore concept consists in a shared variable S that represents the state of the synchronization protocol and/or the preconditions for a process to enter into its critical region. The access primitives of the variable are based on the *test-and-set* instruction. In the entry part of the protocol the process keeps examining a shared variable until it reaches a fixed set of values F. Then it resets the value of the shared variable and control passes to the next statement. This may be written in the following way:

> test S until S is in F
> then S:=Function(S)

The exit routine does not need the test of S, but it can formally be described by the same kind of statement by letting F contain all possible values of the variable S. The test-and-set operation is always atomic. A process that is not allowed to enter its critical region is suspended in a waiting list, thus avoiding the inefficiency of an active wait.

PV Semaphores

The PV semaphore was introduced in [5]; it consists in a shared binary variable S initialized to 1 and an initially empty queue. It can be accessed only by two operations, $P(S)$ and $V(S)$. Their definitions are:

```
procedure P(S: semaphore)
begin
      if S < 1
      then  wait in a queue associated with S
      endif
      S := S − 1
end
```

```
procedure V(S: semaphore)
begin
      S := S + 1
      wake up one of the waiting processes
end
```

PV_{chunk} Semaphores

In this case the semaphore is initialized to a value $t \geq 1$ and a process can ask to decrement it by a value bigger than 1 [36, 17].

The operations on the semaphore are:

```
procedure P(S: semaphore, a:integer)
begin
      if S < a
      then  wait in a queue associated with S
      endif
      S := S − a
end
```

```
procedure V(S: semaphore, b:integer)
begin
      S := S + b
      wake up one of the waiting processes
end
```

If a label c_i is assigned to each process p_i and the corresponding P and V operations use the same value c_i , then a collection of processes can be in their critical sections at the same time if and only if their corresponding labels sum up to no more than the initial value t. The systems of processes synchronizable by the PV_{chunk} primitives are represented by a class of graphs called threshold graphs.

This semaphore has been proved having optimal memory requirements for this class of mutual exclusion graphs [28].

2.2 Graphs

Through this paper we follow the standard graph-theoretic terminology of [25].

Given a graph $G = (V, E)$, $\forall v \in V$, $N_G(v)$ and $N_G[v]$ are the open and closed neighborhood of v in G, respectively. I.e.

$$N_G(v) = \{x | vx \in E\} \quad N_G[v] = N_G(v) \cup \{v\}$$

We define an equivalence relation R on the vertices of G by:

$$x R y \equiv N_G(x) - \{y\} = N_G(y) - \{x\} \quad i.e. \quad N_G[x] = N_G[y]$$

and a partial order on the set of equivalence classes E_a of R

$$E_a \rho E_b \iff E_a \neq E_b \quad and \quad N_G(b) \subseteq N_G[a]$$

ρ is called the *vicinal preorder* associated with G.

A graph $G = (V, E)$ is *split*, shortly $G(K \cup I)$, if it is possible to partition its vertices into two sets K and I such that the subgraphs induced by K and I are a clique and an independent set, respectively.

Definition 1 [2, 17, 25]. The following definitions are equivalent:

A graph $G = (V, E)$ is *threshold* if and only if

- there exists a positive real vertex-labeling c_i, the *threshold labeling*, and a positive real t, the *threshold separator*, such that every subset $I \subseteq V$ is an independent set if and only if the sum of the labels in I is less than or equal to t.
- G does not contain an alternating 4-cycle, i.e. no group of 4 distinct vertices a, b, c, d such that ab, cd are edges in G and bc, da are not edges in G (Fig. 1)[1]. In the following, we will call the graph in Fig. 1 the *forbidden configuration* of a threshold graph.

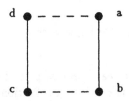

Fig. 1. The forbidden configuration of a threshold graph

- the vicinal preorder associated with G is a linear order.

---

[1] From now on, in the figures, - - - - will show an absent edge.

- G is split and the neighborhoods of the vertices in I are mutually comparable (with respect to set-inclusion).

Now we recall two classes of mutual exclusion graphs that we will show synchronizable by a composition of a fixed number of PV_{chunk} for each process, even if these classes are not 2-threshold.

Definition 2 [10, 25]. The following definitions are equivalent:
A graph $G = (V, E)$ is *matrogenic* if and only if:

- G does not contain the forbidden configuration of Fig. 2.

Fig. 2. The forbidden configuration of a matrogenic graph

- for each couple of vertices x, y if $x(\neg R)y$ then $|N_G[x] \otimes N_G[x]| = 2$.
- its vertices can be partitioned into three sets K, I, C such that: (Fig. 3)[2]

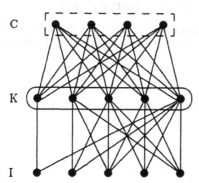

Fig. 3. Structure of a matrogenic graph (the rectangle stands for a crown)

- $K \cup I$ induces a matrogenic split graph in which K is the clique and I the independent set;

---

[2] From now on, in the figures, an oval will stand for a clique.

- C induces a crown, i.e. a pentagon or a matching of size bigger than 2, or an hyperoctahedron (the complement of a perfect matching);
- every vertex in C is connected to every vertex in K;
- no vertices in C are connected to vertices in I.

Definition 3 [30, 25]. A graph $G = (V, E)$ is *matroidal* if and only if it does not contain the forbidden configurations in Fig. 4.

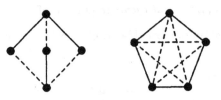

Fig. 4. The forbidden configurations of a matroidal graph

As the forbidden configuration of matrogenic graphs is one of the forbidden configurations of matroidal graphs, the latter is a subclass of the former. Moreover the class of matrogenic graphs properly contains the class of threshold graphs.

Efficient serial and parallel recognition algorithms for threshold and matrogenic graphs are known [29, 26, 3].

The last class of graphs we recall here will be useful for the geometric representation presented in section 3.2.

Threshold graphs are a subclass of interval graphs.

Definition 4 [16, 9, 8]. A graph $G = (V, E)$ is called *interval* if to each vertex $x \in V$ can be assigned an interval $I(x)$ such that $xy \in E \iff I(x) \cap I(y) \neq \emptyset$ (Fig. 5).

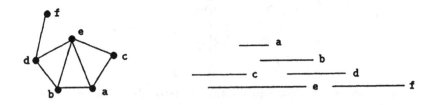

Fig. 5. An interval graph and its interval model

3 Threshold Number of a Graph

A graph $G = (V, E)$ can be *covered* by k graphs $G_1 = (V, E_1), \ldots, G_k = (V, E_k)$ if G is the edge-union of the G_i, i.e. if $E = E_1 \cup \ldots \cup E_k$. The complement of a graph covered by k graphs is the intersection of the complements of the k graphs, i.e. $G' = (V, E')$ where $E' = E'_1 \cap \ldots \cap E'_k$. In the following, when the G_i are threshold graphs we call them *threshold components*.

The *threshold number* of a graph G, $t(G)$, is the minimum number of threshold graphs that cover G, while the *threshold dimension* of G, $tdim(G)$, is the minimum number of graphs that have G as intersection. As the complement of a threshold graph is threshold, the threshold number of a graph is equal to the threshold dimension of its complement. The problem of determining the threshold dimension of a graph G is polynomially reducible to the problem of determining the threshold number of the complement of G, and vice-versa.

Two disjoint edges $e_1, e_2 \in E$ are said to be in *conflict* if the subgraph induced by their 4 endpoints is a forbidden configuration. The *conflict graph* $G^* = (E, E^*)$ of the graph $G = (V, E)$ is the graph that has E as vertex set, and an edge between two vertices if and only if they are in conflict in G. It is immediate to show that the threshold number of a graph is always greater or equal to the chromatic number of its conflict graph.

The next two subsections concern the determination of the threshold number of a graph.

3.1 Threshold Number for $k \geq 3$

Given a graph $G = (V, E)$ a (proper) c-coloring is a partition of the vertices $V = X_1 \cup X_2 \cup \ldots \cup X_c$ such that each X_i is an independent set. All the members of X_i are "painted" with the color i and adjacent vertices will receive different colors. $\chi(G)$ is the smallest possible c for which there exists a c-coloring of G.

$\chi(G)$ is called the *chromatic number* of G. It holds:

Fact 5. Given a graph $G = (V, E)$, it is NP-complete to check for the chromatic number 3 [15].

We recall that a poset is said to have dimension k if it is the intersection of a minimum number k of linear orders, called *linear extensions*.

Fact 6. The 3-dimensional poset recognition is an NP-complete problem [41].

Sketched Proof. It is derived with a reduction from the chromatic number 3 problem. The reduction shows that a graph G can be colored with 3 colors if and only if a poset P built from G has poset dimension $d(P) \leq 3$. ∎

Definition 7. A *chain graph* is a bipartite graph that does not contain $2K_2$ (the graph composed by two disjoint edges) as induced subgraph. The minimum number of chain graphs that cover a bipartite graph B is called *chain dimension* $ch(B)$.

Remark. In the proof of fact 6 it is shown that it is always possible to build from a poset P a bipartite graph $B(P)$ that has chain dimension equal to the poset dimension of P. Therefore, determining if a bipartite graph can be covered with at most 3 chain graphs is NP-complete.

Theorem 8 [41]. *Checking for 3-thresholdness is NP-complete.*

Sketched Proof. It is obtained with a reduction from the chain dimension 3 problem. By adding a clique on one of the two classes of the partition of the graph $B(P)$ we obtain a split graph G that has threshold number $t(G)$ equal to the chain number $ch(B(P))$. ∎

The threshold number 2 case remained open from 1982. In the last 2 years, independently, 2 papers have solved this problem [24, 32].

3.2 Threshold Number for $k = 2$

The underlying ideas of these solutions are very different, the first one is based on the recognition of the geometric model of 2 threshold graphs, while the second is obtained by proving a long-lasting conjecture that affirms:

Conjecture 9 [19]. A graph is 2-threshold if and only if its conflict graph is bipartite.

The interval model of a threshold graph (Fig. 6) is built in the following way:

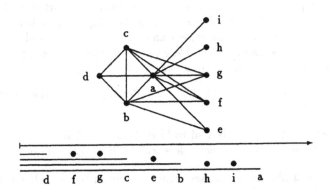

Fig. 6. An interval model of a threshold graph.

– all the intervals associated to the clique vertices share the left (wlog) endpoint;

— all the intervals associated to the independent vertices are points (i.e. have zero-length) at different coordinates.

It is possible to show [24] that the recognition of a threshold dimension 2 graph, i.e. the edge-intersection of two threshold graphs $T_1(K_1 \cup I_1), T_2(K_2 \cup I_2)$, is equivalent to the recognition of a two-dimensional geometric model made of:

— points that are the vertices in $I_1 \cap I_2$;
— segments parallel to the axes and with an endpoint on an axis, that are the vertices in $(I_1 \cap K_2) \cup (K_1 \cap I_2)$;
— rectangles with the edges parallel to the axes and the left-bottom corner in the origin (shown as broken lines in the figure), that are the vertices in $K_1 \cap K_2$.

In Fig. 7 a rectangle model is shown for graph $G = T_x \cap T_y$ with the interval models of the two threshold components T_x, T_y near the corresponding axes.

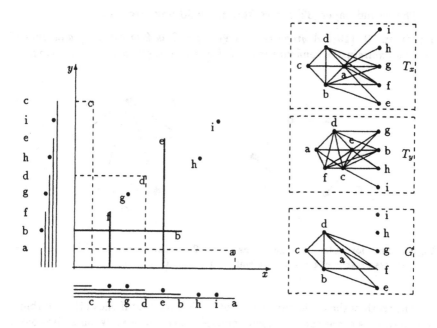

Fig. 7. A rectangle model of a threshold dimension 2 graph $G = T_x \cap T_y$.

Now, we sketch the algorithm that recognizes the threshold rectangle model of a threshold dimension 2 graph [24].

The next steps must be repeated at most $O(n^2)$ times, i.e. the number of maximal cliques in a threshold dimension 2 graph. All the maximal cliques can be found in $O(n^3)$ time for this class of graphs.

Step 1 Let the rectangles set R be the set of vertices of a maximal clique of the graph, and the points and segments sets be the set of isolated vertices of the graph $P = G - R$ and the set of all the vertices of the graph $S = G - R - P$, respectively.

Step 2 Find the relative positions between points, rectangles and segments, by examining their neighborhoods. This relation must be a 2-dimensional poset for this class of graphs.

Step 3 Decompose the poset into the 2 linear orders that correspond to the ordering of the projections of the geometric elements on the axes.

If it is not possible to compute the two linear orders, i.e. the relation is not a 2 dimensional poset, then the chosen maximal clique cannot be the rectangles set of a rectangle model of the graph. Choose a new maximal clique and repeat the process.

If all the $O(n^2)$ maximal cliques are rejected then the graph is not threshold dimension 2.

This algorithm requires $O(n^4)$ time, and can be improved to $O(n^3)$ time [33].

The second paper [32] moves from the following theorem:

Theorem 10 [19]. *A graph G with $\chi(G^*) = 2$ is 2-threshold if and only if it does not contain the configurations of Fig. 8 in respect to the G^* coloring.*

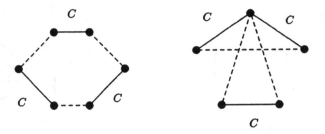

Fig. 8. The forbidden configurations of a 2-threshold graph with $\chi(G^*) = 2$ (—— shows present edges with the same color C in G^*)

The truth of the conjecture is proved by showing that if one of these forbidden configurations is present it is always possible to generate a new bipartite coloring of G^* with less forbidden configurations.

This process produces also the two threshold components, and requires $O(m^2)$ time [32].

4 Synchronization of a General Mutual Exclusion Graph

From the previous section, we saw that if a graph is 2-threshold, then 2 PV_{chunk} semaphores are sufficient to synchronize it. The covering of a general graph requires $O(n)$ threshold components even if one uses algorithms more sophisticated [2] than the trivial one.

In this section we present other approaches for the problem that, unfortunately, seem interesting just from a theoretical point of view because they are correlated either with NP-complete problems or with the dimension of the underlying graph.

4.1 PV-multiple Primitives

The $PV_{multiple}$ set of primitives corresponds to a group of binary semaphores that operate at the same time.

```
procedure P(S₁,...,Sₖ: semaphore)
begin
        if S₁ < 1 ∨ ... ∨ Sₖ < 1
        then  wait in a queue
        endif
        for i := 1 to k do
        Sᵢ := Sᵢ − 1
end
```

```
procedure V(S₁,...,Sₖ: semaphore)
begin
        for i := 1 to k do
        Sᵢ := Sᵢ + 1
        wake up one of the waiting processes
end
```

The test part of the P operation, in this case, is the union of all the tests of the involved semaphores. Taking into account the graph-theoretical interpretation of this system of primitives, it is easy to see that a $PV_{multiple}$ using k semaphores can synchronize a system of processes whose exclusion graph is coverable by k cliques.

Unfortunately, checking if a graph has clique cover number greater or equal than 3 is NP-complete [15].

4.2 PV-general Primitives

In this case we combine multiple PV_{chunk} semaphores, in fact the definition of a $PV_{general}$ is as follows:

```
procedure P(S₁,...,Sₖ: semaphore ; a₁,...,aₖ :integer)
begin
        if S₁ < a₁ ∨ ... ∨ Sₖ < aₖ
        then  wait in a queue
        endif
        for  i := 1 to k do
        Sᵢ := Sᵢ - aᵢ
end
```

```
procedure V(S₁,...,Sₖ: semaphore ; a₁,...,aₖ :integer)
begin
        for  i := 1 to k do
        Sᵢ := Sᵢ + aᵢ
        wake up one of the waiting processes
end
```

The class of exclusion graphs that can be synchronized with a $PV_{general}$ semaphore made of k PV_{chunk} is the class of graphs coverable by k threshold graphs. As we stated above, this recognition problem is NP-complete for fixed $k \geq 3$.

In spite of the correlation with NP-complete problems, semaphores are interesting because it is possible to implement protocols that avoid deadlock and lockout. In the next two subsections we show a possible solution [42].

4.3   A General Protocol Using PV Semaphores

This protocol uses a number of PV semaphores equal to the number of vertices. Each process that must enter its critical region has to check a number of semaphores equal to the size of its closed neighborhood.

- there is a semaphore S_p associated to every vertex p;
- the entry section of each process p consists of a sequence of P operations on the semaphores associated to all vertices in $N_G[p]$, followed by a sequence of V operations on the semaphores associated to all vertices in $N_G(p)$;
- the exit section of each process p consists only of the operation $V(S_p)$.

If a process p is executing its critical region, then only the semaphore S_p is switched off, and therefore all its neighbors are blocked, but in the meantime a non-neighboring process is allowed to enter its critical region.

It is important that the V operations are not all placed in the exit section of the protocol, otherwise when a process is in its critical region the neighboring semaphores are switched off, and a non-neighbor can be blocked if its distance from p is 2. A fixed arbitrary labeling of the processes gives an order to the P operations. To be sure to avoid a deadlock in the protocol, the sequence of V operations is reversed in respect to the P sequence.

The fixed ordering is the trick that avoids a global deadlock during the entry sections of the protocol and that ensures no lockout.

Similar considerations may be done by associating a PV semaphore to every edge (instead than to every vertex).

4.4 A General Protocol Using PV_{chunk} Semaphores

The previous protocol can be extended for many general semaphores. It is interesting to see that in the case of PV_{chunk} it is possible to move all the V operations into the exit section and that this solution allows more concurrency than the previous one.

To each vertex is associated a PV_{chunk} semaphore initialized to the degree of the vertex. The entry protocol of each process p is a list of P operations on the semaphores corresponding to the vertices in the close neighborhood of p, in the order given by a fixed arbitrary labeling of the vertices. The exit protocol is the reversed list of the V operations. In all the P operations on the neighbors of p we subtract 1 from the semaphore, while in $P(S_p)$ we subtract the degree of the vertex.

In this way:

- when p is outside its critical region all its neighbors are allowed to decrement the semaphore without constraining each other;
- if p is inside its critical region it blocks all the neighbors;
- if p is outside its critical region only one neighbor is needed to block p;
- the absence of lockout and deadlock is derived from the fixed labeling.

The drawback of the last two protocols is the fact that in the general case one needs $O(n)$ semaphores and protocols of length $O(n)$ for each process.

5 Synchronization with a Constant Number of PV_{chunk}

As a consequence of the fact that, for a general graph:

- optimal PV_{chunk} synchronization exists when a graph is 2-threshold,
- lockout-free and deadlock-free synchronization exists, even if a linear number of semaphores are necessary,

in this section we investigate a case in which a mutual exclusion graph that is not 2-threshold is synchronizable with a constant number of PV_{chunk} for each node: This case concerns matrogenic (and matroidal) exclusion graphs.

The results affirm that, independently from the dimension of the system (i.e. of the exclusion graph), any process would check sequentially at most four variables by using PV_{chunk} operations [4].

To sketch the idea on which this result is based, we need to recall a split-matrogenic graph characterization theorem.

Theorem 11 [26]. *A split graph G with complete set K and independent set I is split matrogenic if and only if the edges of G can be colored red and black so that:*

1. *The red partial graph is the union of vertex-disjoint pieces, $C_i, i = 1, \ldots, z$. Each piece is either a null graph belonging either to K or to I, or a matching of K' onto I', where $K' \subseteq K$ and $I' \subseteq I$, or an antimatching of K' onto I' (i.e. the bipartite complement of a perfect matching from K' onto I').*

2. *The linear ordering $C_1, \ldots, C_z$ is such that each vertex in K belonging to C_i is linked by a black edge to every vertex in I belonging to $C_j, j = i+1, \ldots, z$ and to no vertex in I belonging to $C_j, j = 1, \ldots, i-1$.*

3. *Any two vertices in K are linked by black edge (Fig. 9).*

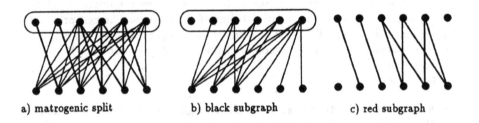

a) matrogenic split   b) black subgraph   c) red subgraph

Fig. 9. Structure of a matrogenic split graph

The idea for synchronizing a system modeled by a split matrogenic graph is to manage the black and the red graph separately. The black graph is a threshold one and then one PV_{chunk} synchronization primitive is applied on it. Further PV operations are needed for the processes of a matching. To manage an antimatching, each process p_i checks its adjacents with two PV_{chunk} primitives managing T_1 and T_2, the two threshold graphs that cover the split graph induced by $I' \cup K'$ (Fig. 10).

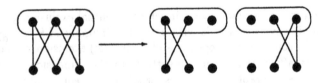

Fig. 10. How to synchronize an antimatching piece

When the matrogenic graph is not split, we have to consider the crown C and its relations with the complete set K in view of its threshold cover. The threshold dimension of the pentagon and the perfect matching are 3 and m, respectively, where m is the number of dually correlated couples of vertices. When the crown is an hyperoctahedron it is proved [4] that we can cover it with three threshold graphs. Table 1 summarizes these results.

Table 1. Threshold number and number of PV_{chunk} in a matrogenic graph

Subgraph	Threshold number	PV_{chunk} per node
matching	m	1
antimatching	2	2
threshold	1	1
pentagon	3	2
hyperoctahedron	3	3
relations C-K	1	1

From Theorem 11 and from Tab. 1 it follows:

Fact 12 [4]. Let the exclusion graph of a system of asynchronous parallel processes be a matrogenic graph. Independently from the dimension of the system (i.e. of the exclusion graph) any process would check sequentially at most four variables by using PV_{chunk} operations.

Notice that the worst case for each vertex in the crown is due to the hyperoctahedron. This consideration justifies the fact that it is not possible to synchronize a matroidal exclusion graph with less than 4 PV_{chunk} per node [35].

6 Conclusion and Open Problems

In this paper the semaphore approach to the synchronization of a system of cooperating processes is surveyed, when the mutual exclusion situation is modeled by a graph. Simple PV or PV_{chunk} primitives are sufficient to synchronize mutual exclusion graphs when they are a clique or a threshold graph, respectively.

Theoretical results on the synchronization of a general mutual exclusion graph are documented, when PV_{chunk} operators are used.

The synchronization of matrogenic and matroidal mutual exclusion graphs with at most 4 PV_{chunk} per node is shown.

In consequence of the fact that protocols using PV_{chunk} operations are easy to implement without lockout and deadlock, it remains much work to do to characterize classes of graphs not 2-threshold, but synchronizable with a constant number of PV_{chunk} semaphores per vertex.

Another open question concerns the efficient dynamic synchronization, i.e. the understanding of how the solution changes when the underlying graph is modified.

Finally other problems such as the minimization of memory requirement in the matrogenic case or the choice of a different from b in the PV_{chunk} operations are also of theoretical and practical interest.

References

1. Eisenberg M. A. and McGuire M. R. Further comments on Dijkstra's concurrent programming control problem. *Comm. ACM*, 15(11), 1972.
2. V. Chvátal and P. L. Hammer. Aggregation of inequalities in integer programming. *Ann. Disc. Math.*, 1:145-162, 1977.
3. S. De Agostino and R. Petreschi. Parallel recognition algorithms for graphs with restricted neighbourhoods. *Int. J. of Foundations of Comp. Sci.*, 1(2):123-130, 1990.
4. S. De Agostino and R. Petreschi. On PV_{chunk} operations and matrogenic graphs. *Int. J. of Foundations of Comp. Sci.*, 3(1):11-20, 1992.
5. E. W. Dijkstra. Cooperating sequential processes. In F. Genuys, editor, *Programming Languages*. Academic Press, New York, 1968.
6. Burns J. E. Symmetry in systems of asynchronous processes. In *Proc. 22nd Annual Symp. on Foundations of Computer Science*, pages 164-174, 1981.
7. Knuth D. E. Additional comments on a problem in concurrent programming control. *Comm. ACM*, 9(5):321-322, 1966.
8. P. C. Fishburn. *Interval Orders and Interval Graphs*. Wiley, New York, 1985.
9. P. C. Fishburn. Intransitive indifference with unequal indifference intervals. *Journal of Math. Psych.*, 7:144-149, 70.
10. S. Földes and P. L. Hammer. On a class of matroid-producing graphs. *Colloq. Math. Soc. J. Bolyai (Combinatorics)*, 18:331-352, 1978.
11. De Bruijn J. G. Additional comments on a problem in concurrent programming control. *Comm. ACM*, 10(3):137-138, 1967.
12. Le Lann G. Distributed systems, towards a formal approach. In *IFIP Congress*, pages 155-160, 1977.
13. Ricart G. and Agrawala A. K. An optimal algorithm for mutual exclusion in computer networks. *Comm. ACM*, 24(1):9-17, 1981. Corrigendum in *Comm. ACM*, **24** (9).
14. Ricart G. and Agrawala A. K. Author's response to 'on mutual exclusion in computer networks' by carvalho and roucariol. *Comm. ACM*, 26(2):147-148, 1983.
15. M. R. Garey and D. S. Johnson. *Computers and intractability. A guide to the theory of NP-completeness*. W. H. Freeman and Co., San Francisco, 1979.
16. P. C. Gilmore and A. J. Hoffman. A characterization of comparability graphs and of interval graphs. *Canad. J. Math.*, 16:539-548, 1964.
17. P. B. Henderson and Y. Zalkstein. A graph-theoretic characterization of the PV_{chunk} class of synchronising primitives. *SIAM J. Comput.*, 6(1):88-108, 1977.
18. Suzuki I. and Kasami T. An optimality theory for mutual exclusion algorithms in computer networks. In *Proc. of the 3rd Int. Conf. On Distributed Computing Systems*, pages 365-370, Miami, 1982.
19. T. Ibaraki and U. N. Peled. Sufficient conditions for graphs to have threshold number 2. In P. Hansen, editor, *Studies on Graphs and Discrete Programming*, pages 241-268. Nort-Holland Publishing Company, Amsterdam, 1981.
20. Lamport L. A new solution of Dijkstra's concurrent programming problem. *Comm. ACM*, 17(8):453-455, 1974.

21. Lamport L. Time, clocks and the ordering of events in a distributed system. *Comm. ACM*, 21(7):558–565, 1978.

22. Peterson G. L. Myths about the mutual exclusion problem. *Inf. Proc. Lett.*, 12(3):115–116, 1981.

23. Peterson G. L. A new solution to Lamport's concurrent programming problem using shared variables. *ACM Toplas*, 5(1):56–65, 1983.

24. Tze-Heng Ma. On the thresold dimension 2 graphs. manuscript, September 1993.

25. N. V. R. Mahadev and U. N. Peled. *Threshold Graphs and Related Topics*. Elsevier Publishers, 1995.

26. P. Marchioro, A. Morgana, R. Petreschi, and B. Simeone. Degree sequences of matrogenic graphs. *Discrete Math.*, 51:47–61, 1984.

27. Carvalho O. and Roucairol G. On mutual exclusion in computer networks. *Comm. ACM*, 26(2):146–147, 1983.

28. E. T. Ordman. Minimal threshold separators and memory requirements for synchronization. *SIAM J. Comput.*, 18(1):152–165, 1989.

29. J. Orlin. The minimal integral separator of a threshold graph. *Ann. Disc. Math.*, 1:415–419, 1977.

30. U. N. Peled. Matroidal graphs. *Discrete Math.*, 20:263–286, 1977.

31. Hehner E. C. R. and Shyamasundar R. K. An implementation of P and V. *Inf. Proc. Lett.*, 12(4):196–198, 1981.

32. Thomas Raschle and Klaus Simon. Recognition of graphs with threshold dimension two. In *The 27th Annual ACM Symposium on Theory of Computing*, Las Vegas, NE, 1995.

33. Thomas Raschle and Andrea Sterbini. 2-threshold graphs can be recognized in $O(n^3)$ time. Private communication.

34. Michel Raynal. *Algorithms for mutual exclusion*. North Oxford Academic, 1986.

35. Andrea Sterbini. *2-Thresholdness and its Implications: from the Synchronization with PV_{chunk} to the Ibaraki-Peled Cojecture*. PhD thesis, University of Rome "La Sapienza", 1994.

36. H. Vantilborgh and A. van Lamsweerde. On an extension of Dijkstra's semaphore primitives. *Inform. Process. Lett.*, 1:181–186, 1972.

37. Dijkstra E. W. Co-operating sequential processes. In F. Genuys, editor, *Programming Languages*, pages 43–112. Academic Press, New York, 1965.

38. Dijkstra E. W. Self-stabilizing systems in spite of distributed control. *Comm. ACM*, 17(11):643–644, 1974.

39. Doran R. W. and Thomas L. K. Variants of the software solution to mutual exclusion. *Inf. Proc. Lett.*, 10(4):206–208, 1980.

40. Kessels J. L. W. Arbitration without common modifiable variables. *Acta Informatica*, 17:135–141, 1982.

41. M. Yannakakis. The complexity of the partial order dimension problem. *SIAM J. Algebraic Discrete Methods*, 3(3):351–358, 1982.

42. K. Yue and R. T. Jacob. An efficient starvation-free semaphore solution for the graphical mutual exclusion problem. *Comput. J.*, 34(4):345–349, 1991.

Task Assignment in Distributed Systems Using Network Flow Methods

IOANNIS MILIS *

INRIA - Sophia Antipolis,
2004 route des Lucioles, B.P. 93, 06902 Sophia Antipolis Cedex, France.
milis@softlab.ece.ntua.gr

Abstract. A common approach to the problem of assigning the tasks of a modular program to the processors of a distributed system relies on its close relation with the multiway cut problem in graphs. All known algorithms exploiting this relation aim to reduce the problem size by extracting partial solutions, using network flow methods. However, for large sized problems, such reductions tend to become useless. In this paper network flow methods are exploited in a different vein and a simple polynomial heuristic algorithm, which exhibits satisfactory computational behavior, is presented.

1 Introduction and Problem Statement

The task assignment problem (TAP) in distributed computing environments consists of assigning the tasks of a modular program to the processors of a distributed system. In such a situation, tasks can in general executed on any of a set of processors but because of different capabilities, speeds or resources of the processors the execution cost/time of a task may vary from one processor to another. Two interacting tasks that are executed on different processors communicate using the system's communication links. In this case a communication cost/time is incurred because of the overhead due to communication protocols and transmission delays. The problem is usually presented in terms of an assignment model that incorporates an optimization criterion as well as some possible constraints. Several variants of optimization criteria have been studied and many results have been appeared in the literature. In this paper we focus on the minimization of the total cost of the program, which is the sum of the task execution costs plus the intertask communication cost, a criterion first introduced by H. Stone [15].

In a minimum execution cost assignment, tasks tend to be assigned to processors on which they have low execution costs. Also, interacting tasks tend to be assigned to the same processor in order to minimize the communication overhead.

---

* Fellow of the European Commission (Program H.C.&M). Present address: National Technical University of Athens (NTUA), Dpt. of Electr. & Comp. Eng., Division of Comp. Sc., Heroon Polytechniou 9, 15773 Zografou - Athens, Greece.

The problem here is to balance these two conflicting objectives by assigning tasks to processors such that the communication overhead is minimized while taking advantage of the specific power of some processors.

Let us consider a modular program consisting of m communicating tasks, $T = \{t_1, t_2, ..., t_m\}$, to be run on the n heterogeneous processors $P = \{p_1, p_2, ..., p_n\}$ of a distributed system. The communication pattern between the tasks can be presented by a task graph $G = (T, E)$, with edges $E = \{(t_i, t_j) |$ data is transferred from (to) t_i to (from) $t_j\}$. It is well known that for scheduling applications the precedence (t_i, t_j) vs. (t_j, t_i) is critical. However, precedence does not affect the minimization criterion considered in this paper and G can be considered to be undirected. A number $c_{i,j}$ on (t_i, t_j) in E shows the amount of data to be transferred.

The communication cost incurred by two communicating modules t_i, t_j, is in general, a function of the processors to which they are assigned, because of the heterogeneity of the communication links between processors. By $d(q, r)$ we denote the cost incurred when a unit of data is transferred between processor p_q and p_r. If two communicating modules are assigned to the same processor p_q, then $d(q, q)$ represents the incurred interference cost. According to the above notation the communication incurred when t_i is assigned to processor p_q and t_j is assigned to processor p_r is equal to $c_{i,j}d(q, r)$.

Communication costs are said to be *uniform* if:
(i) The communication cost between coresident modules is equal to zero, in other words there is no interference cost, i.e. $d(q, q) = 0$, and
(ii) The communication cost incurred by two communicating modules does not depend on the processors to which they are assigned, in other words the communication links between processors are homogeneous, i.e. $d(q, r)$ is constant.
It is clear that in the uniform communications cost case $c_{i,j}$ represents directly the communication cost incurred when two communicated tasks t_i and t_j are assigned in different processors.

Tasks can in general be executed on any one of the processors, but because of the heterogeneity of the processors the cost of executing a task may vary from a processor to another. Thus the execution cost of task t_i on processor p_q depends on p_q and is denoted by $e_i(q)$.

Let us denote an assignment by a vector $X = \{x_1, x_2, ..., x_m\} \in \{1, 2, ..., m\}^m$, where $x_i = q$ means that task t_i is assigned to processor p_q. Then the minimal total cost assignment problem under consideration can be formally stated as:

$$\min_X \mathcal{T}(X) = \sum_{t_i \in T} e_i(x_i) + \sum_{(t_i, t_j) \in E} c_{i,j} d(x_i, x_j).$$

It has been shown in [11] that for $n \geq 3$ the above problem is NP-complete, even in the case of uniform communication costs. Polynomial time optimal algorithms exist for special cases of either the configuration of the distributed system (two-processor systems [15], linear array networks [8]) or the form of the task graph (tree [4], series-parallel [16], k-tree, almost tree [2]). Furthermore branch-and-bound exact algorithms [3], [5], [11], [14] and several heuristics [1], [9], [10], [13] have been, also, proposed in the literature for the general case of the problem.

Most heuristics [1], [9], [10] as well as polynomial algorithms for special cases [8], [15] rely on a close relation between the task assignment problem with uniform communication costs and the multiway cut problem [6]. In Section 2 we present this relation and comment on the previously proposed heuristic algorithms based on this fact. In Section 3 a new algorithm, which relies on a different use of maximum flow methods, is derived. The computational results presented in Section 4 show that the behavior of the proposed algorithm is quite satisfactory. We conclude the paper in Section 5.

2 Partial Assignment Methods

The *Multiway Cut* problem can be defined as follows [6]:

Given a graph $G = (V, A)$, a set $S = \{s_1, s_2, ..., s_n\}$ of n specified vertices or terminals, and a positive weight $w(e)$ for each edge $e \in A$, find a minimum weight set of edges $A' \subseteq A$ such that the removal of A' from A disconnects each terminal from all the others.

It is obvious from this definition that a multiway cut separates the graph G into exactly n disjoint components, each containing a specified node s_i, $i = 1, 2, ..., n$. When $n = 2$ this problem reduces to the well known polynomial "min-cut/max-flow" problem.

In [15], Stone established a close correspondence between the task assignment problem with uniform communication costs, and the multiway cut problem: From the task graph $G = (T, E)$ of a n-processor assignment problem a modified graph $G' = (V', E')$ is constructed. This is done by adding in the original task graph n additional nodes, each one corresponding to a processor p_k, $k = 1, 2, .., n$, as well as edges from each node t_i, $i = 1, 2, ..., m$, of the original task graph to each of the nodes p_k, i.e. $V' = T \cup \{p_k | k = 1, 2, ..., p_n\}$ and $E' = E \cup \{(t_i, p_k) | i = 1, 2, ..., m, k = 1, 2, ..., n\}$. If the weights of the edges of the modified graph are

$$w(t_i, t_j) = c_{i,j} \qquad \text{and} \qquad w(t_i, p_k) = \frac{\sum_{q=1}^n e_i(q)}{n-1} - e_i(k) \qquad (1)$$

then the minimum multiway cut of the modified graph provides a solution to the corresponding assignment problem, i.e. the tasks connected to node p_k are assigned to processor p_k.

The polynomial equivalence between the two problems has been formally shown in [11]. The NP-completeness of the task assignment problem follows from the NP-completeness of the multiway cut problem [6]. It is clear that when $n = 2$

(two-processor systems) the problem is polynomially solvable.

A problem's size can be reduced using the following observation made by Stone: Let a two processor assignment between the processor p_k and the super-processor P_k which is formed by lumping together all the remaining processors. The edges (t_i, P_k) have weights $w(t_i, P_k) = \sum_{q \neq k} w(t_i, p_q)$. Then the tasks assigned to processor p_k by the $p_k - P_k$ two processor assignment, retain this assignment in the optimal n-processor assignment. The iterative application of this process results in a partial assignment which is consistent with the optimal assignment. This result is known as "prefix property" and has a direct analog in the multiway cut problem.

We must distinguish here the above defined multiway cut problem from the similar n-cut problem [7]: Without specified vertices, one is asked to find the minimum weight set of edges, which when deleted separates the graph into exactly n nonempty components. The latter problem is easier from the former, having a polynomial time algorithm for fixed n [7]. However, from Stone's construction, it is clear that this is not useful for our purpose.

Stone's partial assignment is the starting point of most heuristics found in the literature [1], [9], which deal with the tasks that remain unassigned. Lo [9] continues in a similar manner, trying to improve the partial assignment obtained by the Stone process. At each step of her algorithm the tasks assigned in the previous step are removed, a new graph is constructed with updated edge weights and the Stone process is repeated until no task can be assigned in a step. Tasks that remain unassigned are then assigned in a greedy manner. Computational results reported in [9] show that Lo's algorithm found the optimal assignment in 33% of the cases tested, while the deviation of the obtained solution from the optimal one was less than 50% in 90% of the tests. The worst deviation was 170%.

Abraham and Davidson [1] assign all the unassigned tasks to the processor for which $p_k - P_k$ two-cut has the greatest cost. They maintained that this method has a worst case bound of $\frac{2(n-1)}{n}$, as is the bound of the multiway cut problem [6]. Unfortunately, this bound does not hold in the original task assignment problem, because this requires all $w(t_i, p_k)$ weights of equation (1) to be positive. From the equation it is obvious that this is not always true. However, negative weights $w(t_i, p_k)$ can be made positive by adding a fixed number which depends on the execution costs $e_i(q)$. This does not create any problem since the number of edges (t_i, p_k) appearing in every the multiway cut is an invariant because the number of those edges is equal to $n - 1$. Thus, the multiway cut heuristic of [6] can be applied, but the problem is that the bound of $\frac{2(n-1)}{n}$ does not work anymore. This is now bound for a TAP with different execution costs $e_i(q)$ from the original one.

Magirou [10] presents a departure from the basic idea of Stone and reports on a more efficient partial assignment procedure. Instead on lumping all processors, except the one being considered, together, his method replaces them by a processor that is faster in terms of execution cost, say P_k^{min}. The edges (t_i, P_k^{min}) have weights $w(t_i, P_k^{min}) = \min_{q \neq k} e_i(q)$. The prefix property still holds and a partial assignment is, also, obtained by solving n successive $p_k - P_k^{min}$ two processor assignment problems. Computational results in [10] show that Magirou's partial assignment method, for moderate sized problems, has a considerably improved performance over the Stone's original one, obtaining partial assignments with greater cardinality.

Unfortunately, the same results of [10] show that no partial assignment method can be effective for large problems, where both partial assignments tend to become null, as one would expect given the NP-completeness of the problem. The next section suggests the use of maximum flow methods in a different way.

3 A New Algorithm

Partial assignment methods take into consideration the entire set of processors from their initial step and result in a partial assignment. In the first step of the proposed algorithm all the tasks are considered to be assigned to one processor, say p_1, while the remaining processors are added, one by one, at each subsequent step of the algorithm, such that at every step we have a complete assignment on the subset of processors so far examined. Let $X^k = \{x_1^k, x_2^k, ..., x_m^k\}$ denote the assignment vector at the step k of the algorithm. It is clear that initially we have $X^1 = \{\overbrace{1, 1, ..., 1}^{m}\}$ and $x_i^k \leq k$, $1 \leq i \leq m$, $1 \leq k \leq n$.

At each step $k \geq 2$ of the algorithm a modified graph $G_k = (V_k, E_k)$ is constructed. This consists of the original task graph plus two additional nodes, the first one representing the processor p_k, while the second the processors $p_1, p_2, ..., p_{k-1}$, examined so far, denoted by P_k, i.e. $V_k = T \cup \{p_k, P_k\}$. Edges are also added between each task t_i and the nodes p_k and P_k, i.e. $E_k = E \cup \{(t_i, p_k)|1 \leq i \leq m\} \cup \{(t_i, P_k)|1 \leq i \leq m\}$. Let also, C_k be the value of a $P_k - p_k$ two-cut in the modified graph G_k and S_k be the set of tasks associated, in this cut, with the processor p_k.

The algorithm is based in the following lemma:

Lemma 1. *If the weights of the edges of G_k are:*

$$w(t_i, P_k) = e_i(k), \quad 1 \leq i \leq m,$$
$$w(t_i, p_k) = e_i(x_i^{k-1}) + \frac{1}{2} \sum_{x_j^{k-1} \neq x_i^{k-1}} c_{i,j}, \quad 1 \leq i \leq m,$$

$$w(t_i, t_j) = \begin{cases} c_{i,j} & if \ x_i^{k-1} = x_j^{k-1} \\ \frac{1}{2}c_{i,j} & otherwise, \end{cases} \qquad 1 \le i,j \le m,$$

$$then, \ C_k = T(X^k), \ where \ X^k = \begin{cases} x_i^k = k & if \ t_i \in S_k, \\ x_i^k = x_i^{k-1} & otherwise \end{cases} \qquad 1 \le i \le m.$$

Proof. For the sake of simplicity we consider only two communicating tasks t_i and t_j. We distinguish between two cases depending on the assignment of t_i and t_j in the previous $(k-1)$ step of the algorithm:
(i) The two tasks were assigned to the same processor, say p_q, i.e. $x_i^{k-1} = x_j^{k-1} = q, 1 \le q \le k-1$.
(ii) The two tasks were assigned to different processors, say p_q and p_r, i.e. $x_i^{k-1} = q \ne x_j^{k-1} = r, 1 \le q,r \le k-1$.
For each case we consider the four possible cuts-assignments shown in Figures 1(a) and 1(b) respectively. The weights obtained by the equations of the lemma are, also, indicated in the Figure 1.

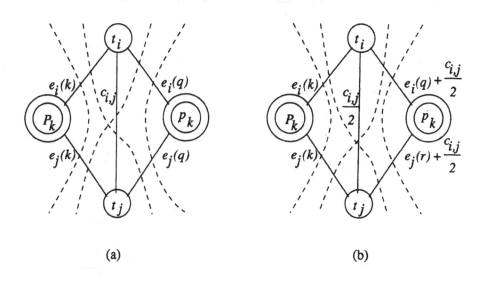

$$(a) \qquad\qquad\qquad\qquad (b)$$

Fig. 1. *Possible assignments of two communicating tasks*

It is easy to verify that the cost of each cut is equal with the cost of the corresponding assignment. Taking in to account that task t_i may communicate with more than one other tasks we obtain the sum in the second equation. $\Box$

In general, each step of our algorithm performs a $p_k - P_k$ maximum flow and results on an assignment X^k by assigning S_k tasks to processor p_k, while the remaining $T - S_k$ tasks remain to the processors to which they were assigned by the previous X^{k-1} assignment, i.e. $X^k = \{x_i^k | x_i^k = k \quad if \quad t_i \in S_k, \quad or \quad x_i^k = x_i^{k-1} \quad otherwise, 1 \leq i \leq m\}$. This is shown by the graphical representation of Figure 2, while the algorithm is summarized as follow:

begin

$$X^1 = \{\overbrace{1, 1, ...1}^{m}\}; \{\text{Initialization}\}$$
    for $k := 2$ **to** n **do**
        begin
            Construct the modified graph G_k according to Lemma 1;
            Apply a maximum flow algorithm to find the set S_k;
            for $i := 1$ **to** m **do if** $t_i \in S_k$ **then** $x_i^k = k$ **else** $x_i^k = x_i^{k-1}$;
    end

end

The complexity of the proposed algorithm is determined by the complexity of the min-cut algorithm used. If the Karzanov's maximum flow algorithm [12] is used, then the complexity of the proposed algorithm becomes $O(nm^3)$.

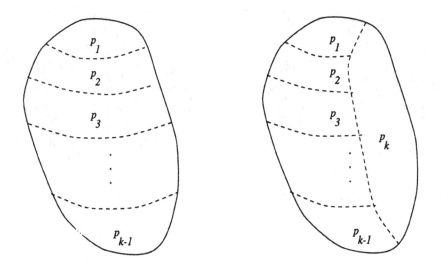

Fig. 2. *A graphical representation of an algorithm's step*

4 Computational Results

To test the behavior of the proposed algorithm, this was run for a significant number of randomly generated problems. In all the cases the obtained solution,

C_{heur}, was compared with the optimal one, C_{opt}, provided from the branch-and-bound algorithm given in [11]. The results of several runs, varying the numbers m, n, as well as the density $s = \frac{2|E|}{m(m-1)}$ of the task graph are shown in Table 1. For every (m, n, s) value twenty randomly generated problems were tested. The table shows:

(a) The number of problems for which the optimum solution was obtained, and
(b) The worst percent deviation, $\frac{C_{heur} - C_{opt}}{C_{opt}} * 100$, between the solution obtained by the proposed algorithm and the optimal one.

Tasks	Processors	Optimal solved problems			Worst deviation (%)		
(m)	(n)						
	Density	1/3	1/2	2/3	1/3	1/2	2/3
4	4	19	18	18	4.0	2.3	15.1
6	4	20	18	19	0.0	2.3	0.8
	6	20	13	16	0.0	10.3	6.2
	4	14	16	15	2.4	3.2	4.3
8	6	16	17	11	4.5	6.3	6.8
	8	15	12	11	12.2	7.1	6.1
	4	16	15	18	2.5	3.2	2.3
10	6	15	10	16	4.3	5.5	13.7
	8	10	12	11	4.3	4.7	8.9
	10	10	7	8	6.7	16.4	12.6
	4	16	14	12	1.6	2.2	3.8
	6	9	8	11	7.5	5.5	13.4
12	8	14	11	7	6.6	15.1	17.3
	10	9	8	11	3.5	7.5	9.3
	12	12	8	10	7.6	15.5	12.5
	4	13	8	11	6.4	3.9	6.1
14	6	9	7	8	4.1	9.8	8.1
	8	8	4	4	4.1	7.2	10.2
	10	5	7	5	9.9	8.7	10.9
	4	13	12	11	9.8	5.6	3.0
16	6	11	8	6	2.1	8.6	5.5
	8	11	5	9	4.1	9.8	9.2
18	4	12	11	11	3.8	6.6	6.3
	6	6	5	7	6.0	10.7	9.3
20	4	11	12	14	3.4	6.5	6.8

Table 1. Computational results

From Table 1 (which includes 1500 examples) we can see that the proposed algorithm attains the optimal assignment in about 53% of all cases. It should also be noted that the ratio $\frac{C_{heur}}{C_{opt}}$ is unexpectedly small. Although one can construct examples with larger deviation, as those shown in the Table 2, we have not envisaged an example for which the proposed algorithm yields deviation greater than 100%.

Tasks (m)	Processors (n)	Deviation (%)
5	5	55.484
10	10	74.516
15	15	80.860
20	20	84.032

Ring task graph with $c_{1,m} = c_{i,i+1} = C = 30$.

$$e_i(q) = \begin{cases} 1 & \text{if } i = q, \\ 2C & \text{otherwise} \end{cases}$$

Table 2. Examples having large deviation.

5 Concluding Remarks

We have presented a simple heuristic algorithm for the TAP which consists of iterative applications of a maximum-flow / minimum cut algorithm and yields to solutions close to optimal. Instead of the common use of network flow methods to obtain partial assignments, the approach behind our algorithm is to improve an initial complete assignment.

Note that all algorithms exploiting network flow methods to investigate the task assignment problem are applicable only in the case of uniform communication costs. In addition, it is not obvious how such an approach could be applied to problems with memory and/or load balancing constraints.

Several interesting questions could be further examined on the performance of the algorithm such as:
a) Does the initial assignment and the numbering of the processors play any significant role on the performance?
b) How well does the iterative application of the whole algorithm perform?

However, it has been shown that the TAP with non uniform communication costs is an ϵ-approximate NP-complete problem [2]. Although we have not encountered an example for which the proposed algorithm yields deviation greater than 100%, an analogous positive or negative result for the case of uniform costs remains an open problem.

Acknowledgment

The author is indebted to Professor C. H. Papadimitriou of UCSD for suggestions and comments on an earlier version of this paper.

References

1. S. G. Abraham and E. S. Davidson, *Task assignment using network flow methods for minimizing communication in n-processor systems*, Center for Supercomp. Res. Develop., Tech. Rep. 598 (1986).
2. D. Fernádez-Baca, *Allocating modules to processors in a distributed system*, IEEE Trans. Soft. Eng. SE-15 (1989) 1427-1436.
3. A. Billionnet, M. C. Costa and A.Sutter, *An efficient algorithm for a task allocation problem*, J.ACM 39 (1992) 502-518.
4. S. H. Bokhari, *A shortest tree algorithm for optimal assignments across space and time in a distributed processor system*, IEEE Trans. Soft. Eng. SE-7 (1981) 583-589.
5. M. S. Cherh, G.H. Chen and P. Liu, *An LC branch-and-bound algorithm for the module assignment problem*, Inf. Proc. Lett. 32 (1989) 61-71.
6. E. Dahlhaus, D. S. Johnson, C. H. Papadimitriou, P. Seymour and M. Yannakakis, *The complexity of multiway cut*, In Proc. 24th ACM STOC (1992) 241-251.
7. A. V. Goldberg and D. S. Hochbaum, *A polynomial algorithm for the k-cut problem*, In Proc. 29th IEEE FOCS (1988) 445-451.
8. C.-H. Lee, D. Lee and M. Kim, *Optimal task assignment in linear array networks* IEEE Trans. Comp. TSC-41 (1992) 877-880.
9. V. M. Lo, *Heuristic algorithms for task assignment in distributed systems*, IEEE Trans. Comp. TSC-37 (1988) 1384-1397.
10. V.F. Magirou, *An improved partial solution to the task assignment and multiway cut problems*, Operations Research Letters 12 (1992) 3-10.
11. V. F. Magirou and J. Z. Milis, *An algorithm for the multiprocessor assignment problem*, Operations Research Letters 8 (1989) 351-356.
12. C. H. Papadimitriou and K. Steiglitz, *Combinatorial Optimization: Algorithms and Complexity*, Prentice Hall, Englewood Cliffs, NJ (1982).
13. C. C. Price and S. Krishnaprasad, *Software allocation models for distributed computing systems*, In Proc. 4th Int. Conf. on Distributed Computing System (1984) 40-48.
14. J. B. Sinclair, *Efficient computation of optimal assignment for distributed tasks*, J. Paral. and Distr. Comput. 4 (1987) 342-362.
15. H. S. Stone, *Multiprocessor scheduling with the aid of network flow algorithms*, IEEE Trans. Soft. Eng. SE-3 (1977) 85-93.
16. D. F. Towsley, *Allocating programs containing branches and loops within a multiple processor system*, IEEE Trans. Soft. Eng. SE-12 (1986) 1018-1024.

Distributed Rerouting in DCS Mesh Networks.*

Dritan Nace and Jacques Carlier

Université de Technologie de Compiègne
URA CNRS 817, HEUDIASYC, BP 529
60205, Compiègne, FRANCE
Dritan.Nace@utc.fr

Abstract

Future telecommunication services will demand a fault-tolerant network with complete survivability. In this context, the reconfiguration of a network in real time has become one of the key issues on network reliability. In continuation of our previous work presented in [3], this paper proposes a fast distributed rerouting algorithm based on a new approach for generation of the restoration paths. The performances of the proposed algorithms are evaluated in terms of restoration efficiency and restoration · time.

keywords: telecommunication networks, distributed rerouting, shortest path, survivability.

1   Introduction.

Survivability must be one of the most important features of today's Telecommunication Networks. It is the aim of the rerouting algorithms to reconfigure the network around the failures such that calls in progress are not dropped and the quality of services is not degraded. Up to now, rerouting algorithms employing centralized control realize restoration times of a few minutes. Many researcher teams are working to improve the restoration time from minutes to sub-seconds by using distributed control algorithms. Classical distributed algorithms have a flooding phase for searching restoration paths. In [3] we have proposed to use precalculated paths. But this is not realistic for large networks. So, we suggest a distributed algorithm which computes the restoration paths in real time. The choice of distributed rerouting deals with some technological advances like the deployment of fibre optic transmission and of reconfigurable DCS (Digital

---

*This work was supported by CNET (Centre National d'Etudes de Télécommunications, France) contract 7970C.

Cross-connect System) nodes, throughout the network. In this paper we will take into consideration DCS Mesh Networks. In such networks, even a single failure causes important losses of traffic but on the other side the DCS nodes permit a faster reaction against the failures. Section 2 will be devoted to main characteristics of the restoral process. A brief description of our previous algorithm will be given in section 3. The distributed rerouting algorithm for DCS mesh Networks will be presented in section 4. Then in section 5 some simulation results will be reported.

2 Main Characteristics of Restoral Process.

The telecommunication traffic is routed according to a matrix of point-to-point demands. Each link has a given capacity which consists of a *working capacity*, used to transmit the traffic of demands routed through this link and a *spare capacity* used for restoration needs in case of element failures. Unlike centralized rerouting where a special node calculates and distributes rerouting tables to other nodes, in the distributed rerouting all nodes collaborate and contribute more or less equally in rerouting the lost traffic. There are two main strategies for the rerouting known as local and end-to-end rerouting. In the first one, the disturbed demands are rerouted between the ends of the failed transmission links. In the second strategy a disrupted demand is rerouted between its origin and its destination. Theoretically, end-to-end rerouting and local rerouting for multiple link failures give multi-commodity flow problems. These flow problems are generally resolved by using alternative routes. In this context, one of the main steps of distributed rerouting algorithms is the route selection. The classical method of route selection employed by most of the proposed restoration algorithms consists of finding the candidate restoration paths by flooding help-messages [4], [8], [10], [13], [16]. The flooding is characterized by broadcasting a help-message firstly by one of the extremities (sender) of the failed link and then by all other nodes receiving these messages, until they reach the other extremity (chooser) of the failed link. In any case, a help-message which has not still reached its destination after crossing a limited number of hops (L), will be rejected, so all existing paths of length inferior or equal to L will be found. The chooser will choose the candidate paths between those found by the help-messages. To decrease the restoration time and reduce the searching area of candidate paths, a double search algorithm [7] and another based on a two-prong approach [4] have also been proposed.

3 A Distributed Rerouting Algorithm Based on Precalculated Paths.

The algorithms utilizing a dynamic method based in exchanging messages for finding the candidate paths, like flooding, penalize the restoration time. Moreover, some of the proposed algorithms cannot always assure an efficient utilization of resources. Theoretically, it is possible that restoration paths having spare capacities still exist at the end of rerouting even though 100% restoration is not achieved. We explain this by the fact that a candidate path which for some reason (e.g. one of links reserved by other messages) cannot be reserved, will never be reconsidered even when this reason is not anymore valid (the reservation is cancelled). These facts incited us to replace the restoration paths found during the flooding by precalculated paths (see [3]). For restoration of the lost traffic between two nodes, the sender keeps a set of restoration paths precalculated before the failure occurs. In case of failure, the sender has to pick paths in the set of stored paths and to send reservation-messages through them. There are two ways to employ this algorithm: the first one precalculates all possible alternative paths between the extremities of the failed element. It corresponds to the algorithm A in the figure 2. But this is not economic in terms of memory space. The second one uses only optimal precalculated paths between the extremities of the failed element. This one corresponds to the algorithm B in the figure 2. These precalculated paths are optimal in the sense that they maximize the rerouted capacity. These paths are obtained by resolving the spare-capacity assignment assuring 100% restoration for single failures on the network. The problem is that they are not sufficient in case of simultaneous failures or sequential failures. These disadvantages justify the idea of computing restoration paths in real time.

4 Another Distributed Rerouting Algorithm for DCS Mesh Networks.

The challenge is to propose a distributed rerouting algorithm where any node (especially a sender) can find locally the shortest path to any other node (chooser) in real time. It corresponds to the algorithm C in the figure 2. For this, we install in each node a shortest path algorithm which will be applied locally by the node. In other words, each sender would be able to compute locally the shortest paths from itself to its respective choosers. This algorithm is the most important module of the proposed distributed restoration algorithm.

4.1 Description of the Distributed Algorithm.

During the description of the algorithm, we will continue to use the same terminology, so terms of sender and chooser will designate the extremities of the failed element, even though they do not correspond exactly to their traditional roles. In fact the sender will play a much more important role than before, it must find the candidate paths by local computations and send reservation-messages. On the other side, the chooser will have only to confirm the reservation routes by sending confirmation-messages back to the sender.

4.1.1 Assumptions.

For the functioning of the algorithm we need to make some classical assumptions:

- Each node is able to detect the failure of its incident links.

- The communication of messages is reliable.

- Each channel assures the communication in both directions.

- A node-failure is considered as failure of its incident links.

4.1.2 Local Rerouting.

When a link-failure occurs, signaling alarms will invoke the sender to initiate the reservation phase of the algorithm.

Reservation Phase.

The sender finds some restoration paths by applying the shortest path algorithm described in section 4.2. Of course, these alternative routes are destinated to reroute the disrupted traffic and the sum of their capacities will not exceed the lost traffic. For each found route, a reservation-message will be created with all necessary information concerning this route. The information carried by a message consists mainly of nodes of the route, its capacity, its type and eventually its priority. After finding an alternative route and creating its corresponding reservation-message, the sender will reserve on its outgoing link through this route and send the message to the other extremity. Each node receiving a message, will keep it in a queuing list till the message will be processed as following. Firstly, the node will reserve on its incoming link and eventually it must resolve the contentions (see below). Secondly it must reserve on its outgoing link and send the message to the following node and so on till it reaches the destination (chooser). If there is no more spare capacity on one of its outgoing links (e.g reserved by other reservation-messages), the message must be sent back to the sender and its information must be updated. So, the type will become "Cancellation" and another information concerning the identity of the "unavailable" link must be added. In this case the node will keep the identity of the sender in a list and will relay the message back through the reserved path. A cancellation-message will cancel the reserved capacities and will inform the sender about the unavailability of the link for not using it during the construction of new restoration paths. In any case, the sender will be prevented by a reactivation-message to update its information if this link becomes available (e.g some reservations are cancelled). Reservation-messages have to reserve the necessary spare capacities till they reach their destinations. In the following paragraph we will explain how this algorithm resolves contentions.

Contention.

In distributed environments, the control of resources is also distributed which can create some delicate situations. The most frequent situation in distributed rerouting is the contention (see fig.1). The resource of a link (its spare capacity), is controlled independently by its two extremity nodes. Following the figure, we can suppose that the nodes I and J attribute the only spare capacity of the link IJ, respectively to message 1 and message 2. When the message 1 (resp. 2) reaches node J (resp. I), this one cannot confirm the requested capacity because it has already reserved it. To resolve the contention we can make use of the identity numbers of extremity nodes by giving the priority to this having the greatest one. In this case, the message sent by the node having the priority will be the "winner" and for the other one, a response-message will be sent back for confirming the contention and cancelling the reservation. In the latter case, a response-message will be relayed by a cancellation-message.

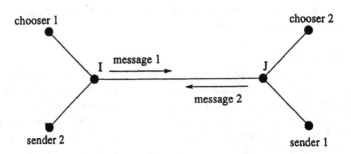

Figure 1: The contention.

Confirmation Phase.
When the chooser will receive a reservation-message, it sends a confirmation message for cross-connecting the links of the reserved route. The procedure is finished when all lost traffic will be restored or there are no more resources for constructing other restoration paths.

4.1.3   End-to-end Rerouting.

Considering the end-to-end rerouting, we have to add another phase in the beginning to liberate the capacities occupied by the disturbed demands. In fact, each of the extremities of the failed link will send release-messages to the extremities of the demands to release the occupied capacities. These released capacities could be recovered or not for restoration needs. The senders receiving these messages will begin rerouting the disrupted traffic in the same way as in local rerouting.

4.1.4   Types of Messages.

In this section we will enumerate all types of messages used in the algorithm.

- A release-message is used for releasing the capacities occupied by the disturbed demands (end-to-end rerouting).

- A reservation-message is used for reserving the spare capacities through a restoration path.

- A cancellation-message is used for cancelling the capacities reserved by a reservation-message on its way back to sender.

- A reactivation-message is used for preventing the senders about changes that have happened and eventually to try another path if it is necessary.

- A confirmation-message is used for the definitive confirmation and for the cross-connection of the links of the path.

- A response-message is used for resolving eventual contentions.

4.2 The Shortest Path Algorithm.

As mentioned before, the senders must be able to find candidate paths for rerouting the traffic of their own demands. For this, each of them must have its own view of the network, as operativity state of network elements, the successors files and a table for spare capacities of links. For a more economic use of resources we have fixed on L (L depends on the network), the maximal length of restoration paths, known also as hop count limit. The successors files are two combined arrays that permit a direct access to successors of each node. These files reflect the topology of the Network, consequently they will be unchanged. Let us denote with C_s the table of spare capacities of links stored in the sender s. Of course, in the beginning any sender has the same table of spare capacities, but during the restoral process, they will update their tables in function of information reported by exchanging messages. Based on the information of their own tables, each sender s can compute exactly the distances to other nodes and reports them to a table denoted by D_s. This computation consists of a classical breadth-first search. This table which is particular to each sender, will keep the key-information for the shortest path algorithm and like the C_s will be updated during the search of shortest paths as shown below. The main advantages of this algorithm is that it is simple and economic in memory space. Notice that a node of the network could be sender for several disturbed demands. In these conditions, using precalculated paths depends on the failure, when the structures needed for this algorithm are valid for any failure and the occupied memory space is insignificant. In the following, we will give a detailed description of the shortest path algorithm.

Description of the shortest path algorithm.
Let suppose that sender s needs to find a candidate restoration path P to node (chooser) t which is the other extremity of the affected demand from s to t.
Let recall that C_s is the table of spare capacities stored in the sender s, D_s is the table of distances from s to other nodes stored in the sender s and L is the maximal allowed length for restoration paths. We say that an arc (i, j) is admissible for the node s if it satisfies the conditions: $D_s[i] = D_s[j] + 1$ and $C_s[i, j] > 0$.

The shortest path algorithm $(s \to t)$;
begin
    $i := t$;
    $c :=$ requested capacity;
      while $(D_s[t] \leq L)$ and $(i \neq s)$
        do
          begin
           if i has an admissible arc then advance(i);
           else retreat(i);
          end;
if $i = s$ then identify-path () *a feasible path is constructed;*
else *there are no more feasible paths;*
end;

    procedure advance(i);
begin
      find the admissible arc (i,j) with maximal capacity;
      $pred(j) := i$ and $i := j$;
end;

    procedure retreat(i);
begin
      $D_s[i] := \min\{D_s[j] + 1$: for all adjacent j of node i with $C_s[i,j] > 0\}$;
      if $i \neq t$ then $i := pred(i)$;
end;
As it can be seen, the procedure *retreat* serves also to adjust the distance table D_s.

    procedure identify-path ();
begin
      identify the path P by using the predecessor indices;
      $c_P := \min\{c, C_s[i,j] : arc(i,j) \in P\}$;
end;

When a restoration path is found, its capacity corresponds to the minimum of the requested capacity and spare capacities of its links. Then the table of spare capacities needs to be updated by subtracting the capacity of the path c_P to those of links of P represented in the table C_s. If the c_P is inferior to the requested capacity c, the procedure of feasible path search has to be repeated till all requested capacity is *covered* or no more resource exists to construct a new path.

Adjustment of the distance table.

Normally the distance table is updated during the procedure retreat() of the algorithm but in case of "insertion" of links, the sender needs to update it explicitly. With "insertion" we understand a link considered before as unavailable when it disposes spare capacities. This happens when the sender receives a reactivation-message. To update the table D, we do not need to recalculate all the table of distances but only those of "successors" of extremities of the "inserted" link.

5   Results and Conclusion.

This section is devoted to report some simulation results and to give some concluding remarks.

5.1   Computer Simulation Results.

Computer simulations were performed in order to test the validity of the algorithm. The algorithm has been tested on a model network deduced from the reference [5]. It has 26 nodes, 43 edges and 177 traffic demands with total working capacities of 645. Spare-capacity is designed to restore perfectly only one link or one node failure with hop limit 9 and the optimal restoration paths are taken by the results of the spare-capacity assignment of this network under the same options as used for the rerouting. Here we will compare the simulation results of three versions of distributed algorithms discussed in this paper. Recall that the algorithm A corresponds to the one using all possible precalculated paths, the algorithm B corresponds to this using optimal precalculated paths and the algorithm C is the one based on the shortest path algorithm described in section 4. In our simulation we have considered end-to-end rerouting for single link failures without recovery of released capacities.

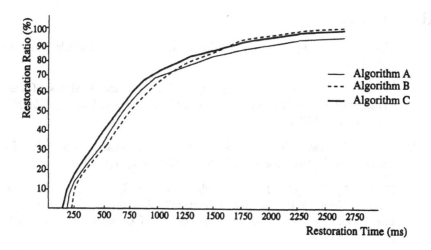

Figure 2: Restoration characteristics of distributed algorithms.

The parameters used for the simulation are:

Message processing time is taken uniformly 3-5 ms;

Message transmission time is taken uniformly 1-2 ms;

Cross-connection time is taken 100ms/per channel with the possibility of parallel cross-connections;

Hop count limit is 9.

As it can be seen by the figure 2, the algorithm C is very close to the algorithm B using optimal restoration paths but in any case it cannot be as efficient as B with 100% restoration for all single link failures. As it is shown in the figure 2, the performances of the algorithm C are better than those of algorithm B. So, the restoration ratio obtained for the algorithm C is 98.14% (only three links are not totally restored), while for algorithm A is 95.81%. On the other hand, the restoration time obtained for the algorithm C is lower than this obtained for algorithm B.

5.2 Conclusion.

A distributed rerouting algorithm based on a new approach of generation of restoration paths was proposed in order to improve the reliability of DCS mesh networks. Main characteristics of the algorithm were discussed and simulation results were reported. We remark that the difference between this algorithm and the others proposed in literature consists mainly in the way that restoration paths are selected and used. Finally we want to underline our conviction that a better combination of the calculation power of nodes with advantages of distributed environments can have an impact in the improvement of reliability of telecommunication networks.

References

[1] Ahuja, R.K., Magnanti, T.L., Orlin, J.B., *Network Flows*, eds. Prentice Hall, (1993).

[2] Bicknell, J., Chow, E., and Syed, S.,"Performance Analyses of Fast Distributed Network Restoration Algorithms" *Proceedings of GLOBE-COM'93*, pp 1596-1600.

[3] Carlier, J., Nace, D., "A New Distributed Restoration Algorithm for Telecommunications Networks", *NETWORKS'94*, pp 377-382.

[4] Chow, E., Bicknell, J., McCaughey, S. and Syed, S.,"A Fast Distributed Network Restoration Algorithm", *Proceedings of 12th International Phoenix Conference on Computers and Communications*, march 1993.

[5] Coan, B.A., Lelet, W.E., Vecchi, M.P., Weinrib, A., Wu, L.T., "Using Distributed Topology Update and Preplanned Configurations to Achieve

Trunk Network survivability", *IEEE Trans. on Reliability, Vol 40, No 4*, 1991.

[6] Doverspike, R., "A Multi-layered Model for Survivability in Intra-lata Transport Networks", *Proceedings of GLOBECOM'91*, pp 2025-2031.

[7] Fujii, H., Yoshikai, N., "Restoration Message Transfer Mechanism and Restoration Characteristics of Double-search Self-healing ATM Networks", *IEEE Journal on Selected Areas on Communications, Vol 12, No 1*, January 1994, pp 149-159.

[8] Grover, W.D., "A Fast Restoration Technique for Networks Using Digital Cross-connect Machines", *Proceedings of GLOBECOM'87*, pp 1090-1095.

[9] Grover, W.D., Venables, B.D., Sandham, J.H., Mine, H.F., "Performances Studies of a Self-Healing Network Protocol in Telecom Canada Long Haul Networks", *Proceedings of GLOBECOM'90*, pp 452-458.

[10] Hasegawa, S., Okanoue, Y.,Egawa, T., Sakauchi, H., "Control Algorithms of SONET Integrated Self-Healing Networks.", *IEEE Journal on Selected Areas on Communications, Vol 12, No 1*, January 1994, pp 110-119.

[11] Kawamura, R., Sato, K., Tokizawa, I., "Self-healing ATM Networks Techniques Based on Virtual Path Concept", *IEEE Journal on Selected Areas on Communications, Vol 12, No 1*, January 1994, pp 120-130.

[12] Kobrinski, H., Azuma, M., "Distributed Control Algorithms for Dynamic Restoration in DCS Mesh Networks: Performance Evaluation", *Proceedings of GLOBECOM'93*, pp 1584-1589.

[13] Komine, H., Chujo, T., Ogura, T., Miyzaki, K., Soejima, T., "A Distributed Restoration Algorithm for Multiple-link and Node Failures of Transport Networks", *Proceedings of GLOBECOM'90*, pp 459-463.

[14] Sakauchi, H., Nichimura, Y., Hasewaga, S., "A Self-healing Network with an Economical Spare-channel Assignment", *Proceedings of GLOBECOM'90*, pp 438-443.

[15] Tsong-Ho, W., Kobrinski, H., Ghosal, D., Lakshman, T.V., "The impact of SONET Digital Cross-connect System Architecture on Distributed Restoration", *IEEE Journal on Selected Areas on Communications, Vol 12, No 1*, January 1994, pp 79-88.

[16] Yang, C.H., Hasewaga, S., "FITNESS: Failure Immunization Technology for Network Service Survivability", *Proceedings of GLOBECOM'88*, pp 1549-1554.

Springer-Verlag
and the Environment

We at Springer-Verlag firmly believe that an international science publisher has a special obligation to the environment, and our corporate policies consistently reflect this conviction.

We also expect our business partners – paper mills, printers, packaging manufacturers, etc. – to commit themselves to using environmentally friendly materials and production processes.

The paper in this book is made from low- or no-chlorine pulp and is acid free, in conformance with international standards for paper permanency.

Lecture Notes in Computer Science

For information about Vols. 1–1053

please contact your bookseller or Springer-Verlag

Vol. 1087: C. Zhang, D. Lukose (Eds.), Distributed Artificial Intelliegence. Proceedings, 1995. VIII, 232 pages. 1996. (Subseries LNAI).

Vol. 1088: A. Strohmeier (Ed.), Reliable Software Technologies – Ada-Europe '96. Proceedings, 1996. XI, 513 pages. 1996.

Vol.. 1089: G. Ramalingam, Bounded Incremental Computation. XI, 190 pages. 1996.

Vol. 1090: J.-Y. Cai, C.K. Wong (Eds.), Computing and Combinatorics. Proceedings, 1996. X, 421 pages. 1996.

Vol. 1091: J. Billington, W. Reisig (Eds.), Application and Theory of Petri Nets 1996. Proceedings, 1996. VIII, 549 pages. 1996.

Vol. 1092: H. Kleine Büning (Ed.), Computer Science Logic. Proceedings, 1995. VIII, 487 pages. 1996.

Vol. 1093: L. Dorst, M. van Lambalgen, F. Voorbraak (Eds.), Reasoning with Uncertainty in Robotics. Proceedings, 1995. VIII, 387 pages. 1996. (Subseries LNAI).

Vol. 1094: R. Morrison, J. Kennedy (Eds.), Advances in Databases. Proceedings, 1996. XI, 234 pages. 1996.

Vol. 1095: W. McCune, R. Padmanabhan, Automated Deduction in Equational Logic and Cubic Curves. X, 231 pages. 1996. (Subseries LNAI).

Vol. 1096: T. Schäl, Workflow Management Systems for Process Organisations. XII, 200 pages. 1996.

Vol. 1097: R. Karlsson, A. Lingas (Eds.), Algorithm Theory – SWAT '96. Proceedings, 1996. IX, 453 pages. 1996.

Vol. 1098: P. Cointe (Ed.), ECOOP '96 – Object-Oriented Programming. Proceedings, 1996. XI, 502 pages. 1996.

Vol. 1099: F. Meyer auf der Heide, B. Monien (Eds.), Automata, Languages and Programming. Proceedings, 1996. XII, 681 pages. 1996.

Vol. 1100: B. Pfitzmann, Digital Signature Schemes. XVI, 396 pages. 1996.

Vol. 1101: M. Wirsing, M. Nivat (Eds.), Algebraic Methodology and Software Technology. Proceedings, 1996. XII, 641 pages. 1996.

Vol. 1102: R. Alur, T.A. Henzinger (Eds.), Computer Aided Verification. Proceedings, 1996. XII, 472 pages. 1996.

Vol. 1103: H. Ganzinger (Ed.), Rewriting Techniques and Applications. Proceedings, 1996. XI, 437 pages. 1996.

Vol. 1104: M.A. McRobbie, J.K. Slaney (Eds.), Automated Deduction – CADE-13. Proceedings, 1996. XV, 764 pages. 1996. (Subseries LNAI).

Vol. 1105: T.I. Ören, G.J. Klir (Eds.), Computer Aided Systems Theory – CAST '94. Proceedings, 1994. IX, 439 pages. 1996.

Vol. 1106: M. Jampel, E. Freuder, M. Maher (Eds.), Over-Constrained Systems. X, 309 pages. 1996.

Vol. 1107: J.-P. Briot, J.-M. Geib, A. Yonezawa (Eds.), Object-Based Parallel and Distributed Computation. Proceedings, 1995. X, 349 pages. 1996.

Vol. 1108: A. Díaz de Ilarraza Sánchez, I. Fernández de Castro (Eds.), Computer Aided Learning and Instruction in Science and Engineering. Proceedings, 1996. XIV, 480 pages. 1996.

Vol. 1109: N. Koblitz (Ed.), Advances in Cryptology – Crypto '96. Proceedings, 1996. XII, 417 pages. 1996.

Vol. 1110: O. Danvy, R. Glück, P. Thiemann (Eds.), Partial Evaluation. Proceedings, 1996. XII, 514 pages. 1996.

Vol. 1111: J.J. Alferes, L. Moniz Pereira, Reasoning with Logic Programming. XXI, 326 pages. 1996. (Subseries LNAI).

Vol. 1112: C. von der Malsburg, W. von Seelen, J.C. Vorbrüggen, B. Sendhoff (Eds.), Artificial Neural Networks – ICANN 96. Proceedings, 1996. XXV, 922 pages. 1996.

Vol. 1113: W. Penczek, A. Szalas (Eds.), Mathematical Foundations of Computer Science 1996. Proceedings, 1996. X, 592 pages. 1996.

Vol. 1114: N. Foo, R. Goebel (Eds.), PRICAI'96: Topics in Artificial Intelligence. Proceedings, 1996. XXI, 658 pages. 1996. (Subseries LNAI).

Vol. 1115: P.W. Eklund, G. Ellis, G. Mann (Eds.), Conceptual Structures: Knowledge Representation as Interlingua. Proceedings, 1996. XIII, 321 pages. 1996. (Subseries LNAI).

Vol. 1116: J. Hall (Ed.), Management of Telecommunication Systems and Services. XXI, 229 pages. 1996.

Vol. 1117: A. Ferreira, J. Rolim, Y. Saad, T. Yang (Eds.), Parallel Algorithms for Irregularly Structured Problems. Proceedings, 1996. IX, 358 pages. 1996.

Vol. 1118: E.C. Freuder (Ed.), Principles and Practice of Constraint Programming — CP 96. Proceedings, 1996. XIX, 574 pages. 1996.

Vol. 1119: U. Montanari, V. Sassone (Eds.), CONCUR '96: Concurrency Theory. Proceedings, 1996. XII, 751 pages. 1996.

Vol. 1120: M. Deza. R. Euler, I. Manoussakis (Eds.), Combinatorics and Computer Science. Proceedings, 1995. IX, 415 pages. 1996.

Vol. 1121: P. Perner, P. Wang, A. Rosenfeld (Eds.), Advances in Structural and Syntactical Pattern Recognition. Proceedings, 1996. X, 393 pages. 1996.

Vol. 1122: H. Cohen (Ed.), Algorithmic Number Theory. Proceedings, 1996. IX, 405 pages. 1996.

Vol. 1123: L. Bougé, P. Fraigniaud, A. Mignotte, Y. Robert (Eds.), Euro-Par'96. Parallel Processing. Proceedings, 1996, Vol. I. XXXIII, 842 pages. 1996.

Vol. 1124: L. Bougé, P. Fraigniaud, A. Mignotte, Y. Robert (Eds.), Euro-Par'96. Parallel Processing. Proceedings, 1996, Vol. II. XXXIII, 926 pages. 1996.

Vol. 1125: J. von Wright, J. Grundy, J. Harrison (Eds.), Theorem Proving in Higher Order Logics. Proceedings, 1996. VIII, 447 pages. 1996.

Vol. 1126: J.J. Alferes, L. Moniz Pereira, E. Orlowska (Eds.), Logics in Artificial Intelligence. Proceedings, 1996. IX, 417 pages. 1996. (Subseries LNAI).

Vol. 1129: J. Launchbury, E. Meijer, T. Sheard (Eds.), Advanced Functional Programming. Proceedings, 1996. VII, 238 pages. 1996.